Viral and Fungal Infections of the Central Nervous System: A Microbiological Perspective

Hiba Sami • Safiya Firoze • Parvez A. Khan
Editors

Viral and Fungal Infections of the Central Nervous System: A Microbiological Perspective

Springer

Editors
Hiba Sami
Department of Microbiology
Jawaharlal Nehru Medical College Hospital
Aligarh, Uttar Pradesh, India

Safiya Firoze
Department of Microbiology
Jawaharlal Nehru Medical College Hospital
Aligarh, Uttar Pradesh, India

Parvez A. Khan
Department of Microbiology
Jawaharlal Nehru Medical College Hospital
Aligarh, Uttar Pradesh, India

ISBN 978-981-99-6444-4 ISBN 978-981-99-6445-1 (eBook)
https://doi.org/10.1007/978-981-99-6445-1

This Springer imprint is published by the registered company Springer Nature Singapore Pte Ltd.
The registered company address is: 152 Beach Road, #21-01/04 Gateway East, Singapore 189721, Singapore

Paper in this product is recyclable.

Dedicated to our Parents and Family Members.

Foreword I

I congratulate the editors to select such a timely topic for their book. This book comprises a broad range of viral and fungal central nervous system infections, including the cause, the pathogenesis, the diagnosis, the therapy, and the prophylaxis. CNS infections continue to be life-threatening conditions that frequently necessitate intensive neurosurgical interventions for accurate diagnosis and treatment, despite advancements in antimicrobial treatments and resuscitation techniques. *Viral and Fungal Infections of the Central Nervous System: A Microbiological Perspective* serves as a refresher for teachers, clinical practitioners, graduate and postgraduate students, and researchers in microbiology. The twenty chapters that follow delve into the intricacies of viral and fungal infections of the CNS, covering a wide range of topics that span from basic microbiology to advanced clinical management. Each chapter is meticulously crafted by experts in the respective fields, ensuring that you receive the most up-to-date and accurate information available. I wish the students and readers a successful educational experience and success to authors and editors.

Principal, JNMCH, AMU,
Aligarh, India

Haris M. Khan

Foreword II

I am delighted to write the foreword for *Viral and Fungal Infections of the Central Nervous System: A Microbiological Perspective*, which reports updated information on central nervous system (CNS) infections of viral and fungal etiologies.

This book serves as a valuable window, connecting the world of medical neurovirology and neuromycology with the world of clinical practice. Medical neuromycology and neurovirology are specialties that have made great strides in recent years and have had a significant impact on clinical outcomes, especially for those with immunocompromised conditions like AIDS.

The availability of a wide range of antifungals, newly discovered antivirals, or immunological-based medicines, supported by successful trials, has improved over the past several decades from a time when there were only a few agents available. The reader will find that all these aspects are covered in this book. The text also briefly covers: approach to a patient with suspected CNS infection; an account of all the causes of CNS infections, including bacterial, parasitic, prion, and postinfectious; immunology aspects of CNS illnesses; role of neuroinfections in immunocompromised states; future diagnostic aspects.

This timely, comprehensive text will plant the roots of knowledge with the final aim to support academicians, neurologists, physicians, and all healthcare personnel to be confident in identifying and managing neuroinfections. I anticipate that healthcare professionals, including medical students, will find this book to be a useful learning tool and reference, which will result in better patient care.

I congratulate all the involved editors and writers for their commendable work, with the hope that it will be widely appreciated by all readers.

Department of Microbiology
and Immunology, College of Medicine
and Health Sciences, Sultan Qaboos University
Muscat, Oman

Meher Rizvi

Preface

There are many different types of microorganisms that may affect the meninges and brain causing central nervous system (CNS) infections. Likewise, the number of CNS illnesses attributed to viruses and fungi is increasing over the globe. Despite the development of novel diagnostic techniques and the accessibility to cutting-edge healthcare technologies, CNS infections are associated with considerable morbidity and mortality worldwide. The importance of this group of diseases for medical care demands the creation of a well-structured resource for recent scholarly literature, which encompasses up-to-date clinical and diagnostic practices as well as novel, prospective advances in the identification, treatment, and prevention of viral and fungal CNS infections.

The book is divided into four main parts. The reader is given profound knowledge on the various etiologies, patho-physiologies, and host responses involved in different CNS infections in the first part. It lists not only the common bacteria, viruses, fungi, parasites, but also the rarer microbes, including prions responsible for neural infections. Syndrome-driven strategies for neuro-infections are also stressed, together with patient history, nervous system examination, cerebrospinal fluid findings, laboratory diagnostics, imaging, and management principles.

The two parts that follow go in depth on CNS illnesses caused by certain viral and fungal pathogens. It includes extensive extracts on neurotropic viruses' long-term effects, mycological neuro-infections, along with the inclusion of lesser-known viruses and fungi that cause neurological disease. In the relevant chapters, subjects including the fundamentals of aggressive antifungal therapy and procedures are actively covered.

The book's final part, which stands out for its originality, is devoted to giving clinicians an understanding of the prospects for effectively identifying and treating viral and fungal neuro-infections in the future in order to address related morbidities.

Neurologists, clinical microbiologists, and medical professionals who work with patients suffering from neurological infections might use this book's clinically focused counsel. The ultimate goal is to shape the reader into someone who is self-assured enough to draw logical conclusions and effectively manage patients with viral or fungal central nervous system infections.

Aligarh, India

Hiba Sami
Safiya Firoze
Parvez A. Khan

Disclosure Statement

The views expressed in the contributed chapters are solely the views of the contributing authors, and the editors bear no responsibility in relation to its contents. The primary author/corresponding author also takes the responsibility for the originality of the contribution and for any form of copyright/plagiarism issues.

Acknowledgments

We submit our work and endeavors to the Almighty as we thank him for bestowing us with the great opportunity of editing this book. We want to express our heartfelt gratitude to all the authors who have contributed to this book's completion. Without their support, dedication, and expertise, this project would not have been possible. Their talent, creativity, and commitment have been truly inspiring. It has been an honor to collaborate with them and connect with their ideas to create a captivating manuscript. We are immensely proud to have been a part of this journey as editors.

Special thanks to Prof. Mohammad Gulrez, the honorable Vice Chancellor of Aligarh Muslim University, India, for his encouragement and continuous support during this publication process. Thanks are due to Prof. Veena Maheshwari, Dean of Jawaharlal Nehru Medical College, and Dr. Nazish Fatima, Chairperson, Department of Microbiology, JN Medical College AMU, Aligarh, India, for their support. We are forever indebted to our mentors, Prof. Haris M Khan, Prof. Indu Shukla, Prof. Mohd Shahid, Dr. Meher Rizvi, and Prof. Mohd Shahid, for guiding us in every phase of life.

Infinite recognition is due towards our institution, Aligarh Muslim University, and the Department of Microbiology, for serving as a backbone and giving us the resolve to complete this assignment. We also want to express immense gratitude to the publishing team for their constant support and direction throughout the process.

Lastly, we would like to express our gratitude to our family, friends, and loved ones who have provided us with unwavering support, encouragement, and understanding throughout the long hours and demanding deadlines. Their presence in our life has been a constant source of inspiration and motivation.

Hiba Sami
Safiya Firoze
Parvez A. Khan

Contents

Editors and Contributors

About the Editors

Hiba Sami is working as an Assistant Professor in the Department of Microbiology, JNMC, AMU, Aligarh, India. She is actively involved in postgraduate and undergraduate teaching learning activities. Her field of interest are Virology, COVID-19, Zoonosis, and Hospital Acquired Infections. She has won the PROF ASHA MATHUR BEST POSTER IN VIROLOGY in UP MICROCON 2016. She underwent molecular training in virology at "ILBS, NEW DELHI." Currently she is the PI of a ICMR-funded project on Brucellosis and Co-PI of ICMR-funded Project On HIV-HBV co-infection. She edited a book titled *Beta-lactam Resistance in Gram-Negative Bacteria: Threats and Challenges* published by Springer Nature in 2022. She has several book chapters and around 55 research publications in international and national journals to her credit Dr. Hiba Sami completed her MBBS from JNMC, AMU, and MD in Microbiology from the same institution in years 2010 and 2013, respectively. She holds life membership of Indian Association of Medical Microbiologist (IAMM) and Hospital Infection Society, India.

Safiya Firoze is currently working as a clinical registrar in the Department of Microbiology, Jawaharlal Nehru Medical College and Hospital, Aligarh Muslim University, India. She received her degree of "Bachelor of Medicine and Bachelor of Surgery" in the year 2016 and the degree of "Doctor of Medicine in Microbiology" in 2021, from JN Medical College and Hospital of Aligarh Muslim University. She has authored/coauthored many works, including book chapters, case studies, research papers, review articles, and systemic analyses for a number of publications. Her writings and researches have addressed a variety of subjects, including but not limited to: antifungal susceptibility testing of dermatophytes; effects of mask use in the COVID-19 era; advanced diagnostics in medical parasitology; biotechnologies for biofilm prevention; antimicrobial-coated implants for orthopedic implant-associated infection prevention. She is renowned among her peers and pupils, for her enthusiastic approach to teaching/learning activities for undergraduate/graduate students. She actively participates in CME activities, intra-departmental seminars, workshops, and has presented at national/international conferences, such as at the 3rd International Conference on Virology Infectious Diseases and COVID-19, UAE, 2021, and the 18th Annual conference of the Association of Medical Microbiologists (UP-UK MICROCON), India, 2023. She was awarded third prize in "Need for COVID-19 Detection" at the Hackathon: Combat COVID-19—2nd World Conference on Advances in COVID-19—Bengaluru Genomics Center, India, and High Altitude Pulmonary and Pathology, Bolivia, 2020. She holds memberships in different professional bodies, including the "Indian Association of Medical Microbiologists," "Association for Professionals in Infection Control and Epidemiology," and the "British Infection Association."

Parvez A. Khan is an Assistant Professor in the Department of Microbiology, JNMCH, AMU, Aligarh, India. He did his MBBS and MD (Microbiology) from JN Medical College and Hospital of Aligarh Muslim University, Aligarh, India, in 2008 and 2011, respectively. He has authored five book chapters and 36 scientific papers in various national and international journals. He holds membership in different professional bodies, including the American Society for Microbiology (ASM), the Indian Association of Medical Microbiologists (IAMM), and the Hospital Infection Society, India.

Contributors

Siraj Ahmad Department of Pharmacology, Jawaharlal Nehru Medical College and Hospital, Aligarh Muslim University, Aligarh, Uttar Pradesh, India

Shariq Ahmed Faculty of Medicine, Department of Microbiology, Jawaharlal Nehru Medical College and Hospital, Aligarh Muslim University, Aligarh, Uttar Pradesh, India

Hala Ali Ebrahim Almahmeed College of Medicine and Medical Sciences, Arabian Gulf University, Manama, Kingdom of Bahrain
Salmaniya Medical Complex, Manama, Kingdom of Bahrain

Shampa Anupurba Department of Microbiology, Institute of Medical Sciences, BHU, Varanasi, India

Nirmaljit Kaur Bhatias Department of Microbiology, Atal Bihari Vajpayee Institute of Medical Sciences and RML Hospital, New Delhi, India

Sidhya Choudhary SMS Medical College, Jaipur, Rajasthan, India

Nafis Faizi Department of Community Medicine, Jawaharlal Nehru Medical College, Aligarh Muslim University, Aligarh, India

Nazish Fatima Department of Microbiology, JNMCH, AMU, Aligarh, UP, India

Abdullah M. Firoze Ahmed Rochester Regional Health, Rochester, NY, USA

Rahma Mohamed Firoze International Medical Center Hospital, Jeddah, Kingdom of Saudi Arabia
UCSI University, Kuala Lumpur, Malaysia

Safiya Firoze Department of Microbiology, Jawaharlal Nehru Medical College Hospital, Aligarh, Uttar Pradesh, India

Aisha Halawani Department of Medical Imaging, King Abdulaziz Medical City-Ministry of National Guard Health Affairs, COM-KSAU-HS-Western Region, Jeddah, Kingdom of Saudi Arabia

Sadia Hassaan Department of Microbiology, Jawaharlal Nehru Medical College and Hospital, Aligarh Muslim University, Aligarh, Uttar Pradesh, India

Jowairiah Hassan Department of Pathology, Jawaharlal Nehru Medical College, AMU, Aligarh, India

Ronni Mol Joji Department of Microbiology, Immunology, and Infectious Diseases, College of Medicine and Medical Sciences, Arabian Gulf University, Manama, Kingdom of Bahrain

Reba Kanungo Department of Microbiology, MAHE, KMC, Mangalore, India
PIMS, Pondicherry, India

Parvez A. Khan Department of Microbiology, Jawaharlal Nehru Medical College Hospital, Aligarh, Uttar Pradesh, India

Saima Khan Shanti Memorial Hospital, Cuttack, Odisha, India

Shayanki Lahiri Department of Computational Medicine and Bioinformatics, University of Michigan Medical School, Ann Arbor, MI, USA

Shalini Malhotra Department of Microbiology, Atal Bihari Vajpayee Institute of Medical Sciences and RML Hospital, New Delhi, India

Samia Masood Department of Medicine, All India Institute of Medical Sciences, New Delhi, India

Rakesh Mishra Department of Neurosurgery, IMS Banaras Hindu University, Varanasi, UP, India

Sumathi Muralidhar Apex Regional STD Centre and SRL – HIV, Vardhman Mahavir Medical College, Safdarjung Hospital, New Delhi, India

Puvaneswaren A. L. Parmasivam Faculty of Medicine and Health Sciences, UCSI University, Kuala Lumpur, Malaysia

Manish Ramesh Patil Department of Neurology, King George's Medical University, Lucknow, Uttar Pradesh, India

S. Patil ARMCH and Research Center, Solapur, India

Kruthika Perumal Department of Microbiology, Sri Ramachandra Medical College and Research Institute, Sri Ramachandra Institute of Higher Education and Research, Chennai, India

Shantanu Prakash King George's Medical University, Lucknow, India

Nadeem Rais Department of Pharmacology, Al-Nakheel Medical Center (NMC Healthcare), Rabigh, Kingdom of Saudi Arabia

Vikash Ranjan Department of Microbiology, Atal Bihari Vajpayee Institute of Medical Sciences and RML Hospital, New Delhi, India

Adil Raza Department of Microbiology, JNMCH, AMU, Aligarh, UP, India

Imran Rizvi Department of Neurology, King George's Medical University, Lucknow, Uttar Pradesh, India

Arghadip Samaddar Department of Neuromicrobiology, National Institute of Mental Health and Neuro Sciences (NIMHANS), Bengaluru, Karnataka, India

Hiba Sami Department of Microbiology, Jawaharlal Nehru Medical College Hospital, Aligarh, Uttar Pradesh, India

Mohd. Shadab Department of Microbiology, Immunology, and Infectious Diseases, College of Medicine and Medical Sciences, Arabian Gulf University, Manama, Kingdom of Bahrain

Mohammad Shahid Department of Microbiology, Immunology and Infectious Diseases, College of Medicine and Medical Sciences, Arabian Gulf University, Manama, Kingdom of Bahrain

Farheen Shaikh Department of Internal Medicine, Juravinski Hospital, Hamilton Health Sciences, Hamilton, ON, Canada

Adesh Shrivastava Department of Neurosurgery, All India Institute of Medical Sciences, Bhopal, MP, India

Suruchi Shukla Department of Microbiology, King George's Medical University, Lucknow, India

Nagarathna Siddaiah Department of Neuromicrobiology, National Institute of Mental Health and Neurosciences (NIMHANS), Bengaluru, Karnataka, India

C. Suman ARMCH and Research Centre, Solapur, India

Karuna Tadepalli Department of Microbiology, All India Institute of Medical Sciences, Bhopal, MP, India

Uzma Tayyaba Department of Microbiology, JN Medical College, Aligarh Muslim University, Aligarh, India

Richa Thakker Healthcare Global Cancer Centre, Mumbai, Maharashtra, India

Kyu Kyu Win Faculty of Medicine and Health Sciences, UCSI University, Kuala Lumpur, Malaysia

Part I

Introduction: An Insight into Central Nervous System (CNS) Infections

Etiology and Epidemiology of Central Nervous System Infections

1

Nafis Faizi and Jowairiah Hassan

Abstract

Infections of the central nervous system (CNS) are of utmost importance because of the various pathogens, emergence and re-emergence of new infections, and the hefty burden they impose on healthcare systems. Despite advances in antimicrobial therapies and resuscitation practices, CNS infections still persist as life-threatening circumstances, often requiring aggressive neurosurgical interventions for proper diagnosis and treatment. CNS infection etiologies vary depending on the geographical location, seasonal time, age group, existing co-morbidities, vaccination status, and the routes of pathogen acquirement. CNS infections predominantly affect low- and middle-income countries (LMICs). CNS infections are caused by a wide range of viruses that can lead to a variety of neurological manifestations, ranging from mild to severe. The etiology of viral CNS infections is attributable to various viruses such as enteroviruses, herpes simplex virus (HSV), varicella-zoster virus (VZV), cytomegalovirus (CMV), Epstein–Barr virus (EBV), and human immunodeficiency virus (HIV). Cryptococcus remains the most common cause of fungal CNS infections, with most cases occurring in sub-Saharan Africa, where it is the leading cause of meningitis. Aspergillosis, histoplasmosis, coccidioidomycosis, and blastomycosis are other common fungal CNS infections.

N. Faizi (✉)
Department of Community Medicine, Jawaharlal Nehru Medical College, Aligarh Muslim University, Aligarh, Uttar Pradesh, India

J. Hassan
Department of Pathology, Jawaharlal Nehru Medical College, Aligarh Muslim University, Aligarh, Uttar Pradesh, India

H. Sami et al. (eds.), *Viral and Fungal Infections of the Central Nervous System: A Microbiological Perspective*, https://doi.org/10.1007/978-981-99-6445-1_1

Keywords

Central nervous system · Meningitis · Virus · Fungi · Epidemiology

1.1 Introduction

Central nervous system (CNS) infections are a set of infections that affect the brain, spinal cord, optic nerves, and the membranes that cover them. CNS infections lead to significant morbidity and mortality, with the persistence of neurological sequelae among survivors. The spread of microorganisms from the blood causes most of these infections. They could be due to the breaching of the blood–brain barrier, head trauma, or even extrinsic contamination, such as in neurosurgery and implants. The epidemiology of CNS infections have changed over time, with new pathogens emerging and the existing ones evolving. These changes are driven by multiple factors, including global travel, migration, climate change, and medical advancements. Despite advances in antimicrobial therapies and resuscitation practices, CNS infections still persist as life-threatening circumstances, often requiring aggressive neurosurgical interventions for proper diagnosis and treatment. CNS infection etiologies vary depending on the geographical location, seasonal time, age group, existing comorbidities, vaccination status, and route of pathogen acquisition. The viral CNS infections predominantly cause meningitis—the infection of meninges surrounding the brain and spinal cord, and, encephalitis—the inflammation of the brain. While encephalitis is diagnosed by an altered state of consciousness, focal deficits, and seizures, often, meningitis is underdiagnosed (Tuppeny 2013). Fungal CNS infections once considered rare, are no longer rare with the increasing use of immunosuppressive therapies and rising fungal resistance. CNS infections are caused by bacteria, mycobacteria, and other parasites as well.

1.2 Epidemiology

CNS infections predominantly affect low- and middle-income countries (LMICs), with an overall combined incidence of 726 cases in low-income countries, 299 in middle-income countries, and 11 in high-income countries per 100,000 people (Robertson et al. 2018). This restricted estimate of incidence selected only five diseases: neurocysticercosis, bacterial meningitis, intracranial abscess, tubercular meningitis/osteomyelitis, and non-tuberculous spinal osteomyelitis. In comparison, the incidence of viral meningitis alone exceeds the combined incidence of all other etiologies, with enteroviruses causing 90% of all viral meningitis (Irani 2008). The global burden of disease reports a reduction in deaths by 21% from 1990 to 2016, but an increase in incidence from 2.50 to 2.82 million in the same time period (Zunt et al. 2018). While viral meningitis is more common than bacterial meningitis, bacterial meningitis is more serious if left untreated.

Viral CNS infections are a significant global public health concern, with an incidence ranging from 0.26 to 17 cases per 100,000 people, varying with age and other factors (McGill et al. 2017). Venkatesan et al. (2013) found that the incidence of viral meningitis in the United States was approximately 0.7 cases per 100,000 people per year, while viral encephalitis occurred at a rate of 1.7 cases per 100,000 people per year. Additionally, Granerod et al. (2010) estimated that the incidence of acute encephalitis syndrome (AES) was approximately 7.4 cases per 100,000 population per year in Europe, with viral etiologies accounting for up to 60% of cases. Viral CNS infections are distributed and reported worldwide, with a predominance of specific subtypes in certain geographies. The burden of viral CNS infections is particularly high in resource-limited settings, where they are associated with high morbidity and mortality rates. For instance, a study by Preux et al. (2017) estimated that the annual incidence of AES in sub-Saharan Africa was approximately 4.3 cases per 100,000 people, with a case-fatality rate of up to 45% in some regions.

Fungal CNS infections were considered rare before the 1970s, reported largely from immunocompromised patients. Fungal CNS infections are now found even in immunocompetent hosts using steroids, cytotoxic drugs, and antibiotics (Chakrabarti 2007). With each passing year, the incidence of fungal CNS infections is steadily increasing owing to a multitude of such factors, despite diagnostic concerns. Fungal central nervous system (CNS) infections are now a significant cause of morbidity and mortality worldwide, particularly in immunocompromised individuals. The burden of fungal CNS infections varies depending on the geographic location, immune status of the host, and the specific causative organism. *Cryptococcosis* is the most common fungal infection affecting the CNS, with an estimated 223,100 cases and 181,100 deaths globally annually (Rajasingham et al. 2017). In the United States, the incidence of cryptococcal meningitis, one of the most common fungal CNS infections, is estimated to be between 0.2 and 0.4 cases per 100,000 people per year (Perfect et al. 2010). Its incidence is higher in sub-Saharan Africa, where HIV/AIDS is endemic, with an estimated incidence of 13 cases per 100,000 people per year (Park et al. 2009). Other fungal pathogens causing CNS infections, such as *Aspergillus* and *Candida* species, are less common, but can also lead to significant morbidity and mortality, especially in immunocompromised patients.

Mycobacterial infections also affect the central nervous system, especially in Southeast Asia. CNS tuberculosis accounts for 1% of all TB cases and TB meningitis is the most common form of CNS TB (Thakur et al. 2018). CNS TB accounts for approximately 5–10% of all extrapulmonary TB (EPTB) cases and approximately 1% of all TB cases. Cerebral malaria caused by *Plasmodium falciparum* disproportionately (>90%) affects children in African regions (WHO 2009).

1.2.1 Epidemiological Triad

The concept of causation has changed drastically since the early days of Henle–Koch's postulates. The mere presence of the agent is not sufficient to cause a disease.

The predispositions are equally important for the causation of a disease. The cause of a disease is defined as "an event, condition, or characteristic that preceded the disease event and without which the disease event either would not have occurred at all or would not have occurred until some later time" (Rothman and Greenland 2005). This sufficient cause model is a conceptual framework used to understand the complex interplay of multiple risk factors in the development of a disease. In this model, a disease is considered to be the result of a combination of one or more necessary and sufficient factors. A necessary factor is one that must be present for the disease to occur, whereas a sufficient factor is one that alone can cause the disease. In the case of cryptococcal CNS infection and immunosuppressant status, Cryptococcus is a necessary factor for the occurrence of infection as it is the causative organism. However, the immune system plays a critical role in the prevention and control of infections. Hence, in a sufficient cause model, the combination of immunosuppression and exposure to *C. neoformans* may be sufficient to cause cryptococcal infection. For example, an HIV-infected person who is also receiving immunosuppressive therapy for another medical condition may be at a higher risk of developing cryptococcal infection if they are exposed to *C. neoformans*. The sufficient cause model highlights the importance of understanding the complex interplay of multiple risk factors in the development of a disease. By identifying the necessary and sufficient factors involved in a disease, researchers and clinicians can develop targeted prevention and treatment strategies to reduce the risk of disease development and improve outcomes in affected individuals.

The classical epidemiological triad comprising agent, host, and environmental factors helps us conceptualize and understand the implications of viral and fungal CNS epidemiology and etiology (Fig. 1.1). Different viruses may show different patterns of transmission and prevalence in different geographic regions. For example, West Nile virus is primarily transmitted through mosquitoes and is most commonly found in parts of the United States, while Japanese encephalitis virus is prevalent in parts of Asia and is primarily transmitted through mosquito bites.

1.3 Viral CNS Infections

Viral infections of the central nervous system (CNS) are caused by a wide range of viruses that can lead to a variety of neurological manifestations, ranging from mild to severe. The etiology of viral CNS infections can be attributed to a variety of viruses such as enteroviruses, herpes simplex virus (HSV), varicella-zoster virus (VZV), cytomegalovirus (CMV), Epstein–Barr virus (EBV), and human immunodeficiency virus (HIV) among others (Venkatesan and Geocadin 2014). Viruses are among the most common causes of meningitis and more common than the combined meningitis cases of all other microorganisms. Focal infections such as viral myelitis are less common. Etiologically, viruses also cause aseptic meningitis which is difficult to distinguish and diagnose (Irani 2008).

Host conditions affect the prevalence of viral CNS infections. In pregnancy, the most common viral CNS infections are Herpes Simplex virus, especially HSV

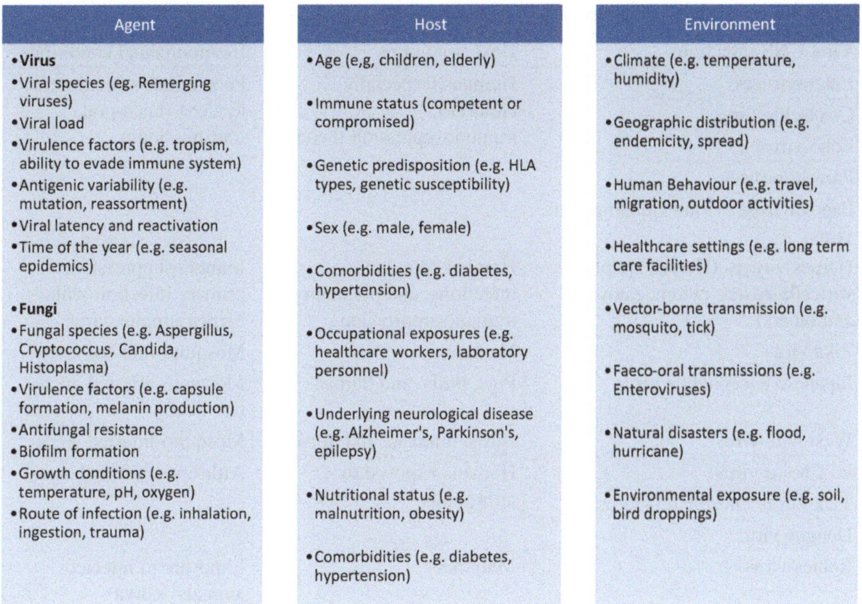

Fig. 1.1 Epidemiological triad for viral and fungal CNS infections

2, Cytomegalovirus, Zika Virus and Rubella virus (Curcio et al. 2020). Arboviral infections such as La Crosse and West Nile virus are more common in certain parts of the world including the United States.

In immunocompromised hosts, the viral CNS infections include Herpes Simplex, Varicella Zoster, Cytomegaloviruses, Epstein–Barr virus, John Cunningham virus, and Human Herpes virus-6 (Walker et al. 2019, Johnston and Guinan 2019). Similarly, causes of viral CNS infections could include poliomyelitis, zika, measles, Japanese encephalitis, and others (Table 1.1).

1.4 Fungal CNS Infections

Cryptococcus remains the most common cause of fungal CNS infections, with most cases occurring in sub-Saharan Africa, where it is the leading cause of meningitis. Aspergillosis, histoplasmosis, coccidioidomycosis, and blastomycosis are other common fungal CNS infections that are associated with significant morbidity and mortality.

Immunocompromised individuals, such as those with HIV/AIDS, solid organ or stem cell transplant recipients, and individuals receiving chemotherapy, are at higher risk for fungal CNS infections (Harvard Health Publishing 2022). Other risk factors include age, diabetes mellitus, and prolonged corticosteroid use.

Prevention and treatment of fungal CNS infections can be challenging due to the limited diagnostic tools and the high mortality rates associated with these infections.

Table 1.1 Viral CNS infections

Viral CNS infections	Host conditions	Environmental conditions
Enteroviruses	Humans (especially children), immunosupression therapy	Poor water, sanitation, and hygiene (Feco-oral transmissions)
Coxsackie virus		
Echo viruses		
Parechoviruses		
Papovaviruses (John Cunningham virus)		
Herpes viruses (Herpes simplex, varicella zoster, cytomegalovirus, and others)	Humans, previous infections, and pregnancy immunosuppression	Immunosuppression, primary infection with herpes simplex virus
Zika virus	Humans	Mosquito-infested areas
Japanese encephalitis virus	Pigs, birds, and humans	Mosquito-infested areas, rice fields
West Nile virus	Birds, humans, and horses	Mosquito-infested areas
La Crosse virus	Humans exposed to arthropods	Arthropod-infested areas
Tick-borne encephalitis virus		
Dengue virus		
Rabies virus	Mammals	Exposure to infected animals' saliva
Poliomyelitis virus	Humans especially children	Poor vaccination coverage, sanitation, and hygiene
Measles, Mumps, Rubella viruses		
Opportunistic infections such as Arenaviruses-Lymphocytic choriomeningitis virus	Reduced CD4 in HIV-infected persons	Poor treatment access or adherence

Early diagnosis and prompt initiation of antifungal therapy are crucial for improving patient outcomes. There are various fungal CNS etiological agents associated with different hosts and environment conditions (Table 1.2).

1.5 Emerging and Re-emerging Infections

Emerging infections are newly recognized or with an increased occurrence within the last few decades and include four distinct categories (Löscher and Prüfer-Krämer 2010).

1. Emerging diagnosis of infectious diseases: old diseases with a recent discovery of the responsible infectious agent.
2. Newly emerging infectious diseases.
3. Re-emerging infectious diseases: reoccurrence or new outbreaks of old infectious diseases with important public health relevance, and
4. Emerging resistance: increasing resistance of infectious agents to antimicrobial substances.

Table 1.2 Fungal CNS infections

Fungal CNS infections	Host conditions	Environmental conditions
Cryptococcosis	Immunocompromised individuals (HIV/AIDS)	Encapsulated yeast found in soil, bird droppings, and decaying wood
Aspergillosis	Immunocompromised individuals (neutropenia, transplantation)	Found in soil and decaying organic matter
Histoplasmosis	Immunocompromised individuals (HIV/AIDS)	Found in soil with high nitrogen content, such as that enriched by bird and bat droppings
Coccidioidomycosis	Individuals living in or traveling to endemic areas (Southwestern United States, Central and South America)	Found in soil in arid and semi-arid regions
Blastomycosis	Immunocompromised individuals (HIV/AIDS)	Found in soil and decaying organic matter

Table 1.3 Emerging and re-emerging viral and fungal CNS infections

Emerging infectious agent	Host	Environmental conditions
Zika virus	Humans, primates	Mosquito-borne
Japanese encephalitis	Humans, birds, pigs	Mosquito-borne
West Nile virus	Birds, humans, horses	Mosquito-borne
Nipah virus	Humans, pigs, bats	Bat-borne, person-to-person transmission
Hendra virus	Humans, horses, bats	Bat-borne, person-to-person transmission
Herpes simplex virus	Humans	Person-to-person transmission
Varicella zoster virus	Humans	Person-to-person transmission
Cryptococcus neoformans	Humans, especially immunocompromised	Tropical climates
Aspergillus spp.	Humans, especially immunocompromised	Tropical climates

Emerging and reemerging infections pose a significant public health threat because of their potential for rapid spread and high mortality rates, many of which include CNS infections. Emerging infections have peaked since the 1980s because of multiple factors, such as climate change and globalization at the planetary level with high population density (megacities), intensive farming, and other factors. The emerging and re-emerging viral CNS infections (Table 1.3), such as Zika virus and West Nile virus have caused large-scale outbreaks in recent years. Many arboviruses, primarily transmitted by mosquitoes, have spread widely because of changes in agricultural practices such as rice and pig farming, urbanization, rise in international travel, urbanization, and other such causes (Chow and Glaser 2014). Arboviruses cause a range of neurological symptoms, including encephalitis, meningitis, and myelitis. Herpesviruses, such as herpes simplex virus and varicella-zoster virus, can also cause CNS infections through person-to-person transmission. These viruses establish latent infections in the sensory ganglia, where they can reactivate and cause viral encephalitis, meningitis, or myelitis. Varicella-zoster virus is a particularly common cause of CNS infections in immunocompromised individuals.

Fungal pathogens, including *Cryptococcus neoformans* and *Aspergillus* spp., are other significant causes of CNS infections (van de Beek et al. 2019). These fungi are commonly found in the environment and can cause meningitis, encephalitis, or brain abscesses in immunocompromised individuals.

1.6 Bacterial, Mycobacterial, and Parasitic CNS Infections

Among the bacterial causes, the most common pathogens are *Streptococcus pneumoniae* and *Neisseria meningitidis* (Bhimraj 2012). *Hemophilus influenzae* type b is also a known cause in children, but its incidence has reduced due to effective vaccination in some settings. In Southeast Asia, tubercular meningitis is common especially in children (Nair et al. 2010). The non-tubercular mycobacterial CNS infections such as *Mycobacterium abscessus* are very rare (Lee et al. 2012). Among the parasitic causes, protozoans such as malaria parasites, toxoplasmosis, and helminths—neurocysticercosis and schistosomiasis—are the most common parasitic organisms causing CNS infections (Garcia 2021). The epidemiological triads for important bacterial, mycobacterial, and parasitic causes of CNS infections are associated with diverse host and environment conditions (Table 1.4).

1.7 Conclusion

The epidemiology and etiology of CNS infections are complex and varied. Understanding the epidemiology, distribution, and determinants is crucial for preventing this rise in viral and fungal CNS infections. There are five significant concerns with the viral and fungal CNS infections:

1. Public Health: CNS infections caused by viruses and fungi are a significant public health concern worldwide. Understanding the epidemiology of these infections, including their incidence, prevalence, and distribution, is essential for developing effective prevention and control strategies.
2. Diagnosis and treatment: Early diagnosis and treatment of CNS infections are critical for preventing long-term neurological sequelae and mortality. Improved understanding of the etiology, pathogenesis, and clinical presentation of viral and fungal CNS infections can aid in their timely diagnosis and management.
3. Emergence of new pathogens: Advances in diagnostic technologies and global travel and migration have led to the emergence and re-emergence of viral and fungal pathogens causing CNS infections. Studying these pathogens is important

Table 1.4 Bacterial, mycobacterial, and parasitic CNS infections

CNS infections	Host conditions	Environmental conditions
Bacterial		
Neisseria meningitidis	Humans, especially unvaccinated	Crowded living conditions, close contact with carriers, and travel to endemic areas
Streptococcus pneumoniae	Humans, particularly the elderly and immunocompromised individuals	Crowded living conditions, exposure to carriers, and comorbidities such as diabetes and HIV
Haemophilus influenzae type b	Children, particularly those under the age of 5	Crowded living conditions, exposure to carriers, and lack of vaccination
Listeria monocytogenes	Neonates, pregnant women, and immunocompromised individuals	Consumption of contaminated food, exposure to contaminated environments, and maternal–fetal transmission
Mycobacterial		
Mycobacterium tuberculosis	Humans, most common in South East Asia	Common in low socio-economic conditions such as overcrowding and immunosuppression
Non-tubercular Mycobacteria	Rare, immunosuppressive conditions	Immunosuppressed conditions
Parasite		
Plasmodium falciparum	Humans affected by cerebral malaria	Tropical and subtropical malarial regions, poor water, sanitation, and poor housing
Taenia solium	Humans (neurocysticercosis)	Poor water and sanitation, consumption of undercooked pork
Schistosoma	Humans, especially children	Exposure to freshwater sources with snails or containing parasitic eggs in feces/urine contaminations
Toxoplasma gondii	Humans, immunocompromised and pregnant women	Contact with cat feces, undercooked meat consumption, hot, humid climates and lower altitudes

for identifying emerging infectious diseases and developing effective strategies for their prevention and control.

4. Drug resistance: The emergence of drug-resistant fungal strains is a growing concern, and understanding the mechanisms of resistance can aid in the development of new antifungal agents. Similarly, the evolution of viral strains can impact the effectiveness of existing antiviral therapies.
5. Interdisciplinary collaboration: The management of CNS infections requires a multidisciplinary approach, involving infectious disease specialists, neurologists, neurosurgeons, and other healthcare professionals. Collaborative research on viral and fungal CNS infections can lead to improved patient outcomes and better public health policies.

For a more complete knowledge of these illnesses on a global scale, advancements in CNS viral and fungal infection detection and research are required.

The prevalence of CNS illnesses will inevitably rise as planetary health continues to be afflicted with climate change and antimicrobial resistance.

References

van de Beek D, Brouwer MC, Tunkel AR (2019) Acute bacterial meningitis. Handb Clin Neurol 160:309–323

Bhimraj A (2012) Acute community-acquired bacterial meningitis in adults: an evidence-based review. Cleve Clin J Med 79(6):393–400

Chakrabarti A (2007) Epidemiology of central nervous system mycoses. Neurol India 55(3):191

Chow FC, Glaser CA (2014) Emerging and reemerging neurologic infections. Neurohospitalist 4(4):173–184

Curcio AM, Shekhawat P, Reynolds AS, Thakur KT (2020) Neurologic infections during pregnancy. Handb Clin Neurol 172:79–104

Garcia HH (2021) Parasitic infections of the nervous system. Continuum (Minneap Minn) 27(4): 943–962

Granerod J, Ambrose HE, Davies NW, Clewley JP, Walsh AL, Morgan D, Cunningham R, Zuckerman M, Mutton KJ, Solomon T, Ward KN (2010) Causes of encephalitis and differences in their clinical presentations in England: a multicentre, population-based prospective study. Lancet Infect Dis 10(12):835–844

Harvard Health Publishing (2022) Fungal infections of the central nervous system. https://www.health.harvard.edu/a_to_z/fungal-infections-of-the-central-nervous-system-a-to-z. Accessed 25 March 2023

Irani DN (2008) Aseptic meningitis and viral myelitis. Neurol Clin 26(3):635–655

Johnston MV, Guinan EC (2019) Viral infections of the nervous system. In: Goldman L, Schafer AI (eds) Cecil medicine, 24th edn. Elsevier, pp 1895–1899

Lee MR, Cheng A, Lee YC, Yang CY, Lai CC, Huang YT, Ho CC, Wang HC, Yu CJ, Hsueh PR (2012) CNS infections caused by *Mycobacterium abscessus* complex: clinical features and antimicrobial susceptibilities of isolates. J Antimicrob Chemother 67(1):222–225

Löscher T, Prüfer-Krämer L (2010) Emerging and re-emerging infectious diseases. In:Krämer A, Kretzschmar M, Krickeberg K, editors. Modern infectious disease epidemiology: concepts, methods, mathematical models, and public health. Berlin: Springer. p. 39–67

McGill F, Griffiths MJ, Solomon T (2017) Viral meningitis: current issues in diagnosis and treatment. Curr Opin Infect Dis 30(2):248–256

Nair N, Wares F, Sahu S (2010) Tuberculosis in the WHO south-East Asia region. Bull World Health Organ 88:164–164

Park BJ, Wannemuehler KA, Marston BJ, Govender N, Pappas PG, Chiller TM (2009) Estimation of the current global burden of cryptococcal meningitis among persons living with HIV/AIDS. AIDS 23(4):525–530

Perfect JR, Dismukes WE, Dromer F, Goldman DL, Graybill JR, Hamill RJ, Harrison TS, Larsen RA, Lortholary O, Nguyen MH, Pappas PG (2010) Clinical practice guidelines for the management of cryptococcal disease: 2010 update by the Infectious Diseases Society of America. Clin Infect Dis 50(3):291–322

Preux PM, Gourmelen J, Druet-Cabanac M, Koffi V, Houinato D (2017) Estimation of the incidence of central nervous system infections in adult West Africans. Ann Trop Med Parasitol 101(4):317–323

Rajasingham R, Smith RM, Park BJ, Jarvis JN, Govender NP, Chiller TM, Denning DW, Loyse A, Boulware DR (2017) Global burden of disease of HIV-associated cryptococcal meningitis: an updated analysis. Lancet Infect Dis 17(8):873–881

Robertson FC, Lepard JR, Mekary RA, Davis MC, Yunusa I, Gormley WB, Baticulon RE, Mahmud MR, Misra BK, Rattani A, Dewan MC (2018) Epidemiology of central nervous system infectious diseases: a meta-analysis and systematic review with implications for neurosurgeons worldwide. J Neurosurg 130(4):1107–1126

Rothman KJ, Greenland S (2005) Causation and causal inference in epidemiology. Am J Public Health 95(S1):S144–S150

Thakur K, Das M, Dooley KE, Gupta A (2018) The global neurological burden of tuberculosis. Semin Neurol 38(2):226–237

Tuppeny M (2013) Viral meningitis and encephalitis. Crit Care Nurs Clin 25(3):363–380

Venkatesan A, Geocadin RG (2014) Diagnosis and management of acute encephalitis: a practical approach. Neurol Clin Pract 4(3):206–215

Venkatesan A, Tunkel AR, Bloch KC, Lauring AS, Sejvar J, Bitnun A, Stahl JP, Mailles A, Drebot M, Rupprecht CE, Yoder J (2013) Case definitions, diagnostic algorithms, and priorities in encephalitis: consensus statement of the International Encephalitis Consortium. Clin Infect Dis 57(8):1114–1128

Walker AC, Gaieski DF, Johnson NJ (2019). How do I manage central nervous system infections (meningitis/encephalitis)? In: Deutschman CS, Neligan PJ, editors. Evidence-based practice of critical care E-Book, p 313. https://www.sciencedirect.com/science/article/abs/pii/B9780323640688000547?via%3Dihub. Accessed April 2023.

World Health Organisation (2009) World malaria report 2009. http://apps.who.int/iris/bitstream/handle/10665/44234/9789241563901_eng.pdf?sequence=1&isAllowed=y. Accessed April 2023

Zunt JR, Kassebaum NJ, Blake N, Glennie L, Wright C, Nichols E, Abd-Allah F, Abdela J, Abdelalim A, Adamu AA, Adib MG (2018) Global, regional, and national burden of meningitis, 1990–2016: a systematic analysis for the Global Burden of Disease Study 2016. Lancet Neurol 17(12):1061–1082

Overview of Infective Syndromes of the Central Nervous System and Its Coverings: Decoding Chameleons and Mimics

2

Safiya Firoze, Hiba Sami, and Parvez A. Khan

Abstract

Infections of the central nervous system (CNS), which include those of the brain's cerebrum and cerebellum, spinal cord, optic nerves, and the membranes that cover them, are medical emergencies that are associated with high rates of morbidity, mortality, or long-term effects that can have detrimental effects on the quality of life of affected individuals. Acute CNS infections may be caused by microorganisms, such as bacteria, viruses, fungi, and parasites, or by trauma caused by fractures at the base of the skull or the cribriform plate, which can cause an opening between the CNS and the adjoining areas, such as sinuses, mastoid, middle ear, or nasopharynx, which may serve as a gateway for infections. Infections in the CNS can manifest as a subarachnoid space infection, such as meningitis, or a parenchymal infection, such as encephalitis, myelitis, or abscess. It is most usually disseminated hematogenously but it can also spread directly from nearby structures (otitis, sinusitis, and dental abscess) as well as invasive and non-invasive trauma (cranial fractures, foreign bodies, and ventricular shunt). CNS infection clinical syndromes may include acute meningitis: characterized by an acute onset of fever, headache, vomiting, meningismus, and impaired mental status in bacterial and viral infections; fast progression over hours to days; subacute or chronic meningitis: found in tuberculosis or fungal infections, with a low-grade fever that develops gradually over weeks; acute encephalitis that is caused by viruses manifests itself in two ways: (a) diffuse: changed mental condition and (b) focal: viral tropism for a single place; and encephalopathy caused by a systemic infection, such as Shigella, typhoid,

S. Firoze · H. Sami (✉) · P. A. Khan
Department of Microbiology, Jawaharlal Nehru Medical College Hospital, Aligarh, Uttar Pradesh, India

H. Sami et al. (eds.), *Viral and Fungal Infections of the Central Nervous System: A Microbiological Perspective*, https://doi.org/10.1007/978-981-99-6445-1_2

malaria, Rickettsia, or endocarditis. Symptoms range from mild to severe and are frequently linked to mood swings.

Keywords

Central nervous system · Meningitis · Encephalitis · Myelitis · Abscess

2.1 Introduction

The brain (cerebellum and cerebrum), spinal cord, optic nerves, and their encasing membranes make up the central nervous system (CNS). The stiff boundaries of the cranium and spinal canal of the vertebral column safeguard these structures. The meninges are three layers of continuous protective tissues that surround the spinal cord and the cerebral cortex, the outermost, gray tissue layer of the brain. The pia mater is the deepest meningeal layer that directly covers the cerebral cortex. The arachnoid and dura mater, respectively, are the terms for the middle and outermost layers (Archibald and Quisling 2013).

2.2 Meningitis

2.2.1 The Meninges: Anatomical and Functional Attributes

The meninges are three layers of continuous protective tissues (Fig. 2.1) that surround the spinal cord and the cerebral cortex, the outermost, gray tissue layer of the brain. The pia mater is the deepest meningeal layer that directly covers the cerebral cortex. The arachnoid and dura mater, respectively, are the terms for the middle and outermost layers of the meninges. A number of compartments, including sinuses for venous drainage, are formed by the dura mater (Archibald and Quisling 2013). These sinuses are entered by arachnoid villi, which are a part of the arachnoid. The Virchow–Robin spaces are connected with the subpial space. The subarachnoid space is not connected to these two compartments, which transmit different vessels to and from the brain parenchyma. While the subarachnoid space is continuous and situated between the pia mater and the arachnoid, the subdural space is situated between the arachnoid and the dura mater. The epidural gap is the space between the dura and the skull. The infection continues to be close to the initiating factor. The majority of infections that spread into the epidural spaces are spread from a nearby source of infection, and the infection remains around the original nidus of infection. Subdural infections, on the other hand, are frequently linked to extracerebral sources, and these infections can also spread broadly inside the subdural compartment, far from the initiating source. The four ventricles of the brain's choroid plexuses continually produce cerebrospinal fluid (CSF). The lateral and third ventricles transport CSF to the fourth ventricle, which is situated between the cerebellum and the midbrain. CSF surrounds the entire central nervous system and travels from the

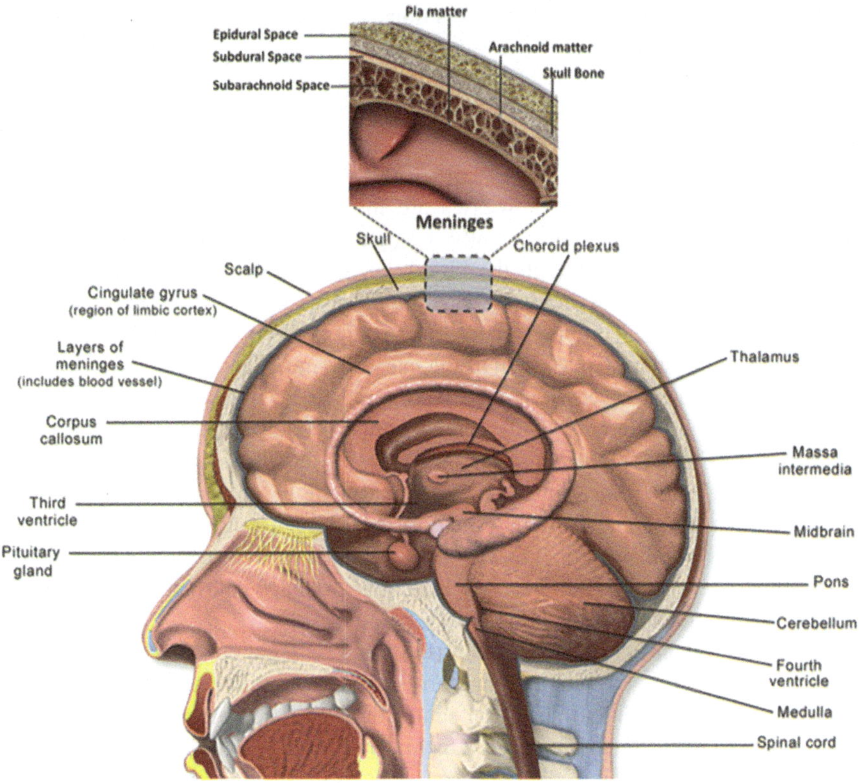

Fig. 2.1 Different layers of meninges

fourth ventricle to the subarachnoid space, arachnoid villi, and the superior sagittal sinus of the dura mater before being reabsorbed into the bloodstream.

Understanding the etiology of CNS infections also requires a solid understanding of the blood supply. The brain and spinal cord have a special capillary supply as the outer layers of endothelial cell layers are merged together. These specialized brain microvascular endothelial cells make up the blood–brain barrier, which isolates the brain and meninges from the circulating blood and prevents the infiltration of microorganisms, toxic agents, and the majority of other compounds while regulating the flow of crucial nutrients and molecules for normal neural function.

As a result, bacteria can infect the central nervous system if they manage to cross the blood–brain barrier. Such breaches can be caused by the organisms themselves, such as *Escherichia coli*, mycobacteria, and spirochetes, or by damage to the blood–brain barrier (such as microhemorrhage or necrosis of the surrounding tissue), the mechanical obstruction of microvessels by parasitized red blood cells, leukocytes, or platelets, excessive production of cytokines that degrade tight junction proteins, or a combination of these factors. The treatment implications are obvious: in order for antimicrobials administered for CNS infections to be successful, they must be able to cross the blood–brain barrier.

2.2.2 Routes of CNS Infection

Meningitis, encephalitis, and abscesses are the three main types of acute CNS infections, and they are all typically brought on by the transmission of the corresponding microbes through the blood. Bacteremia or viremia can be caused by primary infections at more distant anatomic sites, such as the lungs, heart, skin, gastrointestinal tract, or kidney, as well as locations close to or contiguous to the central nervous system (CNS), such as the mastoid, sinuses, or middle ear. The most frequent predisposing diseases in children are middle ear or sinus infections, which cause temporary bacteremia and hematogenous seeding of the central nervous system (CNS) (Kim 2010; Schuchat et al. 1997). Infections with bacteria in the paranasal and otomastoid sinuses frequently result in phlebothrombosis of the nearby cortical veins that drain into them. It is possible for this thrombotic process to enter nearby dural sinuses. A direct passage from the infected sinus to the nearby extra axial spaces or the brain via cortical venous drainage pathways is provided by the phlebothrombosis turning into thrombophlebitis.

When an organism enters the venous sinuses of a patient with bacteremia or viremia, it may cross the blood–brain barrier, pierce the dura as well as arachnoid, and enter the subarachnoid space, infecting the CSF and spreading the infection further all over this anatomic space. Openings between the CNS and the sinuses, mastoid, middle ear, or nasopharynx can result from fractures near the base of the skull or cribriform plate. All of these locations are close to the upper respiratory system; therefore, if there is a CSF leak at one of these locations, respiratory bacteria may track backward up into the subarachnoid space. Intraoperatively, during neurosurgery procedures, an extrinsic contamination of the CNS can happen. In addition, implants or adjunct hardware (such as shunts, ventriculostomies, or external drainage catheters) can colonize and act as foci of infection. Spina bifida and sinus tracts are two examples of congenital abnormalities that can colonize and serve as sources of infections. Through intraneural routes, viruses including polioviruses, herpes simplex virus (HSV), and rabies can travel to the CNS and cause encephalitis (Archibald and Quisling 2013).

2.2.3 Acute Bacterial Meningitis

An inflammatory reaction following pyogenic bacterial invasion of the pia mater, its arachnoid membranes, and the area surrounding the central nervous system is known as bacterial meningitis. Because the subarachnoid space is continuous, this infection frequently affects the full length of the neuraxis, including the brain itself (cerebrum and cerebellum), the spinal cord, optic nerves, and associated encasing membranes. Non-pyogenic microbes (such as mycobacterium or spirochetes like *Leptospira*

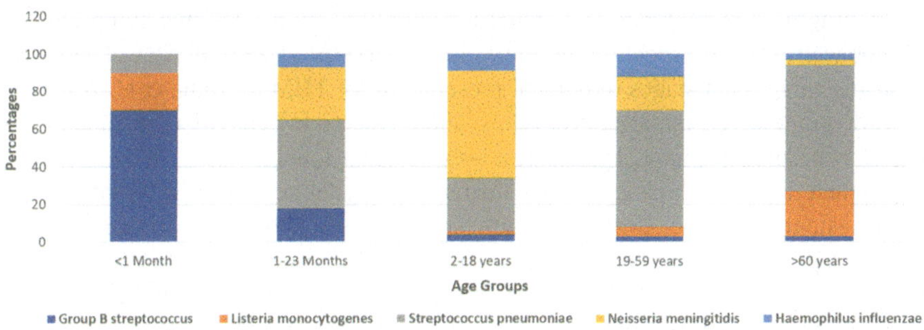

Fig. 2.2 Etiological agents of bacterial meningitis according to different age groups

spp.) are more rarely implicated in pyogenic meningitis and are linked with a prominent, acute inflammatory exudate. Clinically, the illness begins suddenly and progresses over a few hours to a few days to include a fever, headache, irritability, and stiff neck, along with or without specific neurological symptoms (Hsu et al. 2009; Schuchat et al. 1997; Wall et al. 2021).

Streptococcus pneumoniae, Neisseria meningitidis, and *Haemophilus influenzae* type B are the three pathogens most commonly responsible for community-acquired bacterial meningitis. Additionally, specific groups, such as newborns, pregnant women, transplant recipients, and elderly persons, are susceptible to meningitis caused by Gram-negative bacteria like *Escherichia coli* and *Klebsiella pneumoniae* and Group B *Streptococci*, *Listeria monocytogenes*, and *Streptococcus suis* (GBD 2016 Meningitis Collaborators 2018). The age distribution, for the most prevalent agents (Schuchat et al. 1997) in bacterial meningitis, is enumerated in Fig. 2.2.

2.2.4 Tubercular Meningitis

Infection of the meninges with *Mycobacterium tuberculosis* (MTB) bacilli results in tuberculous meningitis (TBM), which presents as extrapulmonary tuberculosis. Droplet inhalation causes the alveolar macrophage to get infected with MTB, which is then transmitted to the host. The lung is where the primary infection first manifests itself, spreading later to the lymph nodes. MTB seeds the meninges in tuberculous meningitis, causing Rich foci, which are sub-ependymal clusters. These foci have the potential to burst into the subarachnoid space, triggering a severe inflammatory reaction that results in meningitis symptoms. This reaction's exudates have the potential to enclose cranial nerves and result in nerve palsies. They can obstruct the flow of cerebral spinal fluid (CSF), which causes hydrocephalus, and entrap blood vessels, resulting in vasculitis. Patients who overcome tuberculous meningitis (TBM) may experience chronic sequelae from these immunological reactions (Slane and Unakal 2023; Thwaites et al. 2000).

Meningitis, tuberculoma, and spinal arachnoiditis are all possible symptoms of Mycobacterium TB infection in the central nervous system (CNS). There are typically three main stages of clinical presentation of TBM (Farinha et al. 2000):

Low-grade fever, malaise, headaches, and a change in behaviors are all signs of the early prodromal phase. Typically, it lasts between 1 and 3 weeks. The meningitic phase then follows, which is distinguished by significant neurologic symptoms, such as persistent headache, urge to vomit, meningismus, fatigue, disorientation, and different presentations of cranial nerve and long-tract indications. In the paralytic phase, confusion is followed by stupor, seizures, coma, and frequently hemiparesis. Untreated sickness frequently ends in death 5–8 weeks after it starts.

2.2.5 Chronic Meningitis

Meningeal inflammation that lasts at least 4 weeks and is accompanied by a CSF pleocytosis is referred to as chronic meningitis (Thakur and Wilson 2018). Atypical bacteria, endemic fungal organisms, and noninfectious etiologies are more frequently responsible for chronic meningitis than community-acquired bacteria and viruses, which are more frequently the causes of acute meningitis. The three most frequent causes of chronic meningitis are neoplasm, TB, and fungal infections.

Chronic meningitis patients with infectious etiology may exhibit symptoms and signs of raised intracranial pressure (ICP), especially when the infection is brought on by *Cryptococcus neoformans* subsp. Fungal meningitis is increasingly recognized in immunocompetent patients as well as those with immunologic diseases taking immunosuppressant drugs or monoclonal antibodies like rituximab. Previously, the disease was thought to only affect those with HIV infection or a hematologic malignancy (Baldwin and Zunt 2014). A list of causes of chronic meningitis (Aksamit 2021; Hildebrand and Aoun 2003; Tan 2003) has been presented in Table 2.1.

2.2.6 Aseptic Meningitis

The phrase "aseptic meningitis" refers to inflammation of the meninges that line the brain caused by a variety of etiologies with sterile cerebrospinal fluid (CSF) culture results. It is determined by CSF pleocytosis of more than 5 cells/mm^3 (Kaur et al. 2023; Tattevin et al. 2019). Although viruses are the most frequent cause of aseptic meningitis, other causes may be categorized as infectious and non-infectious causes.

2.2.7 Meningitis Following Trauma Episodes

Regardless of temporal proximity, post-traumatic meningitis denotes a meningeal infection that is associated with cranio-cerebral trauma. One of the most frequent skull fractures is a basilar skull fracture (BSF) (Das and Pal 2021). Meninges and the

Table 2.1 Causes of chronic meningitis

Infectious causes	Non-infectious	Drugs
Bacterial	Systemic lupus erythematosus (SLE)	NSAIDS
Brucella species	Neoplastic	Intravenous
Francisella tularensis	Sarcoidosis	Immunoglobulins
Actinomyces species	Wegener's Granulomatous Polyangiitis (GPA)	Intrathecal agents
Listeria monocytogenes	Behcet's disease	
Ehrlichia chaffeensis	Central nervous system vasculitis	
Nocardia species	Chemical or drug-induced meningitis	
Tropheryma whipplei (Whipple disease)	Idiopathic	
Spirochetal		
Borrelia burgdorferi		
Treponema pallidum		
Leptospira species		
Mycobacterial		
Mycobacterium tuberculosis		
Fungal		
Cryptococcus neoformans/gattii		
Sporothrix schenckii		
Blastomyces dermatitidis		
Coccidioides immitis		
Others (*scedosporium apiospermum, Paracoccidioides*, dematiaceous molds)		
Parasitic		
Taenia solium		
Angiostrongylus		
Schistosoma		
Toxoplasma gondii		
Acanthamoeba		
Balamuthia mandrillaris		

brain are anatomically close to the inside of the skull; therefore, BSF can cause a dural rupture that causes CSF to leak. BSF frequently involves the frontal and ethmoid bones (Costa et al. 1993). The physical characteristics of the skull bones, especially their thickness and flexibility, play a major role in the degree of deformation and amount of fracture/defect after a skull fracture. The horizontal cribriform plate of the ethmoid bone connects to the frontal bone's ethmoid notch to form a substantial and dense bone complex. However, because of the difference in bone density between these structures, the intersection between the cribriform plate and a comparatively thinner ethmoid labyrinth harboring ethmoidal air cells is especially susceptible to stress. Additionally, frontal sinus involvement is frequent in BSF and even regarded as high risk due to its relationship with a contusion to the anterior part of the frontal lobe and with dural cuts that may result in CSF leakage.

2.3 Space-Occupying Lesions and Associated Infections

An "Intra-cranial space occupying lesion" (ICSOL) may be defined as a mass lesion in the cranial cavity with diverse etiologies, including inflammatory, neoplasm (benign or malignant), parasitic, hematoma, or AV malformation (Hema et al. 2016). The infectious causes of space-occupying lesions involve pyogenic abscess, tuberculosis (Tuberculoma, TB abscess, TB pachymeninigitis), fungal infections (Aspergilloma, Chromomycosis, Zygomycosis, Cryptococcoma, and Candida), parasitic infections (Cysticercosis, Hydatid cyst, and Toxoplasmosis), and viral infections (progressive multifocal leukoencephalopathy caused by JC virus; Santosh et al. 2010).

2.4 Healthcare-Associated Infections of the CNS

Healthcare-associated meningitis or ventriculitis (HCAVM) is defined as "the presence of a positive CSF culture or at least two symptoms (among fever >38 °C, headache, meningeal or cranial nerve signs), along with at least one additional criterion either of abnormal CSF (increased white cells, elevated protein, and decreased glucose), detection of bacterial growth in CSF on Gram Staining, bacteremia, or diagnostic antibody titers" (Ippolito et al. 2022). Common symptoms include fever, new-onset headache, nausea, convulsions, and changes in mental status. There may also be erythema. Intrathecal infusion pumps may contribute to the development of surgical site drainage and wound infections. It can be challenging to diagnose HCAVM associated with the presence of ventriculoperitoneal, ventriculopleural, or ventriculoatrial shunts because of the presentation, which might include symptoms of peritonitis, abdominal discomfort, pleuritis, glomerulonephritis, or bacteremia. The treating physician should be prompted to include HCAVM in the differential diagnosis if there are no other sources of infection that are suspected. The clinical presentation might not be particularly specific because it may be complicated by the underlying neurological disease (e.g., issues that result in a reduced state of consciousness or coma) or because it may be similar to other concurrent conditions (e.g., septic shock from other sources). Shunt infections typically arise as a result of skin flora colonizing the shunt. This could happen during or afterward surgery because of the disintegration of the surgical wound or surrounding skin. Staphylococci are the most common pathogens in these infections, which happen in the first few weeks following shunt implantation. The most common kind of cerebrospinal fluid shunt infection is an early infection with skin flora; approximately, half of all shunt infections are caused by coagulase-negative staphylococci, and approximately, one-third of cases are caused by *Staphylococcus aureus*. *Diphtheroids,* like *Corynebacterium jeikeium* and *Cutibacterium* [previously *Propionibacterium*] acnes, may also be pathogenic.

2.5 Viral Encephalitis and Encephalopathies

Encephalopathy is defined as "an altered consciousness persisting for longer than 24 h, including lethargy, irritability or a change in personality or behaviour" (Granerod et al. 2010). Encephalitis is defined as "encephalopathy and evidence of CNS inflammation, demonstrated by at least two of:

- Fever.
- seizures or focal neurological findings attributable to the brain parenchyma,
- CSF pleocytosis (more than 4 white cells per μL).
- EEG findings suggestive of encephalitis neuroimaging findings suggestive of encephalitis"(Granerod et al. 2010).

Various known causes of Encephalitis (Ellul and Solomon 2018), which may be viral etiology, autoimmune etiology, or due to other causes (Table 2.2).

"Inflammation of the brain parenchyma brought on by a virus" is known as viral encephalitis. It coexists commonly with viral meningitis and is the most prevalent kind of encephalitis. Viruses enter the host outside of the central nervous system and then travel retrogradely from nerve terminals or hematogenously to the spinal cord and brain (Said and Kang 2023). Important causes of viral and fungal encephalitis are discussed in later chapters of this book.

Table 2.2 Important causes of encephalitis

Viral
HSV 1 and 2, VZV, Enteroviruses, Adenoviruses, Measles, Parechovirus, HIV, West Nile virus, Japanese Encephalitis, EBV, CMV, HSV-6 and 7, Mumps, Rubella, t. Louis virus, Eastern equine virus, Western equine virus, Dengue virus, and Rabies virus
Autoimmune
NMDAR antibody encephalitis (ovarian teratoma), LGI-1 antibody encephalitis (thymoma), antibodies against intracellular antigens: anti-Hu (small cell lung tumor), anti-Ma (testicular tumors), anti-GAD, acute disseminated encephalomyelitis, and Bickerstaff's encephalitis
Infective
Bacterial meningitis, TB, opportunistic infections in immunocompromised patients (e.g. crytococcus, toxoplasma, and CMV), Systemic sepsis with encephalopathy
Inflammatory
Vasculitis, systemic lupus erythematosus with CNS involvement, Behçet's disease, and neurosarcoidosis
Metabolic
Hypoglycaemia, hyponatremia, hepatic encephalopathy, and toxins (drugs and alcohol)
Neoplastic
Primary brain tumor (mainly low-grade glioma mirroring CNS inflammation), metastases

2.6 Mycotic Neuro-Invasions

A host's immunological wellness plus a fungal pathogen's degree of virulence play major roles in neuro-invasion (Hernández-Chávez et al. 2017). Mycoses result in considerable morbidity in immunosuppressed hosts; in addition, neurological complications may have lethal effects (Table 2.3) (Góralska et al. 2018). Still, several fungi (e.g., Coccidioides, Cryptococcus, Histoplasma, etc.) may also infect patients who have a functioning immune system (Guarner and Brandt 2011).

Cryptococcus neoformans-associated meningoencephalitis is an exceptionally common mycotic neuro-infection infection globally (Maziarz and Perfect 2016). CNS involvement is frequently linked to diffuse mycosis. Patients exhibiting an aggressive cryptococcosis are predicted to experience full-blown neurological involvement in 67–84% of cases. Likewise, 3–64% of invasive *Candida* infections tend to cause secondary neuro-invasion. Some cases of blastomycosis (40%), disseminated coccidioidomycosis (25%), disseminated histoplasmosis (5–20%), mucormycosis (12%), and invasive cases of severe aspergillosis (4–6%) may also bring about neuropathological sequelae (Góralska et al. 2018).

2.6.1 Blood–Brain-Barrier Interactions in Fungal Invasion

Fungal organisms may reach the central nervous system directly or indirectly by penetrating the blood–brain barrier. Direct seeding can occur via iatrogenic means or via an incident/s of trauma. Fungi can penetrate the blood–brain barrier by three main mechanisms (Table 2.4; Fig. 2.3) (Snarr et al. 2020).

Table 2.3 Overview of mycotic neuro-involvements

Fungal pathogen	Remarks
Aspergillus	Brain abscess, skull/base involvements, infarction/stroke, and dissemination
Absidia	Rhino-cerebral syndromes
Blastomyces	Brain abscess
Candida	Brain abscess, dissemination, meningitis, and meningoencephalitis
Coccidioides	Brain abscess, dissemination, meningitis, and meningoencephalitis
Cryptococcus	Dissemination, meningitis, and meningoencephalitis
Exserohilum	Meningitis
Histoplasma	Brain abscess and meningitis
Mucoromycetes	Brain Abscess, dissemination, and infarction/stroke
Rhizomucor	Rhino-cerebral syndromes
Syncephalastrum	Rhino-cerebral syndromes

Table 2.4 Main mechanisms in fugal blood–brain-barrier penetration

	Mechanism	Remarks
1.	Paracellular	Weakened cell-to-cell adhesions
2.	Transcellular	Adherence to endothelial cells
3.	Trojan Horse	Carried over by phagocytes

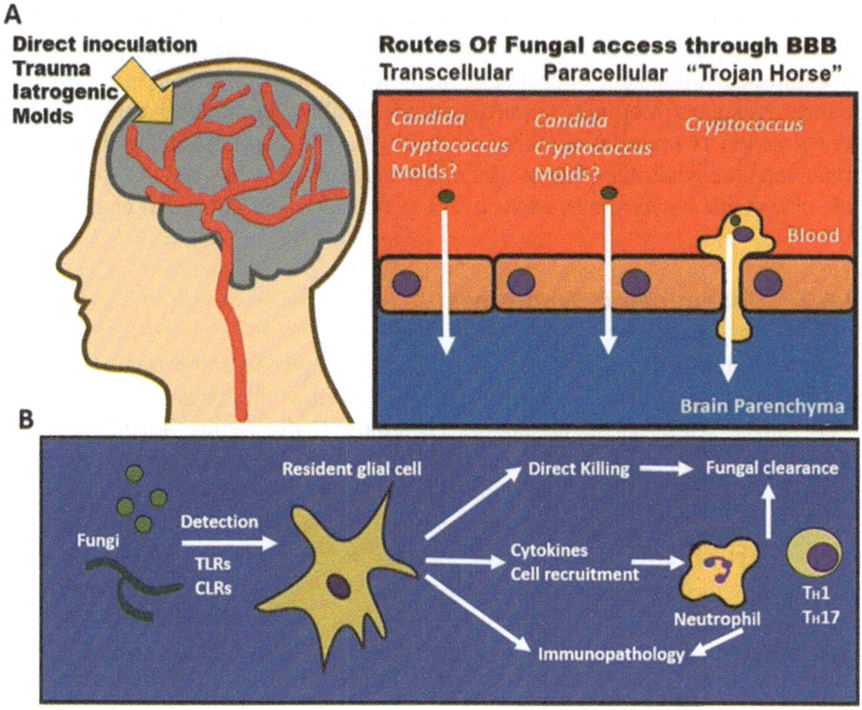

Fig. 2.3 Fugal blood–brain-barrier penetration

The expression of toll-like receptors as well as C-type lectin receptors on astrocytes and microglia enables the detection of invading fungi in the host's CNS (Wu et al. 2023). Subsequently, the glial cells may either actively combat the insult (i.e., fungal clearance from the CNS) or they may bring in particular immune cells by releasing certain chemoattractant peptides and cytokines (Lionakis et al. 2023). It is possible to eradicate the insult with the help of innate (i.e., neutrophils) and adaptive (i.e., T helper 1, 17, etc.) immune system cells. However, resultant severe inflammation plus immunotoxicity may negatively impact the nervous system indefinitely (Bartemes and Kita 2018).

2.6.2 Rhino-Cerebellar Syndromes

A relatively less common condition known as rhino-cerebral mucormycosis, sometimes known as zygomycosis, affects the central nervous system, paranasal sinuses, and the nose (Bhandari et al. 2023). This represents opportunistic fungi that may be isolated among those with compromised immune systems (Hernández-Chávez et al. 2017). A rapid spread of rhino-orbital-cerebral mucormycosis (ROCM) was witnessed in India, among active and post COVID-19 patients (predominantly in diabetic groups) (Dubey et al. 2021).

The filamentous fungi (Table 2.5) spread quickly and vigorously given that they affect people who already have compromised immune systems, leading to a clearly distinguished, fulminant, often fatal condition (Bhandari et al. 2023). In order to reduce associated morbidities and avoid irreversible neurological consequences, timely action is essential (Chikley et al. 2019). The majority of the time, these rhino-cerebral syndromes display acute presentations, but they can also be chronic infections with slower progressions over a period of weeks (Bhandari et al. 2023).

2.6.3 Disseminated Syndromes

Systemic fungal infections and, occasionally, neuro-mycosis may arise from spreading through the bloodstream from distant sites, such as the lungs, gut, or artificial heart valves (Góralska et al. 2018). Therefore, CNS seeding arises via direct propagation from juxta-cranial locations or via hematogenous dissemination (Raman Sharma 2010). Cryptococcosis and coccidioidomycosis are among some of the common disseminated syndromes resulting in CNS fungal infections (Góralska et al. 2018). In addition, meningitis may occur as an unanticipated consequence of disseminated candidiasis (Lionakis et al. 2023). Disseminated blastomycosis, mainly in immunocompromised patients, can present with CNS sequelae, manifesting as meningitis, brain abscess, or epidural abscess (McBride et al. 2017).

2.7 Infections of the Brainstem, Cerebellum, and Spine

Although primary infectious pathologies involving the brainstem are less common, when present, they usually manifest as an abscess or as encephalitis. *Staphylococci*, certain *streptococci*, and *Mycobacterium tuberculosis* are frequent culprits for the development of abscesses in the brainstem (Li et al. 2020). Herpes simplex virus and *Listeria monocytogenes* infection are frequent culprits for brainstem-linked encephalitis (Cunha et al. 2007). During primary brainstem infections, typical brainstem findings occur infrequently and may manifest features of cerebrospinal fluid blockage, as is the case in neurocysticercosis (Archibald and Quisling 2013; Garcia et al. 2014).

Microorganisms, such as John Cunningham virus (JCV), *L. monocytogenes*, and varicella zoster virus (VZV), commonly or primarily attack the cerebellum (Archibald and Quisling 2013; Le Govic et al. 2022). Other such causative viral illnesses include Epstein–Barr virus (EBV), echovirus, and Coxsackievirus infections (Muscat et al. 2017; Muzio 2022). Infections can occasionally produce cerebellar inflammation, which impairs the cerebellum's functional capacity and leads to ataxia. Although bacterial diseases, like Lyme disease, can also produce cerebellar ataxia, viral illnesses like chickenpox are predominantly to blame (Arav-Boger et al. 2002; Betancourt Fursow et al. 2013). This causes an abrupt or rapid bout of ataxia in an individual who was previously well, and it affects youngsters far

Table 2.5 Molds that cause paranasal sinus and orbital infections

Phylum	Zygomycota				Ascomycota	
Class	Zygomycetes				Eurotiomycetes	Dothideomycetes
Order	Mucorales		Eurotiales		Pleosporales	
Family	Mucoraceae	Cunninghamella	Saksenaeaceae	Lichtheimiaceae	Trichocomaceae	Pleosporaceae
Genus/*species*	*Apophysomyces elegans*	*Cunninghamella bertholletiae*	*Apophysomyces elegans*	*Lichtheimia corymbifera (also known as Absidia corymbifera)*	*Aspergillus flavus*	Alternaria
	Mucor ramosissimus		*Apophysomyces trapeziformis*		*Aspergillus fumigatus*	Bipolaris
	Rhizomucor pusillus		*Saksenaea vasiformis*			Curvularia
	Rhizopus oryzae					

more frequently than it does adults. Acute post-infectious ataxia or acute cerebellar ataxia are common names for this type of ataxia (Fogel 2012).

Some relevant points regarding spinal infections:

- It is conceivable that a far-off infection source acts as a nidus for the hematogenous propagation of bacteria toward the spinal cord. Typical antecedent locations include the epidermis and the genitourinary system, but other potential foci have also been identified, including sinusitis, pulmonary, gut, oral infections, septic arthritis, and subacute bacterial endocarditis (Tsantes et al. 2020).
- Spinal infections are increasingly being caused by intravenous drug usage. *Staphylococcus aureus* tends to be the microbe that is most probably responsible for infecting the spine, but among intravenous drug abusers, *Pseudomonas* strains are as frequently responsible (Tsantes et al. 2020).
- Yeast, other fungi, parasitic pathogens, and *M. tuberculosis* are all potential causes of non-pyogenic osteomyelitis of the spine (Moritani et al. 2014; Skaf et al. 2010).
- Following spinal operations, surgical-site infections may occur as a negative outcome. The frequency of such infections in spinal surgeries can be decreased by following rigorous aseptic protocols and by timing pre-op antimicrobial prophylaxis effectively (Aleem et al. 2020).
- Mycotic spinal infections are relatively less common, and they typically affect elderly, diabetic, or immune-compromised people. Alcoholics, leukemic patients, people on chemotherapeutic agents, those suffering from lymphomas, and transplant patients are more vulnerable to such fungus infections (Frazier et al. 2001; Kim et al. 2006).
- The region of the lumbar spine has the highest rate of vertebral osteomyelitis due to the great blood supply there (Graeber and Cecava 2023). Intravenous drug addicts are especially prone to getting cervical spine infections, while in tuberculosis, the thoracic spine tends to be the commoner site (Garg and Somvanshi 2011; Singh et al. 2006).
- The most prevalent kind of spinal infection is spondylodiscitis, which is an infection of the intervertebral disc and the nearby vertebral body. More specifically, primary pyogenic (bacterial) spondylodiscitis, in which the pathogen affects the site(s) by hematogenous dissemination, is the most frequent form of spinal infection (Tsantes et al. 2020).

2.7.1 Epidural Abscess of the Spine

A fairly well-demarcated, pus-filled mass can compress the spinal cord or compromise its blood supply (spinal epidural abscess), inflicting spinal cord damage (Gala and Aswani 2016). A purulent growth within the column's epidural space has the potential to press against the cord, resulting in sensory, motor, and paralytic impairments or if left untreated, even death (Ameer et al. 2023). Morbidity and mortality rates tend to be greater for epidural abscesses of the cervical spine than for

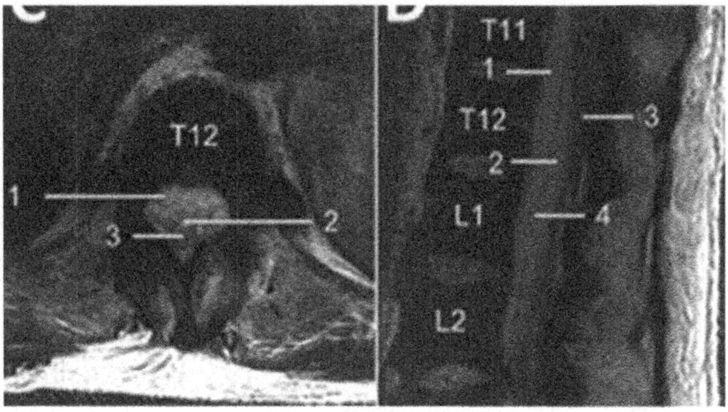

Fig. 2.4 MRI (sagittal and T2-weighted images of thoracic 11 (T11) to Lumbar 2 (L2) vertebra levels) of a patient diagnosed with a methicillin-sensitive *Staphylococcus aureus* infection. There is ventral, 1—epidural and 2—subdural spinal collection with 3—dorsal displacement of conus medullaris. The 4—dura mater can be seen delineating the theca and separating the subdural and epidural spaces. (Image Source—Emerging Infectious Diseases journal—CDC (Jewell et al. 2019)—Copyright Restrictions: NONE; this image is in the public domain and thus free of any copyright restriction)

the more frequently encountered, lumber and thoracic spine abscesses (Sharfman et al. 2020). Although numerous other species, including *Brucella,* Coagulase-negative *Staphylococci, Escherichia coli*, mycobacteria, and *Pseudomonas*, are all documented causes, *Staphylococcus aureus* remains the most reported culprit to date (Fig. 2.4) (Ameer et al. 2023). Methicillin-resistant *S. aureus* (MRSA) is more commonly seen among those who have had spinal procedures, implant insertions, or who have a history of MRSA abscesses (Pi et al. 2023). Individuals who are immunodeficient may present with spinal epidural abscesses linked to unlikely causative microorganisms, including certain fungi (Tsantes et al. 2020).

2.7.2 Neurodegenerative Disorders with Possible Infective Linkages

Neurodegenerative diseases, such as amyotrophic lateral sclerosis (ALS), Alzheimer's disease (AD), and Parkinson's disease (PD), are characterized by progressive neuronal degeneration and functional decline (Katsuno et al. 2018). The fundamental molecular mechanisms behind the pathogenesis of these extremely complicated disorders, which are frequently caused by a number of interrelated hereditary and environmental variables, are still poorly understood (Jellinger 2010).

According to several literary works, it is possible that neuro-infections contribute to the development of neurodegenerative illnesses on multiple fronts (De Chiara et al. 2012; Vigasova et al. 2021; Zhou et al. 2013). ALS, which affects motor neurons, may be caused by enteroviruses and human herpesviruses, according to some serological investigations (Cabrera et al. 2020; Xue et al. 2018). Due to the striking resemblance involving the medical manifestations of Japanese encephalitis virus (JEV) and PD, a JEV-origin for PD has recently been suggested (Ogata et al. 2000; Tadokoro et al. 2018). The flu virus, influenza, has frequently been proposed as a co-factor for PD, dating back to early articles documenting CNS manifestations linked to influenza (Cocoros et al. 2021; Henry et al. 2010; Takahashi and Yamada 2001). A number of research results also suggest that pathogens, mainly herpes simplex virus-1 and *Chlamydia pneumoniae*, play a role in the development of Alzheimer's disease, a complex illness with varying patient-to-patient manifestations (basal cortical neuron, cognition, hippocampus involvement, dementia, etc.) (Breijyeh and Karaman 2020; Harris and Harris 2015; Nicolson and Haier 2009; Wouk et al. 2021). Cytomegalovirus (CMV), EBV, human herpesvirus-6 (HHV-6), HSV, VZV, and now even HHV-7 have been found to be contributors to multiple sclerosis, an inflammatory condition that causes axonal demyelination in the column and brain (Donati and Jacobson 2002).

Transmissible Spongiform Encephalopathies—Prion illnesses, commonly referred to as transmissible spongiform encephalopathies (TSEs), are a collection of deadly CNS illnesses, with the most common theory that they are spread by prions (Moore et al. 2009). Microscopy from affected brain matter reveals numerous cortical holes that resemble a sponge, hence the name. The illness damages CNS function and leads to persistently worsening memory loss, behavioral changes, and difficulties with motion. Human TSEs include Creutzfeldt–Jakob disease (sporadic, familial, iatrogenic, and variant forms), kuru, fatal familial insomnia, and Gerstmann–Sträussler–Scheinker syndrome (Poggiolini et al. 2013). In addition, familial spongiform encephalopathy and variably protease-sensitive prionopathy have also been identified. Having overlapping manifestations, these ailments make up a wide range of disorders (Geschwind 2015; Imran and Mahmood 2011).

2.7.3 Peripheral Nerve Involvement in CNS Infections

Infection may serve as a neglected cause of peripheral nervous system (PNS) disease (Table 2.6) while being very uncommon in comparison to autoimmune, primary inflammatory, or vascular causes. Nevertheless, PNS illness brought on by infection has the potential to result in serious damage to the nervous system, either directly from the microorganism or indirectly via secondary immunological exacerbation. Knowing the variations among multiple illnesses and differentiating between infectious and noninfectious neurological causes may assist in making correct treatment choices and, in certain circumstances, cure or at the very least stop further harm from happening to the patient (Brizzi and Lyons 2014).

2.7.4 Transverse Myelitis, etc.

Transverse myelitis may result from bacterial or viral infections (Table 2.7) that impact the spinal cord. When this inflammatory condition is linked to an infectious etiology, it typically develops following the recovery period (Beh et al. 2013). The

Table 2.6 Microorganisms associated with peripheral nervous system involvement

	Microorganism	
PNS manifestations	Bacteria	Viruses
Acute flaccid paralysis of limb(s)	–	Adenovirus, Enterovirus EV68, Human Immunodeficiency virus (HIV), Poliovirus, Rabies virus, and West Nile virus (WNV)
Acute inflammatory demyelinating polyneuropathy	*Campylobacter jejuni*	Cytomegalovirus (CMV), Epstein-Barr virus (EBV), HIV (early stage), and Varicella zoster virus (VZV)
Axonal sensory polyneuropathy	–	Hepatitis C virus
Compressive radiculopathy	*Mycobacterium tuberculosis*	–
Cranial neuropathy	*Borrelia burgdorferi, Brucella* species, *Corynebacterium diphtheriae, and Mycobacterium tuberculosis*	HIV (early stage), VZV
Demyelinating polyneuropathy	–	WNV
Distal peripheral neuropathy	–	CMV
Distal sensorimotor polyneuropathy	*Corynebacterium diphtheriae and Mycobacterium tuberculosis*	–
Distal symmetric polyneuropathy	*Mycobacterium leprae*	HIV (late stage)
Encephalomyeloradiculitis	–	CMV
Flaccid paralysis	–	WNV
Mononeuropathy (or mononeuritis) multiplex	*Borrelia burgdorferi and Mycobacterium leprae*	CMV, HIV (early stage)
Motor neuropathy/motor neuron disease	–	Human T-lymphotropic virus (HTLV), VZV
Myasthenia gravis-like syndrome	–	HTLV
Myeloradiculitis/ myeloradiculopathy	–	CMV, EBV, VZV
Peripheral neuropathy	*Brucella* species	–
Polyneuropathies with autonomic dysfunctions	–	HTLV
Polyneuropathy	*Borrelia burgdorferi*	–
Postherpetic neuralgia	–	VZV
Radiculopathy	*Borrelia burgdorferi, Brucella* species	–
Sacral radiculitis	–	Herpes simplex virus
Soft palate neuropathy	*Corynebacterium diphtheriae*	–

(continued)

Table 2.6 (continued)

	Microorganism	
PNS manifestations	Bacteria	Viruses
Symmetric descending paralysis	*Clostridium botulinum*	–
Tropical spastic paraparesis	–	HTLV type 1-associated myelopathy

Table 2.7 Microorganisms associated with transverse myelitis

Bacteria	Viruses
Actinomyces species	CMV
Bordetella pertussis	EBV
Borrelia burgdorferi (extremely rare complication)	Echovirus
Clostridium tetani	Enteroviruses (Coxsackievirus, Poliovirus, etc.)
Corynebacterium diphtheriae	Hepatitis B virus
Mycobacterium tuberculosis	Herpes viruses
Treponema pallidum	HIV
Transverse myelitis can also be brought on by bacterial infections of the skin, gastroenteritis, and some strains of pneumonia-causing bacteria.	Influenza virus
	Measles virus
	Mumps virus
	Rubella virus
	WNV
	Zika virus
	Other viruses that do not directly infect the spinal column may nevertheless cause an autoimmune response-associated transverse myelitis.

spinal column can occasionally be affected by parasites and fungi (Beh et al. 2013; Kibiki and Murphy 2006).

Poliomyelitis—The extremely contagious poliomyelitis disease is brought on by the poliovirus, a member of the Picornaviridae family. The main method by which poliovirus spreads from person-to-person is the fecal–oral transmission. For a number of weeks, it can shed in oral secretions, and for a number of months, it can shed in the feces. The virus is capable of destroying the spinal column's anterior horn cells (Mehndiratta et al. 2014).

There are two forms of poliovirus infections: a minor form and a major form (Mehndiratta et al. 2014). The minor-related ailments begin 1–3 days before paralysis presents and are gastrointestinal in nature, including nausea, vomiting, discomfort in the abdomen, cramping, and diarrhea. Additionally, those affected, experience systemic signs, such as headache, fatigue, fever, and sore throat (Wolbert and Higginbotham 2023). The persistence or emergence of any muscular aches and pains indicates that the acute phase has not ended; this phase generally continues for

2–3 weeks but can even last as long as 2 months (Mehndiratta et al. 2014). All poliovirus-related CNS syndromes, such as paralytic poliomyelitis, nonparalytic polio or aseptic meningitis, poliovirus-encephalitis, and bulbar polio, singly or combined, are among the major forms of illness (Mehndiratta et al. 2014).

Countless lives have been affected worldwide by the devastating malformations that have been linked to this illness. The genetic makeup of polioviruses, along with their pathogenicity could only be understood thanks to the persistence and resilience of outstanding scientific researchers during the 1900s (Quarleri 2023). The invention of the oral polio vaccine and the inactivated polio vaccine by Salk and Sabin signaled the beginning of a new era in science. The World Health Organization (WHO) declared the Americas region free from the three forms of wild poliovirus, i.e., types 1, 2, and 3, in 1994. The Western Pacific region followed suit in 2000 and then the European region in June 2002. In 2013, poliovirus was still prevalent in just three nations, namely, Pakistan, Afghanistan, and Nigeria. Polio must be globally eradicated or there will always be a risk of an outbreak (Mehndiratta et al. 2014).

Pott's Spine—Tuberculous spondylitis, referred to as Pott's spine or Pott disease, is a typical extrapulmonary tuberculosis (TB) manifestation. It can cause substantial functional decline and is linked to serious morbidity. Spinal TB is now uncommon in industrialized nations due to the development of anti-tuberculosis medications and better public health practices, yet it remains a serious disease in the developing world (Garg and Somvanshi 2011). Extensive morbidity from spinal involvement of *Mycobacterium tuberculosis*, in the form of serious deformities and lifelong neurological impairment, is conceivable (Rajasekaran et al. 2018). Most people can get their condition under control with therapeutic measures or a combination of therapeutic plus operative regimens (Rasouli et al. 2012).

2.8 Tropical Neuro-Infections

Infections that are common or specific to the subtropics or to the tropics are referred to as tropical diseases (Zumla and Ustianowski 2012). In general, among the more frequent disease, vectors are insects like flies and mosquitoes. The vectors might be carrying a virus, bacteria, or parasite capable of infecting humans as well as animals. The most common way that disease is spread is by the bite of an insect, which transmits the disease-causing agent via subcutaneous blood exchange. Many of these tropical illnesses are not covered by vaccinations, and some may not have efficient treatments (Williams 2023). The presence of extensive wintertime seasons, which drive insects into hibernation, reducing their population, is one reason why these infections are not as common in temperate zones. However, prior to today's knowledge of the causal relationship of different diseases, plenty of these tropical illnesses existed in the Nordic regions and North America around the seventeenth and eighteenth centuries (Rosenau 1925; Zumla and Ustianowski 2012). Expeditions of tropical rainforests by humans, deforestation, increased immigration, global travel, as well as other forms of tropical tourism have all contributed to a spike in the occurrence of such illnesses in non-tropical nations (Lindahl and Grace 2015).

2.8.1 Brain Injury Caused by Neuro-Parasites

Medical manifestations of parasitic CNS infections may vary. It might be challenging to diagnose an illness because its symptoms are frequently modest or ambiguous. It is more probable to identify and treat parasitic neuro-infections when one is acquainted with the fundamental epidemiological traits and distinctive diagnostic imaging results (Carpio et al. 2016) (Table 2.8).

2.8.2 Cerebral Malaria

Cerebral malaria happens to be a particularly serious CNS consequence of *Plasmodium falciparum* disease. It comprises a medical illness marked by coma or other sequelae (Table 2.9); blood films reveal asexual falciparum forms. There is substantial mortality, and those who survive develop sustained brain insult, with lasting cognitive deficits (Idro et al. 2010).

2.8.3 Emerging and Re-emerging Tropical CNS Infections

Globally, as much as 85% of CNS infections are brought about by unidentified causes, with emerging infections (Table 2.10) anticipated to account for a sizable fraction of them. To uncover previously unsuspected infections, clinicians must begin looking beyond the conventional diagnostic algorithms (Ranawaka 2022). Provincial or worldwide outbreaks can be caused by emerging pathogens as they broaden their geographic distribution (such as the Chikungunya virus), often circulating through nonhuman reservoirs (such as the Nipah virus), or developing novel neurovirulence (such as the Chikungunya virus) (Tyler 2009).

2.9 Post-infection Neurological Complications

The relationship between infection and the majority of autoimmune neurological diseases is somewhat complicated. Many individual autoimmune illnesses have been linked to different pathogens, indicating that infections may cause autoimmunity via processes other than just molecular mimicry (Table 2.11) (Blackburn and Wang 2020).

Table 2.8 Overview of CNS lesions caused by parasites (Carpio et al. 2016)

Parasite/disease	CNS lesion(s)	Common medications
Acanthamoeba species, *Balamuthia* species (granulomatous amoebic encephalitis)	Ring-enhancing lesion(s) of gray/white matter, + perilesion edema, mass effect; arterial occlusion and infarction	Fluconazole, flucytosine, pentamidine, sulfadiazine
Echinococcus granulosis (intracranial hydatidosis)	Non-enhancing, thin-walled, globular cysts/small cysts around a bigger cyst, usually in the parietal area	Albendazole
Echinococcus multilocularis (alveolar hydatid disease)	T2WI: contrast-enhanced, globular lesions, ±calcification, +perilesional edema	Albendazole
Naegleria fowlerii, etc. (primary amoebic meningoencephalitis)	T2WI: global edema mostly at the base of the brain, stroke secondary to intracranial tension	Amphotericin B
Paragonimus species (paragonimiasis)	T2WI: 'grape-cluster' ring lesions with perilesion edema at frontal/temporal grey/white matter	Praziquantel
	T1WI: contrast-enhanced lesions with inflammation	
Schistosomiasis (*S. hematobium* favours the spinal column, but other species can infect other parts of the CNS)	T1WI: mass-like lesion with central linear enhancement	Praziquantel
	T2WI: hyperintense lesions, +perilesion edema, mass effect	
	CT: variably enhancing, hyperdense lesion(s), +low-density perilesion edema, mass effect	
Sparganosis	T2WI: edema (increased signal intensities)	Praziquantel
	T1WI/MRI: enhancing/linear mass lesions at frontal/parietal lobes (Rare—cerebellar/brainstem/spinal column involvement)	
Taenia solium (neurocysticercosis)	CT: Solitary lesion/punctate calcification	Albendazole, corticosteroids
	T1WI: 'hole-with-dot' sign, solitary hypointense ring-enhancing lesion (cystic)	
	Intraventricular/subarachnoid/spinal cysts	
Toxoplasma gondii (toxoplasmosis)	Multiple ring-enhancing cerebral lesions, basal ganglia involvement, +cerebral edema	Sulphonamides, pyrimethamine, spiramycin
Trypanosoma brucei (Human African trypanosomiasis)	T1WI: Diffuse grey matter/basal ganglia/internal capsule bilateral hyperintensities	Pentamidine isethionate, corticosteroids, melarsoprol

(continued)

Table 2.8 (continued)

Parasite/disease	CNS lesion(s)	Common medications
Trypanosoma cruzi (Chagas' disease)	T1WI: enhanced, hypointense lesions at frontal/parietal lobes	Benznidazole, corticosteroids
	T2WI: hyperintense perilesion edema, mass effect	

T2WI T2-weighted imaging, *T1WI* T1-weighted imaging, *MRI* magnetic resonance imaging, *CT* computed tomography

Table 2.9 Cerebral malaria's neurocognitive consequences (may be resolving or non-resolving) (Peixoto and Kalei 2013)

Neurocognitive impairments	Attention/executive function/learning/memory abnormalities
Epileptic manifestations	Generalized/tonic clonic seizures
Motor impairments	Spasticity, central hypotonia, cranial nerve palsies, quadriparesis/ hemiplegia/quadriplegia
Movement impairments	Ataxia, dystonia, tremors
Neuropsychiatric manifestations	Abnormal behaviors, attention problems, and hyperactive syndromes
Post malaria neurological syndrome	Abnormal behaviors, acute psychosis, catatonia, hallucinations, inappropriate speech, and seizures
Speech abnormalities	Aphasia, language/pragmatic/vocabulary impairments
Vision abnormalities	Mild (resolving) blindness

Table 2.10 Emerging/reemerging tropical pathogens, which may cause CNS involvement

Emerging/reemerging tropical pathogen	Neurological involvements
Burkholderia pseudomallei	Neurological Melioidosis
Chandipura virus	Acute encephalitis
Chikungunya virus	Encephalitis
	Encephalomyelitis
	Encephalopathy
	Guillain-Barré syndrome
	Meningitis
	Myelopathy
	Myopathy
	Peripheral neuropathy
Dengue virus	Acute disseminated encephalomyelitis
	Cerebellitis
	Cranial/peripheral neuropathy
	Dengue-associated muscle dysfunction
	Encephalitis
	Encephalopathy
	Guillain-Barré syndrome
	Optic neuritis and
	Transverse myelitis
Nipah virus	Encephalitis
Ebola virus	Encephalitis
	Encephalopathy
	Frontal lobe dysfunction
	Meningitis
	Seizures

Table 2.11 Post-infection autoimmune CNS complications (Blackburn and Wang 2020)

Infective pathogen	Autoimmune CNS disorder
Coxsackie B virus	Acute disseminated encephalomyelitis
Dengue virus	
HIV	
Legionella	
Measles	
Mumps	
Mycoplasma pneumonia	
Varicella Zoster virus (VZV)	
Herpes simplex virus	Autoimmune encephalitis
Japanese encephalitis virus	
Mycoplasma pneumoniae	
Severe acute respiratory syndrome coronavirus 2 (SARS-CoV-2)	
West Nile virus (WNV)	
Campylobacter jejuni	Guillain–Barre syndrome
Chlamydia pneumonia	
Cytomegalovirus	
Epstein Barr virus	
Haemophilus influenzae	
Hepatitis A virus (HAV), HEV	
HIV	
Influenza viruses	
Mycoplasma pneumonia	
SARS-CoV-2	
Zika virus	
WNV	Myasthenia gravis
VZV	Neuromyelitis optica spectrum disorders
WNV	Stiff person syndrome
Group A Streptococcus	Sydenham's chorea

References

Aksamit AJ (2021) Chronic meningitis. N Engl J Med 385(10):930–936. https://doi.org/10.1056/NEJMra2032996

Aleem IS, Tan LA, Nassr A, Riew KD (2020) Surgical site infection prevention following spine surgery. Global Spine J 10(1 suppl):92S–98S. https://doi.org/10.1177/2192568219844228

Ameer MA, Knorr TL, Munakomi S, Mesfin FB (2023) Spinal epidural abscess. In: StatPearls. StatPearls Publishing, Treasure Island. http://www.ncbi.nlm.nih.gov/books/NBK441890/

Arav-Boger R, Crawford T, Steere AC, Halsey NA (2002) Cerebellar ataxia as the presenting manifestation of Lyme disease. Pediatr Infect Dis J 21(4):353–356. https://doi.org/10.1097/00006454-200204000-00021

Archibald LK, Quisling RG (2013) Central nervous system infections. In: Textbook of neurointensive care, pp 427–517. https://doi.org/10.1007/978-1-4471-5226-2_22
Baldwin KJ, Zunt JR (2014) Evaluation and treatment of chronic meningitis. Neurohospitalist 4(4): 185–195. https://doi.org/10.1177/1941874414528940
Bartemes KR, Kita H (2018) Innate and adaptive immune responses to fungi in the airway. J Allergy Clin Immunol 142(2):353–363. https://doi.org/10.1016/j.jaci.2018.06.015
Beh SC, Greenberg BM, Frohman T, Frohman EM (2013) Transverse myelitis. Neurol Clin 31(1): 79–138. https://doi.org/10.1016/j.ncl.2012.09.008
Betancourt Fursow Y, Jiménez León JC, Jiménez Betancourt CS (2013) [Acute cerebellar ataxia in childhood]. Medicina 73(suppl 1):30–37
Bhandari J, Thada PK, Nagalli S (2023) Rhinocerebral mucormycosis. In: StatPearls. StatPearls Publishing, Treasure Island. http://www.ncbi.nlm.nih.gov/books/NBK559288/
Blackburn KM, Wang C (2020) Post-infectious neurological disorders. Ther Adv Neurol Disord 13: 1756286420952901. https://doi.org/10.1177/1756286420952901
Breijyeh Z, Karaman R (2020) Comprehensive review on Alzheimer's disease: causes and treatment. Molecules 25(24):5789. https://doi.org/10.3390/molecules25245789
Brizzi KT, Lyons JL (2014) Peripheral nervous system manifestations of infectious diseases. Neurohospitalist 4(4):230–240. https://doi.org/10.1177/1941874414535215
Cabrera JR, Rodríguez-Izquierdo I, Jiménez JL, Muñoz-Fernández MÁ (2020) Analysis of ALS-related proteins during herpes simplex virus-2 latent infection. J Neuroinflammation 17(1):371. https://doi.org/10.1186/s12974-020-02044-4
Carpio A, Romo ML, Parkhouse RME, Short B, Dua T (2016) Parasitic diseases of the central nervous system: lessons for clinicians and policy makers. Expert Rev Neurother 16(4):401–414. https://doi.org/10.1586/14737175.2016.1155454
Chikley A, Ben-Ami R, Kontoyiannis DP (2019) Mucormycosis of the central nervous system. J Fungi 5(3):59. https://doi.org/10.3390/jof5030059
Cocoros NM, Svensson E, Szépligeti SK, Vestergaard SV, Szentkúti P, Thomsen RW, Borghammer P, Sørensen HT, Henderson VW (2021) Long-term risk of Parkinson disease following influenza and other infections. JAMA Neurol 78(12):1461–1470. https://doi.org/10.1001/jamaneurol.2021.3895
Costa H, Cerejo A, Baptista A, Vaz R, Gonçalves M, Guimarães A, Amarante J, Cruz C, Guimarães F (1993) The galea frontalis myofascial flap in anterior fossa CSF leaks. Br J Plast Surg 46(6): 503–507. https://doi.org/10.1016/0007-1226(93)90225-z
Cunha BA, Fatehpuria R, Eisenstein LE (2007) Listeria monocytogenes encephalitis mimicking herpes simplex virus encephalitis: the differential diagnostic importance of cerebrospinal fluid lactic acid levels. Heart Lung 36(3):226–231. https://doi.org/10.1016/j.hrtlng.2007.01.001
Das S, Pal T (2021) Post-traumatic meningitis: case-based review of literature from internists' perspective. Case Rep Acute Med 4(2):41–49. https://doi.org/10.1159/000515540
De Chiara G, Marcocci ME, Sgarbanti R, Civitelli L, Ripoli C, Piacentini R, Garaci E, Grassi C, Palamara AT (2012) Infectious agents and neurodegeneration. Mol Neurobiol 46(3):614–638. https://doi.org/10.1007/s12035-012-8320-7
Donati D, Jacobson S (2002) Viruses and multiple sclerosis. In: Polymicrobial diseases. ASM Press, Washington, DC. https://www.ncbi.nlm.nih.gov/books/NBK2494/
Dubey S, Mukherjee D, Sarkar P, Mukhopadhyay P, Barman D, Bandopadhyay M, Pandit A, Sengupta A, Das S, Ghosh S, Adhikari S, Biswas PS, Pal P, Roy H, Patra N, Das A, Sinha P, Mondal MK, Shrivastava SR et al (2021) COVID-19 associated rhino-orbital-cerebral mucormycosis: an observational study from Eastern India, with special emphasis on neurological spectrum. Diabetes Metab Syndr 15(5):102267. https://doi.org/10.1016/j.dsx.2021.102267
Ellul M, Solomon T (2018) Acute encephalitis – diagnosis and management. Clin Med 18(2): 155–159. https://doi.org/10.7861/clinmedicine.18-2-155
Farinha NJ, Razali KA, Holzel H, Morgan G, Novelli VM (2000) Tuberculosis of the central nervous system in children: a 20-year survey. J Infect 41(1):61–68. https://doi.org/10.1053/jinf.2000.0692

Fogel BL (2012) Childhood cerebellar ataxia. J Child Neurol 27(9):1138–1145. https://doi.org/10.1177/0883073812448231

Frazier DD, Campbell DR, Garvey TA, Wiesel S, Bohlman HH, Eismont FJ (2001) Fungal infections of the spine: report of eleven patients with long-term follow-up. JBJS 83(4):560

Gala FB, Aswani Y (2016) Imaging in spinal posterior epidural space lesions: a pictorial essay. Indian J Radiol Imaging 26(3):299–315. https://doi.org/10.4103/0971-3026.190406

Garcia HH, Nash TE, Del Brutto OH (2014) Clinical symptoms, diagnosis, and treatment of neurocysticercosis. Lancet Neurol 13(12):1202–1215. https://doi.org/10.1016/S1474-4422(14)70094-8

Garg RK, Somvanshi DS (2011) Spinal tuberculosis: a review. J Spinal Cord Med 34(5):440–454. https://doi.org/10.1179/2045772311Y.0000000023

GBD 2016 Meningitis Collaborators (2018) Global, regional, and national burden of meningitis, 1990-2016: a systematic analysis for the Global Burden of Disease Study 2016. Lancet Neurol 17(12):1061–1082. https://doi.org/10.1016/S1474-4422(18)30387-9

Geschwind MD (2015) Prion diseases. Continuum (Minneapolis, Minn.) 21(6 Neuroinfectious Disease):1612–1638. https://doi.org/10.1212/CON.0000000000000251

Góralska K, Blaszkowska J, Dzikowiec M (2018) Neuroinfections caused by fungi. Infection 46(4):443–459. https://doi.org/10.1007/s15010-018-1152-2

Graeber A, Cecava ND (2023) Vertebral osteomyelitis. In: StatPearls. StatPearls Publishing, Treasure Island. http://www.ncbi.nlm.nih.gov/books/NBK532256/

Granerod J, Ambrose HE, Davies NW, Clewley JP, Walsh AL, Morgan D, Cunningham R, Zuckerman M, Mutton KJ, Solomon T, Ward KN, Lunn MP, Irani SR, Vincent A, Brown DW, Crowcroft NS, UK Health Protection Agency (HPA) Aetiology of Encephalitis Study Group (2010) Causes of encephalitis and differences in their clinical presentations in England: a multicentre, population-based prospective study. Lancet Infect Dis 10(12):835–844. https://doi.org/10.1016/S1473-3099(10)70222-X

Guarner J, Brandt ME (2011) Histopathologic diagnosis of fungal infections in the 21st century. Clin Microbiol Rev 24(2):247–280. https://doi.org/10.1128/CMR.00053-10

Harris SA, Harris EA (2015) Herpes simplex virus type 1 and other pathogens are key causative factors in sporadic Alzheimer's disease. J Alzheimers Dis 48(2):319–353. https://doi.org/10.3233/JAD-142853

Hema NA, Ravindra RS, Karnappa AS (2016) Morphological patterns of intracranial lesions in a Tertiary Care Hospital in North Karnataka: a clinicopathological and immunohistochemical study. J Clin Diagn Res 10(8):EC01–EC05. https://doi.org/10.7860/JCDR/2016/19101.8237

Henry J, Smeyne RJ, Jang H, Miller B, Okun MS (2010) Parkinsonism and neurological manifestations of influenza throughout the 20th and 21st centuries. Parkinsonism Relat Disord 16(9):566–571. https://doi.org/10.1016/j.parkreldis.2010.06.012

Hernández-Chávez MJ, Pérez-García LA, Niño-Vega GA, Mora-Montes HM (2017) Fungal strategies to evade the host immune recognition. J Fungi 3(4):51. https://doi.org/10.3390/jof3040051

Hildebrand J, Aoun M (2003) Chronic meningitis: still a diagnostic challenge. J Neurol 250(6):653–660. https://doi.org/10.1007/s00415-003-1101-5

Hsu HE, Shutt KA, Moore MR, Beall BW, Bennett NM, Craig AS, Farley MM, Jorgensen JH, Lexau CA, Petit S, Reingold A, Schaffner W, Thomas A, Whitney CG, Harrison LH (2009) Effect of pneumococcal conjugate vaccine on pneumococcal meningitis. N Engl J Med 360(3):244–256. https://doi.org/10.1056/NEJMoa0800836

Idro R, Marsh K, John CC, Newton CR (2010) Cerebral malaria; mechanisms of brain injury and strategies for improved neuro-cognitive outcome. Pediatr Res 68(4):267. https://doi.org/10.1203/PDR.0b013e3181eee738

Imran M, Mahmood S (2011) An overview of human prion diseases. Virol J 8:559. https://doi.org/10.1186/1743-422X-8-559

Ippolito M, Giarratano A, Cortegiani A (2022) Healthcare-associated central nervous system infections. Curr Opin Anaesthesiol 35(5):549–554. https://doi.org/10.1097/ACO.0000000000001167

Jellinger KA (2010) Basic mechanisms of neurodegeneration: a critical update. J Cell Mol Med 14(3):457–487. https://doi.org/10.1111/j.1582-4934.2010.01010.x

Jewell P, Dixon L, Singanayagam A, Ghani R, Wong E, Coleman M, Pichon B, Kearns A, Russell G, Hatcher J (2019) Severe disseminated infection with emerging lineage of methicillin-sensitive Staphylococcus aureus. Emerg Infect Dis J 25(1):187. https://doi.org/10.3201/eid2501.180684

Katsuno M, Sahashi K, Iguchi Y, Hashizume A (2018) Preclinical progression of neurodegenerative diseases. Nagoya J Med Sci 80(3):289–298. https://doi.org/10.18999/nagjms.80.3.289

Kaur H, Betances EM, Perera TB (2023) Aseptic meningitis. In: StatPearls. StatPearls Publishing, Treasure Island. http://www.ncbi.nlm.nih.gov/books/NBK557412/

Kibiki GS, Murphy DK (2006) Transverse myelitis due to trypanosomiasis in a middle aged Tanzanian man. J Neurol Neurosurg Psychiatry 77(5):684–685. https://doi.org/10.1136/jnnp.2005.079533

Kim KS (2010) Acute bacterial meningitis in infants and children. Lancet Infect Dis 10(1):32–42. https://doi.org/10.1016/S1473-3099(09)70306-8

Kim CW, Perry A, Currier B, Yaszemski M, Garfin SR (2006) Fungal infections of the spine. Clin Orthop Relat Res 444:92. https://doi.org/10.1097/01.blo.0000203451.36522.4c

Le Govic Y, Demey B, Cassereau J, Bahn Y-S, Papon N (2022) Pathogens infecting the central nervous system. PLoS Pathog 18(2):e1010234. https://doi.org/10.1371/journal.ppat.1010234

Li S, Nguyen IP, Urbanczyk K (2020) Common infectious diseases of the central nervous system—clinical features and imaging characteristics. Quant Imaging Med Surg 10(12):2227–2259. https://doi.org/10.21037/qims-20-886

Lindahl JF, Grace D (2015) The consequences of human actions on risks for infectious diseases: a review. Infect Ecol Epidemiol 5:30048. https://doi.org/10.3402/iee.v5.30048

Lionakis MS, Drummond RA, Hohl TM (2023) Immune responses to human fungal pathogens and therapeutic prospects. Nat Rev Immunol 23:433. https://doi.org/10.1038/s41577-022-00826-w

Maziarz EK, Perfect JR (2016) Cryptococcosis. Infect Dis Clin N Am 30(1):179–206. https://doi.org/10.1016/j.idc.2015.10.006

McBride JA, Gauthier GM, Klein BS (2017) Clinical manifestations and treatment of blastomycosis. Clin Chest Med 38(3):435–449. https://doi.org/10.1016/j.ccm.2017.04.006

Mehndiratta MM, Mehndiratta P, Pande R (2014) Poliomyelitis. Neurohospitalist 4(4):223–229. https://doi.org/10.1177/1941874414533352

Moore RA, Taubner LM, Priola SA (2009) Prion protein misfolding and disease. Curr Opin Struct Biol 19(1):14–22. https://doi.org/10.1016/j.sbi.2008.12.007

Moritani T, Kim J, Capizzano AA, Kirby P, Kademian J, Sato Y (2014) Pyogenic and non-pyogenic spinal infections: emphasis on diffusion-weighted imaging for the detection of abscesses and pus collections. Br J Radiol 87(1041):20140011. https://doi.org/10.1259/bjr.20140011

Muscat K, Galea R, Vella M (2017) An adult case of acute EBV cerebellitis. Eur J Case Rep Intern Med 4(1):000519. https://doi.org/10.12890/2016_000519

Muzio BD (2022) Acute cerebellitis | Radiology reference article | Radiopaedia.org. Radiopaedia. https://doi.org/10.53347/rID-35557

Nicolson G, Haier J (2009) The role of chronic bacterial and viral infections in neurodegenerative, neurobehavioral, psychiatric, autoimmune and fatiguing illnesses: part 1. Br J Med Pract 2:20–28

Ogata A, Tashiro K, Pradhan S (2000) Parkinsonism due to predominant involvement of substantia nigra in Japanese encephalitis. Neurology 55:602. https://doi.org/10.1212/WNL.55.4.602

Peixoto B, Kalei I (2013) Neurocognitive sequelae of cerebral malaria in adults: a pilot study in Benguela Central Hospital, Angola. Asian Pac J Trop Biomed 3(7):532–535. https://doi.org/10.1016/S2221-1691(13)60108-2

Pi Y, Gong Y, Jiang J, Zhu D, Tong Y, Jiang L, Zhao D (2023) Extensive spinal epidural abscess caused by Staphylococcus epidermidis: a case report and literature review. Front Surg 10: 1114729. https://doi.org/10.3389/fsurg.2023.1114729

Poggiolini I, Saverioni D, Parchi P (2013) Prion protein misfolding, strains, and neurotoxicity: an update from studies on mammalian prions. Int J Cell Biol 2013:910314. https://doi.org/10.1155/2013/910314

Quarleri J (2023) Poliomyelitis is a current challenge: long-term sequelae and circulating vaccine-derived poliovirus. GeroScience 45(2):707–717. https://doi.org/10.1007/s11357-022-00672-7

Rajasekaran S, Soundararajan DCR, Shetty AP, Kanna RM (2018) Spinal tuberculosis: current concepts. Global Spine J 8(4 suppl):96S–108S. https://doi.org/10.1177/2192568218769053

Raman Sharma R (2010) Fungal infections of the nervous system: current perspective and controversies in management. Int J Surg 8(8):591–601. https://doi.org/10.1016/j.ijsu.2010.07.293

Ranawaka UK (2022) Emerging tropical neurological infections. Clin Med 22(1):18–20. https://doi.org/10.7861/clinmed.2021-0799

Rasouli MR, Mirkoohi M, Vaccaro AR, Yarandi KK, Rahimi-Movaghar V (2012) Spinal tuberculosis: diagnosis and management. Asian Spine J 6(4):294–308. https://doi.org/10.4184/asj.2012.6.4.294

Rosenau MJ (1925) Seasonal prevalence of disease. Sci Mon 20(2):192–214

Said S, Kang M (2023) Viral encephalitis. In: StatPearls. StatPearls Publishing, Treasure Island. http://www.ncbi.nlm.nih.gov/books/NBK470162/

Santosh V, Mahadevan A, Chickabasaviah YT, Bharath RD, Krishna SS (2010) Infectious lesions mimicking central nervous system neoplasms. Semin Diagn Pathol 27(2):122–135. https://doi.org/10.1053/j.semdp.2010.04.004

Schuchat A, Robinson K, Wenger JD, Harrison LH, Farley M, Reingold AL, Lefkowitz L, Perkins BA (1997) Bacterial meningitis in the United States in 1995. Active Surveillance Team. N Engl J Med 337(14):970–976. https://doi.org/10.1056/NEJM199710023371404

Sharfman ZT, Gelfand Y, Shah P, Holtzman AJ, Mendelis JR, Kinon MD, Krystal JD, Brook A, Yassari R, Kramer DC (2020) Spinal epidural abscess: a review of presentation, management, and medicolegal implications. Asian Spine J 14(5):742–759. https://doi.org/10.31616/asj.2019.0369

Singh G, Shetty RR, Ravidass MJ, Anilkumar PG (2006) Cervical osteomyelitis associated with intravenous drug use. Emerg Med J 23(2):e16. https://doi.org/10.1136/emj.2003.012724

Skaf GS, Kanafani ZA, Araj GF, Kanj SS (2010) Non-pyogenic infections of the spine. Int J Antimicrob Agents 36(2):99–105. https://doi.org/10.1016/j.ijantimicag.2010.03.023

Slane VH, Unakal CG (2023) Tuberculous meningitis. In: StatPearls. StatPearls Publishing, Treasure Island. http://www.ncbi.nlm.nih.gov/books/NBK541015/

Snarr BD, Drummond RA, Lionakis MS (2020) It's all in your head: antifungal immunity in the brain. Curr Opin Microbiol 58:41–46. https://doi.org/10.1016/j.mib.2020.07.011

Tadokoro K, Ohta Y, Sato K, Maeki T, Sasaki R, Takahashi Y, Shang J, Takemoto M, Hishikawa N, Yamashita T, Lim CK, Tajima S, Abe K (2018) A Japanese encephalitis patient presenting with parkinsonism with corresponding laterality of magnetic resonance and dopamine transporter imaging findings. Intern Med 57(15):2243–2246. https://doi.org/10.2169/internalmedicine.0337-17

Takahashi M, Yamada T (2001) A possible role of influenza a virus infection for Parkinson's disease. Adv Neurol 86:91–104

Tan TQ (2003) Chronic meningitis. Semin Pediatr Infect Dis 14(2):131–139. https://doi.org/10.1053/spid.2003.127230

Tattevin P, Tchamgoué S, Belem A, Bénézit F, Pronier C, Revest M (2019) Aseptic meningitis. Rev Neurol 175(7–8):475–480. https://doi.org/10.1016/j.neurol.2019.07.005

Thakur KT, Wilson MR (2018) Chronic meningitis. Continuum (Minneapolis, Minn.) 24-(5 Neuroinfectious Disease):1298–1326. https://doi.org/10.1212/CON.0000000000000664

Thwaites G, Chau TT, Mai NT, Drobniewski F, McAdam K, Farrar J (2000) Tuberculous meningitis. J Neurol Neurosurg Psychiatry 68(3):289–299. https://doi.org/10.1136/jnnp.68.3.289

Tsantes AG, Papadopoulos DV, Vrioni G, Sioutis S, Sapkas G, Benzakour A, Benzakour T, Angelini A, Ruggieri P, Mavrogenis AF (2020) Spinal infections: an update. Microorganisms 8(4):476. https://doi.org/10.3390/microorganisms8040476

Tyler KL (2009) Emerging viral infections of the central nervous system. Arch Neurol 66(8): 939–948. https://doi.org/10.1001/archneurol.2009.153

Vigasova D, Nemergut M, Liskova B, Damborsky J (2021) Multi-pathogen infections and Alzheimer's disease. Microb Cell Factories 20(1):25. https://doi.org/10.1186/s12934-021-01520-7

Wall EC, Chan JM, Gil E, Heyderman RS (2021) Acute bacterial meningitis. Curr Opin Neurol 34(3):386–395. https://doi.org/10.1097/WCO.0000000000000934

Williams T (2023) Prevention and treatment of tropical disease. Trop Med Surg 11(2):1–1. https://doi.org/10.35248/2329-9088.23.11:295

Wolbert JG, Higginbotham K (2023) Poliomyelitis. In: StatPearls. StatPearls Publishing, Treasure Island. http://www.ncbi.nlm.nih.gov/books/NBK558944/

Wouk J, Rechenchoski DZ, Rodrigues BCD, Ribelato EV, Faccin-Galhardi LC (2021) Viral infections and their relationship to neurological disorders. Arch Virol 166(3):733–753. https://doi.org/10.1007/s00705-021-04959-6

Wu C, Jiang M-L, Jiang R, Pang T, Zhang C-J (2023) The roles of fungus in CNS autoimmune and neurodegeneration disorders. Front Immunol 13:1077335. https://doi.org/10.3389/fimmu.2022.1077335

Xue YC, Feuer R, Cashman N, Luo H (2018) Enteroviral infection: the forgotten link to amyotrophic lateral sclerosis? Front Mol Neurosci 11:63. https://doi.org/10.3389/fnmol.2018.00063

Zhou L, Miranda-Saksena M, Saksena NK (2013) Viruses and neurodegeneration. Virol J 10(1): 172. https://doi.org/10.1186/1743-422X-10-172

Zumla A, Ustianowski A (2012) Tropical diseases. Infect Dis Clin N Am 26(2):195–205. https://doi.org/10.1016/j.idc.2012.02.007

Clinical Presentation and Diagnosis of CNS Infections Through a Systematic Approach

3

Farheen Shaikh

Abstract

Patients with CNS infections may present with a broad range of vague symptoms and signs; fever, headache, altered mental status, and altered behavior. Traditional physical examination maneuvers, such as Brudzinski's and Kernig's signs, are specific for predicting CSF pleocytosis, although relatively less sensitive. The involvement of parenchyma in patients with encephalitis or brain abscess may experience focal neurologic deficits or seizures. CNS infections are best diagnosed with a combination of neuroimaging and CSF analysis via lumbar puncture. The approach to CNS infections, like in other life-threatening emergency conditions, necessitates keen observation and a high index of suspicion based on the vigorous history and physical examination, which needs to be confirmed with appropriate imaging as well as laboratory evaluation.

Keywords

CNS infections · Altered mental status · Meningeal signs · Neuroimaging · CSF analysis

3.1 Introduction

Timely diagnosis of CNS infections is essential in clinical practice. A high index of suspicion is necessary to both start treatment in a timely manner and to improve patient outcomes. For example, in the case of bacterial meningitis, untreated

F. Shaikh (✉)
Department of Internal Medicine, Juravinksi Hospital, Hamilton Health Sciences, Hamilton, ON, Canada
e-mail: shaikhfa@hhsc.ca

H. Sami et al. (eds.), *Viral and Fungal Infections of the Central Nervous System: A Microbiological Perspective*, https://doi.org/10.1007/978-981-99-6445-1_3

infections are associated with a near 100% mortality (Fitch et al. 2008). With the advent of antibiotic and adjunct steroid administration, these mortality rates are now less than 30% (Ziai and Lewin 2006). The clinician must be swift in their initial examination and diagnostic evaluation.

3.2 Clinical Presentation

An ill patient with a CNS infection can clinically present with many symptoms. These can include headache, fever, focal neurological deficits, seizures, acute confusion, lethargy, neck pain, photophobia, back pain, nausea, vomiting, syncope, coma, as well as dermatological manifestations such as a new rash. Constitutional symptoms can include lymphadenopathy, arthralgias, and myalgias. Focal neurological deficits can virtually involve any sensory and motor function originating both centrally and peripherally. Neurological deficits can also include hemiparesis, cranial nerve palsies, abnormal, or pathologic deep tendon reflexes.

Given the broad spectrum of presenting symptoms as well as causative agents, a definitive diagnosis often takes a few days. Thus, the initial assessment should be on obtaining focused patient histories. It must be noted that there is no specific symptom that allows the recognition of causative agents of CNS infections based solely on history and examination (Mańdziuk and Kuchar, 2023).

3.2.1 Meningitis

Acute meningitis is an infection of the membranes (meninges) covering the brain and spinal cord (Li et al. 2020). It is the most common infectious disease of the CNS (Li et al. 2020).

Fever, neck stiffness, and altered mental status are considered the classic triad of meningitis. Retrospective reviews have found this triad to possess a low sensitivity for the diagnosis of bacterial meningitis (Ziai and Lewin 2006). In a 2004 Dutch study of 696 episodes of adult community-acquired acute bacterial meningitis confirmed by CSF cultures, only 44% of cases had the full triad (van de Beek et al. 2004). Dorsett and Liang (2016) report 99–100% of patients found to have meningitis presented with at least one component of the classic triad (Dorsett and Liang 2016). Individual sensitivities have the following reported ranges (Dorsett and Liang 2016): fever 42–97%; neck stiffness 15–92%; and altered mental status 32–89%.

For instance, for a patient who presents with a headache and neck stiffness, the probability of this patient having meningitis can be as high as 92%. A clinician must hold a high index of suspicion for evaluating CNS infections if two or more of the triad symptoms are noted at presentation.

The onset of symptoms varies depending on the cause of meningitis. For the leading causes of bacterial meningitis, the progression of symptoms can be as short as hours to 1–2 days (Pruitt 1998). An example of this is *S. pneumoniae* meningitis which can lead to a rapid clinical decline with septic shock and death

Table 3.1 Pathogens that cause bacterial meningitis with their common presenting ages

Causative agent	Involved age groups
• *Escherichia coli*	>1 month (m) to <3 years (y)
• Group B *streptococcus*	>1 m to <3 y
• *Haemophilus influenzae*	>3 to <19 y
• *Listeria monocytogenes*	>1 m to <3 m
	>50 y
• *Neisseria meningitidis*	>3 m to 50 y
• *Streptococcus pneumoniae*	>3 m to >50 y

(Fitch et al. 2008). Thankfully, with the advent of Pneumococcal vaccines in 2000, there has been a reduction in mortality in the United States. Comparatively, mortality associated with *S. pneumoniae* meningitis was 0.073 in 2002 and 0.024 in 2008, per 100,000 people, respectively (Castelblanco et al. 2014).

With regard to incidence, some key highlights are important in a physician's clinical acumen. A general age-based distribution of causative agents (Table 3.1) for bacterial meningitis exists (Schuchat et al. 1997).

While *Streptococcus pneumoniae* and *Neisseria meningitidis* are the most common offenders for most cases of bacterial meningitis, it is important to have a high index of suspicion for agents such as *Listeria monocytogenes* which is known to impact infants and elderly patients.

Additionally, meningococcal meningitis due to *Neisseria meningitidis* is particularly worrisome because of its potential to cause epidemics (Castelblanco et al. 2014). Demographic details such as areas of close co-habitation (dorm rooms, prisons, military barracks, shelters) are important to elucidate as well as this can lead to the timely identification of other exposed individuals.

3.2.2 Encephalitis

Clinically, patients with encephalitis can range from subtle deficits to complete unresponsiveness (Whitley and Gnann 2002). The same plethora of neurological manifestations seen in meningitis can be seen in encephalitis patients. Seizures are also common with encephalitis. Patients also classically present with a fever above 38 °C within the last 72 h (Dorsett and Liang 2016).

Of note, meningeal symptoms such as nuchal rigidity may be absent (Fitch et al. 2008). Patients can have a wide range of altered mentation from being confused to obtunded. The hallmark of viral encephalitis is the acute onset of a febrile illness (Whitley and Gnann 2002). Presenting symptoms are often milder than those associated with meningitis.

With viral encephalitis, the presentation is often consistent with the causative agent's predilection for certain CNS cells. For example, in the case of the herpes simplex virus, the infection affected the temporal lobe leading to findings of aphasia, anosmia, temporal lobe seizures, and focal neural deficits (Gilden 2008). Polioviruses in comparison affect motor neurons preferentially.

Encephalitis is most commonly due to viral agents (Whitley and Gnann 2002). Incidence in a particular age group is not as clearly discernible as it is for bacterial agents. However, the immunization status is an important diagnostic clue. For example, vaccination programs have greatly limited the sequalae associated with common childhood viral encephalitis illnesses such as measles, mumps, and rubella (Pruitt 1998). Travel to endemic areas and exposure to vectors and hosts (animal bites) are also important clues in the patient's history.

3.2.3 Brain Abscess

A brain abscess is defined as a localized zone of neurosis with a surrounding membrane within the brain parenchyma as a result of infection (Bokhari and Mesfin, 2023). Etiology can consist of direct local spread or hematogenous spread from peripheral foci of infection. Examples of direct local spread include infections from frontal or ethmoid sinuses, dental infections, recent surgical instrumentation in the head and neck region, and paranasal sinus infection which is reported to account for 30–50% of known causes (Honda and Warren 2009).

CNS abscesses are usually solitary lesions; however certain agents can cause multiple lesions. For example, brain abscesses that are a result of hematogenous spread are associated with multiple infectious collections. *Staphylococcus aureus* and *Viridian streptococci* are the most common pathogens isolated in these cases (Bokhari and Mesfin, 2023).

CNS infections resulting in a brain abscess can present similarly with generalized neurological manifestations. An important distinction is the possibility of mass effect due to the physical space occupied by the abscess. An increase in intracranial pressure (ICP) is what causes a mass effect, which is often difficult to determine in a clinical exam without radiological corroboration. In addition to altered mental status and neurological deficits, patients with increased ICP may also present with nausea and vomiting (Pruitt 1998). Comprehensive ophthalmic exams are essential. Although papilledema is not specific to brain abscesses; it is an important determinant of increased intracranial pressure (ICP). Increased ICP is a relative contraindication to performing a lumbar puncture due to the risk of brain herniation (Honda and Warren 2009).

Fever, headache, and neurological deficits are common. Headaches are noted in approximately 60–70% of patients (Olie et al. 2022). Headaches are usually localized to the area of the infection and can be both gradual and sudden in onset. Fever is the second most common symptom; presenting in up to 50% of patients (Britt et al. 1984). Up to 30% of patients can present with new-onset seizures and are particularly common with frontal lobe abscesses (Olie et al. 2022).

3.3 Physical Examination

A focused physical exam provides necessary diagnostic clues that further prompt specific tests and studies. A patient presenting with symptoms inclusive of a new headache, fever, and altered mental status must immediately prompt the physician to evaluate for CNS infections. The World Health Organization (WHO) has published diagnostic definitions (Fig. 3.1) for the classification of CNS infections (Dubot-Pérès et al. 2019).

3.3.1 Fever

Fever as a solitary vital sign holds a broad differential diagnosis and prompts a comprehensive assessment. It certainly speaks for the systemic immune response our body mounts physiologically against foreign pathogens. For the purposes of diagnosing CNS infections, it is an important finding that needs to be assessed in conjunction with other presenting symptoms.

Fever is defined as a temperature >37.5 °C, typically within 72 h of presentation. The patient may have fevers on the initial presentation as well. However, of importance is recognizing the absence of fevers in certain patient populations.

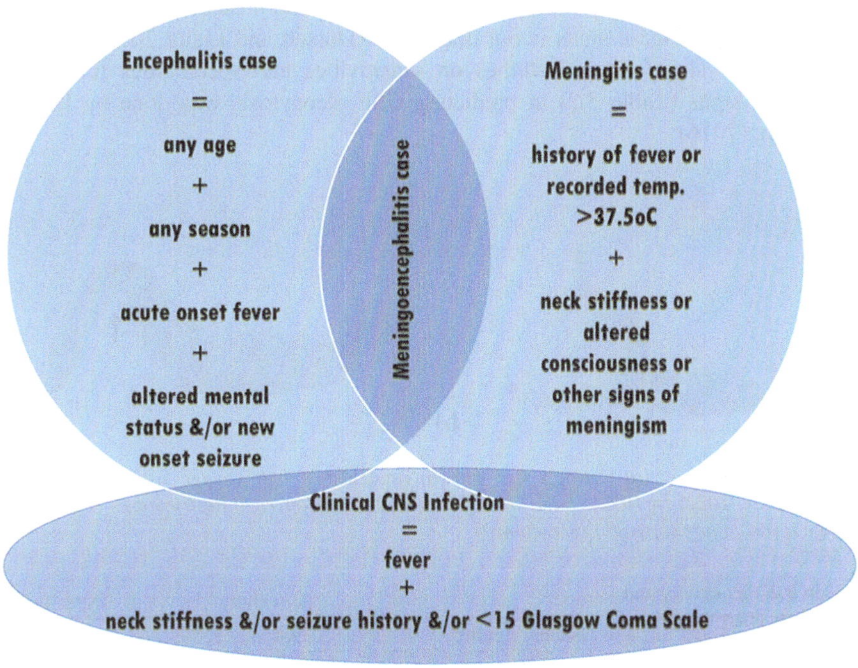

Fig. 3.1 WHO encephalitis and meningitis case definitions

Patients who are elderly, immunocompromised, or have had recent antibiotic treatment may not mount a febrile response (Ziai and Lewin 2006).

A retrospective review done in Laos between 2003 and 2011 ((Dubot-Pérès et al. 2019) noted that out of 622 adults who presented with a fever, 425 had a confirmed CNS infection or approximately 68% of cases. Therefore, the presence or absence of fever as a solitary symptom is not conclusive for a definitive diagnosis of CNS infections.

3.3.2 Classical Maneuvers

In the case of bacterial meningitis, there are two well-known maneuvers conducted as a part of the physical examination: Kernig's sign and Brudzinski's sign. Kernig's sign (Fig. 3.2a), first described in 1882, consists of flexing the patient's neck and then extending the patient's knees. It is considered positive when the maneuver elicits pain while the neck is flexed (Ward et al. 2010). Brudzinski's sign (Fig. 3.2b), first reported in 1909, consists of passive flexion of the neck while the patient is in a supine position. It is considered positive if it results in flexion of the hips and knees (Waghdhare et al. 2010).

The historical significance of these signs in determining meningeal irritation is important to appreciate in clinical practice, however, their utility in modern practice is limited for definitive diagnostic evaluation. When compared with CSF analysis showing pleocytosis, several studies have shown that the presence and more importantly absence of these signs is not diagnostic (Dorsett and Liang 2016). A retrospective analysis of several studies on sensitivities and specificities for classic meningeal signs (Table 3.2) in predicting CSF pleocytosis was done by Dorsett and Liang (2016).

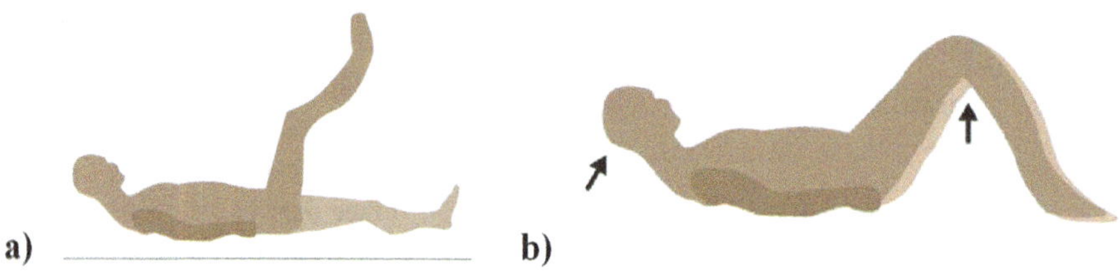

Fig. 3.2 Depiction of (**a**) Kernig's sign and (**b**) Brudzinski's sign. (*Image Source: Emergency Care Journal, 2022; 18:10929; page press* (Siu et al. 2022) [*Copyright Restrictions – NONE; this image is free of any copyright restrictions*])

Table 3.2 Sensitivity and specificity chart for meningeal signs (with 95% confidence intervals)

	Sensitivity	Specificity
Nuchal rigidity	39.4	70.3
Kernig's sign	14.1	92.3
Brudzinski's sign	11.1	93.4

Notably, these physical signs have very low sensitivities which negates the sole reliance on their presence as a diagnostic measure. More importantly, for both Kernig's and Brudzinski's signs, greater than 85% of the time, the diagnosis is missed if the sign is absent. Therefore, much like fever in the absence of neurological symptoms; the presence or absence of these meningeal signs is not diagnostic on its own.

Additionally, it must be noted that when present, both Kernig's and Brudzinski's signs hold high specificities for predicting CSF pleocytosis and should increase clinical suspicion of meningitis (Waghdhare et al. 2010).

3.3.3 Neurological Examination

Patients with CNS infections often present with a neurological complaint or sequalae. As mentioned previously, neurological symptoms can range from headaches to focal neurological deficits, seizures, obtundation, or comatose state. Objective assessment of these patients initially begins with a thorough neurological exam including cranial nerve assessment, peripheral motor, and sensory evaluations as well as alertness and behavior assessments. In addition to the above, a widely used tool in the initial assessment is the Glasgow Coma Scale. It is used to objectively describe the extent of impaired consciousness. The scale (Table 3.3) assesses patients according to three aspects of responsiveness: eye-opening, motor, and verbal responses (Jain and Iverson, 2023).

An abnormal conscious state as graded by the GCS is a strong predictor for poor disease outcomes in CNS infections (Ward et al. 2010). Specifically, scores less than 12 are reported to correlate with disease severity and warrant prompt management (Jain and Iverson, 2023).

In the case of bacterial meningitis, a retrospective review from Lucas et al. (2014) conducted in the Netherlands demonstrated patients presenting with a minimal Glasgow Coma Scale score on admission along with bilaterally absent pupillary

Table 3.3 Glasgow Coma Scale

Motor response (MR)		Verbal response (VR)		Eye opening (EO) response	
	Scoring		Scoring		Scoring
Obeys command	6	Oriented	5	EO spontaneously	4
Localizes pain	5	Confused	4	EO to verbal command	3
Withdraws from pain	4	Inappropriate words	3	EO to pain	2
Flexion response to pain	3	Incomprehensible sounds	2	No EO	1
Extension response to pain	2	No VR	1		
No MR	1				

light responses, bilaterally absent corneal reflexes, or signs of septic shock on admission all died (Lucas et al. 2014). Thankfully, the incidence of such severe neurological compromise was not high in the cases reviewed.

Another retrospective analysis done by van de Beek et al. (2004) showed that patients with pneumococcal meningitis had more severe disease than did patients with meningococcal meningitis, as reflected by a higher frequency of seizures ($P = 0.001$) and focal neurologic deficits ($P < 0.001$) and a lower level of consciousness ($P < 0.001$). Thus, in the case of meningitis itself, the causative agent can pose its own independent risk of mortality and disease severity.

The utility of this grading tool was evaluated in a prospective study by Barsić et al. (1996). They investigated ICU mortality outcomes in critically ill patients with CNS infections vs critically ill patients with other infections using GCS scores. Their analysis suggests a prognostic value for GCS in patients with CNS infections but not in other infectious disease patients. Thus, obtaining initial and serial GCS scores is an important monitoring and assessment tool.

3.3.4 Ancillary signs

A clinician must also pay special attention to ancillary signs and symptoms noted on a physical exam that can provide diagnostic clues with respect to the patient's neurological presentation. In the case of CNS infections, there are numerous etiologies and causative agents. Some unique characteristics apply to certain pathogens. For example, in the case of meningitis: petechial or purpural rash is seen in meningococcal meningitis (Archibald and Quisling 2013). van de Beek et al. (2004) conducted a retrospective review where a petechial rash was seen in 26% of the patients with confirmed meningitis. Therefore, once again, the presence and absence of any constitutional symptom is not diagnostic based on examination alone.

Another example of common ancillary signs is the case of *Haemophilus influenzae* meningitis where ataxia and labyrinthitis are noted. For tuberculosis meningitis, we can commonly see cough, weight loss, night sweats, and cranial nerve deficits (Archibald and Quisling 2013). In the world of viral infections, a morbilliform rash and Koplik's spots are noted in measles encephalitis and parotitis and orchitis are noted in Mumps encephalitis (Archibald and Quisling 2013). Adjunct pathologies such as pleuritis, myocarditis, and myositis are also seen with many infectious agents that pose a risk for CNS infections. Considerable overlap exists in constitutional symptoms and associated prodromes for many infectious agents and a clinician must evaluate all systems and associated exposures or risk factors in every presenting patient.

3.4 Diagnostics

3.4.1 Initial Laboratory Tests

Once a patient has been assessed for suspected CNS infection, diagnostic testing must take place rapidly. Initial laboratory tests include complete blood count with differential and platelet count, erythrocyte sedimentation rate, serum C-reactive protein, blood cultures (drawn before initiating antibiotic therapy), and HIV serology (Bokhari and Mesfin, 2023). Certain infections also demonstrate liver pathology and electrolyte derangements, therefore assessing for these is also recommended. For patients with specific exposures (i.e., known tick bites), obtaining serologies for specific viral agents is also indicated.

3.4.2 Lumbar puncture

A patient with a high index of suspicion for a CNS infection, cerebrospinal fluid analysis (CSF) obtained via a lumbar puncture (LP) is very revealing with respect to the causative agent. Neuroimaging with a prior head CT must be conducted to rule out increased intracranial pressure due to the risk of herniation and death associated with performing LP (Bokhari and Mesfin, 2023). Samples obtained can be employed for culture, Gram stain, serology, histopathology, and polymerase chain reaction.

3.4.3 Neuroimaging

Initial CNS infection diagnostics must include neuroimaging. The two main utilized imaging modalities are computed tomography (CT) and magnetic resonance imaging (MRI).

CT scan is highly sensitive to acute intracranial hemorrhages (parenchymal or extra-axial), calcified lesions, or bony deformities (Li et al. 2020). A CT done with contrast in the setting of a hemorrhagic focus will confound this important finding and therefore the first diagnostic brain imaging should be a CT head without contrast. While findings definitive for CNS infections are more likely to be absent in initial imaging, the presence of mass effect, midline shift, or potential for herniation can still be determined. This must be reviewed prior to undergoing lumbar puncture (Hasbun et al. 2001). Additionally, a non-enhanced CT can also identify fluid collections that may be eligible for CT-guided stereotactic drainage. Serial CT scans can be done with contrast administration including dedicated vascular studies. For example, a CT angiogram of the head and neck can reveal vascular complications of CNS infections impacting arterial or venous structures (Li et al. 2020).

While obtaining a CT scan yields a quick analysis of intracranial pathology, it has its limitations with respect to resolution and detail in comparison to an MRI scan.

MR imaging (Fig. 3.3) is much more sensitive for detecting early changes of CNS infections and depicting various imaging findings during each disease process, due

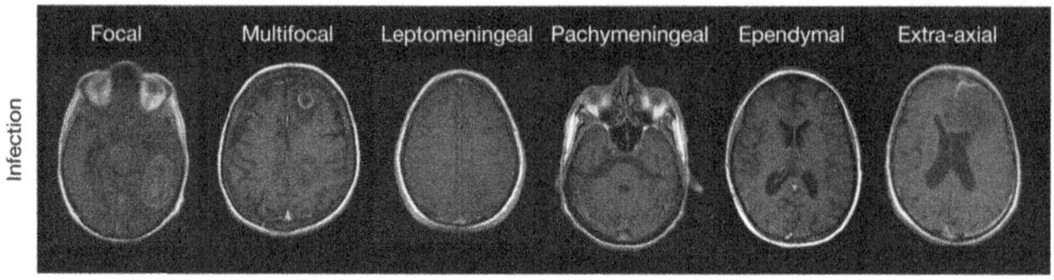

Fig. 3.3 Examples of anatomical imaging classification in MRI head studies with confirmed final diagnosis of CNS infection. (*Image Source: Scientific Reports, 2022; 12:15805* (Lim et al. 2022) [*Copyright Restrictions – NONE; this image is free of any copyright restrictions*])

to its high anatomy resolution, soft tissue differentiation, multi-planar acquisition, and versatile sequences of delineating different characters of pathologic processes of CNS infections (Gilden 2008). Furthermore, in the appropriate patient (i.e., no contrast allergy, appropriate renal function, etc.) the administration of intravenous contrast in MRI increases the sensitivity and specificity of the MRI technique (Li et al. 2020) helping further delineate the pathology, exclude most noninfectious etiologies, and narrow down the differential diagnoses.

Both scan modalities are useful for diagnostics and should be considered in all patients presenting with CNS infections.

For example, in the case of CNS abscess, both CT and MRI provide useful data. Enzmann et al. (1980) reported that CT findings of patchy enhancement in early cerebritis evolve to a rim of enhancement in late cerebritis which later on forms the brain abscess. As cerebritis evolves, a more conspicuous rim-enhancing lesion becomes visible. Can also see satellite daughter lesions or separate adjacent well-located lesions. Vasogenic edema is also noted (Britt et al. 1984). MRI is the imaging modality of choice for diagnosis as well as follow-up of lesions. It allows for greater contrast between cerebral edema and the brain and is also more sensitive for detecting the spread of inflammation into the ventricles and subarachnoid space (Bokhari and Mesfin, 2023).

A 10-year retrospective analysis by Lim et al. (2022) reviewed the most common radiological findings on MRI for CNS infections. A total of 109 patient data were reviewed. Their analysis demonstrated the following: pathological enhancement was seen most frequently (46.8%) followed by hemorrhage (22.0%) and restricted diffusion (19.3%). With respect to disease distribution; multifocal-parenchymal was the most common (34.9%), focal-parenchymal (29.4%), and leptomeningeal (11.9%).

For the purposes of initial assessment, special attention can be paid to positive radiological findings and their association with a specific pathogen. Bacterial CNS infections may have characteristic findings attributable to a single pathogen, however, this is not the case with viral CNS infections. While this is not a comprehensive list, examples of certain pathological distributions of certain viruses are listed in Table 3.4.

Table 3.4 Characteristic MRI findings associated with certain viral pathogens (Gilden 2008)

Cytomegalovirus encephalitis	• Ependymal enhancement around lateral ventricles
Herpes simplex virus encephalitis	• Abnormal signal and edema in left temporal lobe, insult, and cingulate gyrus
	• Typically spares deep nuclear structures
	• Mass effect compressing the left ventricle with uncal herniation
Human immunodeficiency virus infection in the CNS	• Brain atrophy, diffuse white matter attenuation
John Cunningham virus infection of the CNS	• Multifocal and confluent subcortical non-enhancing white matter hyperintensities, extension into the cortical gray matter
Varicella zoster virus vasculopathy	• Ischemia and infarction are more commonly seen in the white matter
	• Predilection for gray-white matter junctions

3.4.4 Cerebrospinal Fluid (CSF) Analysis

Cerebrospinal fluid analysis analysis is the cornerstone of diagnosis and management of CNS infection. Together with presenting neurological symptoms, the presence or absence of positive radiological findings, CSF analysis is the most important diagnostic test for elucidating causative agents.

For patients with suspected CNS infections, timely diagnostics is critical. The current standard of care necessitates obtaining a CT head prior to performing a lumbar puncture. The importance of a CT head prior to lumbar puncture was investigated by Hasbun et al. (2001). Of 78% of patients who underwent CT before LP, 24% had an abnormal finding, and 5% had a mass effect with clinical significance (Hasbun et al. 2001). Even without abnormalities on CT, clinical signs suggestive of increased ICP should caution against the use of LP. As a safety measure, the smallest amount of CSF sampling should be obtained to minimize leakage (Hasbun et al. 2001).

Distinctive CSF findings (Table 3.5) are classically seen in bacterial, viral, and fungal infections of the CNS. In the case of bacterial meningitis, CSF Gram's stain will indicate organisms in 60–90% of cases and positive cultures in 80% of cases (Hrishi and Sethuraman 2019). Additionally, 90% of patients with bacterial meningitis will have a leukocytosis greater than 100. Classically, CSF glucose values will be low in comparison to normal CSF values and protein levels are elevated. Of all the parameters obtained on CSF analysis, a retrospective analysis done by Olie et al. (2022) demonstrated that the presence of CSF leukocytes is the most important predictor of CNS infection, independently.

In terms of diagnostic yield, nearly all CNS infectious patients will have positive cytology noted after 3 LP attempts. Up to 70% can be diagnosed with the first LP (Hrishi and Sethuraman 2019). The yield of Gram stain and culture is significantly reduced if patients have received antibiotics before LP (Bokhari and Mesfin, 2023). Cultures can also take some time to demonstrate growth. Co-administration of

Table 3.5 CSF findings in various CNS infections as distinguished by opening pressure, appearance, protein, glucose levels, Gram stain, and white blood cells (WBC)

	Normal	Bacterial	Viral	Fungal	TB
Pressure (cmH_2O)	5–20	>30	Normal/inc	Normal/inc	Inc
Appearance	Normal clear	Turbid	Clear	Clear	Fibrin web
Protein (g/L)	0.18–0.45	>1	<1	0.1–0.5	0.1–0.5
Glucose (mmol/L)	2.5–3.5	<2.2	Normal	1.6–2.5	1.6–2.5
Gram stain	No microorganism	60–90% positive	Normal	Normal	AFB
WBC per cu mm	<3	10–2000	>100	10–50	100–500
		Neutrophils[a]	Lymphocytes[a]	Lymphocytes[a]	Lymphocytes[a]

TB tuberculosis; *inc* increased, *WBC* white blood cells, *AFB* acid fast bacilli
[a] Predominance

therapy, while diagnostics are being obtained, is the clinical standard of practice in order to decrease morbidity and mortality associated with delay of treatment.

3.5 Summary

The following summarizes the clinical approach to CNS infections. A patient with a suspected CNS infection may present with a range of neurological symptoms, which in isolation are not diagnostic. The classic triad in bacterial meningitis is not often seen in every confirmed case. The combination of fever along with neurological symptoms must alert the clinician to carry out neurological diagnostics. The presentation for patients with encephalitis and brain abscesses can range from vague and chronic symptoms to new-onset seizures and coma. Neuroimaging in the form of a CT head without contrast should be obtained in all patients. Along with serum analysis, CSF analysis via lumbar puncture is an important diagnostic tool that yields the most accurate diagnosis including the identification of the causative agent. All efforts should be made to avoid delay of treatment as CNS infections can result in rapid patient decline and deterioration.

References

Archibald LK, Quisling RG (2013) Central nervous system infections. In: Textbook of neurointensive care. Springer, London, pp 427–517

Barsić B, Marton E, Himbele J, Ravlić Z (1996) Evaluation of the Glasgow Coma Scale score in critically ill infectious disease patients. Infection 24(4):297–300

Bokhari MR, Mesfin FB (2023) Brain abscess [Updated 2022 May 11]. In: StatPearls. StatPearls Publishing, Treasure Island. https://www.ncbi.nlm.nih.gov/books/NBK441841/

Britt RH, Enzmann DR, Placone RC Jr, Obana WG, Yeager AS (1984) Experimental anaerobic brain abscess. Computerized tomographic and neuropathological correlations. J Neurosurg 60(6):1148–1159. https://doi.org/10.3171/jns.1984.60.6.1148. PMID: 6610026

Castelblanco RL, Lee M, Hasbun R (2014) Epidemiology of bacterial meningitis in the USA from 1997 to 2010: a population-based observational study. Lancet Infect Dis 14(9):813–819

Dorsett M, Liang SY (2016) Diagnosis and treatment of central nervous system infections in the emergency department. Emerg Med Clin North Am 34(4):917–942

Dubot-Pérès A, Mayxay M, Phetsouvanh R et al (2019) Management of central nervous system infections, Vientiane, Laos, 2003-2011. Emerg Infect Dis 25(5):898–910

Enzmann DR, Brant-Zawadzki M, Britt RH (1980) CT of central nervous system infections in immunocompromised patients. Am J Roentgenol 135(2):263–267

Fitch MT, Abrahamian FM, Moran GJ et al (2008) Emergency department management of meningitis and encephalitis. Infect Dis Clin North Am 22(1):33–52. https://pubmed.ncbi.nlm.nih.gov/18295682/

Gilden DH (2008) Brain imaging abnormalities in CNS virus infections. Neurology 70(1):84

Hasbun R, Abrahams J, Jekel J, Quagliarello VJ (2001) Computed tomography of the head before lumbar puncture in adults with suspected meningitis. N Engl J Med 345(24):1727–1733. https://doi.org/10.1056/NEJMoa010399. PMID: 11742046

Honda H, Warren DK (2009) Central nervous system infections: meningitis and brain abscess. Infect Dis Clin North Am 23(3):609–623

Hrishi AP, Sethuraman M (2019) Cerebrospinal fluid (CSF) analysis and interpretation in neurocritical care for acute neurological conditions. Indian J Crit Care Med 23(Suppl 2): S115–S119

Jain S, Iverson LM (2023) Glasgow Coma Scale [Updated 2022 Jun 21]. In: StatPearls. StatPearls Publishing, Treasure Island. https://www.ncbi.nlm.nih.gov/books/NBK513298/

Li S, Nguyen IP, Urbanczyk K (2020) Common infectious diseases of the central nervous system-clinical features and imaging characteristics. Quant Imaging Med Surg 10(12):2227–2259

Lim EA, Ruffle JK, Gnanadurai R, Lee H, Escobedo-Cousin M, Wall E, Cwynarski K, Heyderman RS, Miller RF, Hyare H (2022) Differentiating central nervous system infection from disease infiltration in hematological malignancy. Sci Rep 12:15805. https://doi.org/10.1038/s41598-022-19769-2

Lucas M, Brouwer M, van der Ende A, van de Beek D (2014) Outcome in patients with bacterial meningitis presenting with a minimal Glasgow Coma Scale score. Neurol Neuroimmunol Neuroinflamm 1(1):e9

Mańdziuk J, Kuchar EP (2023) Streptococcal meningitis [Updated 2022 May 21]. In: StatPearls. StatPearls Publishing, Treasure Island. https://www.ncbi.nlm.nih.gov/books/NBK554448/

Olie SE, van Zeggeren IE, ter Horst L et al (2022) Seizures in adults with suspected central nervous system infection. BMC Neurol 22:426. https://doi.org/10.1186/s12883-022-02927-4

Pruitt AA (1998) Infections of the nervous system. Neurol Clin 16(2):419–447. https://doi.org/10.1016/s0733-8619(05)70071-2

Schuchat A, Robinson K, Wenger J et al (1997) Bacterial meningitis in the United States in 1995. N Engl J Med 337:970–976. https://www.nejm.org/doi/full/10.1056/nejm199710023371404

Siu V, Lin S, Sivakumar M, Varkey TC (2022) Evidence-based work up and management of non-specific neck pain in the emergency department. Emerg Care J 18. https://doi.org/10.4081/ecj.2022.10929

van de Beek D, de Gans J, Spanjaard L et al (2004) Clinical features and prognostic factors in adults with bacterial meningitis. N Engl J Med 351(18):1849–1859

Waghdhare S, Kalantri A, Joshi R et al (2010) Accuracy of physical signs for detecting meningitis: a hospital-based diagnostic accuracy study. Clin Neurol Neurosurg 112(9):752–757. https://pubmed.ncbi.nlm.nih.gov/20615607/

Ward MA, Greenwood TM, Kumar DR et al (2010) Josef Brudzinski and Vladimir Mikhailovich Kernig: signs for diagnosing meningitis. Clin Med Res 8(1):13–17

Whitley RJ, Gnann JW (2002) Viral encephalitis: familiar infections and emerging pathogens. Lancet 359(9305):507–513. https://doi.org/10.1016/S0140-6736(02)07681-X

Ziai WC, Lewin JJ III (2006) Advances in the management of central nervous system infections in the ICU. Crit Care Clin 22(4):661–694

4 Immune Responses in Infections of the Central Nervous System

Kyu Kyu Win and Puvaneswaren A/L Parmasivam

Abstract

The neurological invasion of pathogens into the central nervous system (CNS) causes inflammatory host–cell interactions, which inexplicably control the pathogens' ability to replicate or to be removed from the host. Undoubtedly, the pathogen's mechanism of entering the CNS determines how the immune response develops. The portal of entry and the types of pathogens determine the exact mechanism of immune response in the CNS. Furthermore, immunological cellular interactions within this kind of ecosystem encourage the expression of inflammatory mediators, which has an acute impact on neuronal function at the cellular level. An increase in neutrophil chemoattractant molecules (like CXCL2), cellular adhesion molecules (ICAM), and complement C5a is observed if pathogen invasion happens. Antigen-presenting cells which are present in both meninges and choroid plexus support ongoing stimulation of T cells upon viral and fungal infections. However, high levels of TH1 cell cytokines, such as tumor necrosis factor, interferon, and IL-1, may potentially cause sustained cognitive function impairment even after the eradication of neurotropic pathogens. Autoimmune diseases in CNS could be better treated once the exact mechanism of molecular mimicry due to infections is clearly understood.

Keywords

Neurotropic pathogen · Immune response · Molecular mimicry · Autoimmune disease

K. K. Win (✉) · P. A/L. Parmasivam
Faculty of Medicine and Health Sciences, UCSI University, Springhill Campus, Port Dickson, Negeri Sembilan, Malaysia
e-mail: kyukyuwin@ucsiuniversity.edu.my; puvaneswaren@ucsiuniversity.edu.my

H. Sami et al. (eds.), *Viral and Fungal Infections of the Central Nervous System: A Microbiological Perspective*, https://doi.org/10.1007/978-981-99-6445-1_4

4.1 Immune Responses in the Cerebrospinal Fluid

The central nervous system (CNS) exists in a strictly confined biological environment as a result of the CNS's evolution over time. Due to the local tissue barrier and a variety of immunosuppressive microenvironments, the CNS has an "immune privilege" status (Engelhardt et al. 2017). The presence of viruses, foreign particles, and tissue in the CNS can therefore be tolerated by the human immune system, up to a certain tolerable threshold (Banks and Erickson 2010). However, it is not necessarily the status quo. Physiological or pathological changes to the CNS that go beyond the point of toleration, however, are regarded as life-threatening occurrences. The CNS appears to have a subtle yet distinct strategy for defending itself against intruders from other bodily systems.

The immunological responses in the CNS have been studied and the science behind immune response is undergoing development to expand our current understanding regarding the immune defence. Based on the condition and functionality of the immune candidates infused inside the cerebrospinal fluid, our understanding of the immune responses can be divided into a number of barriers, which are positioned between the blood and the CNS, in addition to the regular physiological surveillance mechanisms that are in placed to watch over the CNS environment. Blood–brain barrier (BBB), the blood–CSF barrier, and the arachnoid barrier are the three most vital physical barriers within the CNS (Brooks et al. 1983). The anatomical and physiological design of the BBB's functions to restrict or prevent the flow of cells, infections, or any macromolecules into the CNS. By accomplishing this, the interaction between peripheral immune cells and local immune cells in the CNS is restricted. Under certain pathological conditions, the peripheral immune cells have the capacity to pass through the BBB to reach the brain.

In addition to the physical barrier, at the cellular level, CNS has been considered to be an organ, which is inhabited by a limited immune repertoire, such as the parenchymal tissue of the CNS. It contains resident microglia as the main immune cell together with perivascular macrophages (Bailey et al. 2006).

Findings in this field in particular are pointing toward a novel mechanism that may help with immune surveillance within the skull. It has been discovered that CSF which is secreted by the choroid plexus is being transported through the channels within the skull. It suggests that CSF may very well be able to travel via the perivascular spaces of the dural blood vessels, through the channels in the skull, and into the marrow cavities. By leveraging the CSF outflow into the marrow cavities, pathogens could cause the release of immune cells in response to inflammation, resulting in cranial emergency hemopoiesis (Engelhardt et al. 2017). This demonstrates unequivocally that CSF is essential for immunological responses in the CNS. This discovery is believed to provide an explanation regarding how the brain and skull might interact to cause an immune response toward incoming pathogens within the CNS. It is obvious that CSF participation in the immune response has a deeper significance than previously considered.

4.2 T-Cell-Mediated Immune-Regulation

Both resident microglia and perivascular macrophages have a role in inhibiting the infiltration of foreign bodies into the CNS. These cells are part of a non-specific type of immune response against outside intruders. The presence of the specific immune response, also known as the adaptive immune system, is observed in strengthening the nonspecific immune response. The efficiency of the entire spectrum of the immune response in the CNS as a whole have been demonstrated to be increased by this particular form of specific immune response. Within the context of cell-mediated immune response, T cell is the primary immune cell that strives to eradicate any viral infection and obligatory intracellular bacterial infection, together with the B cell (Walsh et al. 2014). T lymphocytes in the CNS will frequently clear viral infections predominantly by non-cytolytic viral clearance processes (Ellwardt et al. 2016). This role of T cells within the perivascular region is enhanced as CNS exhibits an increased expression of T-cell receptors (TCRs) recognition in terms of regulation (Walsh et al. 2014).

Activated T cells move from blood vessels to the subarachnoid space by passing through the stroma of the choroid plexus. This is one way that T cells move from the blood into the CNS. The blood–cerebrospinal fluid barrier that surrounds the choroid plexus stroma is made up of epithelial cells interconnected by tight junctions. Activated T cells are able to cross this barrier. Additionally, recent studies have suggested that the region of the meninges may have a lymphatic system (Walsh et al. 2014). This discovery indicates that T lymphocytes in the CNS may come from additional sources of T cells located within the CNS.

Both CD4+ and CD8+ T lymphocytes were present in meninges, choroid plexus, parenchyma, and the naïve brain compartment. There are two unique ways by which T cells in our CNS function. Certain T-cell subtypes can contribute to repair from one perspective, while another subtype can contribute to damage from another. When certain T-cell subtypes are activated, these two actions are seen simultaneously. T cells have the capacity to concurrently cause neurodegeneration and inflammation when acting as a collective (Louveau et al. 2015). Consequently, depending on the characteristics of the pathogen of interest, a concise and thorough modulation of T-cell-mediated immune response is quite important.

Every pathogen will display a certain protein pattern known as "pattern-associated molecular patterns," which is where the control of T cells comes into play (PAMPs). PAMPs can bind to or adhere to pattern recognition receptors (PRRs) that are present on specific immune cells. This causes the type I interferons (IFN-I) to be expressed as a result, and IFN-I then binds to the IFN-alpha receptor chain 1 (IFNAR1) (Kumar 2019). Any ligand that binds to IFNAR activates the JAK-STAT signaling cascade, which in turn produces the impact of antiviral clearance, one of the numerous functions that IFN-I is assigned (Begolka et al. 2005). IFN can be generated by cells in the CNS ecosystem that are infected with a virus, including astrocytes and microglia (Kumar 2019).

This will have a direct impact on IFN-stimulated gene expression (ISG). The interaction between CXCL10 and CCL2, which is already strengthened, improves the regulation of T cells (Begolka et al. 2005). The combined effects are produced with ISG and both of these important chemokines result in an inflammatory environment. These quicken the process of T-cell recruitment into the CNS. But the primary function of resident myeloid as antigen-presenting cells (APC) is crucial in evoking the inflammatory ecology, which is crucial in activating the stimulation and recruitment of antiviral T cells.

4.3 Cellular and Humoral Immune Responses Against Different Pathogens

Anatomically, the immune system in the CNS is designed to safeguard itself against invading infections. The CNS is further strengthened by cellular heterogeneity, which can prevent and react to numerous neurotropic infections. Despite having robust physical defences already in place, some neurotrophic pathogens including bacteria, viruses, fungi, and protozoa are able to get through the blood–brain barrier (BBB) leading to CNS infections. Meningitis, encephalitis, and myelitis are the results of CNS infections and are categorized clinically based on the affected locations. However, some neurotropic viruses have pathogen-associated molecular patterns (PAMPs) that are particularly important in inducing immune responses inside the CNS. Intrinsic features within PAMP of neurotropic pathogens are crucial in triggering the specific immune response against themselves (Daniels et al. 2014).

Immune response in the CNS is often divided into innate and adaptive immune responses, similar to immune responses induced in other regions of the infected human body (Pedemonte et al. 2006). The initial response includes a strong response from the region-specific local innate immune responses which include a solid antimicrobial response throughout the CNS. Later, the immune response will change or need to employ the adaptive immune response if the infection persists (Kipnis 2016). A particular antigen-presenting mechanism carried out by microglia cells must be initiated to ignite the immune system's adaptive response (Pedemonte et al. 2006).

Astrocytes and microglia, which function as active antigen-presenting cells, will express MHC class II, coupled with B7-1 and B7-2, which are termed "co-stimulatory molecules" and will eventually trigger the activation of several T-cell subtypes. The humoral immune response, which involves the identification and removal of pathogens by antibodies released by B cells, can be stimulated within the CNS by this action alone (Kipnis 2016). The meninge are where the majority of B lymphocytes reside. Additionally, B cells and other immune cells are prevalent in the meninges' dura mater layer, which is part of the meninges. Additionally, recent research suggests that B cells sensitized in the gut may contribute to humoral immunity in the CNS (Wraith and Nicholson 2012). This appears to be a possibility, when specific bacteria enter the blood through the gut, causing bacteraemia. Activation of naïve B cells in the meninges will take place once the pathogen from the blood is able to breach the blood–brain barrier (Klein et al. 2017).

This summarizes all the different modes of immune responses that are present in CNS toward different types of pathogens. This difference in response is due to the various portal of entry and the unique nature of PAMP that is presented to B and T cells.

4.4 Neurotropism

Neurotropism is referring to the ability of microorganisms, especially viruses that can invade and live in neural tissue. Neurotropic viruses are those that have the ability to infect the central nervous system causing diseases with both neurovirulent and neuroinvasive characteristics. Numerous viruses, including the herpes simplex virus, poliovirus, enteroviruses, parechovirus, West Nile virus, Japanese encephalitis virus, measles, and mumps viruses, are classified as neurotropic viruses. Some are quite neurovirulent and neuroinvasive, whilst others are not. In the case of herpes simplex viruses, it is extremely neuroinvasive yet weakly neurovirulent toward the peripheral nervous system. However, at the same time, it is weakly neuroinvasive and highly neurovirulent toward the central nervous system (Abdullahi et al. 2020). This clearly shows the presence of selective neurotropism.

The genomic makeup of the virus, the biological location of the virus, the host immunological state, and a few other environmental factors are among the key determinants of viral infections in the central nervous system leading to a wide range of clinical presentations. However, they are frequently misdiagnosed, and their etiology missed, in part because of inadequate diagnostic tools and in part because of inadequate knowledge of viral biology and epidemiology (Steffen 2019).

The Flaviviridae family includes the West Nile virus, Japanese encephalitis virus, dengue virus, Murray Valley encephalitis virus, and St. Louis encephalitis virus. A flavivirus infection in the CNS affects the anterior horn neurons, substantia nigra, thalamus, and neocortex (Neal 2014). One factor affecting neuroinvasion is the flavivirus E envelope protein's glycosylation, which increases axonal and trans-epithelial transport. Flavivirus uptake into the axon is inhibited by neutralizing antibodies synthesized against the NS and E proteins. CD8+ T cells are vital for the removal of West Nile virus infected cells from the CNS (Neal 2014). The disruption of the blood–brain barrier is observed as a result of the anti-virus response mediated by TLR-3 and TLR-7, allowing virally infected leucocytes and free living virus to enter the CNS (Neal 2014). Semaphorin 7A, low density lipid receptors (LDLR), glycosamino glycans (GAG), heparan sulfate (HSPG), and DC-SIGN are among the numerous cellular virus attachment factors that facilitate flavivirus entry into the host cell. However, these receptors are not specifically expressed for flaviviruses. By destroying tight junction complexes, the flavivirus also penetrates epithelial and endothelial barriers, increasing blood–brain barrier permeability (Neal 2014).

Axonal (either anterograde or retrograde) transport of SARS-CoV-2 into the CNS may occur after it has entered the respiratory tract through the cranial nerves' nerve ends that innervate it. After establishing viremia, the virus will be able to cross the

BBB and/or blood–cerebrospinal fluid barrier (Ludlow et al. 2016). Once the virus successfully enters the CNS, they can cause alterations in neurons that result in neuronal histopathological damages in the cortex and hypothalamus (Gu et al. 2005). In addition to radiological investigations, SARS-CoV-2's neurovirulent potential has been rigorously investigated in post-mortem brain tissue, using vivo animal models during both acute and post-acute stages of COVID-19. The findings from the radiological investigations are pointing toward the presence of edema and microbleeding in the olfactory bulb together with the loss of gray matter in the parahippocampal gyrus, lateral orbitofrontal cortex, and insula (Douaud et al. 2022).

Since SARS-CoV-2 has a variety of underlying neurovirulent disorders, it is possible that more than one mechanism could be responsible for these changes. Furthermore, host factors, such as age, sex, and other underlying diseases, may contribute to complications associated with SARS-CoV-2 infection in CNS. The possible emergence of future variants could make the CNS issues even more complicated. Additionally, immunization could impact the possibility of experiencing various degrees of CNS complications (Bauer et al. 2022). However, it is agreed that the outcome of the CNS complications is influenced in circumstances whereby vaccination is required.

Only few organs, including the brain and spinal cord, can reproduce the poliovirus. This restricted tropism shown by poliovirus may be an effect of organ-specific differences in relation to translation initiation by its internal ribosome entry site (IRES). Viral proliferation in CNS can be eliminated by C-to-U mutation at base 472 in the IRES of the Sabin type 3 poliovirus vaccination strain. This may further decrease neurovirulence (Kauder and Racaniello 2004).

Rabies virus infection occurs either through bites from infected animals or contact with infected saliva or secretions. Subsequently, the virus enters neural axons of sensory and motor nerves via an endosomal transport pathway. The virus then migrates along peripheral nerves using the fast axonal transport system toward CNS at an estimated speed of 8–20 mm/day (Hemachudha et al. 2002). The incubation period for rabies is dependent on the site of inoculation. Furthermore, in patients who have been bitten on the face or neck as opposed to more distant places (such as the arms or legs), the virus will take longer time to reach the central nervous system (Hemachudha et al. 2002).

The progression of the disease is also dependent on certain receptors. It appears that two specific receptors: neural cell adhesion molecule (NCAM) and nicotinic acetylcholine receptor (nAchR) might play a crucial role in concentrating virus particles at the neuromuscular junction and facilitating effective transportation in the intracellular space. Viral replication begins with the transcription of the viral genome by P-L polymerase and subsequent assembly of new viruses occurs mainly in dorsal-root ganglia and anterior-horn cells, once the intact virions have entered the central nervous system. It will result in the formation pathognomonic cytoplasmic inclusions know as Negri bodies. The virus subsequently disseminates throughout the CNS, mostly by cell to cell direct infection, plasma-membrane budding, or trans-synaptic spread, with predominant localization within the brainstem, thalamus, basal ganglia, and spinal cord (Schnell et al. 2010).

The development of immune responses and autoimmune reactions against infected neurons, as well as direct cell death due to virus replication, could add toward compounded neurotoxicity (Hemachudha et al. 2002). It is crucial to note that the massive cytokine production that results from CNS infection has a significant impact on the hippocampus and other limbic-system activities, altering electrical cortical activity, the serotonin metabolism, and hypothalamo–pituitary–adrenal axis (Hemachudha et al. 2002). Rabies virus returns to the periphery later in the disease's progression via intra-axonal transport, with increased tropism toward salivary and lacrimal glands (Mrak and Young 1994).

Based on the examples of neurotropic viruses discussed above, the conclusion that could be derived clearly points toward a complex interplay between various factors, responsible for the pathogenesis of diseases. To have an effective intervention for these infections of the nervous system, molecular, genomic aspect, and neuropathogenesis need to be explored further.

4.5 Immunosuppressive Effects of Pathogens

HIV, measles virus, and cytomegalovirus (CMV) are the viruses that most commonly lead to immunosuppression. They impede hemopoiesis and/or antigen presentation to T cells. The measles virus is a distinct form of a non-persistent, immunosuppressive infection.

In the early years of the twentieth century, Clement von Pirquet noticed that acute measles could temporarily inhibit the results of individuals' tuberculin skin test. This indicates that measles virus infection leads to immunosuppression (Naniche and Oldstone 2000). Following an acute measles infection, this temporary immunosuppression may extend for up to 6 months. High morbidity and mortality rates have commonly been linked to the immunosuppression that comes with measles virus infection (Naniche and Oldstone 2000).

Only a portion of how measles affects the immune system is known. The main site of measles virus replication is the lymphoid tissue, and both B and T cells can become infected via CD150 (Griffin 2021). During acute infection, this leads to the depletion of lymphocytes. However due to increased adaptive immune responses, in the form of higher degree of lymphocytes proliferation and an increased production of specific antibodies and T cell against measles virus are observed. This results in the elimination of the measles virus (Griffin 2021). To develop lifelong immunity, the genetic material of measles should exist in the lymphoid system for longer duration along with germinal center proliferation, generation of antibody-secreting cells, functionally of different populations of T cells, and the maturation of antibody avidity. A decrease in the variety of already-present antibodies, amount of memory, and naïve B cells leads to an increased susceptibility to other infections (Griffin 2021).

Considering HIV infection, it also able to induce severe immunosuppression, which develops slowly and creates an environment that is progressively favorable for opportunistic infections and the reactivation of latent diseases. It is well known that helper T cells are the primary targets of HIV infection, and depletion of these T cells counts leads to AIDS (Macatonia et al. 1990). Antigen-presenting cells' (APC) abnormalities due to HIV infection may potentially trigger the corresponding decrease of CD4 responses at the early stage of the infection. This is observed when the T-cell counts are still within the normal range. Furthermore, it must be noted that dendritic cell (DC) counts in the peripheral blood also decline as the virus progresses in HIV patients (Macatonia et al. 1990). Peripheral major histocompatibility complex (MHC) class II expression is necessary for CD4 T-cell survival. Thus, DCs with high levels of class II expression may help keep CD4 T cells functional. The demise of the CD4 T cells is because fewer functioning DCs are present. Therefore, it is possible that the infection of bone marrow auxiliary cells, which is promoting hemopoiesis, causes the impairment of DC, macrophage, and/or T-cell production in AIDS (Naniche 2000).

HIV, which can infect lymphoid, myeloid, and monocytic progenitor cells, could exacerbate the problem with hemopoietic cell lineage regeneration. In addition to CD4 T lymphocyte depletion, APC malfunction and inability of hemopoietic regeneration have been suggested to contribute to the immunosuppression induced by HIV. According to a theory, apoptosis hinders the immune system's ability to respond in people with HIV. HIV-positive people's CD4 and CD8 T cells are extremely vulnerable to Fas-induced apoptosis. In addition, it has been shown that homing receptor-mediated signaling encourages death in dormant, nonactivated CD4 cells that are HIV-infected (Wang et al. 1999).

Additionally, it has been suggested that HIV's gp120 and tat proteins may have immunosuppressive properties. By blocking CD4 interaction with the APCs, Gp120 attaches to uninfected CD4 cells and causes a nonresponsive state to develop (Fidler and Rees 1999). The soluble viral protein Tat can cause apoptosis in uninfected cells (Cohen et al. 1999: Li et al. 1995). It been proposed that increased levels of some cytokines, such IL-10, can decrease the immune response (Schols and De Clercq 1996).

Another example of a virus that causes immunodeficiency is the human cytomegalovirus. CMV infection in normal immune person results in an acute infection with temporary immunosuppression. Over the course of a few weeks or months, the immunological malfunctions return to the normal state. Following that the virus creates a latent infection with minimal or no symptoms at all. Hence, CMV completely differs from measles and HIV in that it causes a persistent infection with temporary immunosuppression (Naniche 2000).

In peripheral blood mononuclear cells, CMV significantly induces IFN-α/β. These IFN-α/β decreases the oxidative activity and phagocytosis in monocytes leading to the inhibition of functions of monocytes (Noraz et al. 1997). Infection with CMV also results in the suppression of hemopoiesis in the bone marrow. While some wild-type field isolates can only influence stromal cells in the bone marrow, others can directly infect progenitor cells and restrict their proliferation. The reminiscent of HIV infection and stromal cell tropism decrease the production of growth factor G-CSF leading to the inhibition of in vitro progenitor cell colony (Lagneaux et al. 1994; Simmons et al. 1990).

Unlike HIV, which exclusively infects auxiliary support cells, CMV is able to infect both primitive progenitor cells and stromal support cells. Consequently, CMV infection in the bone marrow is responsible for the primary mechanism of CMV-induced immunosuppression. As stated above, these example viruses: measles virus, HIV, and CMV cause systemic immunosuppression but have radically diverse consequences. They affect the body through (a) modifying the early IL-12/Th1/Th2 cytokine equilibrium, (b) impairing macrophage and dendritic cell activities, (c) inhibiting hemopoiesis, and (d) generating proteins that have immunosuppressive actions. The immune system can be destroyed by viruses that target hemopoiesis, like HIV, or they can develop benign latency, like CMV. These numerous sequels are thought to be caused by the complex alterations in the viral life cycle (Naniche 2000).

Another pathogen that tends to infect CNS leading to invasive diseases among immunodeficient person is *Aspergillus fumigatus* which causes 30–95% mortality rate. This fungus is the most important airborne fungus and possesses the virulent component epidithiodioxopiperazine gliotoxin. By hindering functions of neutrophils, gliotoxin decreases innate immunity in patient with invasive aspergillosis. Leukotriene (LT)B4, which is generated by 5-lipoxygenase and LTA4 hydrolase (LTA4H), is a chemoattractant which causes the migration of neutrophils to infection sites. By direct inhibition of LTA4H, gliotoxin reduces the formation of chemoattractant LTB4. This finally leads to interfering neutrophil phagocytic activities in patients with invasive aspergillosis (König et al. 2019).

4.6 Infection-Induced Autoimmune Responses

Our own immune system may at times become reactive to our own self-antigen. This is due to inherent deficiencies of peripheral self-tolerance. The underlying cause could be brought on by an intrinsic genetic defect or by other factors (Bhagavati 2021). With the current scientific understanding, it is a well-established fact, that the majority of organ-specific autoimmune illnesses are prone to involve autoreactive pathology as their root mechanism (Bhagavati 2021). It takes a great deal of observation and persuasive scientific evidence to completely understand the pathophysiological mechanisms involved in each autoimmune illness. This is because a paradigm change concerning how infections lead to autoimmune disorders will eventually influence how people with autoimmune diseases are treated (Ndondo et al. 2022).

One of the most prevalent CNS autoimmune illnesses is multiple sclerosis (MS). It has a fundamental cause that is strongly connected to a person's genetic make-up. However, not all autoimmune reactions are due to a person's genetic composition. Some of them may also be brought on by an infection. In such an instance, viral or certain bacterial infections are most often than not the root causes of autoimmune illnesses (McKechnie et al. 2002). It is thought that viruses could potentially be able to initiate an autoimmune reaction due to the fact that they have been shown to influence immune cells with the objective to defend themselves (Getts et al. 2013).

However, it is yet unknown what causes MS in its specific form. Animal studies showing that myelin-specific CD4+ T lymphocytes can cause clinical sickness similar to that seen in MS in mice have supported the basic idea that MS is an autoimmune disease (Croxford et al. 2002). This is our understanding till today.

It has been observed that young children could get infected with a particular virus at an early age. Due to the fact that MS patients are often diagnosed between 20 and 40, the viruses that may have been the primary causes of the disease may have disappeared by the time, when, the condition presents. Therefore, it is vital to investigate for evidence that might contribute toward virus-induced autoimmunity using MS animal models to further our understanding regarding the relationship between autoimmune and infections (Croxford et al. 2002). Considering the situation of HIV infection, there is a possibility that HIV-1 may hide in resident macrophages and microglial cells in the CNS. This act enables HIV to evade the full immune response, turns into latency, and eventually causes various complications.

This type of mechanism of infection enables the peripheral immune response to be well organized leading to a coordinated cascade of the immune response. The reactions, however, may change into an abnormal result of their own accord based on unanticipated variables and conditions. In this instance the immune response appears to be unable to recognize the difference between self and foreign antigen (Getts et al. 2013). It is understood that the immune system begins to lose its capacity to tolerate self-antigen shortly after an infection. It has been widely accepted that the "molecular mimicry" theory may be the basic theory that could justify the aftereffects of an infection in this context. The observed endpoint includes a cross-reactive reaction involving T-cell response, which in turn results in autoimmune disease in the CNS too (McKechnie et al. 2002).

Another example that should be considered is Theiler's murine encephalomyelitis virus (TMEV). It is a neurotropic mouse virus that exists naturally, which belongs to the Picornaviridae family. A virus-specific CD4 T-cell response develops after TMEV infection, leading to a concentration of the virus-infected macrophages and microglia (Croxford et al. 2002). However, approximately 50 days after infection, the white matter of the neurons in the spinal cord starts to demyelinate, at the same duration, CD4 T-cell responses which is myelin specific begin to come into effect. Clearly, this is a sign of an autoimmune response. This implies the prospect of the "epitope spreading" process, involving a viral-specific response, which in turn becomes a myelin-specific response. This discovery does suggest autoimmunity in the CNS caused by viral infection (Croxford et al. 2002). Even though, the information that we have now is revolving, around "epitope spreading" as the most probable explanation on how a viral infection of the CNS leads to an autoimmune reaction specifically to myelin, the extensively investigated scientific theory, which is thought to be the underlying mechanism of autoreactive T-cell activation, is still molecular mimicry. An earlier observation of a hepatitis B virus peptide-induced T-cell reactivity to the myelin basic protein (MBP) revealed the same underlying mechanism (Croxford et al. 2002). Conclusively, it is obvious that an auto-immune response arises in this situation following an infection, with molecular mimicry serving as the basic mechanism by which a normal immune response changes to an auto-immune response (Croxford et al. 2002).

The West Nile virus (WNV) is another virus that is suspected to be capable of inducing inflammatory illness in the CNS. It is also known to be the neurotropic flavivirus that infects humans and capable of progressively spreading causing symptoms ranging from a simple febrile illness (WNV fever) to a neuroinvasive condition manifesting that includes meningitis and encephalitis (Leis et al. 2014). Bringing together the possibilities of molecular mimicry between WNV antigens and acetylcholine receptor subunits with the length of time that myasthenia gravis (MG) takes to manifest after an acute WNV neuroinvasive illness, it is speculated that the complex process of mimicry may have contributed to the disruption of immunological self-tolerance that brought about MG as observed during WNV infection (Leis et al. 2014).

In conclusion, based on the current findings and postulations, there is great potential for virus and fungus to successfully establish themselves as potent candidates that are capable of inciting autoimmune diseases in the CNS.

References

Abdullahi AM, Sarmast ST, Singh R (2020) Molecular biology and epidemiology of neurotropic viruses. Cureus 12(8):e9674. https://doi.org/10.7759/cureus.9674

Bailey SL, Carpentier PA, McMahon EJ, Begolka WS, Miller SD (2006) Innate and adaptive immune responses of the central nervous system. Crit Rev Immunol 26(2):149–188. https://doi.org/10.1615/critrevimmunol.v26.i2.40

Banks WA, Erickson MA (2010) The blood-brain barrier and immune function and dysfunction. Neurobiol Dis 37(1):26–32. https://doi.org/10.1016/j.nbd.2009.07.031

Bauer L, Laksono BM, de Vrij FMS, Kushner SA, Harschnitz O, van Riel D (2022) The neuroinvasiveness, neurotropism, and neurovirulence of SARS-CoV-2. Trends Neurosci 45(5):358–368. https://doi.org/10.1016/j.tins.2022.02.006

Begolka WS, McMahon EJ, Miller SD (2005) Cytokines and immune regulation in the nervous system. In: Cytokines and the CNS. Routledge, Milton Park, pp 137–162

Bhagavati S (2021) Autoimmune disorders of the nervous system: pathophysiology, clinical features, and therapy. Front Neurol 12:664664. https://doi.org/10.3389/fneur.2021.664664

Brooks BR, Hirsch RL, Coyle PK (1983) Cellular and humoral immune responses in human cerebrospinal fluid. In: Neurobiology of cerebrospinal fluid, vol 2. Plenum Press, New York, pp 263–329

Cohen SS, Li C, Ding L, Cao Y, Pardee AB, Shevach EM, Cohen DI (1999) Pronounced acute immunosuppression in vivo mediated by HIV Tat challenge. Proc Natl Acad Sci U S A 96(19): 10842–10847. https://doi.org/10.1073/pnas.96.19.10842

Croxford JL, Olson JK, Miller SD (2002) Epitope spreading and molecular mimicry as triggers of autoimmunity in the Theiler's virus-induced demyelinating disease model of multiple sclerosis. Autoimmun Rev 1(5):251–260. https://doi.org/10.1016/s1568-9972(02)00080-0

Daniels BP, Holman DW, Cruz-Orengo L, Jujjavarapu H, Durrant DM, Klein RS (2014) Viral pathogen-associated molecular patterns regulate blood-brain barrier integrity via competing innate cytokine signals. mBio 5(5):e01476-14. https://doi.org/10.1128/mBio.01476-14

Douaud G, Lee S, Alfaro-Almagro F, Arthofer C, Wang C, McCarthy P, Lange F, Andersson JLR, Griffanti L, Duff E, Jbabdi S, Taschler B, Keating P, Winkler AM, Collins R, Matthews PM, Allen N, Miller KL, Nichols TE, Smith SM (2022) SARS-CoV-2 is associated with changes in

brain structure in UK Biobank. Nature 604(7907):697–707. https://doi.org/10.1038/s41586-022-04569-5
Ellwardt E, Walsh JT, Kipnis J, Zipp F (2016) Understanding the role of T cells in CNS homeostasis. Trends Immunol 37(2):154–165. https://doi.org/10.1016/j.it.2015.12.008
Engelhardt B, Vajkoczy P, Weller RO (2017) The movers and shapers in immune privilege of the CNS. Nat Immunol 18(2):123–131. https://doi.org/10.1038/ni.3666
Fidler SJ, Rees AD (1999) Antigen presenting cell function in HIV-1 infected patients. Immunol Lett 66(1–3):129–134. https://doi.org/10.1016/s0165-2478(98)00171-0
Getts DR, Chastain EM, Terry RL, Miller SD (2013) Virus infection, antiviral immunity, and autoimmunity. Immunol Rev 255(1):197–209. https://doi.org/10.1111/imr.12091
Griffin DE (2021) Measles immunity and immunosuppression. Curr Opin Virol 46:9–14. https://doi.org/10.1016/j.coviro.2020.08.002
Gu J, Gong E, Zhang B, Zheng J, Gao Z, Zhong Y, Zou W, Zhan J, Wang S, Xie Z, Zhuang H, Wu B, Zhong H, Shao H, Fang W, Gao D, Pei F, Li X, He Z et al (2005) Multiple organ infection and the pathogenesis of SARS. J Exp Med 202(3):415–424. https://doi.org/10.1084/jem.20050828
Hemachudha T, Laothamatas J, Rupprecht CE (2002) Human rabies: a disease of complex neuropathogenetic mechanisms and diagnostic challenges. Lancet Neurol 1(2):101–109. https://doi.org/10.1016/s1474-4422(02)00041-8
Kauder SE, Racaniello VR (2004) Poliovirus tropism and attenuation are determined after internal ribosome entry. J Clin Invest 113(12):1743–1753. https://doi.org/10.1172/jci21323
Kipnis J (2016) Multifaceted interactions between adaptive immunity and the central nervous system. Science 353(6301):766–771. https://doi.org/10.1126/science.aag2638
Klein RS, Garber C, Howard N (2017) Infectious immunity in the central nervous system and brain function. Nat Immunol 18(2):132–141. https://doi.org/10.1038/ni.3656
König S, Pace S, Pein H, Heinekamp T, Kramer J, Romp E, Straßburger M, Troisi F, Proschak A, Dworschak J, Scherlach K, Rossi A, Sautebin L, Haeggström JZ, Hertweck C, Brakhage AA, Gerstmeier J, Proschak E, Werz O (2019) Gliotoxin from Aspergillus fumigatus abrogates leukotriene B(4) formation through inhibition of leukotriene A(4) hydrolase. Cell Chem Biol 26(4):524–534.e525. https://doi.org/10.1016/j.chembiol.2019.01.001
Kumar V (2019) Toll-like receptors in the pathogenesis of neuroinflammation. J Neuroimmunol 332:16–30. https://doi.org/10.1016/j.jneuroim.2019.03.012
Lagneaux L, Delforge A, Snoeck R, Stryckmans P, Bron D (1994) Decreased production of cytokines after cytomegalovirus infection of marrow-derived stromal cells. Exp Hematol 22(1):26–30
Leis AA, Szatmary G, Ross MA, Stokic DS (2014) West nile virus infection and myasthenia gravis. Muscle Nerve 49(1):26–29. https://doi.org/10.1002/mus.23869
Li CJ, Friedman DJ, Wang C, Metelev V, Pardee AB (1995) Induction of apoptosis in uninfected lymphocytes by HIV-1 tat protein. Science 268(5209):429–431. https://doi.org/10.1126/science.7716549
Louveau A, Harris TH, Kipnis J (2015) Revisiting the mechanisms of CNS immune privilege. Trends Immunol 36(10):569–577. https://doi.org/10.1016/j.it.2015.08.006
Ludlow M, Kortekaas J, Herden C, Hoffmann B, Tappe D, Trebst C, Griffin DE, Brindle HE, Solomon T, Brown AS, van Riel D, Wolthers KC, Pajkrt D, Wohlsein P, Martina BEE, Baumgärtner W, Verjans GM, Osterhaus A (2016) Neurotropic virus infections as the cause of immediate and delayed neuropathology. Acta Neuropathol 131(2):159–184. https://doi.org/10.1007/s00401-015-1511-3
Macatonia SE, Lau R, Patterson S, Pinching AJ, Knight SC (1990) Dendritic cell infection, depletion and dysfunction in HIV-infected individuals. Immunology 71(1):38–45
McKechnie NM, Gürr W, Yamada H, Copland D, Braun G (2002) Antigenic mimicry: onchocerca volvulus antigen-specific T cells and ocular inflammation. Invest Ophthalmol Vis Sci 43(2):411–418

Mrak RE, Young L (1994) Rabies encephalitis in humans: pathology, pathogenesis and pathophysiology. J Neuropathol Exp Neurol 53(1):1–10. https://doi.org/10.1097/00005072-199401000-00001

Naniche D (2000) Generalized immunosuppression: individual viruses, intertwined targets. Virology 275(2):227–232. https://doi.org/10.1006/viro.2000.0552

Naniche D, Oldstone MB (2000) Generalized immunosuppression: how viruses undermine the immune response. Cell Mol Life Sci 57(10):1399–1407. https://doi.org/10.1007/pl00000625

Ndondo AP, Eley B, Wilmshurst JM, Kakooza-Mwesige A, Giannoccaro MP, Willison HJ, Cruz PMR, Heckmann JM, Bateman K, Vincent A (2022) Post-infectious autoimmunity in the central (CNS) and peripheral (PNS) nervous systems: an African perspective. Front Immunol 13: 833548. https://doi.org/10.3389/fimmu.2022.833548

Neal JW (2014) Flaviviruses are neurotropic, but how do they invade the CNS? J Infect 69(3): 203–215. https://doi.org/10.1016/j.jinf.2014.05.010

Noraz N, Lathey JL, Spector SA (1997) Human cytomegalovirus-associated immunosuppression is mediated through interferon-alpha. Blood 89(7):2443–2452

Pedemonte E, Mancardi G, Giunti D, Corcione A, Benvenuto F, Pistoia V, Uccelli A (2006) Mechanisms of the adaptive immune response inside the central nervous system during inflammatory and autoimmune diseases. Pharmacol Ther 111(3):555–566. https://doi.org/10.1016/j.pharmthera.2005.11.007

Schnell MJ, McGettigan JP, Wirblich C, Papaneri A (2010) The cell biology of rabies virus: using stealth to reach the brain. Nat Rev Microbiol 8(1):51–61. https://doi.org/10.1038/nrmicro2260

Schols D, De Clercq E (1996) Human immunodeficiency virus type 1 gp120 induces anergy in human peripheral blood lymphocytes by inducing interleukin-10 production. J Virol 70(8): 4953–4960. https://doi.org/10.1128/jvi.70.8.4953-4960.1996

Simmons P, Kaushansky K, Torok-Storb B (1990) Mechanisms of cytomegalovirus-mediated myelosuppression: perturbation of stromal cell function versus direct infection of myeloid cells. Proc Natl Acad Sci U S A 87(4):1386–1390. https://doi.org/10.1073/pnas.87.4.1386

Steffen R (2019) Tick-borne encephalitis (TBE) in children in Europe: epidemiology, clinical outcome and comparison of vaccination recommendations. Ticks Tick Borne Dis 10(1): 100–110. https://doi.org/10.1016/j.ttbdis.2018.08.003

Walsh JT, Watson N, Kipnis J (2014) T cells in the central nervous system: messengers of destruction or purveyors of protection? Immunology 141(3):340–344. https://doi.org/10.1111/imm.12187

Wang L, Chen JJ, Gelman BB, Konig R, Cloyd MW (1999) A novel mechanism of CD4 lymphocyte depletion involves effects of HIV on resting lymphocytes: induction of lymph node homing and apoptosis upon secondary signaling through homing receptors. J Immunol 162(1):268–276

Wraith DC, Nicholson LB (2012) The adaptive immune system in diseases of the central nervous system. J Clin Invest 122(4):1172–1179. https://doi.org/10.1172/jci58648

Part II

Viral Pathogens: Pathogenesis, Pathology, Diagnosis and Treatment

Herpesvirus Infections of the Central Nervous System

5

Sumathi Muralidhar

Abstract

Double-stranded DNA herpesviruses are capable of establishing latency and reactivation. They are subdivided into eight human herpesviruses (HHVs): herpes simplex virus type 1, type 2, varicella-zoster virus, Epstein–Barr virus, cytomegalovirus, HHV-6, HHV-7, and Kaposi sarcoma virus. Most HHVs have the potential to cause serious neurological disease, which may be acute, chronic, monophasic, or recurrent; clinical features vary depending on the herpesvirus involved. During the course of primary infections, or even during reactivations or re-infections, neurological disease in the form of encephalitis, cerebellitis, meningitis, or myelitis may occur. Herpes simplex encephalitis is life-threatening, whereas herpes simplex virus type 2-associated recurrent aseptic meningitis is self-limited, presenting with headache, episodic fever, and meningismus. Polymerase chain reaction testing of cerebrospinal fluid is the method of choice for diagnosis, along with supplementary clinical findings and magnetic resonance imaging (MRI). Cerebrospinal fluid-polymerase chain reaction analysis has brought about an upgrade in the diagnosis of central nervous system (CNS) viral infections, especially those caused by herpesviruses, replacing aggressive brain biopsy techniques. Additionally, quantitating viral DNA levels in the cerebrospinal fluid (CSF) allows prognosis determination. Treatment modalities often involve starting the patient on empiric acyclovir; discontinued only when herpesvirus is ruled out.

S. Muralidhar (✉)
Apex Regional STD Centre & SRL – HIV, Vardhman Mahavir Medical College, Safdarjung Hospital, New Delhi, India
e-mail: drsumathimdhar@vmmc-sjh.nic.in

H. Sami et al. (eds.), *Viral and Fungal Infections of the Central Nervous System: A Microbiological Perspective*, https://doi.org/10.1007/978-981-99-6445-1_5

Keywords

Herpesviruses · Herpes simplex virus · Varicella-zoster virus · Epstein–Barr virus · Cytomegalovirus · Herpes simplex encephalitis · Aseptic meningitis

5.1 Herpesviruses in Central Nervous System Infections

5.1.1 Introduction

Infections affecting the central nervous system (CNS) in humans may be a result of bacteria, viruses, fungi, or parasites. Viruses can affect the CNS in one of many ways.

(a) By directly infecting the central nervous system, viruses may replicate and destroy the cells, as seen in encephalitis. Common etiological pathogens linked to viral encephalitis are herpes viruses (chiefly, herpes simplex virus) and arboviruses, (example, West Nile virus).
(b) By infection limited to the meninges, i.e., meningitis.
(c) Infections elsewhere in the body may trigger the immune system to attack and damage cells around the nerves, leading to a CNS infection; acute disseminated encephalomyelitis, for example.

The inflammatory changes that occur following neuroinfections can affect one or more anatomical regions. Accordingly, meningitis occurs if the meninges are involved, encephalitis in case of brain parenchyma involvement and myelitis if the spinal cord is involved. Multiple regions may also be involved simultaneously, such as meningoencephalitis or encephalomyelitis (Swanson and McGavern 2015).

Encephalitis is parenchymal inflammation of the brain, resulting from either non-infective, infective, or post-infective causes, accompanied by neurologic dysfunction. Approximately, 50% are due to infectious causes (Bradshaw and Venkatesan 2016).

Common encephalitic viruses are: Herpes simplex virus type 1 (HSV-1) and 2 (HSV-2), the non-polio enteroviruses, and arboviruses. Apart from these, the cytomegalovirus (CMV), Epstein–Barr virus (EBV), human herpesvirus 6 (HHV-6), and influenza viruses are also relevant. Humans are the only reservoirs for Herpesviruses. Once infected by herpesviruses, reactivation is possible, especially in immunocompromised individuals, because these viruses remain latent in the host. It is important to diagnose the various herpes virus infections, both clinically and radiologically, at the earliest, because most of them, if treated early and efficiently, have a favorable outcome.

Human herpes viruses (HHV) are a group of viruses that come under the family of Herpesviridae and entail eight species relevant to human infections. The human herpesviruses are mostly neurotropic and cause serious disease of the CNS, with acute or chronic presentations. All eight herpes viruses are capable of establishing

Table 5.1 Classification of viruses in the family-herpesviridae

Subfamily	Common name	Official designation of the virus
Alphaherpesvirinae	Herpes simplex virus-type-1 (HSV-1)	Human herpes Virus-1 (HHV-1)
Alphaherpesvirinae	Herpes simplex virus-type-2 (HSV-2)	Human herpes Virus-2 (HHV-2)
Alphaherpesvirinae	Varicella zoster virus (VZV)	Human herpes Virus-3 (HHV-3)
Gammaherpesvirinae	Epstein-Barr virus (EBV)	Human herpes Virus-4 (HHV-4)
Betaherpesvirinae	Human cytomegalovirus (HCMV)	Human herpes Virus-5 (HHV-5)
Betaherpesvirinae	Human herpes virus −6	Human herpes Virus-6 (HHV-6)
Betaherpesvirinae	Human herpes virus −7	Human herpes Virus-7 (HHV-7)
Gammaherpesvirinae	Kaposi sarcoma-associated herpes virus (KSHV)	Human herpes Virus-8 (HHV-8)

latency after primary infection in the natural host. Latency can be reactivated after varying periods of time, in accordance with the molecular nature of latency in the virus, as well as its type. The herpesviruses can affect the CNS by presenting as cerebellitis, encephalitis, meningitis, or even myelitis. Furthermore, these presentations may be linked to a fallout of primary infection or as an occurrence of viral re-infection or reactivation (Meyding-Lamadé and Strank 2012).

Over the past few decades, considerable importance has been given to herpesviruses, mainly due to their possible role in Alzheimer's disease (AD) and other neuro-degenerative disorders. These ailments are thought of as being linked to not only HSV-1 but also HHV-6 and EBV (Duarte et al. 2019).

The family Herpesviridae along with its subfamilies (Table 5.1) include DNA viruses with icosahedral capsids and a host nuclear-membrane-derived envelope (Muralidhar and Chawla 2019). The genomic constitution is that of a single, linear, double-stranded DNA, encoding 70–200 proteins, which depends on the species. There are three subfamilies in the family Herpesviridae.

5.1.2 General Features for Herpetic CNS Infections

An extensive clinical history along with physical examination are important in making an accurate diagnosis of herpesvirus-associated neuroinfections (Table 5.2). A viral etiology is implicated when there are classic signs and symptoms of meningeal inflammation (including Brudzinski sign, fever, headache, Kernig sign, neck rigidity, photophobia, etc.), along with any characteristic findings pointing toward viral involvement (e.g., vague abdominal symptoms of diarrhea or vomiting, conjunctivitis, herpangina, pharyngitis, rash/s, specific skin lesions, etc.). Focal

Table 5.2 General features and characteristics of human herpes viruses (HHV) of family herpesviridae

Sl. No.	Human herpes virus	Latency site	Primary infection	CNS infection	Laboratory diagnosis	Management
1.	HHV-1	Trigeminal ganglion	Herpes labialis	Encephalitis, *meningitis, and* myelitis	(specific serology HSV-1), viral culture, and PCR of CSF sample	Acyclovir and foscarnet
2.	HHV-2	Sacral and lumbar ganglia	Herpes labialis	Encephalitis, meningitis, and myelitis	(specific serology HSV-2), viral culture, and PCR of CSF sample	Acyclovir and foscarnet
3.	HHV-3	Trigeminal and dorsal root ganglion	Varicella (chickenpox)	Encephalitis, meningitis, cerebellitis, myelitis, Vasculitis, Reye's syndrome, acute cerebellar ataxia, and neuropathy	PCR of CSF sample	Acyclovir and valacyclovir
4.	HHV-4	B cells and epithelial cells	Infectious mononucleosis	Meningitis, encephalitis, ADEM, acute cerebellar ataxia, polyradiculomyelitis, transverse myelitis, GB syndrome, cranial, and peripheral neuropathies	PCR of CSF sample	Acyclovir and ganciclovir
5.	HHV-5	Cells of myeloid lineage	Pneumonitis myocarditis	Meningo-encephalitis, myelitis, and polyradiculopathy	Virus isolation and PCR of CSF sample	Ganciclovir and cidofovir
6.	HHV-6	Monocytes and macrophages and brain tissue	Exanthema subitum	Meningoencephalitis, myelopathy, temporal lobe epilepsy, post-transplant acute limbic encephalitis (PALE), multiple sclerosis, and aseptic meningitis	PCR of CSF sample	Ganciclovir and foscarnet
7.	HHV-7	CD4+ lymphocytes and salivary glands	Exanthema subitum and rubella like illness	Encephalitis and PALE	PCR of CSF sample	Foscarnet and cidofovir
8.	HHV-8		Kaposi's sarcoma	Encephalitis, AIDS-dementia complex, amyotrophic lateral	Serologic assays, immunofluorescence, ELISA,	Cidofovir, foscarnet,

	HHV8 genome is bound to cell DNA in an episome form		sclerosis (ALS), and primary CNS lymphoma.	and Western blot. PCR of CSF sample	ganciclovir, and valganciclovir

PCR polymerase chain reaction, *ELISA* enzyme linked immuno sorbent assay sensory impairments or motor deficits typically indicate encephalitis of viral etiology. Also, when there is a history of exposure to diseased contacts, mosquitoes, rats, and ticks or there are seasonal viral outbreaks, a diagnosis of CNS infection by herpesvirus should be suspected (Autore et al. 2021).

(Note: *There is a difference between Encephalitis and Encephalopathy. Encephalopathy is a broad terminology, referring to a clinical state, defined by changes in behavior, confusion, disorientation, and further cognitive impairments. These changes may be seen not only in encephalitis but also in various other non-inflammatory conditions).*

The presentation of encephalitis can be misleading, with prodromal symptoms favoring a diagnosis of upper respiratory infections. The clinical manifestations of encephalitis may then evolve over a period of days, presenting with fever, headache, seizures, and sometimes focal neurological deficits, which are all common features occurring in many conditions, and not specific to any one disease.

5.1.3 Diagnosis of Herpetic Encephalitis

In general, when investigating a case of suspected herpetic encephalitis in adults, the following tests should be performed:

- Cerebrospinal fluid (CSF) (at least 5 ml to be collected)—bands, cell count with differential count, glucose, Gram's stain, immunoglobulin G index, India ink staining for Cryptococcus lactose, opening pressure, and protein concentration.
- Bacterial cultures as relevant.
- Blood culture.
- Serology (acute and convalescent serum to be tested with titers) VDRL, ELISA for IgM and IgG of relevant organisms (HSV, EBV, etc.)
- Polymerase chain reaction (PCR) test for HSV-1/HSV-2, VZV, enteroviruses, and other possible pathogens.
- For non-neurologic findings, added syndrome-oriented testing ought to be carried out, accordingly (e.g., bronchoalveolar lavage, endobronchial biopsy, skin biopsy, throat swab PCR/culture, etc.).
- Brain MRI may be normal or with specific findings.
- More than 80% of viral encephalitic cases exhibit an abnormal electroencephalogram (EEG) with diffuse slow waves of high amplitude, with or without focal epileptiform activity. It is often necessary to continuously monitor EEG, especially to identify non-convulsive status (da Costa and Sato 2020).
- Blood tests—Blood gas analysis, complete blood count (CBC), differential white blood cell (WBC) count, platelet count, CRP, procalcitonin, serum electrolytes, kidney function test (KFT), liver function tests (LFT), international normalized ratio for prothrombin time (INR), and partial thromboplastin time (PTT).

There are several indications for performing imaging before a lumbar puncture (Autore et al. 2021)

(a) Severe case of depressed mental status.
(b) Papilledema.
(c) Hydrocephalus.
(d) Trauma history or CSF shunt procedure.
(e) Focal CNS impairments.

5.1.4 Treatment of Herpetic Infections of CNS

5.1.4.1 Initial Management

When an emergency case presents with alterations in consciousness, it is of utmost importance to recognize and treat emergency issues first, including hemodynamic and respiratory insufficiency. A quick assessment of other potential reversible associations in encephalopathy should be made, like electrolyte disturbances, hypoglycemic conditions, and so forth, which can be executed easily in an emergency setting. Once these irregularities are handled and the patient is stabilized, proper triaging and assessing for intensive care unit admission should be carried out. A neurological ICU, with a multi-disciplinary team, if available, is recommended, especially for patients with comorbidities, autonomic dysfunctions, or in a coma (Bradshaw and Venkatesan 2016).

5.1.4.2 Empirical Treatment of Encephalitis

Empirical treatment, in all encephalitis cases, must be started right away, while relevant investigations may be carried out alongside. Intravenous acyclovir should be started early at 10 mg/kg eighth hourly and continued for 14–21 days, along with a broad-spectrum antibiotic for bacterial causes, till a definitive diagnosis is made, when it may have to be stopped. This is because bacterial meningoencephalitis is often indistinguishable from herpes virus encephalitis. Till a definite diagnosis is established, the patient has to be followed up very closely (Bradshaw and Venkatesan 2016).

Although there exists an extensive list of different viruses associated with CNS infections, most of them follow similar cascades of pathogenic events. More or less, they involve events such as the infringement of the blood–brain barrier, as well as the release of glutamate, interferon-1, reactive oxygen species, tumor necrosis factor-alpha, and other potentially neurotoxic mediators. It should be noted that these universal mediators may just as well serve as current or future targets for therapeutic agents (Autore et al. 2021).

5.2 Emphasis on Individual HHV Types Causing Infections of the CNS

5.2.1 HHV-1 Also Known as Herpes Simplex Virus Type-1 (HSV-1)

5.2.1.1 Introduction

As a neurotropic agent, HSV-1 exhibits a wide array of medical disorders, ranging from oral and facial skin manifestations to serious infection of the CNS. In adults, it remains the most common reason for severe, life-threatening sporadic, necrotizing encephalitis (85%). The CNS infections of HSV-1 may be labeled as "neonatal HSV neuro-infection," in neonates, or "herpes simplex encephalitis" (HSE) in age groups outside the neonatal age. Apart from this, HSV-1 is implicated as a cause of blinding keratitis, as well as a cause of neonatal disseminated herpes (Mielcarska et al. 2022). Almost 90% of encephalitis in children and adults is due to HSV-1, with a worldwide incidence of 2 and 4 cases/1,000,000, and the majority of these infections occur in over 50-year-olds (Bradshaw and Venkatesan 2016; Shoji et al. 2002).

5.2.1.2 Pathogenesis

Following infection in an uninfected individual, HSV-1 becomes latent in peripheral sensory neurons of the trigeminal or dorsal root ganglion, and the vestibular and facial ganglion, where it remains latent and causes changes in the cellular processes essential for the normal functioning of neuronal cells (Mielcarska et al. 2022). The asymptomatic individuals with HSV-1 infection serve as important reservoirs of the virus, contributing to transmission via means of viral shedding. Recent studies suggest that HSV-1 neuro-infection may be contributory to Alzheimer's disease, wherein it damages the neurons and glial cells and also affects the homeostatic and immune functions. This hypothesis is backed by various incidences involving the detection of AD-linked biomarkers in HSV-1 neuro-infection cases. In addition, advanced age predisposes to HSV-1 reactivation and its entry into the brain (Mielcarska et al. 2022).

The various known routes for HSV-1 to reach the central nervous system include the following:

(a) Olfactory tract via the olfactory epithelium and neurons to the olfactory bulb.
(b) Trigeminal tract—Viral particles can also undergo CNS migration through the anterograde route (along trigeminal neurons) (Swanson and McGavern 2015).
(c) Vertical transmission is also possible after the previous infection of the mother and subsequent HSV-1 infection through the hematogenous route.
(d) After a recurrent HSV-1-orofacial infection, peripheral ganglia reactivation can occur, followed by axonal migration to the CNS.

HSV-1 virions, when present in the CNS, are able to trigger a continuous microglial activation, leading to the emission of large amounts of cytokines and chemokines, as well as substantial neuro-inflammatory events (Campos et al. 2021). These biochemical and morphological changes in the neurons may ultimately end in glial cell and neuronal loss. Acute herpes simplex type 1 encephalitis causes apoptotic or necrotic neuronal cell loss. It involves the frontal lobes, temporal lobes, and certain areas of the cerebral hemispheres (i.e., insular cortex). The immune system clears any virus at the infection site, while those residing within the neurons enter a latent phase, with their DNA existing as episomes. Here, the viral protein is not expressed, except for limited genes (latency-associated transcripts, i.e., LAT) (Mielcarska et al. 2022).

The virus can be reactivated by several stress factors, such as emotional stress, trauma, fever, immunosuppression, ultraviolet rays, hormonal changes, dental operations, etc. (Duarte et al. 2019). It is clear that the immune system serves a crucial part in preventing the reactivation of the virus because the latent virus enters a lytic replication cycle when the immune system is suppressed.

Pathology—Herpes simplex virus enters host tissues via broken skin barriers or mucous membranes, and journeys toward the dorsal root neurons (retrograde axonal transport). It interacts with glycosaminoglycans at cell surfaces, for example, heparan sulfate and cell-adhesion molecules like nectin-I (da Costa and Sato 2020).

Histopathological examination of brain tissue, within 7 days, reveals areas of mononuclear inflammation, and necrotizing processes, characteristically in the frontal and/or temporal lobes, with perivascular inflammation, edema in the affected neurons, as well as microglial engulfment of neurons (Swanson and McGavern 2015).

Associated frontal and temporal lesions may be necrotic, inflammatory, or hemorrhagic. Intracerebral hemorrhage and infarction are both pathologically distinct HSV-associated neuro-infection findings. Infection-associated intracerebral hemorrhage occurring in the temporal region is almost exclusively linked to HSV-1, with morbid consequences (Hauer et al. 2019).

Eosinophilic, intranuclear inclusion bodies are present in the neuron and glial cells. Inclusions containing viral antigen and herpesvirus particles are seen on immunohistochemistry and electron microscopic examination, respectively. Latent states of HSV by PCR analysis have been found in parts of the limbic system such as in the cerebellar amygdala and hippocampus (Shoji et al. 2002; Bradshaw and Venkatesan 2016).

5.2.1.3 Clinical Features

Primary HSE is seen in 30% of cases, while 70% of cases are recurrent infections, with almost all cases due to HSV-1. HSV-1 is highly invasive and capable of accumulation in the dorsal root nerve ganglia (DRG). It is also capable of cerebral cortex, hippocampal, insular, mesial temporal, and orbitofrontal migration. These cases are highly vulnerable to Alzheimer's disease. The majority of adult encephalitic cases are due to HSV-1 infection (Campos et al. 2021). Primary HSV-1 infections usually have a subclinical presentation. When symptomatic, the infections may be gingivostomatitis in younger age groups, or tonsillitis and pharyngitis in adult populations. Eye infections in the form of keratoconjunctivitis are also of importance, as they can lead to corneal scarring and blindness. Herpetic lesions typically begin as vesicles that eventually become shallow ulcers, many a time accompanied by systemic findings of fever, malaise, and myalgia (Muralidhar and Chawla 2019).

HSE has a non-specific clinical picture, with a transitory influenza-like illness or with an abrupt onset. Encephalitis must clinically be suspected once indicators of CNS dysfunction present acutely (in 24–72 h), in the form of headache, seizures, behavioral abnormalities, focal deficits, or coma, along with common systemic manifestations or known risk factors (da Costa and Sato 2020; Meyding-Lamadé and Strank 2012; Shoji et al. 2002).

Complications of HSE—Imperative complications of acute encephalitis, apart from respiratory and circulatory insufficiencies, include cerebral edema, herniation, raised intracranial pressure, and seizures (Bradshaw and Venkatesan 2016).

Differential Diagnosis of HSE—Several clinical conditions have to be actively looked for and ruled out when making a diagnosis of HSE. These include the following:

(a) **Vascular causes**—cerebral venous sinus thrombosis, intracerebral hemorrhage, ischemic stroke, posterior reversible encephalopathy syndrome, reversible vasoconstriction syndrome, subarachnoid hemorrhage, and vasculitis.
(b) **Metabolic disturbances**—electrolyte imbalances, hepatic/renal encephalopathies, hypoglycemic states, mitochondrial encephalopathy-lactic acidosis-stroke-like episodes (MELAS), septic encephalopathy, and Wernicke's encephalopathy.
(c) **Toxic causes**—Alcohol, drugs and lactic acidosis.
(d) **Trauma.**
(e) **Malignancy**—Primary brain tumor with cerebral metastases.
(f) **Epileptic causes**—Status epilepticus (non-convulsive).
(g) **Infective causes**—Other viral encephalitis, other post-infectious encephalitis, bacterial, mycobacterial, and fungal meningitis (Shoji et al. 2002).

5.2.1.4 Laboratory Diagnosis of HSV-1 Encephalitis

(a) Serology—four-fold elevation of HSV antibody (ELISA for IgG and IgM).
(b) Serum/CSF antibody ratio < 20; antibody index $>/= 1.91$.
(c) Polymerase chain reaction—HSV-DNA detection by CSF-PCR is the current gold standard diagnostic test, which is highly reliable and avoids the invasive procedure of brain biopsy. For HSV, it has 100% specificity, along with a sensitivity of more than 95%. In cases with a negative primary test despite high clinical suspicion, the recommendation is to test an additional sample in 2–7 days' time. In view of situations where there is a negative CSF-PCR for HSV initially, which later turns out to be positive, it is always required to give importance to clinical suspicion (James et al. 2009; Shoji et al. 2002).
(d) Viral culture—HSV isolation from CSF.
(e) Chemi-luminescence immune-assay (CLIA).
(f) Cerebrospinal fluid examination—CSF may be hemorrhagic and shows pleocytosis (≥ 5 nucleated cells/ml) with pressure moderately or greatly increased. There may or may not be pleocytosis in the CSF, but lymphocytes

are predominant, and RBCs are frequently seen. The usual HSE-CSF findings comprise (1) moderate lymphocytic pleocytosis of 10–200 cells per mm^3, (2) mildly raised red blood cells, (3) moderately elevated protein (50–100 mg/dl), and (4) mostly normal glucose levels (Shoji et al. 2002).

(g) Electroencephalogram (EEG) studies—Electroencephalography should ideally be executed in all suspected individuals to perceive any generalized/focal epileptiform activities. It might correspondingly aid in distinguishing the origin of specific behavioral alterations, primarily involving psychiatric states or encephalopathy (da Costa and Sato 2020). EEGs might indicate non-specific, slowing periodic lateralizing epileptiform discharges known as PLEDS (da Costa and Sato 2020). The changes in EEG usually involve one side initially and subsequently progress toward the contralateral temporal region.

(h) Brain imaging techniques.
   - Brain computed tomography (CT) scans—CT changes in 80% of patients with HSV show decreased attenuation in the temporal lobe/s or hyper-intensified regions. Whenever there are signs of increased intracranial pressure, a non-contrast CT scan is generally suggested before lumbar puncture (Meyding-Lamadé and Strank 2012).
   - Magnetic resonance imaging (MRI)—MRI is considered the most sensitive/specific radiological investigation in HSE, especially in its initial course, and provides a great account of parenchymal inflammation, demonstrates focal, contrast-enhancing lesions, and helps in recognizing idiopathic inflammatory neurological findings. Brain MRI shows hyperintensity regions on fluid-attenuated inversion recovery (FLAIR) images, T2-weighted (T2W) images, diffusion-weighted images, and T1W images, which may show gadolinium-enhanced cingulate gyrus, insular cortex, orbitofrontal, and/or gadolinium-enhanced temporal lobes (Shoji et al. 2002).

(i) Brain biopsy, although rarely performed these days, is designated for postmortem cases or for highly suspicious, unresponsive cases (Meyding-Lamadé and Strank 2012).

The vital way for early identification and management of HSE includes attaining familiarity with the clinical syndromes and features associated with it. It must also be remembered here that in immunosuppressed patients (e.g., those on chemotherapy, those taking corticosteroids, transplant recipients, or those with human immunodeficiency virus (HIV) infection or other immune-suppressing ailments), the atypical presentations may occur, further complicating the diagnosis, thus needing a strong sense of suspicion (Meyding-Lamadé and Strank 2012).

5.2.2 Treatment

HSE if left untreated has a mortality of 70%, and those who do survive end up with significant neurological deficits, nearly 97%. The first antiviral agent used to treat HSE was idoxuridine, but it was both toxic and ineffective. The next drug to counter

HSE was vidarabine, which was better tolerated and brought down the mortality to 54%, although 86% of the survivors had residual neurologic deficits. Acyclovir is presently the usual treatment of HSE in most parts of the world. It is a guanine analog, acycloguanosine, given in a 3-week course of 30 mg/kg/day (3 divided doses) (James et al. 2009). Empiric treatment with acyclovir for suspected cases (until ruling out of HSE) still remains the recommendation (Autore et al. 2021). However, it is also vital to be mindful of the fact that none of the anti-herpetic drugs can actually cure a latent infection or prevent symptom recurrence and asymptomatic viral shedding.

5.2.3 Human Herpes Virus (HHV)-2 Also Known as Herpes Simplex Virus Type-2

5.2.3.1 Introduction

About 80% of neonatal encephalitis cases transmitted from mothers to newborns are due to HSV-2, which reduces to less than 10% after the neonatal period. Thereafter, the acquisition of HSV-2 infection is mostly through the sexual route. HSE in neonates typically presents clinically around 2 weeks after delivery. In adults, it is the most common cause of uncomplicated genital herpes, although it can occasionally affect the CNS and cause meningitis, radiculomyelitis, and rarely encephalitis (Meyding-Lamadé and Strank 2012). Determination of adult HSV-2 infection is made by various history points, e.g., age of sexual debut, race, sexual partners, female gender, positivity for other sexually transmitted infections, and the presence of HIV or other infections.

5.2.3.2 Pathogenesis

Primary infection often begins at mucocutaneous surfaces with retrograde viral migration and latency of its genome in peripheral sensory ganglia. Subsequently, episodic reactivation occurs when there is antegrade communication to nerve endings and mucocutaneous surfaces. HSV-2 affects several parts of the CNS, including brain matter, brainstem, cranial nerves, nerve roots, retina, spinal column, and other parts of the neuroaxis. Primary HSV-2 associated aseptic meningitis cases may be seen in 13% males and 36% females (Hauer et al. 2019). Serious cerebrovascular-related consequences of HSV neuroinfections can occur, which are mostly comorbidities, irrespective of gender, age, and immunosuppression status.

5.2.3.3 Clinical Features

Herpes simplex encephalitis due to HSV-2 accounts for 1.6–6.5% in adults, especially in immunocompromised hosts. HSV-1 affects the orbitofrontal and/or temporal lobes, but HSV-2 prefers to infect the brainstem. It remains latent in the sacral and lumbar ganglia. HSV-2 encephalitis can occur without features of meningitis. Alterations in awareness, cranial neuropathies, hemiparesis, as well as hemisensory loss, are some of the CNS manifestations. Lumbosacral or thoracic ascending myelitis may occur in immunocompromised hosts. Acute retinal necrosis, which is characterized by red eyes, periorbital pain, decreased vision, episcleritis, keratotic precipitates, retinal vasculitis, and retinal detachment, is one of the eye syndromes that meningoencephalitis can afflict. In infants and neonates, the HSV-2 infection may present in a localized or disseminated form, either of which can lead to encephalitis, with a more serious prognosis than HSE due to HSV-1. Meningitis and myelitis linked to genital herpes are more prevalent in adults, and HSV-2 as an opportunistic infection accompanying AIDS can cause acute encephalitis and acute necrotizing myelopathy. HSV-2 causes Mollaret's meningitis, which is a type of recurrent aseptic meningitis, although benign, and characterized by recurrent, 2- to 5-day meningitis bouts, which may occur several weeks to years apart.

5.2.3.4 Diagnosis

(a) Tzanck smear—To check for multinucleated giant cells, a swab is obtained from a lesion's base and stained with Giemsa.
(b) Immunofluorescence and immunoperoxidase stains—Here, antibodies to early viral proteins are detected. This is more sensitive than the Tzanck smear.
(c) Viral culture—specimens used for HSV-2 culture include vesicle scraping, genital swabs, or fluid. The virus is grown on human tissue culture; syncytia development and intranuclear Cowdry type A inclusion bodies in host cells are signs of viral growth. Though specific, this is not a very sensitive method of diagnosis.
(d) Serology—Intrathecal monitoring of HSV-2 antibodies is useful after the acute infection stage. In general, the serological tests are not very sensitive and are useful only in differentiating between HSV-1 and 2.
(e) Brain imaging—In the early stages of HSE, CT and MRI scans of the head reveal diffuse edema, while later stages show calcifications, cystic encephalomalacia, and cerebral atrophy.
(f) Electroencephalography (EEG)—There is slow background and paroxysmal discharges in the EEG.
(g) Cerebrospinal fluid—As in all CNS infections of viral etiology, CSF analysis shows normal glucose, raised protein levels, and lymphocytic pleocytosis, which is not quite specific.
(h) PCR—It can detect HSV-2 in cerebrospinal fluid, serum, and other tissues. It is currently the gold standard for the identification of HSV neuroinfections since it is quick, specific, and sensitive for herpes simplex virus type 1 and 2.

5.2.3.5 Treatment

Currently, intravenous administration of acyclovir sodium at a dose of 10 mg/kg every 8 h for 2–3 weeks is advised for treating HSE in adults and children older than three months. Neonatal patients receive a dose of 20 mg/kg, 8 hourly, for a 2–3 week course.

Age, level of awareness at presentation, duration, severity of encephalitis, and HSV load are some of the variables that influence HSE therapy and prognosis. If HSV-2 DNA is still present in the CSF post 3 weeks of treatment, acyclovir (intravenous) medication is given for an additional 2 weeks. PCR of a cerebrospinal fluid sample is done again for HSV-DNA. Acyclovir sodium is used as therapy for HSV-2 meningitis at 5–10 mg/kg three times per day. Acyclovir is used to manage myelitis or radiculitis in a similar manner (10 mg/kg three times each day). In patients with HSV-2 myelitis, it is recommended that glucocorticosteroid be given in high-dose IV therapy along with acyclovir, to decrease the risk of ascending myelitis, as a complication (Shoji et al. 2002). Drug resistance to acyclovir in HSV-1 and -2 can occur due to thymidine kinase mutations and can be treated with valacyclovir. Acyclovir-resistant HSV also responds to cidofovir and foscarnet sodium (Berger and Houff 2008).

5.2.4 Human Herpes Virus (HHV)-3 Varicella Zoster Virus

5.2.4.1 Introduction

Varicella zoster virus (VZV) in the young (as varicella infection) has decreased considerably since there is a reliable vaccination available. VZV CNS infections occur in about 1% of varicella cases, a few days or weeks before or after the appearance of rash. These CNS infections may follow the cranial nerve dermatome or may present as cerebellar ataxia or encephalitis. Immunocompromised patients are clearly at increased risk.

5.2.4.2 Clinical Features

Among the CNS infections caused by VZV, VZV meningitis is the most common, followed by encephalitis, myelitis, Reye's syndrome, acute cerebellar ataxia, and neuropathy (Shoji et al. 2002). Clinically, VZV Encephalitis often presents as an acute or subacute delirium and generally non-specific features, such as fever, headache, meningism, ataxia, and seizures. There is a lot of variation regarding the areas of the brain that are impacted, the extent of the CNS impairment, and the cognitive functions that are involved (Meyding-Lamadé and Strank 2012). Postherpetic neuralgia, which lingers long after the cutaneous signs have disappeared, is the major consequence. The peripheral and central nervous systems are both susceptible to acute neurological problems (Meyding-Lamadé and Strank 2012). In HIV-positive individuals, there may be difficulty in diagnosis because the CNS infections may be a part of the opportunistic infections, in the absence of the typical skin eruptions (zoster sine herpete).

5.2.4.3 Diagnosis

The clinical diagnosis of varicella is relatively easy when typical skin lesions are present. Laboratory diagnosis involves cerebrospinal fluid analysis, which exposes a slight mononuclear pleocytosis along with normal glucose and protein levels.

As in HSV infections, real-time PCR is the favored test, for use in the diagnosis of VZV DNA in CSF. Quantitative PCR is more useful than qualitative PCR. Compared to individuals with other VZV-related CNS disorders, those suffering from encephalitis or meningitis generally have greater viral loads. This might be useful for anticipating the course of VZV-meningoencephalitis in HIV-infected persons as well as evaluating therapeutic efficacy (Meyding-Lamadé and Strank 2012).

Massive and minor ischemic infarcts or hemorrhagic infarcts are visible on brain MRI, in cortical or subcortical grey and white matter. The likelihood of VZV involvement is highly considered when abnormalities, particularly at the gray-white matter junctions, are noted.

5.2.4.4 Treatment

Routine vaccination of children aged 12–15 months, and healthy persons aged 13 years and above, is known to minimize infections of CNS and complications in later life. Treatment of varicella is essentially symptomatic. Herpes zoster management involves antiviral therapy, which should be started within 72 h of the rash's emergence. Valacyclovir, which is known to be superior to acyclovir (8 days: per oral 800 mg 5 times a day), is advised for 1 week, per orally as 1 g three times a day (Meyding-Lamadé and Strank 2012). If steroid treatment is indicated, it should be for a short duration of 3–5 days, only to minimize side effects (Meyding-Lamadé and Strank 2012).

5.2.5 Human Herpes Virus (HHV)-4 Also Known as Epstein–Barr Virus (EBV)

5.2.5.1 Introduction

Epstein–Barr virus is responsible for infectious mononucleosis (IM) as well as neuroinfections. Among the most prolific viruses, it affects over 90% of people within the first 10 years of life and lasts the entire person's lifetime. Primary infection is usually asymptomatic and occurs through infected saliva, in childhood, as well as up to 40% of adults and teenagers, leading to infectious mononucleosis. Meningitis is a rare but serious side effect of EBV infection and typically develops during the initial illness.

5.2.5.2 Pathogenesis

Unlike CNS infections caused by HSV, neurological diseases caused by EBV have distinct etiology. Brain biopsies of EBV-encephalitis cases fail to reveal EBV-proteins or nucleic acids, unlike the presence of HSV proteins in encephalitic brain tissue of HSVE cases. Thus, an autoimmune mechanism has been attributed to EBV-associated CNS disease (Meyding-Lamadé and Strank 2012).

It is important to note that EBV establishes latency only in lymphoid cells and not in neurons. Hence, upon reactivation, the extra-neural locations are likely to become infected with the virus, and infected lymphocytes will then disseminate the EBV to the CNS (Meyding-Lamadé and Strank 2012).

5.2.5.3 Clinical Features

A sore throat with fever and adenopathy, or even splenomegaly, are common symptoms of infection with Epstein–Barr virus, which is typically a self-limiting, lymphoproliferative condition. The CNS infections may occur before, during, or after IM and can be grouped into two categories—CNS infections occurring with primary or reactivated EBV, and occurrence of a chronic-active EBV involvement. Acute cerebellar ataxia, acute disseminated encephalomyelitis (ADEM), cranial/peripheral neuropathy, encephalitis, Guillain–Barré syndrome, meningitis, poly-radiculomyelitis, and transverse myelitis are among the medical conditions that may affect the central nervous system (Meyding-Lamadé and Strank 2012). The involvement of the cerebellum is more common in EBV CNS infections. Other presentations, although uncommon, include gait abnormalities, extrapyramidal involvement, Reye syndrome, improper antidiuretic hormone secretion, or acute fatty liver disease along with encephalopathy. Cases of chronic encephalitis or recurrent meningitis have also been described (Shoji et al. 2002; Meyding-Lamadé and Strank 2012).

5.2.5.4 Diagnosis and Treatment

Traditional diagnosis of EBV infections involves serological tests to detect both specific and non-specific antibodies to the virus. The heterophile antibody detection by the conventional Paul-Bunnel test, along with clinical features and the presence of atypical lymphocytes, is fairly diagnostic for Epstein–Barr virus involvement.

In addition to alterations in serology that are indicative of acute EBV infection, PCR analysis for CSF-DNA can be used to identify neurological disorders associated with EBV. While CT scans may be normal, MRI may nevertheless detect lesions since it is so sensitive, especially on T2W imaging. With brain-MRI, basal ganglia can display focal abnormalities that could aid in differential diagnosis. In cases of encephalitis, EEGs usually reveal generalized slowness with sporadic episodes (Meyding-Lamadé and Strank 2012).

Treatment—For the treatment of EBV infections of CNS, acyclovir and corticosteroids are recommended. Some groups recommend acyclovir usage, or ganciclovir, for managing EBV-encephalitis (da Costa and Sato 2020).

5.2.6 Human Herpes Virus (HHV)-5 Also Known as Human Cytomegalovirus (HCMV)

5.2.6.1 Introduction

CMV is known to cause congenital maternal–fetal infection, with asymptomatic maternal infection leading to symptomatic cases in newborns. CMV causing acquired CNS infection is rare in healthy children but can cause severe disease and complications in immunocompromised hosts, with high residual adverse effects and mortality.

5.2.6.2 Pathogenesis and Clinical Features

Congenital cytomegalovirus infection's neuropathogenesis is due to the dissemination of CMV through the CSF, including the brain's endothelial cells, the contiguous astrocytes, and the choroidal epithelial cells. Viral replication in these cells leads to intranuclear inclusion bodies, like all herpes viruses and lysis of cells. Immune-mediated damage is also implicated in the antigen–antibody complex deposition (James et al. 2009).

Clinical Features—CMV infection rarely causes CNS infections in adults. However, when it does affect the CNS, the cases may present as myelitis, encephalitis (CMVE), or radiculopathies. Diagnosis of CMVE is very difficult because there are no clinical features unique to this condition, and it may present with non-specific signs, such as confusion, seizures, or disorientation. In addition, it may be clinically indistinguishable from HIV dementia (Meyding-Lamadé and Strank 2012). Ventriculoencephalitis or encephalopathy is seen in immunocompromised hosts, such as AIDS patients. Cases of CMV lumbosacral radiculoneuritis have also been reported, which include progressive flaccid paralysis of limbs, areflexia, and recto-urinary impairment.

5.2.6.3 Diagnosis and Treatment

No typical laboratory or neuroimaging finding exists for CMVE. Once cerebral fluid pleocytosis and CMV-PCR test are confirmed, the condition is diagnosed.

Treatment—CMV infections are treated with IV ganciclovir at 15 mg/kg per day every 12 h, for 6 weeks (Shoji et al. 2002). As CMV infection is self-limited and usually tends to resolve, antiviral treatment is generally not recommended for immunocompetent children. However, a 12-month maintenance therapy with valganciclovir 15 mg/kg/dose twice a day, per orally, is advised for immunocompetent kids with severe symptomatic infections (Autore et al. 2021). With this course, hearing outcomes are improved in neonates having symptomatic congenital cytomegalovirus neuroinfections, presenting as aberrant CSF for age, chorioretinitis, hearing loss, intracranial calcifications, and/or microcephaly. One of the limitations of antiviral treatment of congenital CMV infection is that the damage to CNS has already been done in utero before the condition is recognized. That is why, preventive efforts are of paramount importance (James et al. 2009).

5.2.7 Human Herpes Virus (HHV)-6 Also Known as HHV-6

5.2.7.1 Introduction

Among the most pervasive herpes viruses, HHV-6 infects humans at a prevalence of close to 100%. HHV-6 type A (HHV-6A) and type B (HHV-6B) are the two different types that make up this virus. Although HHV-6A is not linked to any serious human illnesses, HHV-6B is the source of the widespread childhood illness, exanthema subitum, often known as the sixth disease, or as roseola infantum. It often affects children between the ages of 1 and 2. It is characterized by a fever of 2–3 days and, in 25–30% of newborns, a non-vesicular, cutaneous rash that is primarily on the trunk and back. Via saliva from mother to child, or among children, is the most likely method of transmission.

5.2.7.2 Pathogenesis and Clinical Features

HHV-6 is also known to establish latency, like all other herpes viruses, which is in monocytes or macrophages, as well as in brain tissue. Reactivation commonly occurs in immunocompromised patients but may also be seen in immunocompetent hosts.

Clinical Features—The primary infection with HHV-6B, exanthema subitum, usually presents with fever and rash. The CNS is affected as a complication of this primary infection, which may present as encephalopathy, meningitis, meningoencephalitis, seizures, or even hemiplegia. A seemingly rare complication of exanthema subitum in healthy children is aseptic meningitis, which generally has a good prognosis (Autore et al. 2021).

In adults and elderly patients, HHV-6 may affect the CNS and cause epilepsy, myelopathy, temporal lobe epilepsy, post-transplant acute limbic encephalitis (PALE), and multiple sclerosis without skin lesions (Shoji et al. 2002). HHV-6 virus establishes latency in the cells of medial temporal lobes, and hence, the clinical features of PALE include abnormal temporal lobe EEG readings, significant anterograde amnesia, or even seizures (Meyding-Lamadé and Strank 2012).

Differential diagnosis—On MRI, HHV-6 CNS infections have to be differentiated from HSVE, other viral encephalopathies, lymphoproliferative disorders affecting the CNS, and drug-induced encephalopathies. Radiological findings of mesial temporal lobe region association along with an immunocompromised state, particularly in post-transplant recipients on prophylactic acyclovir maintenance, are the distinguishing features that allow for a definitive diagnosis of HHV-6 involvement (Meyding-Lamadé and Strank 2012).

5.2.7.3 Diagnosis

CSF, in HHV-6 infections of CNS, shows mild pleocytosis. MRI abnormalities include bilateral, non-enhancing, medial temporal lobe T2W, or FLAIR hyperintensity images, sharply defined by para-hippocampal gyrus sparing. A major diagnostic tool, however, is CSF-PCR for HHV-6-DNA detection, although the findings of this test have to be interpreted with caution. This is because the HHV-6 virus is known to get integrated into host chromosomes; hence, this chromosomal integration has to be distinguished from primary infection and reactivation of the virus (Meyding-Lamadé and Strank 2012). Verifying that CSF-PCR positivity is followed by a negative serum PCR result will help to solve this issue (da Costa and Sato 2020).

5.2.7.4 Treatment

In immunocompetent children, there is really no need to treat HHV-6 infections, as these are self-limiting and recovery is the norm. However, in immunocompromised hosts, there can be life-threatening complications of CNS infections, which need immediate action. Acyclovir is not effective in treating this condition, due to the absence of the thymidine kinase enzyme. Therefore, ganciclovir is utilized as first-line, in HHV-6-encephalitis, either by itself or in synergistic combination with foscarnet (da Costa and Sato 2020). It should be noted that acyclovir/ganciclovir as a prophylaxis is already given to individuals receiving immunosuppressive medication to stop any reactivation of HSV and CMV, respectively. In these circumstances, ganciclovir may also work as prophylaxis to stop HHV-6 reactivation (Meyding-Lamadé and Strank 2012). There is no effective vaccine available for the prevention of HHV-6 infections.

5.2.8 Human Herpes Virus (HHV)-7 Also Known as HHV-7

5.2.8.1 Introduction

HHV-7 is a ubiquitous virus, present worldwide, mostly acquired in infancy. Seventy percent of youngsters get the illness via droplets or breast milk, by the age of four. Due to its amino acid resemblance with that of several HHV-6 proteins and properties, the clinical features and pathogenicity are similar in both these viral infections. This virus reactivates in around 20% of recipients of solid-organ transplants, resulting in infection.

5.2.8.2 Pathogenesis

In CD4+ T cells, HHV-7 develops both active and latent infection. It very easily creates continuing infection within the salivary glands. Other tissues implicated in the latency of this virus are the tonsils, skin, mammary glands, lungs, kidneys, and liver. It partakes a narrower tissue tropism as compared to HHV-6. In immunocompromised patients, it can reactivate either alone or with HHV-6B and CMV viruses. Human herpesvirus 7 infects, yet does not multiply in macrophages and monocytes in vitro. In Kaposi sarcoma lesions, HHV-7 has been found to affect CD34+ hematopoietic progenitor cells, CD68+ macrophages, and monocytes. Additionally, it can be present in breast milk.

5.2.8.3 Clinical Features

Sporadic viral encephalitis is known to occur with HHV-7, HHV-6A, as well as HHV-6B, especially in immunocompromised individuals, when serious encephalitis and generalized symptoms can be present. Risk factors for reactivation of HHV-7 leading to encephalitis comprise repeated hematopoietic stem cell transplantations, unmatched cord blood transfusions, immune suppression, and cytotoxic drugs. The condition of post-transplantation acute limbic encephalitis (PALE), which is linked to significant morbidity, is perhaps the most severe symptom. Occasionally, primary HHV-7 infection can invade the brain and induce febrile convulsions. HHV-7 is also known to cause infantile hemiplegia together with skin rash (Shoji et al. 2002).

5.2.8.4 Diagnosis and Treatment

Both HHV-7 and HHV-6 have similar growth characteristics, notably their cytopathic outcomes in cell culture, which include the production of refractile, ballooned, multinucleated giant cells. In comparison to HHV-6, HHV-7 tends to grow at a slower pace and is not as cytopathic.

To culture human herpesvirus 7 from a continuous CD4+ lymphoblastic cell line, peripheral blood mononuclear cells, or cord blood, primary phytohemagglutinin-stimulated CD4+ T cells are needed.

Enzyme-linked immunosorbent assays (ELISA) or indirect immunofluorescence assays are used to identify antibodies in serological screening for Human herpesvirus 7 in children. The absence of antibodies makes it more likely that the existence of the viral-DNA in blood will be indicative of an acute HHV-7 infection. In adults, the recognition of viral protein appears to be considerably more specific than viral DNA, in determining the cause of encephalitis to be due to HHV-7.

Treatment—Treatment of HHV-7 is best accomplished with cidofovir and foscarnet in vitro, although ganciclovir inhibits viral replication.

5.2.9 Human Herpes Virus 8 (HHV-8) Also Known as Kaposi Sarcoma-Associated Herpes Virus (KSHV)

5.2.9.1 Introduction

All types of Kaposi sarcoma (KS), classic, AIDS related, endemic, transplant related, primary effusion lymphoma (PEL), even solid organ variants, and lymphoproliferative, multicentric Castleman's disease (MCD), are etiologically linked to HHV-8. In general, HHV-8 does not cause CNS infections, except when there is immunosuppression in the host, wherein several complications can occur.

5.2.9.2 Pathogenesis and Clinical Features

In the latent state, the HHV8 genome is tethered to the cellular DNA as an episome.

Clinical Features—KSHV infection is prevalent in individuals all over the world and is associated with a rare cancer called Kaposi sarcoma. The lesions of this malignancy may involve the skin, nose, throat, or other body tissues. KSHV also causes certain types of lymphoma.

In terms of neurological syndromes, HHV-8 is linked to primary CNS lymphomas, amyotrophic lateral sclerosis, and AIDS–dementia complex. As a part of any differential diagnosis for unexplained viral encephalitis, this virus should always be taken into consideration, especially in immunocompromised persons (Shoji et al. 2002).

5.2.9.3 Diagnosis and Treatment

The majority of the time, serologic analyses like immunofluorescence, ELISA, and Western blots are used in laboratories to determine whether a patient has HHV-8 infection. PCR of CSF samples is the test of choice for diagnosing HHV-8 encephalitis. Minimal KSHV titers, as determined by PCR, do not always imply that a condition is linked to KSHV infection (Meyding-Lamadé and Strank 2012).

Treatment—Anti-HSV agents that are known to be effective against KSHV are cidofovir, foscarnet, ganciclovir, and valganciclovir. They act by reducing HHV8 viremia and thus contribute to controlling diseases presenting with high levels of KSHV replication.

References

Autore G, Bernardi L, Perrone S, Esposito S (2021) Update on viral infections involving the central nervous system in pediatric patients. Children 8:782

Berger JR, Houff S (2008) Neurological complications of herpes simplex virus type 2 infection. (reprinted). Arch Neurol 65(5):596. https://jamanetwork.com/on. Accessed 2 May 2010

Bradshaw MJ, Venkatesan A (2016) Herpes simplex virus-1 encephalitis in adults: pathophysiology, diagnosis, and management. Neurotherapeutics 13:493–508. https://doi.org/10.1007/s13311-016-0433-7

Campos EMN, Rodrigues LD, Oliveira LF (2021) Júlio César Claudino dos Santos. Dementia and cognitive impairment in adults as sequels of HSV-1-related encephalitis a review. Dement Neuropsychol 15(2):164–172. https://doi.org/10.1590/1980-57642021dn15-020002

da Costa BK, Sato DK (2020) Viral encephalitis: a practical review on diagnostic approach and treatment. J Pediatr 96(S1):12–19

Duarte LF, Farías MA, Álvarez DM, Bueno SM, Riedel CA, GonzÁlez PA (2019) Herpes simplex virus type 1 infection of the central nervous system: insights into proposed interrelationships with neurodegenerative disorders. Front Cell Neurosci 13:46. https://doi.org/10.3389/fncel.2019.00046

Hauer L, Pikija S, Schulte EC, Sztriha LK, Nardone R, Sellne J (2019) Cerebrovascular manifestations of herpes simplex virus infection of the central nervous system: a systematic review. J Neuroinflammation 16:19. https://doi.org/10.1186/s12974-019-1409-4

James SH, Kimberlin DW, Whitley RJ (2009) Antiviral therapy for herpesvirus central nervous system infections: neonatal herpes simplex virus infection, herpes simplex encephalitis, and congenital cytomegalovirus infection. Antivir Res 83(3):207–213. https://doi.org/10.1016/j.antiviral.2009.04.010

Meyding-Lamadé U, Strank C (2012) Herpesvirus infections of the central nervous system in immunocompromised patients. Ther Adv Neurol Disord 5(5):279–296. https://doi.org/10.1177/1756285612456234

Mielcarska MB, Skowronska K, Ewski ZW, Toka FN (2022) Disrupting neurons and glial cells oneness in the brain—the possible causal role of herpes simplex virus type 1 (HSV-1) in Alzheimer's disease. Int J Mol Sci 23:242. https://doi.org/10.3390/ijms23010242

Muralidhar S, Chawla R (2019) Lippincott illustrated reviews-microbiology. South Asian edition. Wolters Kluwer, Chennai

Shoji H, Wakasugi K, Miura Y, Imaizumi T, Kazuyama Y (2002) Herpesvirus infections of the central nervous system. Jpn J Infect Dis 55:6–13

Swanson PA, McGavern DB (2015) Viral diseases of the central nervous system. Curr Opin Virol 11:44–54. https://doi.org/10.1016/j.coviro.2014.12.009

6 Enterovirus-Associated Meningoencephalitis and Enteroviruses in Patients with Acute Encephalitis

Aisha Halawani, Saima Khan, Samia Masood, and Safiya Firoze

Abstract

Enteroviruses are from the family of Picornaviridae, consisting of hundreds of serotypes, all having single positive-stranded RNA genome. The Enterovirus group comprises of 12 species, including 4 human species: A to D. Encephalitis and meningoencephalitis are infrequent presentations of enteroviral infection, but various enterovirus serotypes, coxsackievirus serotypes, and echovirus serotypes are reported in epidemics in the Southeast Asia region and some European countries. Enteroviruses mostly enter via faeco-oral routes and present with asymptomatic or mild diseases. However, they are also known to present as biphasic prodromal disease, with neurological involvement often beginning as invasion in the anterior horn cells, or even as progression to the brainstem, cerebellum, midbrain, or motor cortex, causing paralysis from neuronal death. More so, enterovirus encephalitis can present as fever with headache, altered sensorium, acute onset muscle flaccidity, hyporeflexia, meningeal signs, and myoclonic jerks. The diagnosis of enteroviral neurological illness is 'definitive' when it is detected by cerebrospinal fluid polymerase chain reaction or culture, along with detection by polymerase chain reaction from throat, rectal

A. Halawani (✉)
Department of Medical Imaging, King Abdulaziz Medical City-Ministry of National Guard Health Affairs-Western Region, Jeddah, Kingdom of Saudi Arabia

S. Khan
Green City Hospital and cancer center, Lucknow, Uttar Pradesh, India

S. Masood
Department of Medicine, All India Institute of Medical Sciences, New Delhi, India

S. Firoze
Department of Microbiology, Jawaharlal Nehru Medical College Hospital, Aligarh, Uttar Pradesh, India

H. Sami et al. (eds.), *Viral and Fungal Infections of the Central Nervous System: A Microbiological Perspective*, https://doi.org/10.1007/978-981-99-6445-1_6

swabs, and serum. Diagnosis is 'probable', if enterovirus is detected in polymerase chain reaction analysis of throat and rectal swabs. If it is detected in either throat or in rectal swab polymerase chain reaction tests, it is denoted as 'possible' enterovirus infection. There is no definitive treatment for enteroviruses, although, intravenous immunoglobulins and ribavirin have shown some promising outcomes in patients diagnosed with enteroviral encephalitis.

Keywords

Enteroviruses · Encephalitis · Meningoencephalitis · Cerebrospinal fluid

6.1 Pathogens

6.1.1 Overview of Enteroviruses

The genus, *Enterovirus*, EV, from the *Picornaviridae* family, is classified by the ICTV, International Committee on Taxonomy of Viruses, into a total of 12 species, of which seven are notable human pathogens (Lei et al. 2016). Human infection-associated EVs comprise of coxsackieviruses (CV-A, CV-B), numbered enteroviruses (EV-A71, EV-D68, etc.), echoviruses (ECV), human rhinoviruses (HRV), and even polioviruses (PV) (Jubelt and Lipton 2014) (Table 6.1).

These ubiquitous RNA-viruses are often times involved in human infections of varying spectra; some with limited time frames of illness, and some presenting as fulminant, non-reversible changes in different organs or systems. These viruses can cause diseases ranging from neurological illnesses, including paralysis, inflammatory reactions, and even morbid conditions contributing to death. Enteroviruses have the potential to cause outbreaks of poliomyelitis disease. Moreover, they tend to cause morbid conditions in the form of encephalitis, meningitis, and vesicular stomatitis in susceptible individuals. But these viruses may be responsible for simpler conditions too, such as the common cold (Nikonov et al. 2017; Rotbart 2000).

6.1.2 Viral Structure

The genomic component of an EV comprises of a positive-sense, single-strand ribonucleic acid of up to 8000 nucleotides with an open reading frame- ORF of a 5′ untranslated region- UTR and a 3'-UTR end (Table 6.2).

The 3′-UTR has a poly (A) tail (pseudoknot) and the 5′-UTR has a 40S ribosome subunit-binding site, known as the IRES (internal ribosomal entry site), for cap-independent translation. The polyprotein precursor with its regions: P1, P2, and P3, is encoded by the ORF (Fig. 6.1). In the initial phase, as an EV infects a cell, viral 3C proteinase cleaves between P2 and P3 of this precursor protein, while 2A proteinase cleaves at the P1-P2 junction (Racaniello 2016). Eventually, mature

Table 6.1 Picornaviridae classification (depiction only of clinically relevant genera and species) (Simmonds et al. 2020; Zell et al. 2017 Viral structure

Order: Picronavirales ⇒ FAMILY: Picornaviridae (i.e., Picornaviruses)							
Subfamily	Caphthovirinae		Ensavirinae			Heptrevirinae	Paavivirinae
Genera	Aphthovirus	Cardiovirus	Enterovirus			Hepatovirus	Parechovirus
SPECIES (including subspecies and genotypes)	**Foot and mouth disease viru**, etc.	**Cardioviruses** A–F	**Enterovirus A:** Coxsackievirus A2–8, A10, A12, A14, A16; enterovirus A71, A76, A89–92, A99, A114, A119–125 **Enterovirus B**: Coxsackievirus B1–6, A9; Echovirus 1–9, 11–22, 24–27, 29–33; Enterovirus B69, B73–75, B77–88, B93, B97, B98, B100, B101, B106, B107, B110–114; **Enterovirus C:** Poliovirus 1,2,3; Coxsackievirus A1, A11, A13, A17–22, A24; Enterovirus C95, C96, C99, C102, C104, C105, C109, C113, C116–118 **Enterovirus D:**	**Enterovirus E:** (E1–5); Enterovirus F: (F5–7); **Enterovirus G:** (G1–20); **Enterovirus H:** (H1); **Enterovirus I:** (I1, I2); **Enterovirus J:** (J1, J103, J108, J112, J115, J121); **Enterovirus K:** (K1, K2); **Enterovirus L:** (L1);	**Rhinovirus A:** (A1–2, A1B, A7–13, A15–16, A18–25, A28–34, A36, A38–41, A43, A45–47, A49, A51, A53–68, A71, A73–78, A80–82, A85, A88–90, A94, A96, A100–108); **Rhinovirus B:** (B3–6, B14, B17, B26–27, B37, B42, B48, B52, B69–70, B72, B79, B83–84, B86, B91–93, B97, B99–104); **Rhinovirus C:** (C1–51, C54–57)	**Hepatovirus A** (hepatitis A virus); **Hepatoviruses** B-I	**Parechovirus A:** Human parechovirus 1–18; **Parechovirus B:** Ljunganvirus 1–6; **Parechoviruses C-F**

(continued)

Table 6.1 (continued)

Order: Picronavirales ⇒ FAMILY: Picornaviridae (i.e., Picornaviruses)							
Subfamily	Caphthovirinae		Ensavirinae			Heptrevirinae	Paavivirinae
			Enterovirus D68, D70, D94, D111, D120; Human rhinovirus 87;				

Table 6.2 Genomic components of Enteroviruses (and other Picornaviruses)

Viral genome	(+) sense; single-stranded RNA (non-segmented); 6.7 to 10.1 kb
Virion	30–32 nm; icosahedral capsid; non-enveloped

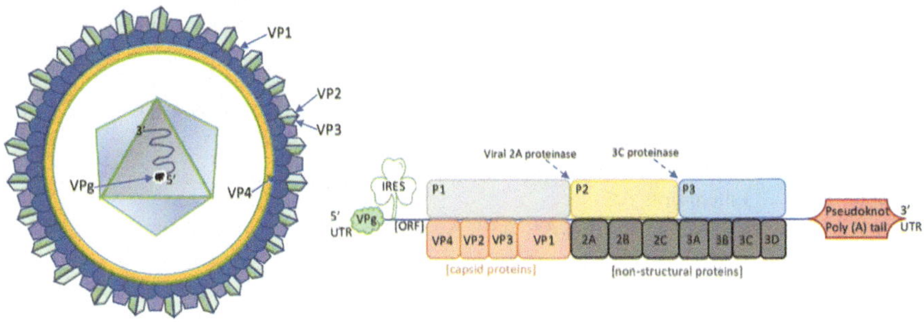

Fig. 6.1 Basic Enterovirus Virion and Genome

EV proteins are processed from this polyprotein precursor, consisting of structural capsid proteins: VP1, VP2, VP3, VP4, non-structural proteins: 2A, 2B, 2C, 3A, 3B, 3C, and 3D. The modulation of proteins involved with apoptosis, innate immunity, poly-adenylation, ribonucleic acid processing and translation is aided by 2A and 3C proteinases, in particular, which in turn, profoundly, effects the infected host cells (Kräusslich et al. 1987; Lei et al. 2010, 2002; Wang et al. 2013; Weng et al. 2009).

6.1.3 Epidemiological Profile

Non-polio enteroviruses (NPEV) may give rise to an array of syndromes, ranging from common cold, haemorrhagic conjunctivitis, hand-foot-mouth disease, herpangina, meningitis, myocarditis, neonatal sepsis, as well as paediatric fever-rash illnesses. This chapter provides a well-constructed insight on non-polio enteroviral CNS infections; encephalitis and meningitis, in particular.

Meningitis and encephalitis are most commonly caused by viral aetiologies, out of which NPEV remain the prominent cause (Michos et al. 2007; Romero 2002; Rotbart 2000). They have been suggested to cause 10% of viral encephalitis cases (Calleri et al. 2017). NPEVs are also responsible for 50–80% aseptic meningitis cases in adults (Han et al. 2016). Compared to meningitis of other viral or bacterial aetiologies, EV meningitis often has a milder course, with occasional serious illness reserved for the early childhood populations, and those with weakened immune systems.

Amongst non-polio enteroviruses, several subspecies of Enterovirus A (e.g., EV-A71, CV-A10), B (e.g., CV-A9, CV-B5, ECV-6, ECV-9), C (e.g., CV-A11, CV-A13), and D (e.g., EV-D68) have been implicated in CNS infections, resulting in a variety of neurological complications, namely, aseptic meningitis, acute flaccid paralysis, and encephalitis (Table 6.3).

Table 6.3 Neurological infections caused by NPEVs (B'Krong et al. 2018; Fan and Liu 2019; Messacar et al. 2018; Pallansch et al. 2013; Sun et al. 2019; Suresh et al. 2018)

CNS infection	Commonly implicated serotypes	
Aseptic meningitis	Enterovirus A: Coxsackievirus A2, A5–7, A10, A16; Enterovirus A71 Enterovirus B: Coxsackievirus B1–6, A9; Echovirus 1–4, 6, 7, 9, 11–21, 24, 25, 27, 29–31;	Enterovirus C: Coxsackievirus A11, A13, A17, A22, A24; Enterovirus D: Enterovirus D68, D70
Encephalitis	Enterovirus A: Coxsackievirus A2, A6, A10, A16; Enterovirus A71 Enterovirus B: Coxsackievirus B1, B2–5, A9; Echovirus 3–7, 9, 11, 13, 14, 16–19, 21, 24, 25, 27, 30, 33;	Enterovirus C: Coxsackievirus A11, A13; Enterovirus D: Enterovirus D68, D70
Acute flaccid paralysis	Enterovirus A: Coxsackievirus A2–7, A10, A12, A14, A16; Enterovirus A71, A76, A90; Enterovirus B: Coxsackievirus B1–6, A9; Echovirus 1–7, 9, 11–22, 24–27, 29–33; Enterovirus B73–75, B77, B79–81, B85–88, B93, B97, B100, B106, B107	Enterovirus C: Coxsackievirus A1, A11, A13, A17, A20–22, A24; Enterovirus C96, C99, C109; Enterovirus D: Enterovirus D68, D70, D94

Enteroviral transmission occurs almost exclusively via faeco-oral routes (Nikonov et al. 2017; Rotbart 2000). There are, however, reports of person-to-person transmission of enteroviruses via respiratory droplets and also via contact with enterovirus-contaminated objects. EVs reside in the gut of infected persons, and are shed in their faeces. They are quite stable and capable of living in environments outside of the human body as well. Contact with faecal matter can lead to the contamination of hands and surfaces with enteroviruses. Viral particles that are ingested and/or in contact with mucosal membranes lead to EV infection. Improved hand hygiene can thereby reduce the spread of enteroviruses (Romero 2002; Rotbart 2000).

Infected individuals can be asymptomatic for long periods, during which enterovirus shedding potentiates risks of giving rise to epidemics in different time frames and countries, making it difficult to trace the initial source (Nikonov et al. 2017). Even while enterovirus infections are often contracted, very few of those infected go on to acquire meningitis or encephalitis. Also, there is very little chance of close contact-associated spread of EV meningitis. Enteroviral infections in general may occur on a perennial basis in the subtropics or tropics, and in temperate climates they tend to have seasonal surges during the summers or early autumns (Romero 2002; Rotbart 2000). Infected populations are predominantly children and infants, with epidemic or sporadic occurrences. Even though numerous enterovirus serotypes can cause encephalitis or meningitis, only a few are actually widespread (Greenberg 2003).

6.2 Etiopathogenesis

6.2.1 Routes of Central Nervous System Invasion

Enteroviruses which cause meningoencephalitis most commonly infect the human through the faecal-oral route and rapidly replicate in the gastrointestinal tract. However, there are few exceptions like for example enteroviruses, EV-D68, they spread via respiratory secretion and can cause respiratory infection. After initially infecting the first exposed area, the viruses easily gain access to the central nervous system via various pathways, which are not mutually exclusive (Rhoades et al. 2011; Huang and Shih 2015).

6.2.1.1 Through the Bloodstream

First, the neurotropic enteroviruses take the course of bloodstream to reach the CNS. Normally, there is a highly selective semipermeable blood brain barrier (BBB) which restricts the viral particle spread from brain's blood vessels to the CNS. However, if the central nervous system's microvascular endothelial cells (BMECs) are infected the integrity of the BBB is largely compromised. The cytokines which are locally produced during the time of infections also poses a big threat to the BBB.

6.2.1.2 Through the Peripheral Circulating Immune Cells (Trojan Horse Route)

In this route, the immune cells which are circulating peripherally carry intracellular viruses (Tabor-Godwin et al. 2010). As a well-established fact, that brain has an active immune surveillance system comprising numerous non-specific leukocytes like lymphocytes and phagocytes to be involved into the meninges and cerebrospinal fluid (Forrester et al. 2018). Additionally, it has been demonstrated that cerebrospinal fluid (CSF) has an appropriate number of trafficking mononuclear cell types, with T cells accounting for the majority 90%, B lymphocytes for a minor 5%, monocytes for the remainder, and dendritic cells for the smallest percentage, i.e., less than 1% (Ransohoff and Engelhardt 2012). These cells become vehicles for viruses to enter the CNS after being infected. The enterovirus is released from myeloid cells upon entrance into the CNS, where it then infects the neuroglial cells and neurons. Recent studies have shown that the sialomucin membrane protein hPSGL1, which is produced on the surface of leukocytes, can bind to EV-A71, exhibiting leukocytic infection (Nishimura et al. 2013).

6.2.1.3 Through Nerves in the Periphery Via Retrograde Axonal Transport and Trans-Synaptic Propagation

Peripheral nerves, which are accessible to enteroviruses through retrograde axonal transport and trans-synaptic propagation, are another possible route for them to enter the CNS (Gromeier et al. 1996; Chen et al. 2007; Ong et al. 2008). Transport via axons is an important cellular process in neurons which is required for the movement to and from the cell body of synaptic vesicles, lipids, proteins, and other organelles including mitochondria, lysosomes, endosomes, and autophagosomes. Few neurotropic viruses have the ability to hijack the retrograde axonal transport to directly invade and infect the central nervous system. The viral particles which are endocytose in the terminal end of axon are moved in retrograde direction through dynein-mediated vesicular transport towards the cell body without uncoating (Ohka et al. 2009). The event of uncoating takes place upon arrival of the motor neuron at the cell body (Chen et al. 2007; Ong et al. 2008; Hixon et al. 2017). The ability of EV-A71 to directly infect the brainstem via cranial nerves has been shown, which is intriguing and implies that for CNS infiltration, not only does the virus use the motor portions of spinal nerves, but also the cranial nerves (Tanet al. 2014).

6.2.2 Cell Receptors for Virus Entry

There are numerous entry mechanisms and receptors used by enteroviruses to invade and infect the host cell. EnteroVirus-71 has been conclusively shown to use several receptors, including P-selectin glycoprotein ligand-1, sialylated glycans, and Scavenger receptor B2 (Nishimura et al. 2013, Yamayoshi et al. 2014, Yang et al. 2015). Some of the enteroviruses are capable of using multiple receptors to invade the host cell. The first barrier for the virus's entrance is determined by the receptor expression on the targeted cells. An infection may become less likely during differentiation, as the amount of viral receptor diminishes. This is what was concluded in a recent study that specifically linked reduced coxsackievirus and adenovirus receptor (CAR) expression in differentiated-primary neurons, to a decrease in infection (Ahn et al. 2008).

6.2.3 Tropism

Tropism is distinct for each enterovirus and is mainly determined by a number of host and viral factors. As elaborated above, neurotropic enteroviruses invade CNS and cause neurological disorders. There is evidence of viral dissemination into the CNS sporadically. There is also evidence that interferons (IFN) which are the innate immune antiviral activities are essential for virus tropism (Wessely et al. 2001; Ida-Hosonuma et al. 2005). EV-A71 invades the nervous system and the areas infected are very distinct. Encephalitis of the brainstem is the commonest neurological presentation of EV-A71 infection. The lesions produced by the virus are profoundly found in the brainstem and is located in various parts of medulla oblongata like the ventral, medial, and caudal areas. (Kao et al. 2004). The spinal column, cerebellum, and cortex may also present with a few lesions. In severe EV-A71 infection, the CNS exhibits significant histo-morphological changes that are characterized by inflammatory damage that specifically cause heart failure and neurogenic pulmonary oedema. It has been determined that the medullary neurons are liable for the development of neurogenic pulmonary oedema (Davison et al. 2012). According to recent investigations in postmortem, EV-A71 can also affect neurons and produce neuronal degeneration, triggering inflammatory responses in the afflicted area and resulting in encephalitis (Yan et al. 2000; Khong et al. 2012; Yao et al. 2012; Feng et al. 2016). It is interesting to note that while EV-A71 can infect neurons, it seems to primarily target astrocytes and neural progenitor cells. The ability to undergo mitosis, that could prove essential for virus replication, is a shared characteristic for the two cell types (Yu et al. 2015). Cognitive, learning, memory, and other such functions depend on neural progenitor cells which are thought of as the cells which give rise to neuroglial and neuronal cells. Therefore, the loss of neural progenitor cells brought on by an EV infection may result in long-term or permanent neurological problems (Chang et al. 2007).

6.2.4 Enteroviruses and Autophagy

Enteroviruses have been shown to greatly benefit and induce the degradation at cellular levels. This process is commonly known as autophagy (Huang et al. 2009; Suhy et al. 2000, Wong et al. 2008). They utilize the autophagosome membrane for replication of the virus as a scaffold. In an experimental work, scientists inferred that induction of autophagy was seen in rat-primary neurons due to raised viral replications (Yoon et al. 2008). Autophagy plays a vital role in preventing neuronal cellular damage (Alirezaei et al. 2015).

6.2.5 Persistent Infection

Although enteroviruses are proved to be cytolytic and the caused disease by the infection is short-lived, numerous studies are now showing association with lifelong disorders which are permanent. The cause of this is not clear yet but is hypothesized that this persistence of enteroviral infection may occur due to the presence of infected viral RNA and protein in the affected tissues at stages of disease after acute infection (Chapman and Kim 2008). Due to lack of proofreading capacity in RNA polymerases, the enteroviruses have high mutation rates thereby generating a variety of mutants to invade the immune system. As CNS is inaccessible to immune surveillance, it makes it vulnerable to persistent infection.

6.3 Diagnosis

6.3.1 Presenting Features

Encephalitis, meningitis, myelitis, and neuritis are a few of the important clinical presentations of enteroviral invasion of the central nervous system. In general terms, encephalitis refers to brain parenchymal inflammation characterized by signs of neurologic impairment in the form of clinical, laboratory, or imaging findings. Aseptic meningitis, on the other hand, refers to sudden development of meningeal warning signs, with/without fever, and pleocytosis on CSF biochemistries, as well as negative bacterial cultures, along with no evidence of parenchymal involvement. Owing to the structure of the nervous system, patients with meningitis generally always have concomitant involvement of the brain parenchyma (meningoencephalitis), and in some cases, there may also be involvement of the spinal cord (encephalomyelitis) or nerve roots (encephalomyeloradiculitis). The individual pathogen and the host's immunological status have a significant impact on the clinical spectrum of neurologic dysfunction as well as the prognosis (Mandell et al. n.d.) In spite of an extensive workup, the cause of encephalitis remains elusive in up to approximately 60% of the population. Enterovirus is the most common cause of aseptic meningitis in children as well as adults; however, the data regarding enteroviral encephalitis in adults vis à vis the presentation and outcomes is sparse (Fischer et al. 2022; Glaser et al. 2003; Hasbun et al. 2017). Most of the reported cases of enteroviral encephalitis in adults are centred around outbreaks involving the paediatric population, with common culprits implicated including echovirus 30 and enterovirus-A71 (Peigue-Lafeuille et al. 2002; Sapkal et al. 2009; Solomon et al. 2010), and cases are generally diagnosed and reported retrospectively.

There are two known types of infection-related encephalitis: primary and post- or para-infectious. Direct central nervous system invasion plus neuronal damage, which frequently also affects the grey matter, conduces to a primary encephalitis. While the symptoms of a post- or para-infectious encephalitis are similar to those of a primary encephalitis, these infections do not directly invade the CNS, and instead, the neurologic effects are a result of the host's immune response, which frequently affects the white matter (Lewis and Glaser 2005).

Data shows that the onset of enteroviral encephalitis may be gradual or abrupt, and there may be a brief prodrome of fever and chills, with severe headache being the predominant complaint. Meningismus may be present, varying from mild to severe, depending on the extent of meningeal involvement. Kernig and Brudzinski signs are present in only about one-third of patients. Other symptoms may include photophobia, nausea, vomiting, diarrhoea, and myalgia (Peigue-Lafeuille et al. 2002). Pharyngitis and other symptoms suggestive of upper respiratory tract infection are commonly present. Quite often, there may be a biphasic pattern of presentation, resembling poliomyelitis, with an initial prodromal phase comprising of fever, upper respiratory tract symptoms, and myalgias, a phase of defervescence of symptoms for a few days followed by an abrupt relapse with the trifecta of fever, headache, and meningism (Bernit et al. 2004).

General physical examination may herald a few clues, for example the presence of herpangina, or hand-foot-and-mouth disease may indicate coxsackievirus; however, most often the findings would be too non-specific for the clinician to clinch a specific viral organism.

The neurologic manifestations of enteroviral encephalitis may be broadly considered under four headings: behavioural alterations (i.e., agitation, altered mentation, hallucinations, personality changes, psychosis, etc.), cognitive decline (e.g., acute memory problems), focal CNS conditions (i.e., anomia, dysphasia, hemianopia, hemiparesis, etc.), and seizures, depending on the site of involvement (Steiner et al. 2005). The most frequent focal neurologic manifestations include ataxia, aphasia, hemiparesis, myoclonus, cranial nerve palsies, and seizures. Involvement of centres for autonomic control may manifest as a loss of vasomotor tone and temperature control, and rarer presentations include the syndrome of inappropriate secretion of antidiuretic hormone (SIADH) or diabetes insipidus owing to hypothalamic dysfunction. Other presentations may include brainstem involvement or even involvement of the anterior horn cell, resulting in acute flaccid paralysis (Mandell et al. n.d.).

A severe, and often fatal form of brainstem encephalitis (rhombencephalitis) is known to be caused by enterovirus-A71 and, occasionally, other EV serotypes. It is also accompanied by secondary cardiac symptoms, such as neurogenic pulmonary oedema and cytokine storm (Hamaguchi et al. 2008; Huang et al. 1999). Countries of the Asia-Pacific Rim belong to one of the hotspots for EV-A71 encephalitis, with records of multiple large outbreaks over the past 10–15 years (Solomon et al. 2010). Principally affecting toddlers and infants, typically neurological manifestations are preceded by hand-foot-and-mouth disease or herpangina. This is followed by the development of progressive myoclonic jerks, tremors, and ataxia; mortality is reported to be as high as 19% (Chan et al. 2003; Ho et al. 1999).

Those who have immunodeficiencies (whether congenital or acquired) have shown a notable predisposition for the acquisition and development of serious EV infections which can present as acute or even chronic infection involving multiple organ systems. Such individuals have been described as having a wide range of CNS consequences, like brainstem involvement, cerebellar manifestations, cranial nerve palsies, encephalopathy, extrapyramidal defects, frontal dementia, and localized (focal) cortical involvement. There is a high prevalence of cochlear nerve involvement when it comes to cranial nerve palsies in these patients (Wagner et al. 2021). Echovirus has also commonly been implicated in the setting of encephalitis in patients with hypogammaglobinaemia (Prentice et al. 1985).

6.3.2 Differential Diagnosis

The constellation of signs and symptoms associated with viral encephalitis and meningitis, including fever, nuchal rigidity, and headache is not specific, and the various differentials include bacterial meningitis, cerebritis, brain abscess, subdural and epidural empyemas, and septic cerebral venous or sinus thrombosis, which should accordingly be ruled out by imaging and further testing. The most important differential to rule out would be bacterial meningitis, as some cases of inadequately treated bacterial meningitis patients may resemble viral encephalitis, especially with regards to the CSF parameters (Steiner et al. 2005).

Viral aetiologies of encephalitis in immunocompetent patients typically include Epstein–Barr virus, herpes simplex virus, and varicella-zoster virus; arboviruses are associated with epidemics of encephalitis. Other differentials to consider would be leptospirosis, borreliosis, Lymes disease, lymphocytic choriomeningitis virus, and acute human immunodeficiency virus infection, which account for most of the remaining cases of infectious aseptic meningitis (Glaser et al. 2003).

6.3.3 Laboratory Diagnostics

Pleocytosis and other inflammatory alterations in the cerebral fluid, along with EV detection by polymerase chain reaction (PCR) analysis, are used for diagnosing enteroviral encephalitis and meningitis (Logan and MacMahon 2008). EV-induced CNS infections, however, may be difficult to diagnose in the presence of unusual clinical manifestations, particularly among elderly persons, as well as in circumstances with lack of cerebrospinal fluid pleocytosis (Ihekwaba et al. 2008; Valcour et al. 2008; Wang et al. 2014). Concurrent testing from locations other than the cerebrospinal fluid may help with diagnosis; however, EV is produced by around 7–8% of healthy controls during viral seasons and is shed from various body sites for several weeks after infection has cleared up (Table 6.4). Moreover, genotyping of enteroviruses with phylogenetic analysis is already a common practice, particularly during epidemics (Savolainen-Kopra et al. 2011; Torok et al. 2017). Enteroviruses spread through the faeco-oral route or less commonly, through respiratory routes, and symptoms can appear anywhere post the incubation period of 3–21 days (Harvala et al. 2018).

Clinical samples should be collected in accordance with clinical symptoms (Table 6.5). Ideally, they should be transported directly to the lab in viral transport media (VTM) and if need be, stored at 4 °C for up to 24 h.

6.3.3.1 Cerebrospinal Fluid (CSF) Biochemistry

To obtain cerebrospinal fluid for a definite diagnosis, a lumbar puncture is carried out. The CSF biochemical analysis typically displays normal glucose levels; although reduced glucose levels may be seen in 15% of patients (Greenberg 2003). White blood cell (WBC) counts in the CSF of aseptic meningitis cases are somewhat higher. Lymphocyte and other mononuclear cell predominance (pleocytosis) is found in more than 66% of infections (Table 6.6). However, in some patients, abundant polymorphonuclear leukocytes in the CSF may also be detected within the initial 6–48 h. Also, whilst the level of protein is normal to slightly elevated, the level of glucose may occasionally be slightly lowered (Chia 2018).

Table 6.4 Laboratory Techniques to diagnose EV Meningitis and/or Encephalitis (Harvala et al. 2018; Storch 2000; Ye et al. 2013)

Laboratory techniques	Comments
Cerebrospinal fluid (CSF) biochemistry	Definitive diagnosis may be obtained by CSF analysis plus CSF polymerase chain reaction (PCR)
Electron microscopy	Reserved for research and further morphological identification/characterization
Immunohistochemistry	Reserved for research studies
Molecular methods:	
i. In situ hybridization	Reserved for research studies
ii. Nucleic acid sequence-based amplification (NASBA)	Up to 100% sensitivity
iii. Reverse transcription-polymerase chain reaction (RT-PCR) [nested, multiplex, probe-based, etc.]	PCR has higher diagnostic value than EV culture;
iv. RT-PCR is more sensitive (95–100%) than culture; ≥ 97% specific	
v. Real-time RT-PCR (rtRT-PCR)	Sensitivity of up to 100%; specificity of ≥96%
Serology:	Routine serology testing for acute EV infection diagnosis is NOT recommended due to subpar EV detection standards of most of these tests
i. Antigen detection	
ii. Enteroviral IgM antibodies	
iii. Enzyme immunoassays	
iv. Enzyme-linked immunosorbent assay (ELISA)	
v. Fluorescent antibody staining	
vi. Immuno-peroxidase antibody staining	
vii. Neutralization assays	
Viral culture (Cell/tissue culture)	NOT sensitive (~30–70%); NOT for routine diagnostic use, but may be used for further EV characterization at national level

Table 6.5 Advised Sample types for EV Meningitis/Meningoencephalitis Laboratory Diagnosis (Ye et al. 2013)

Clinical Sample	Remarks
Cerebrospinal fluid (CSF)	Viral RNA in CSF is detectable by PCR in almost all EV meningitis cases, but detection is inconsistent in EV encephalitis cases
Other sterile sites: Serum/urine/vesicular fluid/collection at autopsy	More reliable than non-sterile sites
Respiratory sample; throat/nasal swabs; Faeces sample	There is prolonged viral excretion in faecal samples and throat, but its detection does not directly imply any etiological link, i.e., may merely imply coincidental carriage

Table 6.6 Typical CSF Findings in EV Aseptic Meningitis and/or Encephalitis (M. Tille 2021)

	Findings	Normal Ranges
Glucose	Normal/slight ↓ [30–45 mg/dL]	45–100 mg/dL [glucose CSF: Serum of 0.6 or 50–70% of normal blood glucose value]
Leukocytes (mm^3)	2–1000	0–5
Opening pressure	Normal/slight ↑	< 180 mmH_2O
Predominant cell type	↑ lymphocytes	None
Protein	Normal/slight ↑ [50–100 mg/dL]	15–50 mg/dL

6.3.3.2 Electron Microscopy

Electron microscopy (EM) is reserved mostly for research purposes in highly professional and designated research facilities. Morphological characteristics can be discovered and studied using EM or transmission electron microscopy as the basis for virus identification using thin-section, negative staining, or cryo-EM technique (with cell or brain tissue specimens). Unfortunately, it has a low sensitivity and needs at least 10^6 virions per milligram of specimen to be visible under a microscope (Hussin et al. 2022).

6.3.3.3 Immunohistochemistry

Immunohistochemistry and other histology-based techniques are mostly reserved for research studies. For instance, immunohistochemistry of encephalitic brain matter may be done, using different antibodies to look for significantly stained cytoplasm of affected neurons or microglia. Patented anti-CV-B polyclonal antibodies, EV-A71 mouse monoclonal antibodies, and mouse monoclonal antibodies to conserved EV VP1 are some of the antibodies that may be used (Dourmashkin et al. 2012).

6.3.3.4 Molecular Methods

Polymerase Chain Reaction—Culture identification of enteroviruses in meningitis or encephalitis are not sensitive (around 30%), because of low EV titres in cerebrospinal fluid. For EV meningitis and encephalitis, reverse transcription-polymerase chain reaction (RT-PCR) tests are much more sensitive (up to 100%) and over 94–97% specific (Chia 2018; Torok et al. 2017). Due to their sensitivity, specificity and quick turnaround time, RT-PCR and real-time reverse transcription-polymerase chain reaction (rtRT-PCR) tests targeting the 5′ non-coding regions should be employed for EV infection diagnosis. Yet, it is crucial to guarantee that the technique being used is regularly updated and can identify all types of enteroviruses. Every laboratory that conducts EV testing ought to be accredited (Harvala et al. 2018). It is quite effective at identifying EV RNA in CSF samples, with several studies reporting a 100% sensitivity with 97% specificity. To put it simply, the gold standard for diagnosing neurological enteroviral infections has been supplanted by EV RT-PCR since it has a far higher sensitivity than culture techniques (DeBiasi and Tyler 2004).

This test is authorized for the diagnosis of enterovirus meningitis, and results are available within 24 h. But in patients presenting with myalgic encephalomyelitis or chronic fatigue syndrome, only 30% have EV identified by PCR test of their blood specimens. The yield is less reliable for other bodily fluids such faeces samples, respiratory secretions, and blood (Chia 2018).

For other chronic EV infections, PCR is not thought to be sensitive. The likelihood of EV RNA or gene being detected in blood by PCR is minimal since the enteroviruses are swiftly eliminated from circulating blood. With specialized methods and repetitive testing, EV RNA can be discovered in almost 30% of whole blood specimens obtained from individuals presenting with persistent enteroviral infection (Chia 2018). Also, False-negative reports can sometimes be produced because of improper CSF handling or collection during late phases of EV illness (Ye et al. 2013).

Nucleic acid sequence-based amplification—Even though PCR tests are widely accessible at viral diagnostic institutes, nucleic acid sequence-based amplification (NASBA) is another molecular technique which can be explored as a good option for efficient detection and amplification of EV sequences in a variety of clinical samples, including cerebrospinal fluid (Fox et al. 2002). It is an in-vitro, isothermal, transcription-based amplification technique that has been converted into standardized kits for use in diagnostic labs where RT-PCR technology is not accessible. NASBA doesn't require certain specialized equipment like thermal cyclers (DeBiasi and Tyler 2004).

In situ hybridization—This technology is applied for EV positive-sense RNA detection in formalin-fixed paraffin-embedded specimens with the aid of designated probes. It enables anatomical localization and serotype determination of viruses (Laiho et al. 2015). In situ hybridization (ISH) techniques, when used alongside reverse transcription quantitative real-time polymerase chain reaction, may provide useful information on the EV infection and on potential targets for antivirals, paving ways for further discoveries (Salmikangas et al. 2020). Nucleic acid hybridization, however, is time-consuming and occasionally insensitive enough for diagnosis, for instance, in EV RNA detection in cerebrospinal fluid; they are yet to be standardized and reserved for research purposes only.

6.3.3.5 Serology

Only some numbered enteroviruses, including echoviruses- 6, 7, 9, 11, 30, and coxsackie viruses B1 through B-6 can be detected using serology-based diagnostic methods. These tests are unable to distinguish between the other known enteroviruses. It should be noted, a negative EV-serology test result does not always indicate that enterovirus is absent (Chia 2018). Serologic assays do not perform well in cases of acute enteroviral infections and have little use in cases of chronic EV illnesses (Nasri et al. 2007).

Serotyping mostly has no bearing on how a patient is managed. The establishment of EV immunoassays has been impeded by the lack of a broadly shared antigen. Although findings of monoclonal antibodies which cross-react with several EV serotypes are encouraging, additional research is necessary to discover whether those findings have any clinical significance (Ye et al. 2013). Simply put, serology-based assays are reserved for specialized or public health research lab facilities and only used for specific indications, such as enterovirus serotyping (Peaper and Landry 2014).

The most frequently used techniques for identifying antibodies to enteroviruses are microneutralization assays. In between acute and convalescent periods of enteroviral disease, a serological investigation may show a rise in the level of antibodies that neutralize enteroviruses. In these circumstances, serum samples from both acute as well as convalescent cases must be taken with a spacing of not less than 4 weeks apart. A fourfold or larger rise of antibody titre levels between acute and convalescent samples can be used to retroactively diagnose an acute EV infection. Initially in the course of illness, serum IgM antibodies to CV-B groups can frequently be found, but positive results are not serotype-specific. Antibody titres of 1: 160–320 or higher can indicate recent infections. Nonetheless, serologic diagnosis is time-consuming and is usually impracticable in the clinical context since it requires collecting samples of both acute and convalescent time frames (Chia 2018; Sandoni et al. 2022).

Commercial lab facilities provide type-specific immunoassays which measure antibodies for only some enterovirus serotypes; however, due to cross reactivity and subpar standards, these tests are seldom useful.

Acute enteroviral illnesses really aren't confirmed using serological techniques like neutralization assays or enzyme-linked immunosorbent assays (ELISA). Although there are immunoglobulin M and G assays for the identification of enteroviral infections, their clinical value is constrained due to antigen cross-reactivity between different serotypes (Anwar 2022).

Serological tests can take 2 weeks to complete, rendering sluggish diagnosis and clinical irrelevance. As direct sample isn't really possible in the diagnosis of EV cardiomyopathy, myocarditis, or pericarditis, serological testing may be implicated. As a result, antibody detection may be helpful in these cases. It is possible to measure neutralizing antibodies by testing them against certain enteroviruses (Anwar 2022).

The multiplicity of serotypes of enteroviruses hinders serologic methods. Consequently, it is challenging to distinguish enteroviral neuro-illness from bacterial meningitis with just these methods. As a result, unnecessary tests and interventions involving empirical IV antibiotics are performed often, and hospital stays are prolonged (DeBiasi and Tyler 2004).

6.3.3.6 Viral Culture

Based upon the affected site, EV may be isolated from blood, CSF, or stool. Yield increases if sampling is done from multiple sites. The VP1 gene sequencing approach or neutralizing assays (CDC approved ones) utilizing type-specific antisera may be employed to further determine serotypes of EVs recovered by this method (Chia 2018). Both viral culture as well as shell vial culture require considerable time to conduct, are relatively insensitive and are unreliable since they depend on the existence of viable enterovirus.

Viral cultures are unable to accurately define several EV strains due to improperly collected, handled, or processed specimens, or because the cell lines being employed may be inherently insensitive. When the EV is present at low titres in samples such as cerebrospinal fluid, viral culture might take as long as 8 days for cytopathic effect (CPE) to manifest, and several coxsackievirus-A types will not thrive in cell culture. Also, even though the culture time for shell vial culture with monoclonal antibodies has been shortened compared to tube culture, it is less sensitive than traditional culture (Ye et al. 2013). Neutralizing antibodies, relatively low viral loads at time of diagnosis and the fact that some enterovirus serotypes are inherently uncultivable are all likely to play a role in the viral culture's insensitivity. In most cases, as much as 8 days are needed for the tissue culture method for enterovirus isolation from cerebrospinal fluid. The labour-intensive process of EV culture necessitates cultivation on numerous cell lines. Although extended excretion (4 and 16 weeks, respectively) from both sites can occur after enteroviral disease, faeces sample or throat swab cultures give only circumstantial proof of aetiology in the presence of meningitis or encephalitis (DeBiasi and Tyler 2004). Whereas a CPE normally takes 3–8 days to develop in a cerebrospinal fluid EV culture, PCR data are available within 24 h (~5 h), reducing the turnaround time compared to viral culture in all cases. Hence, polymerase chain reaction tests are preferred since the expenses are comparable (Ye et al. 2013).

6.3.4 Imaging

Patients with enterovirus infection usually present with picture of rhombencephalitis. The spinal and cranial nerves involvement is common. Imaging findings are variable based on patient's symptoms. Imaging can be normal even in positive CSF analysis. Here are the most commonly reported imaging findings in this group of patients.

Computed Tomography (CT)—Fast and feasible specially in emergency department, it is helps in patient's screening and to exclude other mimickers like acute infarction, haemorrhage, hydrocephalus, brain herniation, and tumours. CT scan can be normal in most of cases. If there is a large parenchymal involvement by encephalitis, this can appear as an area of hypodensity at the affected brain. MRI is the second step used for confirming findings and disease characterization.

MRI Brain—The commonest MRI findings are rhombencephalitis; it commonly appears as a hyperintensities on fluid-attenuated inversion recovery images (FLAIR) (Fig. 6.2) and T2-weighted images, involving posterior aspect of brainstem, along with dentate or cerebellar regions. MRI may show normal findings in some patients. Cranial nerve palsies commonly are associated with specific coxsackievirus and adenovirus receptors (CAR) in the cochlea, which are linked with viral docking, seen as abnormal enhancement of the affected cranial nerve (Excoffon et al. 2006; Venail et al. 2007; Wagner et al. 2021). Rare locations that have been reported are the hippocampus, thalamus, putamen, cerebral region along with subcortical region, and

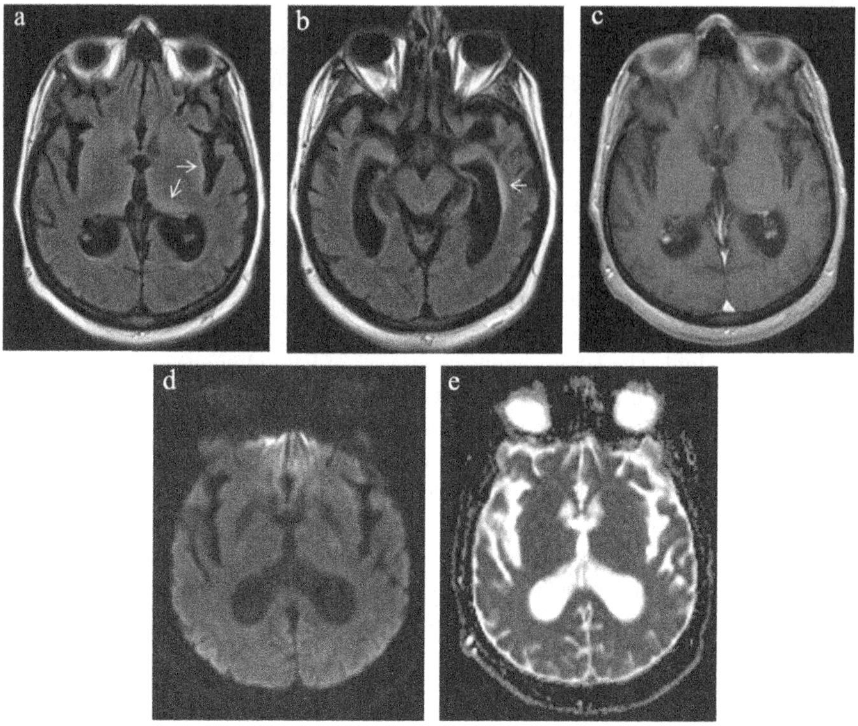

Fig. 6.2 (**a**, **b**) Axial FLAIR images show an abnormal hyperintensities at bilateral Pulvinar, insular ribbon (**a**) and left periventricular WM (**b**) (white arrows), no corresponding enhancement (**c**), or diffusion restriction (**d**, **e**). *Image Source: Courtesy of King Abdulaziz Medical City- Western Region, Jeddah, KSA [Copyright Restrictions – NONE; this image is free of any copyright restrictions]*

corpus callosum in patients with EV71 infection (Jang et al. 2012; Lian et al. 2012; Zeng et al. 2012).

MRI Spine—Patients with spinal involvement usually had acute flaccid paralysis, which appears in MRI as an ipsilateral anterior horn signal changes with or without enhancement. It usually affects lower cervical/upper thoracic cord, or cauda equina nerve roots based on level of involvement. It can be unilateral or bilateral (Chen et al. 2001; Chonmaitree et al. 1981; Chumakov et al. 1979; Kornreich et al. 1996; Melnick 1984).

Electroencephalography (EEG)—Most of viral infections show a picture of encephalitis in EEG. It remains challenging to find a specific pattern of encephalitis in EEG (Rubiños and Godoy 2020). Electroencephalography may demonstrate periodic discharges (unilateral/bilateral), electrical discharges, and generalized slow waves (generalized or focal), which are able to show a non-specific characteristics pattern (Halperin 2017; Rubiños and Godoy 2020; Sutter et al.

2015). The associated neurological changes are mostly regional in nature and less commonly associated with global brain involvement.

6.4 Management and Treatment

6.4.1 General

Most enteroviral infections are self-limiting, requiring only supportive care, with the exception of enteroviral encephalitis, meningitis, myocarditis, neonatal infections, and infections in B-cell deficient patients. Every acute encephalitis case necessitates hospitalization with access to intensive therapy unit, and mechanical ventilation should be prompt, depending on the severity of symptoms. Strict monitoring of the patient's fluids, a stringent lookout for any warning signs of deep vein thrombosis or aspiration pneumonia along with their prevention, medical management of increased intracranial pressure, and any secondary bacterial infections, are all part of the management strategy. Antiepileptics are indicated if the patient develops seizures. Secondary complications including cerebral infarction, cerebral venous thrombosis, SIADH, aspiration pneumonia, and disseminated intravascular coagulopathy are frequent and contribute significantly to morbidity and mortality.

Impending uncal herniation or elevated intracranial pressure in encephalitis that is unresponsive to medicinal therapy (steroid and mannitol) are indications for surgical decompression (Steiner et al. 2005).

6.4.2 Antivirals

There is currently no approved antiviral medication available for the treatment of severe enteroviral infections, including encephalitis and meningitis. Pleconaril, an antiviral that prevents enteroviral replication, was developed as a result of the molecular characterization of enteroviruses. By binding to the viral protein capsid, it prevents enteroviral attachment and uncoating. At doses of 0.1 μg/mL or less, it has wide antiviral effects on enteroviruses, with antiviral effectiveness against more than 90% of the frequently circulating serotypes (Sawyer 2002), although it is currently not licensed for use.

6.4.3 Newer Agents

Owing to a lack of specific enteroviral agents in our repertoire, intravenous immunoglobulins (IVIG) are frequently utilized as a therapeutic as well as prophylactic measure in enteroviral encephalitis. Immunodeficient patients consistently have better outcomes with IVIG, as shown by the encouraging response of intraventricular or intravenous administration of IVIG in patients with X-linked hypogammaglobinaemia (Dwyer and Erlendsson 1988; McKinney et al. 1987).

Enteroviral encephalitis in patients with iatrogenic hypogammaglobinaemia, as seen in patients subjected to rituximab therapy, which is a B-cell depleting therapy, has also been associated with a positive response to IVIG administration; however, its use in patients who are immunocompetent is currently debatable (Schilthuizen et al. 2010; Wagner et al. 2021). Patients with para-infectious enteroviral encephalitis are good candidates for immunomodulatory therapy including corticosteroids and IVIG (Pillai et al. 2015).

6.4.4 Complications/Prognosis

Patients with enteroviral encephalitis, especially adults, are seen to less frequently have severe disease and have better outcomes as compared to other causes of viral encephalitis. These patients also have a shorter length of hospital stay and less severe morbidity; however, there are differences in outcomes based on the infective serotype (Fowlkes et al. 2008).

While a majority of enteroviral infections behave like acute febrile illness with a short-lived course and their natural history is specific for conclusion with either recovery or acute worsening, recent data have also uncovered the presence of persistent infection, as well as an association with lifelong disorders like amyotrophic lateral sclerosis, schizophrenia, type 1 diabetes mellitus, and post-viral cardiomyopathy (Chen et al. 2020). Enteroviral encephalitis in children has also been linked to autism spectrum disorder and attention deficit hyperactivity disorder (ADHD) (Chou et al. 2015; Marques et al. 2014). EV-A71 encephalitis has been associated with long-term sequelae like limb weakness and atrophy, as well as long-term behavioural issues in children (Huang et al. 1999).

Patients with X-linked agammaglobulinemia are prone to develop chronic enteroviral meningoencephalitis of agammaglobulinemia (CEMA), marked by prolonged enteroviral encephalitis. It is characterized by diverse neurologic manifestations with subsequent involvement of multiple organ systems, most commonly associated with echoviruses, and has a guarded prognosis with regards to morbidity and mortality (McKinney et al. 1987).

6.4.5 Prevention

In response to the extensive EV-A71 outbreaks that have been detected in the Far East, a phase III research for an EV-A71 vaccine recently concluded successfully against EV-A71-associated hand, foot, and mouth disease, with up to 90% or more efficacy outcomes observed. (Li et al. 2014; Zhu et al. 2013, 2014). Further efficacy outcomes remain to be seen.

References

Ahn J, Jee Y, Seo I, Yoon SY, Kim D, Kim YK et al (2008) Primary neurons become less susceptible to coxsackievirus B5 following maturation: the correlation with the decreased level of CAR expression on cell surface. J Med Virol 80:434–440. https://doi.org/10.1002/jmv.21100

Alirezaei M, Flynn CT, Wood MR, Harkins S, Whitton JL (2015) Coxsackievirus can exploit LC3 in both autophagy-dependent and -independent manners in vivo. Autophagy 11:1389–1407

Anwar Z (2022) Picornaviridae: Enterovirus. In: Rezaei N (ed) Encyclopedia of infection and immunity. Elsevier, Oxford, pp 154–191. https://doi.org/10.1016/B978-0-12-818731-9.00154-3

B'Krong NTTC, Minh NNQ, Qui PT, Chau TTH, Nghia HDT, Do LAH, Nhung NN, Van Vinh Chau N, Thwaites G, Van Tan L, van Doorn HR, Thanh TT (2018) Enterovirus serotypes in patients with central nervous system and respiratory infections in Viet Nam 1997–2010. Virol J 15:69. https://doi.org/10.1186/s12985-018-0980-0

Bernit E, de Lamballerie X, Zandotti C, Berger P, Veit V, Schleinitz N, de Micco P, Harlé JR, Charrel RN (2004) Prospective investigation of a large outbreak of meningitis due to echovirus 30 during summer 2000 in Marseilles, France. Medicine (Baltimore) 83:245–253. https://doi.org/10.1097/01.md.0000133623.27828.27

Calleri G, Libanore V, Corcione S, De Rosa FG, Caramello P (2017) A retrospective study of viral central nervous system infections: relationship amongst aetiology, clinical course and outcome. Infection 45:227–231. https://doi.org/10.1007/s15010-017-0993-4

Chan KP, Goh KT, Chong CY, Teo ES, Lau G, Ling AE (2003) Epidemic hand, foot and mouth disease caused by human enterovirus 71, Singapore. Emerg Infect Dis 9:78–85. https://doi.org/10.3201/eid1301.020112

Chang LY, Huang LM, Gau SS, Wu YY, Hsia SH, Fan TY et al (2007) Neurodevelopment and cognition in children after Enterovirus 71 infection. N Engl J Med 356:1226–1234. https://doi.org/10.1056/nejmoa065954

Chapman NM, Kim KS (2008) Persistent coxsackievirus infection: enterovirus persistence in chronic myocarditis and dilated cardiomyopathy. Curr Top Microbiol Immunol 323:275–292. https://doi.org/10.1007/978-3-540-75546-3_13

Chen CS, Yao YC, Lin SC, Lee YP, Wang YF, Wang JR et al (2007) Retrograde axonal transport: a major transmission route of Enterovirus 71 in mice. J Virol 81:8996–9003. https://doi.org/10.1128/jvi.00236-07

Chen CY, Chang YC, Huang CC, Lui CC, Lee KW, Huang SC (2001) Acute flaccid paralysis in infants and young children with enterovirus 71 infection: MR imaging findings and clinical correlates. AJNR Am J Neuroradiol 22:200–205

Chen B-S, Lee H-C, Lee K-M, Gong Y-N, Shih S-R (2020) Enterovirus and encephalitis. Front Microbiol 11:261. https://doi.org/10.3389/fmicb.2020.00261

Chia J (2018) Diagnose & treat [WWW Document]. Enterovirus Foundation, San Francisco, CA

Chonmaitree T, Menegus MA, Schervish-Swierkosz EM, Schwalenstocker E (1981) Enterovirus 71 infection: report of an outbreak with two cases of paralysis and a review of the literature. Pediatrics 67:489–493

Chou I-C, Lin C-C, Kao C-H (2015) Enterovirus encephalitis increases the risk of attention deficit hyperactivity disorder. Medicine (Baltimore) 94:e707. https://doi.org/10.1097/MD.0000000000000707

Chumakov M, Voroshilova M, Shindarov L, Lavrova I, Gracheva L, Koroleva G, Vasilenko S, Brodvarova I, Nikolova M, Gyurova S, Gacheva M, Mitov G, Ninov N, Tsylka E, Robinson I, Frolova M, Bashkirtsev V, Martiyanova L, Rodin V (1979) Enterovirus 71 isolated from cases of epidemic poliomyelitis-like disease in Bulgaria. Arch Virol 60:329–340. https://doi.org/10.1007/BF01317504

Davison DL, Terek M, Chawla LS (2012) Neurogenic pulmonary edema. Crit Care 16:212

DeBiasi RL, Tyler KL (2004) Molecular methods for diagnosis of viral encephalitis. Clin Microbiol Rev 17:903–925. https://doi.org/10.1128/CMR.17.4.903-925.2004
Dourmashkin RR, Dunn G, Castano V, Mccall SA (2012) Evidence for an enterovirus as the cause of encephalitis lethargica. BMC Infect Dis 12:136
Dwyer JM, Erlendsson K (1988) Intraventricular gamma-globulin for the management of enterovirus encephalitis. Pediatr Infect Dis J 7:S34
Excoffon KJDA, Avenarius MR, Hansen MR, Kimberling WJ, Najmabadi H, Smith RJH, Zabner J (2006) The Coxsackievirus and adenovirus receptor: a new adhesion protein in cochlear development. Hear Res 215:1–9. https://doi.org/10.1016/j.heares.2006.02.009
Fan Y-K, Liu Y-P (2019) Magnetic resonance imaging features of pediatric Coxsackievirus encephalitis. J Belg Soc Radiol 103:6. https://doi.org/10.5334/jbsr.1685
Feng M, Guo S, Fan S, Zeng X, Zhang Y, Liao Y et al (2016) The preferential infection of astrocytes by Enterovirus 71 plays a key role in the viral neurogenic pathogenesis. Front Cell Infect Microbiol 6:192. https://doi.org/10.3389/fcimb.2016.00192
Fischer TK, Simmonds P, Harvala H (2022) The importance of enterovirus surveillance in a post-polio world. Lancet Infect Dis 22:e35–e40. https://doi.org/10.1016/S1473-3099(20)30852-5
Forrester JV, Mcmenamin PG, Dando SJ (2018) CNS infection and immune privilege. Nat Rev Neurosci 19:655–671. https://doi.org/10.1038/s41583-018-0070-8
Fowlkes AL, Honarmand S, Glaser C, Yagi S, Schnurr D, Oberste MS, Anderson L, Pallansch MA, Khetsuriani N (2008) Enterovirus-associated encephalitis in the California encephalitis project, 1998–2005. J Infect Dis 198:1685–1691. https://doi.org/10.1086/592988
Fox JD, Han S, Samuelson A, Zhang Y, Neale ML, Westmoreland D (2002) Development and evaluation of nucleic acid sequence based amplification (NASBA) for diagnosis of enterovirus infections using the NucliSens basic kit. J Clin Virol 24:117–130. https://doi.org/10.1016/s1386-6532(01)00241-4
Glaser CA, Gilliam S, Schnurr D, Forghani B, Honarmand S, Khetsuriani N, Fischer M, Cossen CK, Anderson LJ (2003) In search of encephalitis etiologies: diagnostic challenges in the California encephalitis project, 1998—2000. Clin Infect Dis 36:731–742. https://doi.org/10.1086/367841
Gromeier M, Alexander L, Wimmer E (1996) Internal ribosomal entry site substitution eliminates neurovirulence in intergeneric poliovirus recombinants. Proc Natl Acad Sci USA 93:2370–2375. https://doi.org/10.1073/pnas.93.6.2370
Greenberg SB (2003) Viral meningitis and encephalitis. Office Pract Neurol 2003:466–469. https://doi.org/10.1016/B0-44-306557-8/50066-6
Halperin JJ (2017) Chapter 18 - diagnosis and management of acute encephalitis. In: Wijdicks EFM, Kramer AH (eds) Handbook of clinical neurology, critical care neurology part I. Elsevier, Amsterdam, pp 337–347. https://doi.org/10.1016/B978-0-444-63600-3.00018-0
Hamaguchi T, Fujisawa H, Sakai K, Okino S, Kurosaki N, Nishimura Y, Shimizu H, Yamada M (2008) Acute encephalitis caused by Intrafamilial transmission of enterovirus 71 in adult. Emerg Infect Dis 14:828–830. https://doi.org/10.3201/eid1405.071121
Han S-H, Choi H-Y, Kim J-M, Park K-R, Youn YC, Shin H-W (2016) Etiology of aseptic meningitis and clinical characteristics in immune-competent adults. J Med Virol 88:175–179. https://doi.org/10.1002/jmv.24316
Harvala H, Broberg E, Benschop K, Berginc N, Ladhani S, Susi P, Christiansen C (2018) Recommendations for enterovirus diagnostics and characterisation within and beyond Europe. J Clin Virol 101:11–17. https://doi.org/10.1016/j.jcv.2018.01.008
Hasbun R, Rosenthal N, Balada-Llasat JM, Chung J, Duff S, Bozzette S, Zimmer L, Ginocchio CC (2017) Epidemiology of meningitis and encephalitis in the United States, 2011–2014. Clin Infect Dis 65:359–363. https://doi.org/10.1093/cid/cix319
Hixon AM, Yu G, Leser JS, Yagi S, Clarke P, Chiu CY et al (2017) A mouse model of paralytic myelitis caused by Enterovirus D68. PLoS Pathog 13:e1006199. https://doi.org/10.1371/journal.ppat.1006199

Ho M, Chen ER, Hsu KH, Twu SJ, Chen KT, Tsai SF, Wang JR, Shih SR (1999) An epidemic of enterovirus 71 infection in Taiwan. Taiwan enterovirus epidemic working group. N Engl J Med 341:929–935. https://doi.org/10.1056/NEJM199909233411301

Huang CC, Liu CC, Chang YC, Chen CY, Wang ST, Yeh TF (1999) Neurologic complications in children with enterovirus 71 infection. N Engl J Med 341:936–942. https://doi.org/10.1056/NEJM199909233411302

Huang HI, Shih SR (2015) Neurotropic Enterovirus infections in the central nervous system. Viruses 7:6051–6066. https://doi.org/10.3390/v7112920

Huang SC, Chang CL, Wang PS, Tsai Y, Liu HS (2009) Enterovirus 71-induced autophagy detected in vitro and in vivo promotes viral replication. J Med Virol 81(7):1241–1252

Hussin S, Chua A, Al-Talib H, Sekaran S (2022) An overview of laboratory diagnosis of central nervous system viral infections. J Pure Appl Microbiol 16:2225–2245. https://doi.org/10.22207/JPAM.16.4.34

Ida-Hosonuma M, Iwasaki T, Yoshikawa T, Nagata N, Sato Y, Sata T et al (2005) The alpha/beta interferon response controls tissue tropism and pathogenicity of poliovirus. J Virol 79:4460–4469. https://doi.org/10.1128/jvi.79.7.4460-4469.2005

Ihekwaba UK, Kudesia G, McKendrick MW (2008) Clinical features of viral meningitis in adults: significant differences in cerebrospinal fluid findings among herpes simplex virus, varicella zoster virus, and enterovirus infections. Clin Infect Dis 47:783–789. https://doi.org/10.1086/591129

Jang S, Suh S, Ha SM, Byeon JH, Eun B-L, Lee YH, Seo HS, Eun S-H, Seol H-Y (2012) Enterovirus 71-related encephalomyelitis: usual and unusual magnetic resonance imaging findings. Neuroradiology 54:239–245. https://doi.org/10.1007/s00234-011-0921-8

Jubelt B, Lipton HL (2014) Chapter 18 - Enterovirus/picornavirus infections. In: Tselis AC, Booss J (eds) Handbook of clinical neurology, neurovirology. Elsevier, Amsterdam, pp 379–416. https://doi.org/10.1016/B978-0-444-53488-0.00018-3

Kao SJ, Yang FL, Hsu YH, Chen HI (2004) Mechanism of fulminant pulmonary edema caused by Enterovirus 71. Clin Infect Dis 38:1784–1788. https://doi.org/10.1086/421021

Kornreich L, Dagan O, Grunebaum M (1996) MRI in acute poliomyelitis. Neuroradiology 38:371–372. https://doi.org/10.1007/BF00596591

Kräusslich HG, Nicklin MJ, Toyoda H, Etchison D, Wimmer E (1987) Poliovirus proteinase 2A induces cleavage of eucaryotic initiation factor 4F polypeptide p220. J Virol 61:2711–2718. https://doi.org/10.1128/JVI.61.9.2711-2718.1987

Khong WX, Yan B, Yeo H, Tan EL, Lee JJ, Ng JK et al (2012) A non-mouse-adapted Enterovirus 71 (EV71) strain exhibits neurotropism, causing neurological manifestations in a novel mouse model of EV71 infection. J Virol 86:2121–2131. https://doi.org/10.1128/JVI.06103-11

Laiho JE, Oikarinen S, Oikarinen M, Larsson PG, Stone VM, Hober D, Oberste S, Flodström-Tullberg M, Isola J, Hyöty H (2015) Application of bioinformatics in probe design enables detection of enteroviruses on different taxonomic levels by advanced in situ hybridization technology. J Clin Virol 69:165–171. https://doi.org/10.1016/j.jcv.2015.06.085

Lei X, Liu X, Ma Y, Sun Z, Yang Y, Jin Q, He B, Wang J (2010) The 3C protein of enterovirus 71 inhibits retinoid acid-inducible gene I-mediated interferon regulatory factor 3 activation and type I interferon responses. J Virol 84:8051–8061. https://doi.org/10.1128/JVI.02491-09

Lei X, Xiao X, Wang J (2016) Innate immunity evasion by enteroviruses: insights into virus-host interaction. Viruses 8:22. https://doi.org/10.3390/v8010022

Lewis P, Glaser CA (2005) Encephalitis. Pediatr Rev 26:353–363. https://doi.org/10.1542/pir.26-10-353

Li M-L, Hsu T-A, Chen T-C, Chang S-C, Lee J-C, Chen C-C, Stollar V, Shih S-R (2002) The 3C protease activity of enterovirus 71 induces human neural cell apoptosis. Virology 293:386–395. https://doi.org/10.1006/viro.2001.1310

Li R, Liu L, Mo Z, Wang X, Xia J, Liang Z, Zhang Y, Li Y, Mao Q, Wang J, Jiang L, Dong C, Che Y, Huang T, Jiang Z, Xie Z, Wang L, Liao Y, Liang Y, Nong Y, Liu J, Zhao H, Na R, Guo L, Pu J, Yang E, Sun L, Cui P, Shi H, Wang J, Li Q (2014) An inactivated enterovirus

71 vaccine in healthy children. N Engl J Med 370:829–837. https://doi.org/10.1056/NEJMoa1303224
Lian Z, Huang B, He S, Liang C, Guo Y (2012) Diffusion-weighted imaging in the diagnosis of enterovirus 71 encephalitis. Acta Radiol 53:208–213. https://doi.org/10.1258/ar.2011.110407
Logan SAE, MacMahon E (2008) Viral meningitis. BMJ 336:36–40. https://doi.org/10.1136/bmj.39409.673657.AE
Mandell, Douglas, and Bennett's Principles and P - 9780323482554 [WWW Document] (n.d.) US Elsevier Health. URL https://www.us.elsevierhealth.com/mandell-douglas-and-bennetts-principles-and-practice-of-infectious-diseases-9780323482554.html. Accessed 28 Mar 2023
Marques F, Brito MJ, Conde M, Pinto M, Moreira A (2014) Autism Spectrum disorder secondary to enterovirus encephalitis. J Child Neurol 29:708–714. https://doi.org/10.1177/0883073813508314
McKinney RE, Katz SL, Wilfert CM (1987) Chronic enteroviral meningoencephalitis in agammaglobulinemic patients. Rev Infect Dis 9:334–356. https://doi.org/10.1093/clinids/9.2.334
Melnick JL (1984) Enterovirus type 71 infections: a varied clinical pattern sometimes mimicking paralytic poliomyelitis. Rev Infect Dis 6(Suppl 2):S387–S390. https://doi.org/10.1093/clinids/6.supplement_2.s387
Messacar K, Asturias EJ, Hixon AM, Van Leer-Buter C, Niesters HGM, Tyler KL, Abzug MJ, Dominguez SR (2018) Enterovirus D68 and acute flaccid myelitis—evaluating the evidence for causality. Lancet Infect Dis 18:e239–e247. https://doi.org/10.1016/S1473-3099(18)30094-X
Michos AG, Syriopoulou VP, Hadjichristodoulou C, Daikos GL, Lagona E, Douridas P, Mostrou G, Theodoridou M (2007) Aseptic meningitis in children: analysis of 506 cases. PLoS One 2:e674. https://doi.org/10.1371/journal.pone.0000674
Nasri D, Bouslama L, Pillet S, Bourlet T, Aouni M, Pozzetto B (2007) Basic rationale, current methods and future directions for molecular typing of human enterovirus. Expert Rev Mol Diagn 7:419–434. https://doi.org/10.1586/14737159.7.4.419
Nishimura Y, Lee H, Hafenstein S, Kataoka C, Wakita T, Bergelson JM et al (2013) Enterovirus 71 binding to PSGL-1 on leukocytes: VP1-145 acts as a molecular switch to control receptor interaction. PLoS Pathog 9:e1003511. https://doi.org/10.1371/journal.ppat.1003511
Nikonov OS, Chernykh ES, Garber MB, Nikonova EY (2017) Enteroviruses: classification, diseases they cause, and approaches to development of antiviral drugs. Biochemistry (Mosc) 82:1615–1631. https://doi.org/10.1134/S0006297917130041
Ong KC, Badmanathan M, Devi S, Leong KL, Cardosa MJ, Wong KT (2008) Pathologic characterization of a murine model of human Enterovirus 71 encephalomyelitis. J Neuropathol Exp Neurol 67:532–542. https://doi.org/10.1097/NEN.0b013e31817713e7
Ohka S, Sakai M, Bohnert S, Igarashi H, Deinhardt K, Schiavo G et al (2009) Receptor-dependent and –independent axonal retrograde transport of poliovirus in motor neurons. J Virol 83:4995–5004. https://doi.org/10.1128/JVI.02225-08
Pallansch M, Oberste M, Whitton J (2013) Enteroviruses: polioviruses, coxsackieviruses, echoviruses, and newer enteroviruses. In: Knipe DM, Howley P (eds) Fields virology. Lippincott Williams & Wilkins, Pennsylvania, PA
Peaper DR, Landry ML (2014) Chapter 5 - laboratory diagnosis of viral infection. In: Tselis AC, Booss J (eds) Handbook of clinical neurology, neurovirology, vol 123. Elsevier, Amsterdam, pp 123–147. https://doi.org/10.1016/B978-0-444-53488-0.00005-5
Peigue-Lafeuille H, Croquez N, Laurichesse H, Clavelou P, Aumaître O, Schmidt J, Maillet-Vioud M, Henquell C, Archimbaud C, Bailly J-L, Chambon M (2002) Enterovirus meningitis in adults in 1999-2000 and evaluation of clinical management: enterovirus and meningitis in adults. J Med Virol 67:47–53. https://doi.org/10.1002/jmv.2191
Pillai S, Tantsis E, Prelog K, Ramanathan S, Webster R, Ouvrier RA, Kesson A, Brilot F, Dale RC (2015) Confirmed enterovirus encephalitis with associated steroid-responsive acute disseminated encephalomyelitis: an overlapping infection and inflammation syndrome. Eur J Paediatr Neurol 19:266–270. https://doi.org/10.1016/j.ejpn.2014.12.015

Prentice RL, Dalgleish SAG, Gatenyb N, Loblay PA, Wade RHS, Kappagoda N, Basten A (1985) Central nervous system echovirus infection in bruton's x-linked hypogammaglobulinemia. Aust NZ J Med 15:443–445. https://doi.org/10.1111/j.1445-5994.1985.tb02770.x

Racaniello VR (2016) Picornaviridae: the viruses and their replication. Basicmedical key. https://basicmedicalkey.com/picornaviridae-the-viruses-and-their-replication/. Accessed 20 Dec 22

Ransohoff RM, Engelhardt B (2012) The anatomical and cellular basis of immune surveillance in the central nervous system. Nat Rev Immunol 12:623–635. https://doi.org/10.1038/nri3265

Rhoades RE, Tabor-Godwin JM, Tsueng G, Feuer R (2011) Enterovirus infections of the central nervous system. Virology 411:288–305. https://doi.org/10.1016/j.virol.2010.12.014

Romero JR (2002) Diagnosis and Management of Enteroviral Infections of the central nervous system. Curr Infect Dis Rep 4:309–316. https://doi.org/10.1007/s11908-002-0023-1

Rotbart HA (2000) Viral meningitis. Semin Neurol 20:277–292. https://doi.org/10.1055/s-2000-9427

Rubiños C, Godoy DA (2020) Electroencephalographic monitoring in the critically ill patient: what useful information can it contribute? Med Intensiva 44:301–309. https://doi.org/10.1016/j.medine.2019.06.008

Salmikangas S, Laiho JE, Kalander K, Laajala M, Honkimaa A, Shanina I, Oikarinen S, Horwitz MS, Hyöty H, Marjomäki V (2020) Detection of viral −RNA and +RNA strands in enterovirus-infected cells and tissues. Microorganisms 8:1928. https://doi.org/10.3390/microorganisms8121928

Sandoni M, Ciardo L, Tamburini C, Boncompagni A, Rossi C, Guidotti I, Garetti E, Lugli L, Iughetti L, Berardi A (2022) Enteroviral infections in the first three months of life. Pathogens 11:60. https://doi.org/10.3390/pathogens11010060

Sapkal GN, Bondre VP, Fulmali PV, Patil P, Dadhania V, Ayachit VM, Gangale D, Kushwaha KP, Rathi AK, Chitambar SD, Mishra AC, Gore MM (2009) Enteroviruses in patients with acute encephalitis, Uttar Pradesh, India. Emerg Infect Dis 15:295–298. https://doi.org/10.3201/eid1502.080865

Savolainen-Kopra C, Paananen A, Blomqvist S, Klemola P, Simonen M-L, Lappalainen M, Vuorinen T, Kuusi M, Lemey P, Roivainen M (2011) A large finnish echovirus 30 outbreak was preceded by silent circulation of the same genotype. Virus Genes 42:28–36. https://doi.org/10.1007/s11262-010-0536-x

Sawyer MH (2002) Enterovirus infections: diagnosis and treatment. Semin Pediatr Infect Dis 13:40–47. https://doi.org/10.1053/spid.2002.29756

Schilthuizen C, Berenschot HWA, Levin M-D (2010) Enteroviral encephalitis in a patient with a marginal zone lymphoma treated with rituximab. Neth J Med 68:221

Simmonds P, Gorbalenya AE, Harvala H, Hovi T, Knowles NJ, Lindberg AM, Oberste MS, Palmenberg AC, Reuter G, Skern T, Tapparel C, Wolthers KC, Woo PCY, Zell R (2020) Recommendations for the nomenclature of enteroviruses and rhinoviruses. Arch Virol 165:793–797. https://doi.org/10.1007/s00705-019-04520-6

Solomon T, Lewthwaite P, Perera D, Cardosa MJ, McMinn P, Ooi MH (2010) Virology, epidemiology, pathogenesis, and control of enterovirus 71. Lancet Infect Dis 10:778–790. https://doi.org/10.1016/S1473-3099(10)70194-8

Steiner I, Budka H, Chaudhuri A, Koskiniemi M, Sainio K, Salonen O, Kennedy PGE (2005) Viral encephalitis: a review of diagnostic methods and guidelines for management. Eur J Neurol 12:331–343. https://doi.org/10.1111/j.1468-1331.2005.01126.x

Storch GA (2000) Diagnostic virology. Clin Infect Dis 31:739–751. https://doi.org/10.1086/314015

Sun Y, Miao Z, Yan J, Gong L, Chen Y, Chen YJ, Mao H, Zhang Y (2019) Sero-molecular epidemiology of enterovirus-associated encephalitis in Zhejiang Province, China, from 2014 to 2017. Int J Infect Dis 79:58–64. https://doi.org/10.1016/j.ijid.2018.11.002

Suresh S, Forgie S, Robinson J (2018) Non-polio enterovirus detection with acute flaccid paralysis: a systematic review. J Med Virol 90:3–7. https://doi.org/10.1002/jmv.24933

Sutter R, Kaplan PW, Cervenka MC, Thakur KT, Asemota AO, Venkatesan A, Geocadin RG (2015) Electroencephalography for diagnosis and prognosis of acute encephalitis. Clin Neurophysiol 126:1524–1531. https://doi.org/10.1016/j.clinph.2014.11.006

Suhy DA, Giddings TH, Kirkegaard K (2000) Remodeling the endoplasmic reticulum by poliovirus infection and by individual viral proteins: an autophagy-like origin for virus-induced vesicles. J Virol 74:8953–8965

Tabor-Godwin JM, Ruller CM, Bagalso N, An N, Pagarigan RR, Harkins S et al (2010) A novel population of myeloid cells responding to coxsackievirus infection assists in the dissemination of virus within the neonatal CNS. J Neurosci 30:8676–8691. https://doi.org/10.1523/JNEUROSCI.1860-10.2010

Tille PM (2021) Meningitis and other infections of the central nervous systems. In: Bailey & Scott's diagnostic microbiology. Elsevier

Torok E, Moran E, Cooke F (2017) Enterovirus (including Coxsacckie). In: Oxford handbook of infectious diseases and microbiology. Oxford University Press, United Kingdom, pp 430–432

Valcour V, Haman A, Cornes S, Lawall C, Parsa AT, Glaser C, Yagi S, Tihan T, Bhatnagar J, Geschwind M (2008) A case of enteroviral meningoencephalitis presenting as rapidly progressive dementia. Nat Rev Neurol 4:399–403. https://doi.org/10.1038/ncpneuro0804

Venail F, Wang J, Ruel J, Ballana E, Rebillard G, Eybalin M, Arbones M, Bosch A, Puel J-L (2007) Coxsackie adenovirus receptor and ανβ3/ανβ5 integrins in adenovirus gene transfer of rat cochlea. Gene Ther 14:30–37. https://doi.org/10.1038/sj.gt.3302826

Wagner JN, Leibetseder A, Troescher A, Panholzer J, von Oertzen TJ (2021) Characteristics and therapy of enteroviral encephalitis: case report and systematic literature review. Int J Infect Dis 113:93–102. https://doi.org/10.1016/j.ijid.2021.10.002

Wang AY, Machicado JD, Khoury NT, Wootton SH, Salazar L, Hasbun R (2014) Community-acquired meningitis in older adults: clinical features, etiology, and prognostic factors. J Am Geriatr Soc 62:2064–2070. https://doi.org/10.1111/jgs.13110

Wang B, Xi X, Lei X, Zhang X, Cui S, Wang J, Jin Q, Zhao Z (2013) Enterovirus 71 protease 2Apro targets MAVS to inhibit anti-viral type I interferon responses. PLoS Pathog 9:e1003231. https://doi.org/10.1371/journal.ppat.1003231

Weng K-F, Li M-L, Hung C-T, Shih S-R (2009) Enterovirus 71 3C protease cleaves a novel target CstF-64 and inhibits cellular polyadenylation. PLoS Pathog 5:e1000593. https://doi.org/10.1371/journal.ppat.1000593

Wessely R, Klingel K, Knowlton KU, Kandolf R (2001) Cardioselective infection with coxsackievirus B3 requires intact type I interferon signaling: implications for mortality and early viral replication. Circulation 103:756–761. https://doi.org/10.1161/01.cir.103.5.756

Wong J, Zhang J, Si X, Gao G, Mao I, McManus BM, Luo H (2008) Autophagosome supports coxsackievirus B3 replication in host cells. J Virol 82:9143–9153

Yamayoshi S, Fujii K, Koike S (2014) Receptors for Enterovirus 71. Emerg Microbes Infect 3:e53. https://doi.org/10.1038/emi.2014.49

Yan JJ, Wang JR, Liu CC, Yang HB, Su IJ (2000) An outbreak of Enterovirus 71 infection in Taiwan 1998: a comprehensive pathological, virological, and molecular study on a case of fulminant encephalitis. J Clin Virol 17:13–22. https://doi.org/10.1016/s1386-6532(00)00067-6

Yang Y, Salayandia VM, Thompson JF, Yang LY, Estrada EY, Yang Y (2015) Attenuation of acute stroke injury in rat brain by minocycline promotes blood-brain barrier remodeling and alternative microglia/macrophage activation during recovery. J Neuroinflammation 12:26. https://doi.org/10.1186/s12974-015-0245-4

Yao PP, Qian L, Xia Y, Xu F, Yang ZN, Xie RH et al (2012) Enterovirus 71-induced neurological disorders in young gerbils, Meriones unguiculatus: development and application of a neurological disease model. PLoS One 7:e51996. https://doi.org/10.1371/journal.pone.0051996

Ye F, Cui M, Khasawneh RH, Shibata R, Wu J, Sharaan M, Zhang DY (2013) Molecular Virology. In: Cheng L, Zhang DY, Eble JN (eds) Molecular genetic pathology. Springer, New York, NY, pp 655–731. https://doi.org/10.1007/978-1-4614-4800-6_25

Yoon SY, Ha YE, Choi JE, Ahn J, Lee H, Kweon HS, Lee JY, Kim DH (2008) Coxsackievirus B4 uses autophagy for replication after calpain activation in rat primary neurons. J Virol 82 (23):11976–11978

Yu J, Zhang L, Ren P, Zhong T, Li Z, Wang Z et al (2015) Enterovirus 71 mediates cell cycle arrest in S phase through non-structural protein 3D. Cell Cycle 14:425–436. https://doi.org/10.4161/15384101.2014.980631

Zell R, Delwart E, Gorbalenya AE, Hovi T, King AMQ, Knowles NJ, Lindberg AM, Pallansch MA, Palmenberg AC, Reuter G, Simmonds P, Skern T, Stanway G, Yamashita T (2017) ICTV virus taxonomy profile: picornaviridae. J Gen Virol 98:2421–2422. https://doi.org/10.1099/jgv.0.000911

Zeng H, Wen F, Gan Y, Huang W (2012) MRI and associated clinical characteristics of EV71-induced brainstem encephalitis in children with hand-foot-mouth disease. Neuroradiology 54:623–630. https://doi.org/10.1007/s00234-011-0979-3

Zhu F-C, Meng F-Y, Li J-X, Li X-L, Mao Q-Y, Tao H, Zhang Y-T, Yao X, Chu K, Chen Q-H, Hu Y-M, Wu X, Liu P, Zhu L-Y, Gao F, Jin H, Chen Y-J, Dong Y-Y, Liang Y-C, Shi N-M, Ge H-M, Liu L, Chen S-G, Ai X, Zhang Z-Y, Ji Y-G, Luo F-J, Chen X-Q, Zhang Y, Zhu L-W, Liang Z-L, Shen X-L (2013) Efficacy, safety, and immunology of an inactivated alum-adjuvant enterovirus 71 vaccine in children in China: a multicentre, randomised, double-blind, placebo-controlled, phase 3 trial. Lancet 381:2024–2032. https://doi.org/10.1016/S0140-6736(13)61049-1

Zhu F, Xu W, Xia J, Liang Z, Liu Y, Zhang X, Tan X, Wang L, Mao Q, Wu J, Hu Y, Ji T, Song L, Liang Q, Zhang B, Gao Q, Li J, Wang S, Hu Y, Gu S, Zhang J, Yao G, Gu J, Wang X, Zhou Y, Chen C, Zhang M, Cao M, Wang J, Wang H, Wang N (2014) Efficacy, safety, and immunogenicity of an enterovirus 71 vaccine in China. N Engl J Med 370:818–828. https://doi.org/10.1056/NEJMoa1304923

Encephalitic Arboviruses

7

Suruchi Shukla and Shantanu Prakash

Abstract

Togaviridae, Flaviviridae, Bunyaviridae, and Reoviridae families of Arboviruses are known to affect the neurological system. Mosquitoes, ticks, biting flies, mites, and nits transmit arboviruses to vertebral hosts during blood feedings. While the majority of arboviral infections do not cause neuro-invasive illness, they are among the most serious infectious threats to the central nervous system's health. Arboviruses replicate in peripheral tissues, cause viremia, penetrate the CNS, replicate in neurons, and disseminate to other neuron populations, potentially resulting in encephalitis. Apart from encephalitis, they can also cause meningitis, myelitis, encephalomyelitis, neuritis, and myositis. Which, in some situations, can result in long-term central nervous system abnormalities. This is seen in Zika virus-related prenatal brain abnormalities(e.g., microcephaly), as well as West Nile virus-induced synaptic dysfunctions that can persist long after infection and result in cognitive deficits. Hence, when evaluating a febrile patient with neurological symptoms, arboviruses are relevant differentials to investigate.

Keyword

Japanese Encephalitis Virus · Dengue virus · Arbovirus · Laboratory Diagnosis of encephalitis

S. Shukla (✉)
Department of Microbiology, King George's Medical University, Lucknow, India

S. Prakash
King George's Medical University, Lucknow, India

H. Sami et al. (eds.), *Viral and Fungal Infections of the Central Nervous System: A Microbiological Perspective*, https://doi.org/10.1007/978-981-99-6445-1_7

7.1 Introduction

Encephalitis is inflammation of the brain parenchyma caused by various pathogens and an autoimmune response, resulting in morbidity, mortality, and permanent neurological disability in both adults and children. The development of a coherent case definition for encephalitis is further complicated by the clinical overlap between encephalitis and encephalopathy. These terms are often used synonymously in the literature, but they can represent different pathophysiological processes. Encephalopathy is a clinical condition of altered mental status characterized by confusion, disorientation, behavioral changes, or other cognitive impairment with or without inflammation of brain tissue (Granerod and Crowcroft 2007). The inflammation causes the brain to swell, which can lead to headaches, stiff necks, photophobia, inner confusion, and seizures. The disease affects 1.9–14.3 people per 100,000 population each year and causes an average of 20,258 hospitalizations per year. According to reports from different countries, it continues to cause 5639.3% of the deaths among the affected patients. The condition can affect anyone but is more common in younger people. In fact, in an algorithmic testing panel, the specific cause of the encephalitis remains unknown in about 30–40 cases (Vora et al. 2014).

Encephalitis by viral pathogens is the most common form of encephalitis and often co-occurs with viral meningitis. Viruses enter the host outside of the central nervous system and then enter the spinal cord and brain via hematogenous spread or retrograde from nerve endings. Vaccinations against measles, mumps, rubella, and chickenpox have reduced the incidence of encephalitis from these diseases, but other infections can also cause encephalitis. Contagious encephalitis is generally caused by a viral infection, most commonly Arboviruses, Herpes simplex infections types 1 and 2, Varicella zoster infection, and Enteroviruses (Jmor et al. 2008; Palus et al. 2014).

7.2 Diagnostic Criteria of Encephalitis (Refer to Table 7.1)

Confirmed encephalitis requires one of the following (Venkatesan and Geocadin 2014; Glaser et al. 2006):

1. Pathologic confirmation of brain inflammation consistent with encephalitis;
2. Defined pathologic, microbiologic, or serologic evidence of acute infection with a microorganism strongly associated with encephalitis from an appropriate clinical specimen (for examples, see references;
3. Laboratory evidence of an autoimmune condition strongly associated with encephalitis.

Fever is common in patients with acute encephalitis but is nonspecific. The requirement for objective documentation of fever within a limited period of 72 hours after hospitalization was chosen to rule out secondary health-related infections. It is known that fever can occur as a result of a number of infections

Table 7.1 Diagnostic criteria of encephalitis

Major Criterion:
(personality change) lasting ≥24 h with no alternative cause identified. Patients presenting to medical attention with altered mental status (defined as decreased or altered level of consciousness, lethargy or
Minor Criteria (2 required for possible encephalitis; ≥3 required for probable or confirmed encephalitis):
Documented fever ≥38 ° C (100.4 °F) within the 72 h before or after presentation
Generalized or partial seizures not fully attributable to a preexisting seizure disorder
New onset of focal neurologic findings
CSF WBC count ≥5/cubic mm
Abnormality of brain parenchyma on neuroimaging suggestive of encephalitis that is either new from prior studies or appears acute in onset
Abnormality on electroencephalography that is consistent with encephalitis and not attributable to another cause

outside the central nervous system that can cause encephalopathy, as well as noninfectious diseases that resemble encephalitis. It is also recognized that fever can be variable and therefore patients with infectious encephalitis may not have an objective fever at the time of clinical evaluation Furthermore, immunosuppressed patients with encephalitis may not mount a fever.

Seizures associated with encephalitis can be generalized, indicating global CNS dysfunction, or focal, indicating a localized process. Subclinical seizures may also occur, which may be a cause of altered sensory perception. Seizures associated with high temperature are relatively common in young children and, when isolated, do not require testing for encephalitis. The main requirement for at least 24 h of altered mental status was chosen to exclude the postictal state that occurs in patients with febrile seizures.

CSF pleocytosis indicates an inflammatory process in the brain parenchyma, meninges, or both (meningoencephalitis). However, the absence of CSF pleocytosis does not rule out encephalitis. In particular, it is known that the liquor in immunocompromised patients can be cell-free. Conversely, in inflammation confined to the meninges, the CSF profile may be indistinguishable from that in patients with encephalitis. However, in most cases of encephalitis, the absolute leukocyte count is $<1000/\text{mm}^3$ and lymphocytes typically predominate.

To ensure adequate sensitivity of the definition, the group defined CSF pleocytosis as ≥ 5 WBC/mm^3. In cases where there are large numbers of red blood cells in the CSF, such as with a traumatic lumbar puncture, the following formula may allow correction of the WBC count: True CSF WBC = actual CSF WBC— (WBC in blood X RBC in CSF)/RBC in blood. Notably, rules for adjusting leukocytes in blood-contaminated CSF have not been well validated.

Neuroimaging plays a critical role in evaluating patients with suspected encephalitis as it can aid in the diagnosis of a specific etiology or identify alternative disorders that resemble encephalitis. Magnetic resonance imaging (MRI) is the radiological procedure of choice for evaluating patients with suspected encephalitis.

Several studies have confirmed that MRI is superior to computed tomography (CT) for detecting CNS abnormalities.

MRI can be helpful in determining the etiology because the localization of the inflammation to specific pathogens (e.g., temporal lobe involvement in patients with herpes simplex virus encephalitis) or to an autoimmune phenomenon (e.g., demyelination in patients with acute disseminated encephalomyelitis). A non-contrast CT scan is most useful to assess the safety of performing a spinal tap and to rule out alternative diagnoses such as a subarachnoid hemorrhage. We recognize that MRI or CT may not be available in resource-constrained environments. In this case, the diagnosis of encephalitis must be based on clinical and laboratory criteria (Glaser et al. 2006; Weil et al. 2002).

The EEG abnormalities reported in encephalitis range from nonspecific generalized slowing to characteristic patterns indicative of distinct entities, including repetitive sharp wave complexes over the temporal lobe or periodic lateralizing epileptiform discharges in HSV-1 and bilateral synchronous periodic sharp and slow waves associated with subacute sclerosing panencephalitis. EEG abnormalities are often nonspecific and may be due to medication or metabolic disorders. The EEG can identify epileptiform discharges when there are no clinical signs of seizure activity (subclinical or nonconvulsive status epilepticus) as a cause of drowsiness.

7.2.1 Viral Encephalitis Pathogenesis

The first step in triggering viral encephalitis is to disrupt the Blood Brain Barrier. Once the virus enters the CNS, the first line of defense consists of microglia, which represent the CNS's primary innate immune response. In the brain, microglia cells play an essential role in maintaining homeostasis by supporting neurons through the secretion of neurotrophins and growth factors, in addition to their contribution to the immune response against pathogens. Microglia are susceptible to infection by multiple viruses, including WNV, JEV, DENV, ZIKV, HIV, and HSV-1 (Jhan et al. 2017). Astrocytes are one of the most abundant cell types in the brain and play multiple roles in brain homeostasis, such as regulating ionic balance and neurotransmitters, supporting neuronal synapses, and maintaining Blood Brain Barrier permeability. Similarly activated astrocytes can secrete factors that allow recruitment of immune cells to the injured area, thus promoting amplification of neuroinflammation. Astrocytes are tolerant to infection by WNV, JEV, ZIKV, TBEV, HSV-2, HIV, hRSV, and SARS-CoV-2. Depending on their location in the brain, neurons can have different functions. For example, neurons near the blood–brain barrier help regulate blood flow and secrete factors that promote angiogenesis, among other things. The function of the glutamatergic pyramidal neurons in the hippocampus is mainly associated with learning, memory, and emotions. For several viruses including WNV, JEV, ZIKV, TBEV, HSV, HIV, influenza virus, hRSV and SARS-CoV-2, neurons are the primary target of infection in the brain. Infection of neurons promotes cell damage, cell death, secretion of cytokines that induce immune cell recruitment, and changes in neurocognitive processes (Kofler and Wiley 2011; Bohmwald et al. 2021).

7.3 Arbovirus Causing Encephalitis (Refer to Table 7.2)

Arboviruses are classified into three families in the current virus classification system, namely Togaviridae, Flaviviridae, and Bunyaviridae. Other groups of arboviruses include the Rhabdoviridae, Reoviridae, and Asfarviridae; They all play minimal or no pathogenic role in the origin of human infections. Encephalitis is caused by certain arboviruses that are transmitted by vectors such as mosquitoes, ticks, and other insects or animals (Salimi et al. 2016; Vasilakis and Tesh 2015). The group of arboviruses that cause encephalitis are:

- Japanese encephalitis virus (JEV).
- West Nile virus (WNV).
- Dengue virus (DV).
- La Crosse virus.
- St. Louis Encephalitis Virus (SLE).
- Tick-Borne Encephalitis Viruses.
- California group viruses such as Lacrosse virus.
- Eastern equine encephalitis (EEE).
- Western equine encephalitis (WEE).
- Venezuelan Encephalitis Virus (VEV).
- Powassan virus.
- Zika
- Chikungunya.

Each of these viruses is more common in certain parts of the United States. Travelers abroad are most at risk of contracting Japanese encephalitis and tick-borne encephalitis. Japanese encephalitis is transmitted by mosquitoes. It is mainly found in India, China, Japan, Korea, and eastern Russia. It is also less common in the Republic of China (Taiwan), Singapore, and Hong Kong. In all these areas, Japanese encephalitis is mainly a rural disease. In 1999, an outbreak of West Nile encephalitis occurred in the New York area; West Nile virus is closely related to St. Louis encephalitis virus and is common in Africa, Asia, and the Middle East. Emerging infections such as Zika, Chikungunya, and Powassan viruses may also contribute to this trend. Other infectious microorganisms such as bacteria, rickettsia such as typhoid, fungi, and parasites can also, although rarely, cause encephalitis (Cdc 2001; Emedicine n.d.).

Arboviral infections typically occur from June to October when mosquitoes are active. Few species of mosquito carry and transmit these arboviruses, and usually only a small proportion of these mosquitoes are carriers of the virus. Another virus, eastern equine encephalitis (EEE), has been found in rare cases in birds and horses in the United States. Tick-borne encephalitis is a viral infection of the central nervous

Table 7.2 Arboviral encephalitis etiology

Arboviral family/genus	Virus name	Genome	Vector	Geographical distribution
Flaviviridae/ flavivirus	Japanese Encephalitis	Single stranded positive sense RNA	Mosquito	Asia
	West Nile Virus	Single stranded positive sense RNA	Mosquito	Asia, Africa, South America
	Dengue Virus	Single stranded positive sense RNA	Mosquito	Asia, Africa, America, Europe, Pacific Islands
	Zika virus	Single stranded positive sense RNA	Mosquito	Asia, Africa, America, Europe, Pacific Islands
	Tick borne Encephalitis	Single stranded positive sense RNA	Ticks	Assia nad Europe
	St. Louis Encephalitis	Single stranded positive sense RNA	Mosquito	America
	Murray Valley Encephalitis	Single stranded positive sense RNA	Mosquito	Australia
	Powassan	Single stranded positive sense RNA	Ticks	Europe, Canada, North United States
Togaviridae/ Alpha virus	Chikungunya	Single stranded positive sense RNA	Mosquito	Asia, Africa, America, Europe, Pacific Islands
	Eastern Equine Encephalitis	Single stranded positive sense RNA	Mosquito	America
	Western Equine Encepahilitis	Single stranded positive sense RNA	Mosquito	America
	Venezuelan Equine Encephalitis	Single stranded positive sense RNA	Mosquito	Central and South America
Reoviridae	Colorado tick fever	Double stranded RNA	Ticks	Western Canada and United States
Bunyaviridae	La Crosse encephalitis California encephalitis	Single stranded negative sense RNA	Mosquito, small midgets	United States Asia and Europe

system. Transmission occurs through the bite of certain ticks. Humans can become infected through the bite of infected Ixodes ricinus ticks. This often happens in people who visit or work in forests, fields, or pastures. You can also get it from eating unpasteurized dairy products from infected cows, goats, or sheep. Infection with an arbovirus occurs only through the bite of an infected arthropod, such as a mosquito. These diseases are not transmitted from person to person. The main vector (transmitter) of SLE and WNV is the northern house mosquito (Culex pipiens). The northern house mosquito breeds in small stagnant bodies of water (such as ditches, roadside catch basins) and containers — such as discarded tin cans, flower urns, old tires, buckets, and other containers — that contain water. Mosquitoes become infected with St. Louis encephalitis virus and WNV after biting infected birds. The mosquito that transmits California encephalitis (LaCrosse) is the tree hole mosquito (Aedes triseriatus) (WHO n.d.-a). Found in forested areas, the tree hole midge breeds in discarded tires and other objects filled with water.

7.3.1 Japanese Encephalitis Virus

Japanese encephalitis virus JEV is the most important cause of viral encephalitis in both Asia and India. Transmission of JEV is endemic in 24 countries in the WHO Southeast Asia and Western Pacific region, putting more than 3 billion people at risk of infection. It is a mosquito-borne flavivirus and belongs to the same genus as dengue, yellow fever, and West Nile viruses. The first case of Japanese encephalitis virus (JE) disease was documented in Japan in 1871. The annual incidence of clinical disease varies in both within and in endemic countries, with outbreaks ranging from <1 to >10 per 100,000 population or more. A review of the literature estimates that there are nearly 68,000 clinical cases of JE each year worldwide, with approximately 13,600 to 20,400 deaths. JE mainly affects children. Most adults in endemic countries have natural immunity after childhood infection, but individuals of any age can be affected. Although symptomatic Japanese encephalitis (JE) is rare, the mortality rate in people with encephalitis can be as high as 30%. Permanent neurological or psychiatric sequelae can occur in 30–50% of people with encephalitis. JEV is transmitted to humans through bites of infected Culex mosquitoes (mainly Culex tritaeniorhynchus). Humans do not develop sufficient viremia after infection to infect feeding mosquitoes. The virus exists in a cycle of transmission between mosquitoes, pigs, and/or waterfowl (enzootic cycle). The disease occurs mainly in rural areas

7.3.1.1 WHO Recommended Case Definition for Japanese Encephalitis (JE) (WHO n.d.-a)

7.3.1.1.1 Clinical Case Definition

Clinically, a case of acute encephalitis syndrome (AES) is defined as a person of any age, at any time of year, with the acute onset of fever and a change in mental status (including symptoms such as confusion, disorientation, coma, or inability to talk) AND/OR new onset of seizures (excluding simple febrile seizures). Other early clinical findings can include an increase in irritability, somnolence, or abnormal behavior greater than that seen with usual febrile illness.

7.3.1.1.2 Case Classification

Suspected case: A case that meets the clinical case definition for AES. Suspected cases should be classified in one of the following four ways.

Laboratory-confirmed JE: A suspected case that has been laboratory-confirmed as JE.

Probable JE: A suspected case that occurs in close geographic and temporal relationship to a laboratory-confirmed case of JE, in the context of an outbreak.

Acute encephalitis syndrome – other agent: A suspected case in which diagnostic testing is done and an etiological agent other than JE virus is identified.

Acute encephalitis syndrome – unknown: A suspected case in which no diagnostic testing is done, or in which testing identified no etiological agent, or in which the test results were indeterminate.

Sometimes, it may be difficult to differentiate JE from those caused by other viruses, bacteria, etc., as clinical signs of JE are indistinguishable from other causes of AES. Under such circumstances, laboratory confirmation is essential for accurate diagnosis of JE. Confirmation of a suspected or probable case of JE would require the support of a well-equipped laboratory to test blood and cerebrospinal fluid (CSF) for the same.

WHO recommends testing for JEV-specific IgM antibodies in a single sample of cerebrospinal fluid (CSF) or serum using an IgM capture ELISA. Testing of a CSF sample is preferred to reduce the false-positivity rate due to previous infection or vaccination. Surveillance of the disease is mostly syndromic in acute encephalitis syndrome. Confirmatory laboratory testing is often performed at dedicated Sentinel sites, and efforts are being made to expand laboratory-based monitoring. Case-based surveillance is established in countries that effectively control JE through vaccination. There is no antiviral treatment for the disease. Treatment focuses on relieving severe clinical symptoms and helping the patient overcome the infection. Safe and effective vaccines are available to prevent JE. (WHO n.d.-a; Turtle and Solomon 2018) WHO recommends that JE vaccination be included in national immunization plans in all areas where JE disease is recognized as a public health problem. Safe and effective JE vaccines are available to prevent disease. WHO recommends strong JE prevention and control activities, including JE immunization in all regions where the disease is a recognized public health priority, and strengthening surveillance and reporting mechanisms. Although the number of confirmed JE cases is low, vaccination should be considered where there is a suitable environment for transmission of the JE virus. There is little evidence of a reduction in the burden of JE disease by interventions other than vaccination of humans. Therefore, vaccination of humans should be given priority over vaccination of pigs and mosquito control measures.

Four main types of JE vaccines are currently used: inactivated mouse brain vaccines, inactivated Vero cell vaccines, live attenuated vaccines, and live recombinant (chimeric) vaccines. In recent years, the China-made live attenuated vaccine SA14–14-2 has become the most widely used vaccine in endemic countries and was prequalified by the WHO in October 2013. Cell culture-based inactivated vaccines and live recombinant vaccine based on the yellow fever vaccine strain have also been approved and prequalified by the WHO. In November 2013, Gavi opened a funding window to support JE vaccination campaigns in eligible countries. To reduce the risk of JE, all travelers to Japanese encephalitis endemic areas should take precautions to avoid mosquito bites. Personal prevention measures include the use of mosquito repellent, long-sleeved clothing, coils, and vaporizers. Travelers staying in JE endemic areas for extended periods are advised to be vaccinated prior to travel. Major JE outbreaks occur every 2–15 years (WHO n.d.-a; Cdc. gov n.d.).

JE transmission increases during the rainy season when vector populations increase. However, so far there is no evidence of increased JEV transmission after major floods or tsunamis. The spread of JEV into new areas is related to agricultural development and intensive rice cultivation supported by irrigation programs.

WHO recommends and supports implementation of JE vaccination in all regions where the disease is a recognized public health priority. It provides technical support for JE surveillance, JE vaccine rollout, and large-scale JE vaccination campaigns, and evaluation of JE vaccine efficacy and programmatic impact.

7.3.2 Chikungunya Virus

Chikungunya virus is an alphavirus (genus Alphavirus, family Togaviridae) transmitted to humans primarily by Aedes mosquitoes and occasionally mother-to-child transmission. The word Chikungunya comes from the Makonde language spoken in Tanzania and Mozambique and means "that which bends"; this refers to the debilitating arthralgia that often occurs in the acute phase of infection, along with fever, myalgia, headache, and rash. Although the first outbreaks were described in the 1960s, the virus was not recognized as a major public health problem until 2004, when it caused explosive outbreaks in the tropics. Serious complications of chikungunya infection, including neurological disorders, are increasingly recognized. Classically, alphaviruses are divided into two groups: the Old World viruses, including Sindbis, Onyongnyong, and Ross River viruses, which cause a predominantly arthritic syndrome, and the New World viruses, which include Eastern, Western, and Venezuelan equine encephalitis viruses, responsible for outbreaks of encephalitis. Chikungunya virus is now recognized as a cause of arthritic and neurological diseases throughout the tropics.

Dengue fever and Zika are also arthropod-borne viruses (arboviruses) transmitted by Aedes mosquitoes like chikungunya, but are flaviviruses (genus Flavivirus, family Flaviviridae). All 3 arboviruses initially cause a fever-arthralgia-intoxication syndrome and are associated with neurological complications (Mehta et al. 2018).

Neurological disorders following chikungunya virus infection were first reported in an outbreak in 1964 in Madras, India. Four cases of chikungunya virus have been described and confirmed serologically or by virus isolation. Two presented with a meningoencephalitic picture (delirium or coma and signs of meningeal irritation with neck stiffness and Kernig's signs, sluggish pupillary response, etc.), one with acute flaccid paralysis and elevated CSF protein suggestive of Guillain–Barr syndrome (GBS) and one with transient dysarthria. Since then, neurological manifestations have been reported throughout the Indian Ocean, southern Asia, the

Pacific Islands, southern Europe, the Caribbean, and South America, ranging from mild behavioral disturbances to severe acute syndromes of both the CNS and peripheral nervous system. (Mehta et al. 2018; Webb et al. 2022)

Diagnosis is based on clinical, radiological, serological, and molecular assays. Although not yet commercially available, there is hope that a vaccine against chikungunya is on the horizon. Two phase 1 clinical studies to date have shown a good safety and immunogenicity profile. A recent study testing an insect-specific alphavirus as a vaccine platform showed promising results in mice and macaques, including immunogenicity after a single dose (Webb et al. 2022).

7.3.3 Dengue Virus

Dengue infection is the most important tropical viral disease in the world today. According to the World Health Organization, 50 million symptomatic dengue infections occur annually, which is a major public health problem, especially in Southeast Asia and the Western Pacific. Dengue viruses (DENV) are single-stranded RNA arboviruses with four serological types (DENV 14) and belong to the Flaviviridae family. (WHO n.d.-b) Clinical signs of dengue infection vary and range from simple myalgia, arthralgia, headache, dengue fever (DF), dengue hemorrhagic fever (DHF), dengue shock syndrome (DSS) to neurological dengue fever, manifested as encephalopathy, encephalitis, encephalomyelitis, myelitis and brachial neuritis, Guillain-Barre syndrome, hypokalemic palsy, viral myositis, and rare opsoclonus-myoclonus syndrome. Dengue infection with acute encephalopathy was first reported in 1976 by Sanguansermsri et al. Since then there have been reports from several Southeast Asian countries. Although the basic pathophysiology of central nervous system involvement in dengue infection remains unclear, in the reported cases the encephalopathy has been attributed to cerebral edema, anoxia, hemorrhage, hyponatremia, liver failure, and release of toxic substances. Various animal experiments and clinical studies indicate a neurotrophic potential of DENV that can lead to encephalitis. Detection of viral antigen in brain autopsy specimens, dengue-specific immunoglobulin M antibody (IgM-Ab), and positive reverse transcriptase PCR (RT-PCR) in cerebrospinal fluid (CSF) support the hypothesis of neuro-invasion during acute dengue infection (WHO n.d.-b; Li et al. 2017).

Both patients and doctors are struggling as the symptoms of dengue fever and COVID-19 overlap. Even after using rapid dengue tests, misdiagnosis of COVID-19 as dengue fever has been reported. The consequences of COVID-19 and dengue misdiagnosis are relevant and may include ineffective patient management, potentially resulting in preventable patient death, and unsuccessful prevention strategies including rapid patient isolation (in the case of COVID-19) and vector control (in the case of Dengue fever). Therefore, before any treatment, there should be diagnostic (clinical, hematological, and biochemical) accuracy. Due to the lack of the above literature on early confirmatory clinical and laboratory diagnosis of DENV, we planned this study. The aim of this study was to investigate the incidence, predictive and prognostic factors of dengue encephalitis (DE) against the background of DENV infection (Li et al. 2017).

Dengue encephalitis is rare and occurs as a result of direct infection of neurons by the dengue virus. Dengue encephalitis is believed to be benign but can sometimes be fatal. The role of an antiviral agent in such cases needs to be further defined because of the extensive parenchymal involvement and possible adverse outcome. Clinically presents with decreased consciousness, headache, seizures, disorientation and behavioral symptoms. Radiological Imaging features alone are not diagnostic. Bilateral thalamic involvement with positive CSF IgG/IgM for dengue fever virus is diagnostic of dengue encephalitis. A polymerase chain reaction (PCR) is required to confirm viral RNA as antibodies are not always visible. Differential Diagnoses of dengue encephalitis is Japanese encephalitis which depicts bilateral thalamic involvement with hemorrhagic foci, involving the temporal lobe and brainstem is usually present and is very rare in dengue fever. Chikungunya and herpetic encephalitis is another differential diagnosis. Chikungunya encephalitis on MRI shows hyperintense T2W white matter lesions with restricted diffusion. No bleeding or involvement of the basal ganglia was reported whereas herpetic encephalitis shows bilateral and asymmetric involvement. Basal ganglia and thalami are typically spared. Treatment is mostly supportive. Ribavarin is a newer, promising drug that inhibits viral replication; however, further validation is needed in dealing with the virus (Li et al. 2017; Carod-Artal et al. 2013).

7.3.4 Zika Virus

Zika virus (ZIKV) is an arthropod-borne virus (arbovirus) of the genus Flavivirus and family Flaviviridae. ZIKV was first isolated from a nonhuman primate in 1947 and from mosquitoes in Africa in 1948. Human ZIKV infections were sporadic for half a century before emerging in the Pacific and Americas. ZIKV is usually transmitted through the bite of infected mosquitoes. The clinical presentation of Zika fever is nonspecific and can be misdiagnosed as other infectious diseases, particularly those caused by arboviruses such as dengue fever and chikungunya. Before the large outbreak in French Polynesia in 2013 and 2014, when severe neurological complications were reported and in Brazil there was a dramatic increase in congenital malformations (microcephaly) suspected to be related to ZIKV; till that time ZIKV infection was only associated with minor illnesses. Zika virus infection during pregnancy is a cause of microcephaly. During pregnancy, a baby's head size proportionately increases with fetus brain tissue. Microcephaly can occur because a baby's brain did not develop properly during pregnancy or stops growing after birth. (Ferrari-Marinho et al. 2022)

Laboratory diagnosis of Zika fever is based on virus isolation or detection of ZIKV-specific RNA. Serological diagnosis is complicated by cross-reactivity between members of the Flavivirus genus. The adaptation of ZIKV to an urban cycle involving humans and native mosquito vectors in tropical areas where dengue fever is endemic suggests that the emergence of ZIKV infections may be underestimated. There is a high potential for ZIKV emergence in urban centers in the tropics that are infested with competent mosquito vectors such as Aedes aegypti and Aedes albopictus. ZIKA virus infection has been continuously reported from many states of India where dengue disease is already a known and endemic entity (Shukla et al. 2023).

7.4 Clinical Manifestation of Arboviral Encephalitis

Arboviral infections can be asymptomatic or result in disease of varying severity, sometimes with central nervous system (CNS) involvement. When the CNS is affected, clinical syndromes can occur, ranging from febrile headache to aseptic meningitis to encephalitis, and are usually indistinguishable from similar syndromes caused by other viruses. Arboviral meningitis is characterized by fever, headache, neck stiffness, and pleocytosis. Arboviral encephalitis is characterized by fever, headache, and an altered mental state ranging from confusion to coma, with or without additional signs of brain dysfunction (e.g., paresis or palsy, cranial nerve palsies, sensory deficits, abnormal reflexes, generalized convulsions, and abnormal movements). The symptoms of WNV, SLE and LaCrosse encephalitis virus are similar. If the infection is severe, severe headache, high fever, muscle pain, stiffness in the neck, problems with muscle coordination, disorientation, convulsions, and coma can occur rapidly. Symptoms usually appear five to 15 days after an infected mosquito bite (Jmor et al. 2008; Cdc 2001).

Although anyone can be infected with an arbovirus, JEV and dengue commonly affect children and adolescents, while WNV and SLE usually occur in people over the age of 50. Most patients make a full recovery, although in rare cases severe infection can result in neurological damage or death. Infection with California (LaCrosse) encephalitis virus is most common in children. The disease is generally milder than JE, dengue, WNV, and SLE, and fatalities are rare. However, studies suggest that some children with California encephalitis virus (LaCrosse) may experience persistent neurological problems. Infection with an arbovirus can confer some immunity to that specific virus but not to other arboviruses (Cdc 2001).

7.5 Diagnosis

7.5.1 Case Classification

7.5.1.1 Probable

An encephalitis or meningitis case occurring during a period when arboviral transmission is likely, and with the following supportive serology: (1) a single or stable (less than or equal to twofold change) but elevated titer of virus-specific serum antibodies; or (2) serum IgM antibodies detected by antibody-capture EIA but with no available results of a confirmatory test for virus-specific serum IgG antibodies in the same or a later specimen.

7.5.1.2 Confirmed

An encephalitis or meningitis case that is laboratory confirmed.

Closely related arboviruses show serological cross-reactivity, positive results from serological tests with antigens from a single arbovirus can be misleading. In certain circumstances (e.g., in areas where two or more closely related arboviruses are present, or in the case of imported arbovirus diseases), it may be epidemiologically important to attempt to closely isolate the infecting virus by conducting cross-neutralization tests with an appropriate battery related viruses to locate viruses. This is important, for example, to determine whether detected antibodies to St. Louis encephalitis virus are not the result of West Nile virus (or dengue virus) infection or vice versa, in areas where both are present, viruses occur. Although pathological examination of brain tissue is considered the gold standard diagnostic test for this syndrome. However, because of the potential morbidity associated with invasive neurosurgical intervention, it is rarely performed premortem. In the absence of pathological confirmation, encephalitis was previously defined on the basis of selected clinical, laboratory, electroencephalographic, and imaging features. Arbovirus infection is usually diagnosed by a blood test or cerebrospinal fluid test. A doctor will try to relieve the symptoms of the disease, but there is no specific treatment or cure for these diseases.

7.5.1.3 Laboratory Criteria for Diagnosis of Arboviral encephalitis (Venkatesan and Geocadin 2014):

For most arboviruses, serological testing of serum and CSF is preferable to molecular testing because the peak of viraemia typically occurs before the onset of symptoms. For example, in West Nile Virus (WNV) patients associated with neuroinvasive disease, CSF-PCR is relatively insensitive (57%) compared to detecting WNV IgM in CSF. The cumulative percentage of seropositive patients increases by approximately 10% per day during the first week of illness, suggesting the need for repeat testing when patients with initially negative results are strongly suspected of having disease. In particular, arbovirus IgM antibodies may be persistently detectable in serum and, more rarely, in CSF many months after an acute infection and may therefore not be indicative of a current infection. Therefore, when possible, documentation of acute infection by seroconversion and/or a fourfold or greater increase in titer using paired sera is recommended.

- Fourfold or greater change in virus-specific serum antibody titer, OR.
- Isolation of virus from or demonstration of specific viral antigen or genomic sequences in tissue, blood, cerebrospinal fluid (CSF), or other body fluid, OR.
- Virus-specific immunoglobulin M (IgM) antibodies demonstrated in CSF by antibody-capture enzyme immunoassay (EIA), OR.
- Virus-specific IgM antibodies demonstrated in serum by antibody-capture EIA and confirmed by demonstration of virus-specific serum immunoglobulin G (IgG) antibodies in the same or a later specimen by another serologic assay (e.g., neutralization or hemagglutination inhibition).
- PRNT.

- Molecular real time PCR.
- Serotyping and genotyping.

7.6 Treatment

Young children and the elderly are most often seriously ill. There are no proven treatment options for arboviral encephalitis. Most people recover from the disease, but permanent brain problems and death can result.

7.7 Preventive strategies (Olliaro et al. 2018)

Because the mosquitoes that transmit arboviruses thriving in small pools of water, removing potential breeding grounds is the most effective form of disease prevention. Here are a few suggestions:

- Remove or empty water from old tires, tin cans, buckets, barrels, bottles, or other places where mosquitoes might breed. Be sure to check clogged gutters and flat roofs that may have poor drainage. Make sure cisterns, sumps, septic tanks, fire barrels, rain barrels, and dumpsters are tightly covered with a lid or 16-mesh screen.
- Empty plastic wading pools at least once a week and store indoors when not in use. Swimming pools should be properly maintained; if not used, pools should be drained and kept dry during mosquito season.
- Change the water in bird baths, plant saucers, and trays weekly.
- Store boats covered or upside down, or remove rainwater weekly.
- Empty your pet's water bowl and refill daily.
- Level the ground around your home so water can run off and not collect in low spots. Fill in holes or depressions near your home that accumulate water.
- Fill in tree rot holes and hollow stumps that hold water.
- Stock ornamental water gardens with fish (e.g., minnows, "mosquito fish," or goldfish) that eat mosquito larvae.
- Small pools of water can be treated for mosquito larvae with "Bti," a bacterial insecticide. Many hardware stores carry Bti briquettes (such as donut-shaped Mosquito Dunks) for this purpose. Be sure to follow the insecticide label exactly.
- Keep weeds and grass cut short; adult mosquitoes look for these shady places to rest during the hot daylight hours. If adult mosquitoes are present in high weeds along the edge of a yard, that location can be sprayed with an appropriately labeled insecticide.
- Be sure screens in homes and buildings are intact and tight-fitting to prevent the entry of mosquitoes. Use a flyswatter or household spray to kill mosquitoes, flies or other insects that get into buildings.
- Some mosquito control methods are not very effective. Bug zappers and anti-mosquito buzzers (or sound devices) are not effective in controlling biting

mosquitoes. Various birds and bats will eat mosquitoes, but there is little scientific evidence that this reduces mosquitoes around homes.

- Some communities conduct community-wide mosquito abatement programs. Whenever possible, the primary effort of such programs should be identification of mosquito-breeding sites, followed by removal or treatment of these sites with an insecticide used for control of mosquito larvae.
- Use mosquito repellents sparingly on exposed skin. An effective repellent will contain 20% to 30% DEET (N,N-diethyl-meta-toluamide). Higher concentrations may cause side effects, particularly in children. Avoid applying repellents to the hands of children and do not use repellents on children under 3 years of age. Follow package instructions carefully.
- If participating in outdoor activities when mosquitoes are biting, wear protective clothing (shoes, socks, shirt, and long pants). For additional protection from mosquitoes, use an insect repellent. The Environmental Protection Agency (EPA) registers products for use as mosquito repellents. Products containing DEET, picaridin, oil of lemon eucalyptus (PMD), or IR3535 typically provide reasonably long-lasting protection when applied to skin and clothing. The CDC also recommends the use of permethrin-containing products on clothing, shoes, bed nets, and camping gears. The efficacy and duration of protection may vary among products and types of mosquitoes. If you start to get mosquito bites, reapplication maybe needed.

References

Bohmwald K, Andrade CA, Kalergis AM (2021) Contribution of Pro-Inflammatory Molecules Induced by Respiratory Virus Infections to Neurological Disorders. Pharmaceuticals 14:340

Carod-Artal FJ, Wichmann O, Farrar J, Gascón J (2013) Neurological complications of dengue virus infection. Lancet Neurol 12:906–919

Cdc (2001). https://ndc.services.cdc.gov/case-definitions/arboviral-encephalitis-or-meningitis-2001/

Cdc. gov (n.d.). https://www.cdc.gov/japaneseencephalitis/vaccine/index.html

Emedicine (n.d.). https://emedicine.medscape.com/article/234159-overview

Ferrari-Marinho T, De Marchi LR, Caboclo LO (2022) Clinical neurophysiology of zika virus encephalitis. J Clin Neurophysiol 39(4):259–264

Glaser CA, Honarmand S, Anderson LJ et al (2006) Beyond viruses: clinical profiles and etiologies associated with encephalitis. Clin Infect Dis 43:1565–1577

Granerod J, Crowcroft NS (2007) The epidemiology of acute encephalitis. Neuropsychol Rehabil 17(4–5):406–428

Jhan M-K, Tsai T-T, Chen C-L, Tsai C-C, Cheng Y-L, Lee Y-C et al (2017) Dengue virus infection increases microglial cell migration. Sci Rep 7:00182–Z

Jmor F, Emsley HC, Fischer M, Solomon T, Lewthwaite P (2008) The incidence of acute encephalitis syndrome in Western industrialised and tropical countries. Virol J 5:134

Kofler J, Wiley CA (2011) Microglia: key innate immune cells of the brain. Toxicol Pathol 39:103–114

Li G-H, Ning Z-J, Liu Y-M, Li X-H (2017) Neurological manifestations of dengue infection. Front Cell Infect Microbiol 7:449

Mehta R, Gerardin P, de Brito CAA, Soares CN, Ferreira MLB, Solomon T (2018) The neurological complications of chikungunya virus: a systematic review. Rev Med Virol 28(3):e1978

Olliaro P, Fouque F, Kroeger A, Bowman L, Velayudhan R, Santelli AC, Garcia D, Skewes Ramm R, Sulaiman LH, Tejeda GS, Morales FC, Gozzer E, Garrido CB, Quang LC, Gutierrez G, Yadon ZE, Runge-Ranzinger S (2018) Improved tools and strategies for the prevention and control of arboviral diseases: a research-to-policy forum. PLoS Negl Trop Dis 12(2):e0005967

Palus M, Bílı T, Elsterová J, Langhansová H, Salát J, Vancová M et al (2014) Infection and injury of human astrocytes by tick-borne encephalitis virus. J Gen Virol 95:2411–2426

Salimi H, Cain MD, Klein RS (2016) Encephalitic arboviruses: emergence, clinical presentation, and neuropathogenesis. Neurotherapeutics 13(3):514–534

Shukla S, Yadav PD, Patil S, Jain A, Aggarawal V (2023) Atypical clinical presentation of Zika virus from India. Indian J Pathol Microbiol 52:213. https://www.ijpmonline.org/preprintarticle.asp?id=361290

Turtle L, Solomon T (2018) Japanese encephalitis — the prospects for new treatments. Nat Rev Neurol 14:298–313

Vasilakis N, Tesh RB (2015) Insect-specific viruses and their potential impact on arbovirus transmission. Curr Opin Virol 15:69–74

Venkatesan A, Geocadin RG (2014) Diagnosis and management of acute encephalitis: a practical approach. Neurol Clin Pract 4(3):206–215

Vora NM, Holman RC, Mehal JM, Steiner CA, Blanton J, Sejvar J (2014) Burden of encephalitis-associated hospitalizations in the United States, 1998-2010. Neurology 82(5):443–451

Webb E, Michelen M, Rigby I, Dagens A, Dahmash D, Cheng V, Joseph R, Lipworth S, Harriss E, Cai E, Nartowski R, Januraga PP, Gedela K, Sukmaningrum E, Cevik M, Groves H, Hart P, Fletcher T, Blumberg L, Horby PW, Jacob ST, Sigfrid L (2022) An evaluation of global Chikungunya clinical management guidelines: a systematic review. EClinicalMedicine 54: 101672

Weil AA, Glaser CA, Amad Z, Forghani B (2002) Patients with suspected herpes simplex encephalitis: rethinking an initial negative polymerase chain reaction result. Clin Infect Dis 34:1154–1157

WHO (n.d.-a). https://www.who.int/news-room/fact-sheets/detail/japanese-encephalitis

WHO (n.d.-b). https://www.who.int/news-room/fact-sheets/detail/dengue-and-severe-dengue

8 Neurological Illnesses Involved in Human Immunodeficiency Virus and Human T cell Lymphotropic Virus Infections

Manish Ramesh Patil and Imran Rizvi

Abstract

Human immunodeficiency virus (HIV) is a virus from the Lentivirus subgroup, of the Retroviridae family. Neurological involvement is frequent in pediatric human immunodeficiency virus infection and acquired immunodeficiency syndrome (AIDS). Direct invasion of the central nervous system by HIV type 1 (HIV-1) may result in HIV-1 associated encephalopathy, categorized as normal neurological findings, static encephalopathy, or as progressive encephalopathy. Aseptic meningitis, acute encephalitis (HIV-associated neurological disorder, HAND), and polyneuropathy typically occur earlier in the illness, whereas AIDS dementia complex often presents later. HIV/AIDS-associated encephalopathy is linked to HIV's tropism for macrophages or microglial cells, and for lymphocytes (CD4+). In HIV infection with neurologic manifestations, a CSF analysis and neuroimaging is mandatory for ruling out other opportunistic infections. The Retroviridae family also comprises of the Human T cell lymphotropic virus, HTLV. CNS involvement is principally from immunologically mediated insult. HTLV type 1 has a classical neurogenic presentation of myelopathy, labeled as HAM, HTLV-1-associated myelopathy, or as TSP, tropical spastic paraparesis. However, HAM/TSP is not the only neurological outcome that can result from HTLV-1 infection. Cases of complicated encephalitis have been reported in patients afflicted with HTLV type 1 infections.

Keywords

Human immunodeficiency virus · HIV-associated neurocognitive disorder · Myelopathy · Human T cell lymphotropic virus

M. R. Patil · I. Rizvi (✉)
Department of Neurology, King George's Medical University, Lucknow, Uttar Pradesh, India

H. Sami et al. (eds.), *Viral and Fungal Infections of the Central Nervous System: A Microbiological Perspective*, https://doi.org/10.1007/978-981-99-6445-1_8

Abbreviations

ADC	AIDS dementia complex
AFB	Acid fast bacilli
AKT	Anti Koch's therapy
ANI	Asymptomatic neurological involvement
CALAS	Cryptococcal capsular antigen with latex agglutination
cART	Combination anti-retroviral therapy
CBNAAT	Cartridge-based nucleic acid amplification test
CM	Cryptococcal meningitis
CMV	Cytomegalovirus
CNS	Central nervous system
CSF	Cerebrospinal fluid
CT	Computed tomography
DSP	Distal symmetric polyneuropathy
EBV	Epstein Barr virus
ELISA	Enzyme linked Immunosorbent assay
FDG PET	Fluorodeoxyglucose positron emission test
HAART	Highly active anti-retroviral therapy
HAD	HIV-associated Dementia
HAND	HIV-associated neurocognitive disorders
HHV6	Human herpes virus 6
HIV	Human immunodeficiency virus
HIV-VM	Human immunodeficiency virus-vacuolar myelopathy
HTLV	Human T cell lymphotropic virus
IHDS	International HIV dementia scale
IRIS	Immune reconstitution inflammatory response
IVIG	Intravenous immunoglobulin
JC virus	John Cunnigham Virus
LFA	Lateral flow assay
MMSE	Mini mental status examination
MND	Mild neurocognitive disorder
MND	Motor neuron disease
MoCA	Montreal cognitive assessment
MRI	Magnetic resonance imaging
NHL	Non-Hodgkin's lymphoma
PCNSL	Primary CNS lymphoma
PCR	Polymerase chain reaction
PLWHA	People living with HIV AIDS
PML	Progressive multifocal leukoencephalopathy
SPECT	Single photon emission computed tomography
SSRI	Selective serotonin reuptake inhibitor

Neurological involvement is observed very frequently in retroviral infections like HIV and HTLV. The spectrum is quite broad and their comprehension is important due to high prevalence of HIV. In this era of HAART, there is a significant increase in longevity among PLWHA. These neurological observations are also increasing. They involve neurological systems at almost all the levels. In this chapter, we try to see them in an organized fashion. This in no way would attempt be an extensive record but intends to be a practical one.

Outline

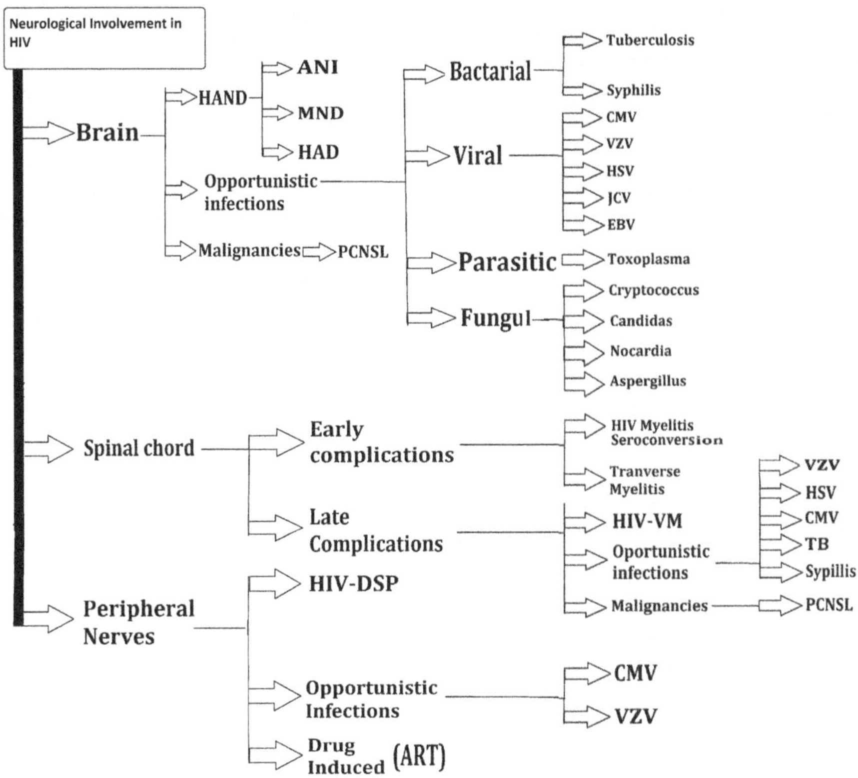

8.1 Brain

Central nervous system comprises of brain and spinal cord. Involvement of CNS is seen from very early stages of HIV infection.

8.1.1 HAND (HIV-Associated Neurocognitive Disorders)

Cognitive involvement is very common among patients with HIV, even with invent of HAART, approximately half of the patients do suffer from cognitive involvement. This cognitive involvement may be due to the presence of HIV virus inside the central nervous system itself or due to the various opportunistic infections. The former type of cognitive involvement is referred as HIV-associated neurocognitive disorders.

Symptoms Patients suffering from HAND may be asymptomatic (ANI, i.e., asymptomatic neurological involvement) which can be demonstrated only with the detailed neuropsychological tests. Or they may have minimal affection so that daily activities are mostly unaffected (MND—mild neurocognitive disorder). The severe form where day-to-day activities are affected due to cognitive limitations is HAD (HIV-associated dementia). ADC (AIDS dementia complex) was the term used in pre-cART era, for slowly progressive form of subcortical dementia where attention, concentration, cognitive, and motor slowing were the main features (Clifford and Ances 2013). Now with HAART, executive functions (planning, organizing, monitoring, and correcting the tasks), language, memory (cortical functions) are the more commonly affected domains of higher mental functions. Initial mild personality changes, fall in occupational and social performances are better and earlier noticed by the family members. So, interviewing family members may provide deeper insight to the problem. HAND include this spectrum of asymptomatic to severe cognitive impairment seen in HIV patients.

Diagnosis Versions of MMSE (mini mental status examination), MoCA (Montreal cognitive assessment), CogState tools and The International HIV Dementia scale (IHDS), etc: the limitations to these are that they are time-consuming and less sensitive to very mild cognitive involvements. Neuroimaging (MRI) may show cerebral atrophy and signs suggestive of (mostly symmetric) white matter involvement.

Treatment No definitive therapy is available for HANDs at present. HAART is the mainstay of the treatment for viral control. Various adjuvant therapies have been tried like SSRI, memantine and are of questionable benefit.

8.1.2 Opportunistic Infections

The opportunistic infections involving CNS are bacterial like tuberculosis; viral like CMV, JC, Virus, etc: parasitic like toxoplasma, etc., or fungal like cryptococcus, nocardia, etc.

8.1.2.1 Tuberculous Meningitis

The risk of tuberculosis infection increases from 10 to 30% depending upon the stages of HIV over lifetime as compared to HIV uninfected population. There is inflammation, increased pro-inflammatory cytokines, free radicles in chronic HIV infection. So, there is reduction of glutathione which normally scavenges free radicles, which possibly increases susceptibility for CNS tuberculosis (Wilkinson et al. 2012). CNS tuberculosis may present as tuberculous meningitis, tuberculomas, and tuberculous abscess. The spinal cord involvement is also common as myelitis and arachnoiditis. It is pragmatic to investigate every patient of tuberculous meningitis for HIV with ELISA and not to content with mere card tests. In India, tuberculous meningitis is a common presenting illness of HIV infection. The course observed may be more fulminant as compared to the immune-competent individuals.

Symptoms Fever, headache, vomiting, blurring of vision, double vision, alteration of sensorium, seizure, limb weakness, cranial nerve paresis.

Signs neck stiffness, Kernig's sign (when trying to extend a knee of a lower limb flexed at hip and knee, elicits pain and discomfort at neck due stretching of the inflamed meninges), Brudzinski's sign, cranial nerve palsy, papilledema, limb paresis, etc.

Investigations CT/MRI brain with contrast-may show meningeal post contrast enhancement, tuberculomas, abscess, infarcts, hydrocephalus (communicating or noncommunicating depending on whether the CSF flow is affected at arachnoid granulations or between the ventricles of the brain) (Image 8.1). CSF evaluation may

Image 8.1 MRI (T1 GAD enhanced) brain image of patient with TBM, showing leptomeningeal (pia and arachnoid mater) enhancement. And communicating hydrocephalus

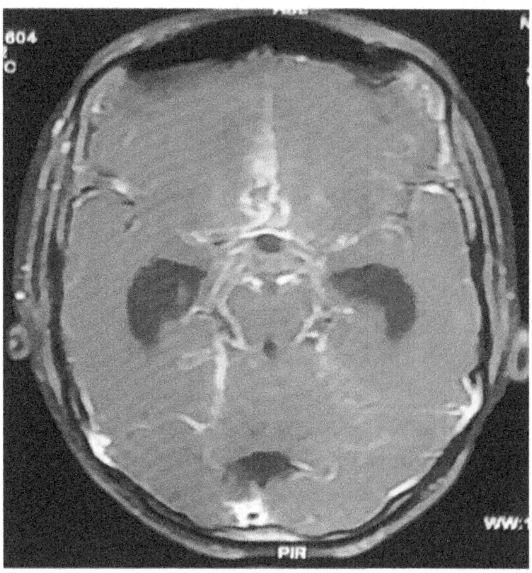

show elevated cell count (mostly lymphocytes), elevated protein, and low sugar levels. AFB staining may show acid fast bacteria, CSF CBNAAT may be positive.

Treatment AKT (anti Koch's therapy) (first line: Isoniazid, Rifampicin, Ethambutol, pyrazinamide, Streptomycin), second line drugs used in case of drug resistance or drug-related side effects. Adjuvant use of steroids. Treatment of tuberculosis in HIV needs special caution as simultaneous administration of AKT and ART may lead to paradoxical deterioration in clinical status due to hyperactivation of the immune response leading to IRIS (Immune reconstitution inflammatory response).

8.1.2.2 CNS Toxoplasma Infection

Toxoplasma is obligate intracellular parasite. Infection is acquired mainly via contaminated water and food or vertical transmission. Tachyzoites are the rapidly replicating and bradyzoites are slow replicating forms. There is constant transition between these forms. Few of these intermediates may evade the host immune response leading a possible persistent infection. Brain is the main organ where bradyzoites encyst, other being skeletal muscle, cardiac muscles, spinal cord, and retina. Vast necrosis of midline and periventricular structures, i.e., Corpus callosum, septum pellucidum, fornix, and basal ganglia were observed during autopsy of the affected patients. (Horowitz et al. 1983) Seroprevalence for toxoplasma among immunocompetent and immunocompromised population is reported to be between 15–67% in different studies. Geographical differences are hypothesized to be the cause for these differences (Basavaraju 2016). Immunocompromised hosts are highly susceptible to toxoplasma mainly through reactivation of the latent infection. Basal ganglial contrast (central or ring) enhancing lesions in such patients should raise suspicion for toxoplasma and guide further evaluation.

Symptoms May be associated with fever, headache, altered sensorium, psychosis, dementia, seizure, focal neurological deficit as limb weakness, sensory deficit, etc.: other extracranial toxoplasma manifestations like chorioretinitis (diminution of vision, blurring of vision, scotomas, eye pain, etc.), pneumonitis like illness may accompany.

Signs Focal neurological deficits may be elicited. Chorea if present is considered as pathognomic.

Diagnosis In suspected cases (mostly immunocompromised individuals with CD4 count <100/μl) if there is positive toxoplasma serology (IgM and IgG ELISA, CFT), symptoms and signs suggesting meningoencehalic involvement (neck stiffness, Kernig's sign, or focal deficit and suggestive neuroimaging; diagnosis of CNS toxoplasmosis is contemplated. Positive response (takes approximately 2–4 weeks) to the treatment may be considered as diagnostic evidence. Biopsy of the lesion provides definitive diagnosis, but may not be feasible in each suspected case.

Radiological diagnosis can be done with CT brain and MRI. Out of these, contrast enhanced MRI offer better visualization and sensitivity of the early lesions. Hypodensities, ring or central enhancing lesions with perilesional edema mainly affecting basal ganglia, cortical, and cerebellar regions are noted. 'Eccentric target sign' where nodule is noted at the rim of the enhancing ring on contrast enhanced T1 weighted imaging is highly specific but unusual radiological sign of toxoplasma. The 'concentric ring sign' with alternating hypo and hyperintense rings may be seen on T2 weighted imaging.

Treatment Sulfadiazine, pyrimethamine, folinic acid, and clindamycin are the first choices for the treatment. Other drugs which may be used are atovaquone, azithromycin, clarithromycin, and dapsone are used. Chemoprophylaxis with co-trimoxazole or dapsone and pyrimethamine is advisable for immunocompromised patients.

8.1.2.3 CNS Fungal Infections

The fungi capable of causing CNS pathologies fall under three major morphological categories like yeasts, dimorphic fungi, and molds; these morphological characteristics are important determinants of their clinical manifestations. Small Yeasts (e.g., Cryptococcus, histoplasmosis, sporotrichosis, blastomycosis, coccidiomycosis) cause leptomeningeal involvement. Whereas large yeasts like candida involve brain parenchyma predominantly manifesting as abscesses and granulomas. Those with large branched hyphae like Aspergillus, Zygomycosis invade blood vessels causing strokes or involve orbit, paranasal sinuses, and skull bones (Nathan et al. 2021).

8.1.2.3.1 Cryptococcal Meningitis (CM)

CM is accountable for approximately 15% of HIV AIDS-related deaths globally (Rajasingham et al. 2017). It mostly manifests as subacute meningoencephalitis, occurs due to inhalational infection by fungi of group basidiomycetes; encapsulated yeasts found in soil contaminated by bird (mainly pigeon) droppings (most common pathogenic species being *C. neoformans*, *C. gattii*). And is prevalent in population where HIV is rampant. It is also seen in patients with cell-mediated immunodeficiency due to various other causes like cytotoxic drugs, in organ transplant recipients, and in patients with hypogammaglobulinemia, liver disease, Cushing's syndrome, even immunocompetent individuals may suffer from CM.

Cryptococcus is the most common cause of fungal meningitis in humans both immunocompromised and immunocompetent (*C. gattii* mostly affects immunocompetent individuals) (Pyrgos et al. 2013). From lung and lymph nodes, these fungi spread to various organs through hematogenous and lymphatic route. Access to the subarachnoid space is achieved via transcellular migration across ependymal cells. As there is lack of complement mediated and soluble ant-cryptococcal factors in brain, there is predilection for CNS affection by cryptococci. As CM carries very high mortality in untreated cases, it should be considered as differential in all cases of lymphocytic meningitis (Williamson et al. 2016).

Symptoms Headache (most common and prominent feature), fever (may not be a feature in chronic course), alteration of sensorium, seizures, focal neurological deficits (cranial nerve palsies), visual disturbances.

Signs Neck stiffness, Kernig's sign may be present. papilledema, focal deficits.

Diagnosis CSF evaluation (lymphocytic pleocytosis, elevated protein and decreased glucose levels), India ink test of CSF have good sensitivity, CSF culture, CSF Ag LFA (lateral flow assay), detection of cryptococcal capsular antigen with latex agglutination (CALAS).

MRI brain may show dilated Virchow Robbin spaces, pseudocysts, contrast enhancing cryptococcomas (mainly involving basal ganglia) (though enhancing lesions are less common in immunocompromised individuals), less frequently leptomeningeal enhancement may be noted. Vascular infarcts, may be seen, etc.

Treatment Amphotericin B, flucytosine for induction phase, fluconazole, and flucytosine for consolidation phase and fluconazole for maintenance phase.

8.1.2.4 Opportunistic Viral Infections

8.1.2.4.1 Cytomegalovirus

Cytomegalovirus is the most important opportunistic virus causing neurologic complications in HIV patients. Other important ones are Epstein Barr virus (mainly its role in CNS lymphoma in HIV patients), varicella zoster virus, HHV-6, and JC virus (causing Progressive Multifocal Leukoencephalopathy). CMV is a human herpes virus, infects humans through close contacts, sexual intercourse, perinatally, through blood products and organ transplantations. Most CNS complications are caused by through the reactivation of prior infection due to immunodeficiency. Median CD4 count in CMV encephalitis is <20 cells/microliter. Encephalitis, myeloradiculitis, and neuritis are main neurologic disease manifestations seen in CMV infection. Focal cell aggregation and or necrosis; ventriculoencephalitis, which is the most distinctive clinical reflection of CMV encephalitis, presents with focal or diffuse ependymal inflammation and periventricular necrosis. Aggressive CMV-dependent dementia is also a noted clinical presentation of CMV.

Symptoms CMV encephalitis—Confusion, lethargy and progressive dementia. Focal neurological deficits and cranial neuropathies are commonly seen as opposed to HAND where focality is not a feature. Internuclear ophthalmoplegia, ataxia, and vertigo are also frequently noticed.

Signs Retinitis, pneumonitis, esophagitis, colitis, adrenalitis, etc., other signs of systemic involvement are commonly found. Along with this alteration of sensorium and evidence of dementia on dementia assessment tools. Focal neurological deficits, cranial neuropathies and other peripheral neuropathies, and myeloradiculopathy might be noted.

Diagnosis CSF evaluation may suggest moderate rise in protein levels, polymorphic pleocytosis (more so in myeloradiculitis rather than ventriculomyelitis), positive Polymerase chain reaction for CMV is highly sensitive. Neuroimaging (MRI with contrast) may show cerebral atrophy, ventricular dilatation with periventricular contrast enhancement.

Treatment Antivirals like ganciclovir, foscarnet, and cidofovir.

8.1.2.4.2 Herpes Zoster Virus

Symptoms trigeminal nerve and multidermatomal radicular involvement is frequent with immunodeficiency. Radicular pain, dysesthesia, reddish papules, vesicles involving dermatomal distribution is seen.

Encephalitis with necrotizing vasculitis may present with altered sensorium, focal deficits and seizures, meninomyeloradiculitis, and retrobulbar optic neuritis are other reported manifestations.

Diagnosis CSF evaluation, CSF PCR. MRI Brain

Treatment Injectable acyclovir, steroids may be used.

8.1.2.4.3 PML (Progressive Multifocal Leukoencephalopathy)

It occurs due to infection of CNS oligodendrocytes by JC virus. It is seen in immunocompromised hosts like HIV infection, malignancies (especially lymphoid), patients on immunomodulator agents.

Signs and Symptoms Neurological symptoms develop slowly over weeks. Neuropsychological symptoms include affection of attention, concentration, cognitive and motor slowness (subcortical dementia), executive dysfunctions, visual disturbances, and motor deficits. Its course remains progressive, ultimately leading to death in months.

Diagnosis CSF evaluation, presence of JC virus in CSF by PCR, biopsy of brain.

MRI brain in early-stage show areas of patchy demyelination (non-contrast enhancing, asymmetric T1 hypo and T2 hyperintensities with no perilesional oedema or mass effect) in subcortical white matter (which progressively becomes more confluent and necrotic.

Treatment cART in HIV-affected patients. Steroids are used in PML IRIS patients with impending brain herniation (Weissert 2011).

8.1.2.5 Primary CNS Lymphoma (PCNSL)

Among the AIDS defining cancers (Kaposi sarcoma, PCNSL, NHL non-Hodgkins's lymphoma, Burkitt's lymphoma, and cervical cancer), PCNSL is frequent CNS involving cancer (Rubinstein et al. 2014). Approximately 15% of NHLs in HIV patients present as PCNSL. Mostly these are related to Epstein bar virus and thought to be due to ineffective immunoregulation. Supratentorial solitary or multiple mass lesions is a common presentation.

Symptoms and Signs Suggestive of raised intracranial tension (headache, vomiting, and blurring of vision), focal neurological deficits, seizures, alteration of sensorium, cranial nerve palsies.

Diagnosis Biopsy of the lesion with histopathological evaluation even though very sensitive diagnostic test. The yield drops vastly with the use of steroids. CSF cytology and flow cytometric immunophenotyping. CSF EBV DNA PCR (as HIV-associated PCNSL is consistently associated with EBV). FDG PET, SPECT.

MRI brain may show solitary or multiple hypo to isointense lesions on T1 weighted images. They are mostly located in periventricular regions with homogeneous contrast enhancement in immunocompetent patients whereas are located at cortical or subcortical regions with ring or heterogeneous less avid enhancement in immunocompromised patients. Periventricular location with subependymal and leptomeningeal spread is mostly observed.

Treatment Whole brain radiotherapy, high dose methotrexate-based chemotherapy.

8.2 Spinal Cord

Spinal cord injury mediated by HIV is mostly by indirect mechanisms like immune modulation, degeneration, and opportunistic infections and neoplasms. The pathological mechanism may involve necrosis, demyelination, or vasculitis. Acute transverse myelitis is seen mostly in early stages of seroconversion, whereas Vacuolar myelopathy and opportunistic infections are noted in late stages of illness.

Signs and Symptoms may present with difficulty in walking (weakness in lower limbs -spastic paraparesis, incoordination while walking-sensory ataxia, autonomic disturbances—bowel/bladder involvement etc.).

8.2.1 Diseases Affecting Spinal Cord in Early Stages of HIV

Primary HIV-Associated Transverse Myelitis Is noted in early stages of seroconversion. Mostly accompanied by constitutional features like fever, malaise, and backache, which gets followed by symptoms of acute transvers myelitis. Biopsy suggests specific multinucleated giant cells or microglial nodules. CSF may suggest lymphocytic pleocytosis. MRI may show T2 hyperintensities (dorsal spine is mostly involved). Steroid pulse therapy mostly shows clinical improvement.

Immune Mediated Transverse Myelitis Mimics CNS demyelinating disorder like multiple sclerosis and Neuromyelitis Optica spectrum disorder. With brain and spinal cord demyelinating lesions may also present with optic neuritis (painful loss of vision and painful eye movements). There is absence of characteristic multinucleated giant cells or microglial nodules and JC virus. Clinical picture suggestive of longitudinally extensive transverse myelitis may be seen. cART and steroids are used for the treatment.

HIV-Associated Motor Neuron Disease (MND) Presents with symptoms of both upper motor neuron (weakness, hyperreflexia, hypertonia) and lower motor neuron (weakness, atrophy, fasciculations) disorder. Even though exact pathology is unknown, histopathology of HIV-associated MND shows pathology other than ALS (amyotrophic lateral sclerosis). Those with HIV-associated MND present earlier and progress rapidly as compared to ALS. cART, intravenous IVIG show stabilization and clinical reversal.

8.2.2 Spinal Cord Involvement in Late and Uncontrolled HIV Stages

Vacuolar myelopathy (HIV-VM) is a very common entity seen due to direct viral pathology (Wuliji et al. 2019). Its prevalence is quite high up to 22–55% and does not carry good prognosis. There is formation of myelin sheath vacuoles and lipid-filled macrophage infiltration predominantly in lateral and posterior parts of the spinal cord with subsequent cord atrophy. This is mainly noted in lower thoracic spinal cord. Advanced disease may show complete demyelination, axonal degeneration, and astrocytic gliosis (Mongezi et al. 2021).

The viral particles were cultured from the spinal cord tissue, but immunohistochemical studies did not demonstrate them in the vacuoles.

HIV-VM is a diagnosis of exclusion. It has got subacute onset and slow progression. The possible causes of spinal cord involvement like infectious, inflammatory, demyelinating, compressive, and metabolic has to be considered and looked for through neuroimaging and CSF evaluation. There is no definitive treatment at present. c ART and symptomatic therapy is mostly provided (Robinson-Papp et al. 2019).

In humans, tropical spastic paraparesis and Japanese myelopathy are due to HTLV infection.

PCNSL is frequently associated with spinal cord involvement. The disease harbors poor prognosis.

CMV myeloradiculitis may present as acute transvers myelitis. When presenting with polyradiculitis may show LMN (lower motor neuron) type of weakness with bowel and bladder involvement. May mimic GBS (Guillain Barre syndrome).

Herpes simplex virus (HSV) sacral myeloradiculitis presents with perinium numbness with bowel bladder incontinence and progressive ascending paresis. And this too may mimic GBS. HSV may also cause severe necrotizing myelitis in HIV.

Varicella zoster may case radicular involvement. Retrograde spread may cause myelitis and encephalitis. It carries poorer prognosis.

Syphilis (Treponema Pallidum) and tuberculosis (Mycobacterium Tuberculosis) frequently cause extensive spinal cord and radicular involvement.

8.3 Peripheral Nerves

Peripheral neuropathy is one of the commonest complication seen in HIV patients. As the longevity among HIV patients is increasing with the advent of HAART, the prevalence of neuropathies is also increasing. Even though distal symmetric neuropathy is more frequent presentation, virtually all forms of peripheral neurological involvement can be seen in HIV. These include acute and chronic inflammatory demyelinating polyradiculopathies, mononeuropathies, mononeuropathy multiplex, cranial neuropathies, autonomic neuropathies, and ALS like motor neuropathies. And it is very difficult to distinguish between those due to direct viral injury and those due to antiviral drugs (Kaku and Simpson 2014). In pre-ART era, low CD4 count and viral load was associated with HIV DSP. But nowadays this correlation is not well established (Morgello 2004). Other factors like diabetes, hypertriglyceridemia, use of statins, older age, height, etc., play a more important role.

References

Basavaraju A (2016) Toxoplasmosis in HIV infection: an overview. Trop Parasitol 6(2):129. https://doi.org/10.4103/2229-5070.190817

Clifford DB, Ances BM (2013) HIV-associated neurocognitive disorder. Lancet Infect Dis 13(11): 976–986. https://doi.org/10.1016/s1473-3099(13)70269-x

Horowitz SB, Bentson JR, Frank Benson D, Davos I, Pressman BD, Gottlieb M (1983) CNS toxoplasmosis in acquired immunodeficiency syndrome. Arch Neurol 40(10):649–652. https://doi.org/10.1001/archneur.1983.04050090085015

Kaku M, Simpson DM (2014) HIV neuropathy. Curr Opin HIV AIDS 9(6):521–526. https://doi.org/10.1097/coh.0000000000000103

Mongezi T, Sibi J, Jerry G, Lourdes de Fátima I-V, Tozama D, Humberto FS (2021) Atypical HIV-vacuolar myelopathy: a case report. Eur J Med Res 26(1):13. https://doi.org/10.1186/s40001-021-00483-0

Morgello S (2004) HIV-associated distal sensory polyneuropathy in the era of highly active antiretroviral therapy. Arch Neurol 61(4):546. https://doi.org/10.1001/archneur.61.4.546

Nathan CL, Emmert BE, Nelson E, Berger JR (2021) CNS fungal infections: a review. J Neurol Sci 422:117325. https://doi.org/10.1016/j.jns.2021.117325

Pyrgos V, Seitz AE, Steiner CA, Prevots DR, Williamson PR (2013) Epidemiology of cryptococcal meningitis in the US: 1997–2009. PLoS ONE 8(2):e56269. https://doi.org/10.1371/journal.pone.0056269

Rajasingham R, Smith RM, Park BJ, Jarvis JN, Govender NP, Chiller TM, Denning DW, Loyse A, Boulware DR (2017) Global burden of disease of HIV-associated cryptococcal meningitis: an updated analysis. Lancet Infect Dis 17(8):873–881. https://doi.org/10.1016/s1473-3099(17)30243-8

Robinson-Papp J, Nmashie A, Pedowitz E, George MA, Sharma S, Murray J, Emma, Lawrence SJ, Machac J, Heiba S, Kim-Schulze S, Navis A, Roland BC, Morgello S (2019) The effect of pyridostigmine on small intestinal bacterial overgrowth (SIBO) and plasma inflammatory biomarkers in HIV-associated autonomic neuropathies. J Neurovirol 25(4):551–559. https://doi.org/10.1007/s13365-019-00756-9

Rubinstein PG, Aboulafia DM, Zloza A (2014) Malignancies in HIV/AIDS. AIDS 28(4):453–465. https://doi.org/10.1097/qad.0000000000000071

Weissert R (2011) Progressive multifocal leukoencephalopathy. J Neuroimmunol 231(1-2):73–77. https://doi.org/10.1016/j.jneuroim.2010.09.021

Wilkinson KA, Torok ME, Schwander S, Meintjes G (2012) HIV-associated tuberculosis 2012. Clin Dev Immunol 2012(2012):950964. https://doi.org/10.1155/2012/950964

Williamson PR, Jarvis JN, Panackal AA, Fisher MC, Molloy SF, Loyse A, Harrison TS (2016) Cryptococcal meningitis: epidemiology, immunology, diagnosis and therapy. Nat Rev Neurol 13(1):13–24. https://doi.org/10.1038/nrneurol.2016.167

Wuliji N, Mandell MJ, Lunt JM, Merando A (2019) HIV-associated vacuolar myelopathy and HIV-associated dementia as the initial manifestation of HIV/AIDS. Case Rep Inf Dis 2019(2019):3842425. https://doi.org/10.1155/2019/3842425

Neurological Complications of Measles and Mumps

9

Abdullah M. Firoze Ahmed and Rahma Mohamed Firoze

Abstract

Measles and mumps are two vaccine-preventable viral diseases that can have serious neurological complications. Four types of encephalitis are associated with measles infection—primary measles encephalitis, postinfectious encephalomyelitis, subacute sclerosing panencephalitis (SSPE), and measles inclusion body encephalitis (MIBE). Primary measles encephalitis usually occurs during the exanthem phase of measles. Postinfectious encephalomyelitis is an autoimmune disorder with acute disseminated encephalomyelitis (ADEM) occurring soon after the measles rash and associated with high mortality. SSPE is a rare, rapidly progressive, and fatal complication occurring usually 6–10 years after measles infection. MIBE is similar to SSPE but occurs in immunocompromised hosts usually within 9 months of measles infection. Meanwhile, the neurological complications of mumps are usually benign. Aseptic meningitis is the most common extrasalivary manifestation of mumps with usually complete recovery within a few days. Mumps-associated encephalitis today is rare. Sensorineural hearing loss is possible with both measles and mumps. The most effective prevention strategy for neurological complications of measles and mumps is vaccination. This is evident by the decline in these neurological complications in regions with high vaccination rates. Low rates of vaccinations in certain regions and the rise of vaccine hesitancy continue to make the eradication of measles and mumps challenging.

A. M. Firoze Ahmed (✉)
Rochester General Hospital, Rochester, NY, USA
e-mail: abdullah.firozeahmed@rochesterregional.org

R. M. Firoze
International Medical Center Hospital, Jeddah, Saudi Arabia

H. Sami et al. (eds.), *Viral and Fungal Infections of the Central Nervous System: A Microbiological Perspective*, https://doi.org/10.1007/978-981-99-6445-1_9

Keywords

Measles · Mumps · Neurological complications · Postinfectious encephalitis · Subacute sclerosing panencephalitis

9.1 Measles and Its Neurological Complications

9.1.1 Introduction

Measles, also known as rubeola, is a highly contagious, yet preventable viral disease. Before the introduction of anti-measles vaccines, measles would result in 30 million cases and 2 million deaths globally (World Health Organization 2017). By the age of 15 years, more than 95% of children would have had measles infection.

Measles was first documented by the Persian physician Abu Bakr Al-Razi as a distinct disease from smallpox around the ninth century (World Health Organization 2023a). Then in 1757, the Scottish physician Francis Home proved its presence in blood. It would not be until 1963 that a vaccine would be publicly introduced for the prevention of measles and 1971 that Dr. Maurice Hilleman would combine the individual measles, mumps, and rubella vaccines to create the live attenuated "MMR" vaccine.

Between 2000 and 2021, vaccination prevented 56 million deaths globally from the measles virus (Minta et al. 2022). Of note, the COVID-19 pandemic briefly hampered vaccination efforts globally with vaccination rates for one dose of measles vaccine dropping from 86% in 2019 to 83% in 2020 and 81% in 2021 (Minta et al. 2022).

The Centers for Disease Control and Prevention (CDC) in the United States advises that every child receive the MMR vaccine with the first dose between 12 and 15 months; and then, a booster between 4 and 6 years of age (McLean et al. 2013). In the United Kingdom, the first dose of MMR is recommended at 12 months and the second at 3 years and 4 months (NHS 2020). The World Health Organization (WHO) in its 2017 position paper recommends the first dose of anti-measles vaccine at 9 months in areas with high cases of measles and infant mortality secondary to it (World Health Organization 2017). Most countries have their own measles vaccination schedule recommendations based on their public health infrastructure and local incidence of measles.

Despite the time-proven efficacy and benefit of the measles vaccine, the number of deaths and complications from this preventable disease remains high. This is due to variable factors from lack of access to vaccines to a rise in vaccine hesitancy. In 2018 alone, the WHO reported more than 140,000 deaths from the measles virus (World Health Organization 2023b).

9.1.2 Virology

The measles virus (MeV), which belongs to the Paramyxoviridae family and *Morbillivirus* genus, is responsible for measles disease. It is a single-stranded, negative-sense RNA virus (Ryan 2022). MeV comprises about 16,000 nucleotides encoding six structural proteins—nucleoprotein (N), phosphoprotein (P), large protein (L), matrix protein (M), hemagglutinin protein (H), and fusion protein (F) (Griffin 2014). There are also two nonstructural proteins—V and C. The genome is encapsulated helically by the N proteins and packaged with P and L proteins necessary for the RNA-dependent RNA polymerase (RdRp) complex (Ryan 2022; Watanabe et al. 2019). This core is then enveloped by a host-derived lipid membrane with the M protein lying on the inner side. H and F proteins can be found on the surface of the lipid envelope forming glycoprotein spikes.

Humans are the only known hosts for the wild-type measles virus. Due to mainly genetic variability in the H and N genes, up to 24 genotypes of the wild-type measles virus have been identified (Beaty and Lee 2016). However, the limited variability in the surface glycoproteins (H and F) allows for only one known serotype with long-term immunity achieved from either primary infection or vaccination. This is why measles vaccines derived from the now-extinct wild-type genotype A continue to confer immunity against all genotypes of measles (Griffin 2014; Moss 2017). In public health, genotyping can help in outbreaks by tracking the infection, determining the country of origin, and distinguishing infection from wild-type measles virus versus vaccination (CDC 2022a).

9.1.3 Pathogenesis

Measles is a highly contagious viral disease as reflected by its high basic reproduction number—$R_0 = 12–18$, where *R* naught (R_0) reflects the average number of uninfected and unvaccinated people one person with measles can infect (Maldonado and Shetty 2018a). One person with measles can infect up to 90% of their close nonimmune contacts and are usually highly infectious four days prior to and 4 days after the measles rash appears (CDC 2020; World Health Organization 2023b). The measles virus (MeV) is transmitted through infected respiratory droplets and aerosolized particles which may remain contagious in the air or on surfaces for up to two hours (Gastañaduy and Goodson 2023).

The viral envelope surface proteins—hemagglutinin (H) and fusion (F)—play an important role in the pathogenesis of measles infection (Rota et al. 2016). The H protein allows for binding to the host cell via the signaling lymphocytic activation molecule (SLAM) receptor found mainly on immune cells. The F protein then facilitates the fusion of the viral lipid envelope with the host cell's plasma membrane. This allows for the viral genetic core and RdRp complex to enter the host cell's cytoplasm where replication occurs. The matrix (M) protein then helps with assembling the replicated virus and budding of new virions. The F protein plays a role in also allowing lateral cell-to-cell viral transmission as well as fusion between cells (Cherry and Lugo 2019). Meanwhile, the V and C proteins have been shown to play a role in evading the host's immune system (Rota et al. 2016).

When the measles virus first enters the respiratory tract, it appears that it is unable to directly infect the respiratory epithelial cells as they lack SLAM receptors. Instead, the virus is believed to first infect the $SLAM^+$ antigen-presenting cells (APCs) like dendritic cells and macrophages in the respiratory tract (Laksono et al. 2016). These APCs likely extend their arm-like projections between the tight epithelial cellular junctions coming in contact with the MeV (Noyce and Richardson 2012). Once infected via the SLAM receptor, these immune cells allow for MeV replication within the local lymph nodes. From there, the MeV disseminates throughout the lymphoid tissues in the body that have high concentrations of $SLAM^+$ immune cells (Griffin 2014; Rota et al. 2016). This results in viremia occurring in the first week of infection (during the incubation period) with the virus spreading throughout the body including lymphoid tissues, spleen, lungs, kidney, and skin (Griffin 2014). The infected $SLAM^+$ immune cells likely then infect the respiratory epithelial cells on the basolateral side via the cellular adhesion molecules—nectin 4—using them as a receptor for entry (Mühlebach et al. 2011; Noyce and Richardson 2012; Singh et al. 2016). The respiratory epithelial cells then transmit the virus laterally from cell to cell via the nectin 4 cellular adhesions. The matrix (M) protein of the MeV helps with the assembly and budding of the new virions apically from these infected respiratory cells allowing for transmission to other susceptible hosts. The fusion (F) protein facilitates the fusion of infected cells in the epithelium creating giant multinucleated cells that can be identified in urine and secretions from the eyes and nose (Griffin 2014).

9.1.3.1 Neuropathophysiology

How the measles virus enters the central nervous system (CNS) remains unclear. However, MeV RNA has been found in some brain tissue samples in those with neurological complications of measles (Griffin 2014). The viral RNA antigens isolated from brain tissue appear to show mutations when compared to the wild-type virus. It is believed that the MeV may cross the blood–brain barrier through infected immune cells via the Trojan horse method. Once in the CNS, it likely spreads laterally via cell-to-cell transmission. In some CNS complications, intranuclear and cytoplasmic inclusion bodies may also be seen in the neurons reflecting hyperfusogenic activity by the mutated MeV (Sato et al. 2018; Watanabe et al. 2019). Given that SLAM and nectin 4 are normally not expressed by neurons, it is likely that some other receptors or mechanisms may play a role in the pathogenesis of the CNS.

9.1.4 Clinical Manifestations

9.1.4.1 Classic Measles

Measles infection can be clinically split into three distinct periods—incubation, prodromal, and exanthem (Cherry and Lugo 2019).

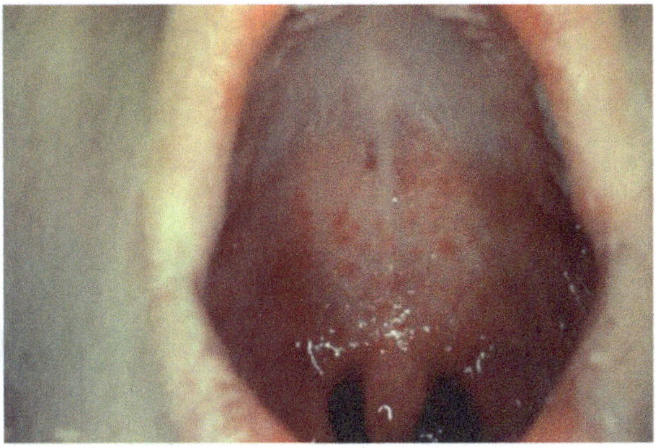

Fig. 9.1 Koplik spots can be seen on the mucosa of the soft palate and oropharynx in this patient with measles. Image source: Centers for Disease Control and Prevention—Public Health Image Library—ID#3187 (Eichenwald 1958)—copyright restrictions: none; this image is in the public domain and thus free of any copyright restriction

The incubation period starts with the measles virus entering the respiratory (or in some cases conjunctival) mucosa and leads to viremia. It lasts on average about 10 days but can be as long as three weeks (Gastañaduy et al. 2021; Riedel et al. 2019).

The short prodromal period then follows lasting about 2–4 days and is characterized by fever with a gradual increase in temperature of up to 39.5–40.5 ° C (Gastañaduy et al. 2021). This is accompanied by one or more of the classic three Cs—cough, coryza (rhinitis), and conjunctivitis.

In about 60% of infected individuals, Koplik spots (Fig. 9.1) on buccal and sometimes vaginal mucosa may be seen 1–2 days prior to the onset of the measles rash (Cherry and Lugo 2019; Perry and Halsey 2004). Koplik spots (the enanthem of measles) are considered pathognomonic for measles and may allow for earlier diagnosis. They are characterized as blue-white spots on an erythematous mucosal base usually seen in the mouth opposite the lower molars.

The exanthem period begins about 14 days after the initial exposure to the measles virus and is characterized by the measles rash (Gastañaduy et al. 2021). The rash (Fig. 9.2) starts as an erythematous maculopapular eruption involving the forehead and behind the ears and spreads centrifugally along the trunk to the extremities (Perry and Halsey 2004; Riedel et al. 2019). Over a few days, the rash may become confluent in some places, especially the face (Cherry and Lugo 2019). The first few days of the rash are commonly associated with high fevers, pharyngitis, and lymphadenopathy. The measles rash lasts for about 5–7 days fading in the same cranial to caudal fashion as it had spread. Meanwhile, the cough may last for about 10 days more.

9.1.4.2 Modified Measles

Modified measles occurs in a subset of individuals that are partially immune to the measles virus and presents with milder symptoms (Cherry and Lugo 2019). The incubation period can be longer (up to 21 days) with a shortened prodromal period (Perry and Halsey 2004; Riedel et al. 2019). Koplik spots may not occur, and the respiratory symptoms and rash may be milder. Modified measles can occur in infants

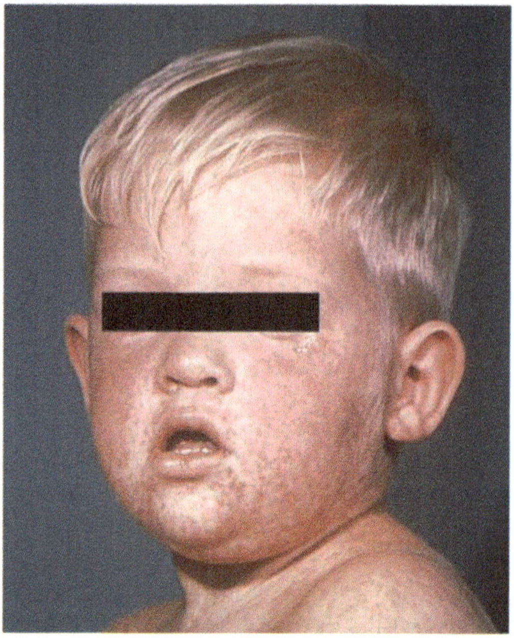

Fig. 9.2 A boy diagnosed with measles exhibiting the measles rash on day 3 of the exanthem period. Image source: Centers for Disease Control and Prevention—Public Health Image Library—(Measles Photos 2023)—copyright restrictions: none; this image is in the public domain and thus free of any copyright restriction

with maternal measles antibodies transferred transplacentally, individuals with prior immunoglobulin administration for measles postexposure prophylaxis, or more commonly from secondary vaccine failure (Cherry and Lugo 2019).

9.1.4.3 Atypical Measles

Atypical measles is another clinical presentation that is rare today. It was primarily seen in individuals that were vaccinated with the inactivated measles vaccine in the United States from 1963 to 1968 (Perry and Halsey 2004). Clinical presentation was with higher fevers, atypical rash spreading from extremities to the trunk, and pulmonary involvement from hilar lymphadenopathy to pneumonia (Cherry and Lugo 2019).

9.1.5 Diagnosis

Any person exhibiting a fever, rash, and one or more of the three C symptoms—cough, coryza, or conjunctivitis—should be suspected of measles (World Health Organization 2022). Laboratory diagnosis is needed for confirmation of a measles case to help distinguish it from other diseases with rashes such as rubella or parvovirus B19 (World Health Organization 2017). Reverse transcriptase polymerase chain reaction (RT-PCR) analysis for MeV from nasopharyngeal, oral, or urine specimens, as well as the detection of MeV-specific immunoglobulin M (IgM) antibodies in serum samples, are typically used for laboratory-based confirmatory testing (World Health Organization 2017). False negatives with IgM testing may occur in those with secondary vaccine failure (Cherry and Lugo 2019). A significant rise in anti-measles immunoglobulin G (IgG) titers between the acute and convalescent phase may be also used for establishing a laboratory diagnosis (Gastañaduy and Goodson 2023). Isolation of the measles virus in cell culture is rarely done for the purpose of establishing measles infection.

The CDC advises healthcare professionals to collect two samples when measles is suspected: serum to detect anti-measles IgM antibodies and throat or nasopharyngeal swabs for RT-PCR analysis (CDC 2022b). A urine sample may also be collected to detect MeV RNA. As IgM antibodies tend to peak within 1–3 weeks from the appearance of the measles rash, in about 25% of individuals they may not be detectable if checked within 3 days of rash onset (Moss 2017). In this case, if RT-PCR testing was either not done or negative, repeating IgM antibody testing within ten days is recommended by the CDC.

9.1.6 Management

Supportive care is the primary approach to treating measles infection. It may include rehydration and nutritional support as well as targeting specific symptoms such as antipyretics for fever and antitussives for cough. Antibiotics may be indicated in those that develop superimposed bacterial infections like pneumonia (Moss et al. 2009). Given that measles is extremely contagious, those infected should isolate for at least four days after rash onset (CDC 2022c).

The WHO recommends giving oral vitamin A to acutely infected children with measles, as vitamin A deficiency has been linked to recovery delays and higher post-measles complication risks (World Health Organization 2017). Meanwhile, the CDC recommends vitamin A only for hospitalized cases of severe measles in children (CDC 2022c).

9.1.7 Neurological Complications

Most people infected with measles usually experience complete recovery within 10–14 days of rash onset. Up to 30% may experience complications from measles infection including otitis media, pneumonia, diarrhea, laryngitis, myocarditis, pericarditis, and encephalitis (Cherry and Lugo 2019; Gastañaduy et al. 2021). Encephalitis from measles is one of the most feared complications as it can sometimes be fatal. Ever since the discovery and implementation of measles vaccination, the incidence of these neurological complications (Table 9.1) has significantly dropped globally. There is no specific therapy available for measles-associated encephalitis, making prevention through vaccination key.

Measles-associated encephalitis can be classified into four diseases:

Table 9.1 Neurological complications of measles

	Incidence (per cases of measles)	Possible etiology	Onset of encephalitis	Typical CSF findings	T2-weighted MRI findings
Primary measles encephalitis	1:1000	Direct CNS invasion by MeV	Exanthem period of measles	Lymphocytic pleocytosis, elevated protein, and normal glucose	Scattered bilateral hyperintensities in brain
Postinfectious encephalomyelitis	1:1000	Autoimmune; demyelination; molecular mimicry	Within 10 days to weeks after measles rash	Normal or with lymphocytic pleocytosis, elevated protein, and normal glucose	Multiple, bilateral, asymmetric hyperintensities in deep white matter in brain and spinal cord (can be seen in gray matter too)
Subacute sclerosing panencephalitis	4–11: 100,000	Direct CNS invasion by defective MeV	6–10 years after primary measles infection	Normal CSF analysis, elevated anti-MeV IgG titers	Normal to multifocal cortical lesions to diffuse cerebral atrophy
Measles inclusion body encephalitis	Unknown	Direct CNS invasion by defective MeV, immunocompromised host	Within 9 months of primary measles infection or measles vaccination	Normal CSF analysis, initially no anti-MeV antibodies detected	Normal to multifocal cortical lesions to diffuse cerebral atrophy

CNS central nervous system, *CSF* cerebrospinal fluid, *MeV* measles virus, *MRI* magnetic resonance imaging

1. Primary measles encephalitis.
2. Postinfectious encephalomyelitis.
3. Subacute sclerosing panencephalitis.
4. Measles inclusion body encephalitis.

9.1.7.1 Primary Measles Encephalitis

9.1.7.1.1 Epidemiology

Primary measles encephalitis (PME) is a form of encephalitis that usually occurs during the exanthem (rash) period of the measles infection. The incidence of PME is predominantly based on older literature putting it to be about 1 per 1000 cases of measles (La Bocetta and Tornay 1964). It should be noted that in some reports, primary measles encephalitis and postinfectious encephalomyelitis have been grouped as one entity.

9.1.7.1.2 Clinical Presentation

The onset of primary measles encephalitis is usually within a few days of the rash developing during the exanthem period (Cherry and Lugo 2019). Presenting symptoms may include fever, altered mental status, convulsions, irritability, focal neurological symptoms, and coma (La Bocetta and Tornay 1964). Infrequently, primary measles encephalitis may present without a rash in immunocompetent hosts as Zeng et al. reported 12 such cases in the literature (Zeng et al. 2016).

9.1.7.1.3 Diagnosis

Clinical diagnosis can be challenging and usually dependent on establishing concurrent measles infection and encephalitis. The differential diagnosis for encephalitis is broad and may include viral, bacterial, fungal, inflammatory, toxin, or tumorous etiologies.

Encephalitis can be clinically suspected using the criteria outlined by the Consensus Statement of the International Encephalitis Consortium. It requires altered mental status and the onset of at least 2 or 3 of the following—fever, convulsions, focal neurological abnormalities, pleocytosis in the cerebrospinal fluid (CSF), abnormalities on neuroimaging, or abnormal electroencephalogram (EEG) changes (Venkatesan et al. 2013). Lumbar puncture, EEG, and neuroimaging (preferably magnetic resonance imaging) should be performed.

In primary measles encephalitis, the CSF analysis usually reveals lymphocytic pleocytosis, slightly elevated protein, and normal glucose levels (Cherry and Lugo 2019; La Bocetta and Tornay 1964). MeV may also be detected in the CSF in some cases (Zeng et al. 2016). EEG may show nonspecific changes such as generalized slowing (Gibbs et al. 1959). Neuroimaging with magnetic resonance imaging (MRI) may reveal nonspecific scattered bilateral hyperintensities on T2-weighted imaging (Carmo et al. 2019; Lee et al. 2003).

9.1.7.1.4 Treatment

Treatment for primary measles encephalitis is largely supportive with symptomatic management such as antiepileptics for seizures and osmotically active agents for severe cerebral edema (Buchanan and Bonthius 2012). Ribavirin has been used in some cases with mixed results.

9.1.7.1.5 Prognosis

Death can occur in about 15% of cases and with 25% having permanent neurological sequelae like deafness or intellectual disability (Buchanan and Bonthius 2012; Cherry and Lugo 2019; La Bocetta and Tornay 1964).

9.1.7.1.6 Neuropathophysiology and Neuropathology

Primary measles encephalitis is believed to possibly be due to direct CNS infection by the MeV. The underlying neuropathophysiology remains unclear and controversial. Initially, it was believed that these cases of primary measles encephalitis may actually have been secondary to the autoimmune demyelinating process of acute disseminated encephalomyelitis (ADEM). However, studies have reported viral antigen being detected in brain tissue suggesting direct invasion of the brain by the measles virus (Cherry and Lugo 2019). Neuropathology findings may include significant perivascular infiltration, edema, hemorrhage, demyelination, and intranuclear and intracytoplasmic inclusion bodies (Adams et al. 1966).

9.1.7.2 Postinfectious Encephalomyelitis

9.1.7.2.1 Epidemiology

Acute disseminated encephalomyelitis, often referred to as "ADEM," is a poorly understood neurological autoimmune disorder that is characterized by demyelination of the central nervous system. It typically occurs following a systemic viral illness, hence sometimes referred to as "postinfectious" encephalomyelitis. The most commonly implicated viral infections include measles, mumps, rubella, varicella, and influenza (Noorbakhsh et al. 2008). Other implicated infections include Epstein-Barr virus, herpes simplex virus, human immunodeficiency virus, group A beta-hemolytic streptococcus, *Legionella pneumophila*, *Salmonella typhi*, *Mycoplasma pneumoniae*, and *Rickettsia rickettsii* to name a few (Noorbakhsh et al. 2008). The term acute postinfectious measles encephalomyelitis is also sometimes used when referring to ADEM associated with measles.

About 75% of children with ADEM have a preceding (usually upper respiratory tract) infection, and about a third may have no identified cause (Cole et al. 2019; Menge et al. 2005; Noorbakhsh et al. 2008; Tenembaum et al. 2002). ADEM is a rare autoimmune disorder that is seen mostly among children and has an incidence of about 0.2–0.8 per 100,000 children (Cole et al. 2019; Leake et al. 2004; Noorbakhsh et al. 2008). The median age of onset is around 5 years, and a slight male predominance has been reported in some studies (Cole et al. 2019; Johnson et al. 1984; Tenembaum et al. 2002). The incidence of ADEM following measles infection is around 1 per 1000 cases and based on older reports (Griffin 2014).

9.1.7.2.2 Clinical Presentation

In individuals with measles infection, ADEM usually presents within 10 days after the onset of the exanthem phase, but in some cases may occur weeks after (Perry and Halsey 2004). It often begins with fever starting again, headaches, and malaise which after a few days are followed by neurological symptoms. These may include encephalopathy from altered mental status to coma, cranial nerve deficits, acute hemiparesis, cerebellar ataxia, extrapyramidal symptoms, seizures, and other motor and sensory deficits (Johnson et al. 1984; Noorbakhsh et al. 2008; Tenembaum et al. 2002). Spinal column involvement in the form of acute transverse myelitis may also occur resulting in loss of bladder and bowel control (Buchanan and Bonthius 2012).

9.1.7.2.3 Diagnosis

ADEM is a diagnosis of exclusion that is usually made based on clinical history, CSF findings, and neuroimaging after other diagnoses have been excluded. Differential diagnosis for ADEM is quite broad, but the help of neuroimaging with MRI can help further distinguish diagnoses like multiple sclerosis, viral or bacterial encephalitis, viral or bacterial meningitis, vasculitis like primary angiitis of the central nervous system (PACNS), toxic leukoencephalopathies, or tumors like astrocytoma (Hemingway 2020; Pohl et al. 2016).

The CSF analysis may appear normal or with slight pleocytosis, mildly increased protein concentration, and normal glucose levels (Griffin 2014; Hemingway 2020). CSF testing should also be done to rule out infectious etiologies. EEG is usually abnormal with nonspecific findings of generalized slowing (Cole et al. 2019).

Neuroimaging is essential for establishing a clinical diagnosis of ADEM with magnetic resonance imaging (MRI) being the ideal modality as computed tomography (CT) scans initially may be unremarkable. Fluid-attenuated inversion recovery (FLAIR) and T2-weighted imaging often reveal demyelination with multiple, bilateral, large, asymmetric, and hyperintense lesions in the deep white matter and subcortical regions of the CNS (Noorbakhsh et al. 2008; Tenembaum et al. 2002). Gray matter lesions can be observed in the basal ganglia and in the thalamus. The spinal column may also be involved with clinical transverse myelitis symptoms (Pohl et al. 2016).

While there are no defined criteria for identifying ADEM in adults, the International Pediatric Multiple Sclerosis Study Group (IPMSSG) has suggested a diagnostic criteria for diagnosing ADEM in the pediatric population. It includes children presenting with unexplained noninfectious encephalopathy and multifocal clinical CNS involvement secondary to a presumed inflammatory or demyelinating event that is supported by MRI findings (Krupp et al. 2007). While ADEM is usually monophasic, in some cases it may be recurrent or multiphasic; these have also been defined by the IPMSSG. Brain biopsy may be done in some cases when a clinical diagnosis is not able to be established.

9.1.7.2.4 Treatment

Given that ADEM is believed to be essentially an autoimmune disorder, the general consensus is to treat with immune-modulating therapies. Treatment regimens are largely based on observational studies. The initial choice of treatment is usually high-dose intravenous glucocorticoids followed by an oral glucocorticoid taper over 4–6 weeks (Hemingway 2020; Pohl et al. 2016). Intravenous immunoglobulin and in refractory cases—plasma exchange—may also be considered.

9.1.7.2.5 Prognosis

Mortality associated with acute postinfectious measles encephalomyelitis (ADEM occurring following a measles infection) is high and has been reported to be between 10 and 40% in prior studies (Griffin 2014; Johnson et al. 1984; Noorbakhsh et al. 2008). Morbidity is also high in these individuals with up to a third having residual neurological deficits like cognitive impairment, hemiparesis, behavioral problems, and epilepsy (Cherry and Lugo 2019; Noorbakhsh et al. 2008).

This is in contrast to nonmeasles-associated ADEM where up to 75% may have a complete recovery (Noorbakhsh et al. 2008). In one study of 84 pediatric patients with ADEM but without known primary measles infection, 89% had complete neurological recovery or mild neurological impairment without disability (Tenembaum et al. 2002).

9.1.7.2.6 Neuropathophysiology and Neuropathology

The exact underlying mechanism of acute postinfectious measles encephalomyelitis remains unclear. Some hypothesize that it is a result of molecular mimicry between the measles virus itself and the myelin proteins leading to this exaggerated immune response rather than direct viral CNS invasion (Laksono et al. 2016; Noorbakhsh et al. 2008). Genetic predisposition has also been implicated in possibly playing a role in this likely autoimmune process (Griffin 2014).

Histopathological findings of white and gray matter include perivascular inflammation and demyelination with axons being relatively spared (Noorbakhsh et al. 2008). Viral proteins, RNA, and inclusion bodies are not seen suggesting again no direct infectious invasion of the central nervous system. Histopathological findings of ADEM have been found to be analogous to those seen in experimental autoimmune encephalomyelitis (EAE) animal models.

9.1.7.2.7 Postvaccination Encephalomyelitis

Rare incidences of ADEM have been reported after vaccination, in which case they can be referred to as "postvaccination" encephalomyelitis. Whether vaccines such as MMR truly cause ADEM remains controversial. The Vaccine Safety Datalink reviewed 64 million vaccine doses that were administered between 2007 and 2012 in the United States. Eight cases of ADEM and seven cases of transverse myelitis were found to have occurred within the 5–28 days of the defined "primary exposure window" (Baxter et al. 2016). Vaccines like MMR were not found to have an increased risk of ADEM in this study. The Tdap vaccine, however, did show an attributable risk of 0.4 cases of ADEM per million vaccine doses (Baxter et al. 2016).

It should be noted that this was based on only 2 cases of ADEM being reported in the post-Tdap vaccination group. Vaccinations were not found to be associated with an increased risk of transverse myelitis.

In some studies, the incidence of post-measles vaccination encephalitis has been reported to be around 1 case of encephalitis per 1,000,000 live measles vaccines. In these studies, however, encephalitis was not defined as ADEM specifically. Regardless, in comparison to the primary wild-type measles infection risk of 1 case of ADEM per 1000 measles infections, the advantages of measles immunization clearly outweigh any possible risks.

9.1.7.3 Subacute Sclerosing Panencephalitis (SSPE)

9.1.7.3.1 Epidemiology

Subacute sclerosing panencephalitis (SSPE) is a rare, rapidly progressing, late-onset, and fatal neurological complication of measles. SSPE onset prior to vaccination was around 6–10 years after primary measles virus infection but could occur even more than 10 years after (Griffin 2014). The median age of onset is around 9 years with a range of 5–15 years usually (Campbell et al. 2007; Dyken 1985; Maldonado 2008). There is a gender predilection with SSPE affecting boys more than girls with about a 3:1 ratio (Campbell et al. 2007; Jabbour et al. 1972). A higher risk of developing SSPE has also been associated with primary measles infection at an age of less than 2 years (Bellini et al. 2005; Jabbour et al. 1972).

Since the introduction of the measles vaccine, SSPE incidence has significantly decreased and is reportedly around 4–11 SSPE cases per 100,000 cases of measles infection (Campbell et al. 2007; Griffin 2014; Leung and Marlow 2022). The prevaccination incidence of SSPE is unclear. There is also great variability in the reported incidence of SSPE among different countries with it being quite rare in those with high rates of vaccination against measles.

In the United States between 1960 and 1974, the risk of developing SSPE following measles infection was determined to be on average 8.5 cases of SSPE per million cases of measles with the first measles vaccine being licensed in 1963 (Dyken et al. 1982). The incidence of SSPE in those under 20 years of age was about 0.61 per million population in 1970 and with increased vaccination having dropped to 0.35 in 1975 and 0.06 in 1980 (Dyken et al. 1982). By the late 1980s, only 1–2 cases of SSPE per year were reported, thanks to increased vaccination coverage (Bellini et al. 2005). Between 1989 and 1991, there was a resurgence of measles infection; Bellini et al. estimated the risk of developing SSPE to be much higher during this period at about 22 cases of SSPE per 100,000 cases of measles and suggested possible underreporting of true SSPE incidence (Bellini et al. 2005; Campbell et al. 2007).

In South India from 1983 through 1987, the annual incidence of developing SSPE was estimated to be much higher between 2.14 to 21 cases of SSPE per 1 million population (Saha et al. 1990). It should be noted that the Universal Immunization Program (UIP) began in 1985 in India with about 45% reported measles coverage by 1987 (Government of India 2005). This higher incidence of SSPE in India and other countries may be attributable to high measles infection rates and measles infection occurring at a younger age. For example, in India, 60% of measles infections occurred in those younger than two years of age (Cherry and Lugo 2019).

9.1.7.3.2 Clinical Presentation

Initial symptoms of SSPE can be subtle like progressively worsening school performance, inattentiveness, changes in personality, and behavioral problems that can be initially misdiagnosed as psychiatric disorders (Cherry and Lugo 2019; Perry and Halsey 2004). These initial symptoms are classified as stage I of IV of SSPE and associated with 0–30% disability lasting for less than 6 months and in some cases for years (Dyken 1985).

This is followed by stage II which includes characteristic myoclonic jerking movements of the head, trunk, and limbs occurring repetitively and eventually every 5–10 s (Dyken 1985). Seizures, loss of coordination, spasticity, extrapyramidal symptoms such as chorea-athetosis and cognitive decline including dementia occur (Jabbour et al. 1969). This stage lasts for about 3–12 months and is associated with 31–55% disability (Dyken 1985).

Neurological symptoms continue to deteriorate further in stage III with myoclonic jerks giving way to increased rigidity and decerebrate posturing (Dyken 1985; Jabbour et al. 1969). Stage III lasts around 3–18 months and is associated with 55–80% disability (Dyken 1985).

Stage IV is defined by mutism, loss of cerebral cortex function, flaccidity or spasticity, autonomic dysfunction, a vegetative state, and then death (Dyken 1985; Jabbour et al. 1969). SSPE is rapidly progressive and may result in death in months to a few years. Seizures and death can occur at any stage. Ocular involvement may occur in 50% of SSPE cases and includes papillitis, macular degenerative changes, chorioretinitis, optic neuritis, and blindness (Colpak et al. 2012).

9.1.7.3.3 Diagnosis

The modified Dyken's criteria is often used to make a clinical diagnosis of SSPE (Dyken 1985; Gutierrez et al. 2010). It requires that both major criteria and at least one minor criteria be met.

The two major criteria include the following:

1. Typical or atypical clinical presentation—not all presenting with SSPE may be noted or recognized to be going through the previously described 4 stages of SSPE. Typical presentations may include a clinical course that is acute and rapidly progressive, subacute and progressive, chronic and progressive, or chronic relapsing and remitting. Atypical presentations may include those with prolonged stage I so that the diagnosis is missed, seizures, or presentation at an unusual age like during infancy or adulthood (Dyken 1985). Most present with an acute or subacute progressive form of SSPE (Dyken 2001).
2. Elevated anti-MeV IgG antibody titers in CSF.

The four minor criteria include the following:

1. Typical EEG findings—discussed further below.
2. Cerebrospinal fluid globulin levels comprising more than 20 percent of total CSF protein—this includes gamma globulins with or without oligoclonal bands.
3. Brain biopsy—demonstrating histopathological findings consistent with SSPE.
4. Molecular diagnostic testing—detection of wild-type measles virus with mutations.

Classic EEG findings of SSPE are usually seen during the myoclonic stage of SSPE—stage II. These are characterized as bilaterally synchronous periodic bursts of two to four high-amplitude slow waves occurring every 2–20 s (Ekmekci et al. 2005; Markand and Panszi 1975). However, other EEG findings may also be seen ranging from normal in stage I to low amplitude and disorganized in stage IV (Dyken 1985). Many different types of nonclassical or atypical EEG findings such as prolonged discharges, periodic bursts that are asynchronous, or lateralized periodic bursts have been reported in the literature making EEG alone insufficient to rule out SSPE (Ekmekci et al. 2005; Markand and Panszi 1975).

On neuroimaging, CT and MRI may initially be normal in SSPE at the time of presentation. However, with disease progression, T2-weighted images on MRI may reveal hyperintensities usually involving the periventricular and subcortical white matter and diffuse cerebral atrophy (Anlar et al. 1996). MRI findings may or may not correlate with the clinical neurological symptoms. Differential diagnosis can be broad and include psychiatric disorders, seizure disorders, viral encephalitis, autoimmune encephalitis, multiple sclerosis, brain tumors, or progressive rubella panencephalitis (Garg 2008).

9.1.7.3.4 Treatment

Treatment for subacute sclerosing panencephalitis is largely supportive such as the use of antiepileptic drugs for managing seizures or antispasmodic drugs for spasms. Some medications used for management based on mostly small clinical studies or case reports with varying degrees of effect include immunomodulator drugs (isoprinosine, interferon alpha, interferon beta, amantadine, intravenous immunoglobulin) and antiviral drugs (ribavirin, favipiravir, and remdesivir) (Samia et al. 2022).

9.1.7.3.5 Prognosis

SSPE is a fatal disease that usually results in death within 1–3 years of onset (Buchanan and Bonthius 2012). The only preventative measure is vaccination.

9.1.7.3.6 Neuropathophysiology and Neuropathology

The exact mechanism by which the measles virus leads to subacute sclerosing panencephalitis remains unclear. MeV RNA detected from brain tissue of SSPE cases appears to be "defective" or "mutated" compared to the wild-type measles virus (Maldonado and Shetty 2018a).

It is believed that the wild-type measles virus, through an unknown mechanism (perhaps via the Trojan horse method) enters the central nervous system during the initial primary measles infection. An immune response is then mounted by the host but does not completely eliminate the virus from the CNS (Griffin 2014). The measles virus is able to replicate within the neurons and likely spreads by the use of unknown receptors or mechanisms.

The M protein which is responsible for budding and viral shedding appears to be highly mutated in SSPE. This suggests that the MeV does not rely on budding as a mechanism of CNS infection and persistence. MeV in cell culture with mutations in the F protein tail has been found to be hyperfusogenic (Rima and Paul Duprex 2005; Watanabe et al. 2019). This may enable the virus to propagate from cell to cell in the CNS with viral persistence for years until the onset of SSPE. It is unclear why a latent period exists between the initial exposure to the measles virus and SSPE onset. Also, what triggers the onset of SSPE is not known.

Histopathological findings of tissue samples from brain biopsy or postmortem analysis may reveal perivascular inflammation with infiltration of lymphocytes and plasma cells, astrocytosis, presence of intranuclear and intracytoplasmic MeV inclusion bodies in glial and neuronal cells, and diffuse cortical atrophy (Bellini et al. 2005; Jabbour et al. 1969; Ryan 2022). MeV antigens can often be identified from brain tissue samples demonstrating that SSPE is a complication of measles involving a defective measles virus strain that has managed to evade the host's immune responses.

9.1.7.3.7 Immunization and SSPE

There has been some controversy regarding whether the live attenuated measles vaccine could directly cause SSPE. This is because there have been some instances of SSPE being diagnosed in those without a history of prior measles infection but a history of measles vaccination. The current evidence does not support a causation link between SSPE and the measles vaccine (World Health Organization 2005). It is possible that these cases attributed to postvaccination SSPE are actually related to wild-type primary measles infection. Bellini et al. analyzed measles virus RNA from brain tissue samples of 11 patients with SSPE that had been referred to the US Centers for Disease Control and Prevention (Bellini et al. 2005). All of these samples revealed genotypes consistent with wild-type measles virus including the six cases with a history of measles vaccine but no prior history of measles infection or rash reported. Campbell et al. reported 23 cases of SSPE from the literature in vaccinated individuals all revealing nonvaccine, wild-type genotype virus (Campbell et al. 2007). If SSPE is caused by the measles vaccine the isolated virus should be the vaccine genotype and not the wild type. This suggests that reports of postvaccination SSPE may actually be from exposure to wild-type measles virus with either the diagnosis being missed or perhaps having had subclinical symptoms. The benefits of vaccination outweigh any potential risk of SSPE and have made it a rare complication in regions with high vaccination rates.

9.1.7.4 Measles Inclusion Body Encephalitis

9.1.7.4.1 Epidemiology

Measles inclusion body encephalitis (MIBE) is another rare, rapidly progressive, and fatal complication of measles. It occurs in immunocompromised individuals with impaired cell-mediated immunity like acute lymphoblastic leukemia, other malignancies, human immunodeficiency virus (HIV) infection, and transplantation (Griffin 2014). MIBE may also occur in immunocompromised individuals following MMR vaccination with onset of about four to nine months after vaccination (Bitnun et al. 1999; Gastañaduy et al. 2021).

The actual incidence of MIBE is unclear. Mustafa et al. in 1993 published a case series of 33 patients diagnosed with MIBE in which the mean age of diagnosis was around six years (Mustafa et al. 1993). MIBE appears to be similar to SSPE with underlying measles virus persistence, except that the host in MIBE is immunocompromised.

9.1.7.4.2 Clinical Presentation

Measles inclusion body encephalitis presents weeks to months after exposure to the measles virus (Cherry and Lugo 2019). Initial measles infection may be missed at the time of presentation due to the absence of symptoms or only mild symptoms occurring due to the host's impaired immune responses. The enanthem of measles (Koplik spots) may be absent, and the exanthem of measles (rash) may be mild or have an atypical presentation (Hughes et al. 1993). Most cases of MIBE present with altered mental status and intractable seizures without any fevers (Griffin 2014; Mustafa et al. 1993). Other neurologic symptoms such as hemiparesis, ataxia, aphasia, visual disturbances, dysphagia, emotional lability, and hypertension may occur (Mustafa et al. 1993).

9.1.7.4.3 Diagnosis

At presentation, CSF findings in MIBE are usually normal with undetectable anti-measles virus antibodies (Mustafa et al. 1993). EEG is nondiagnostic and may reveal focal or generalized epileptiform discharges or generalized slowing (Buchanan and Bonthius 2012). MRI brain imaging initially may be normal and eventually reveal nonspecific multifocal cortical lesions (Aldecoa et al. 2020; Buchanan and Bonthius 2012; Lytvyn et al. 2020; Rodriguez et al. 2020). Definitive diagnosis is usually made on brain biopsy or postmortem analysis of the brain with the measles virus being identified (Griffin 2014).

9.1.7.4.4 Treatment

There is no approved treatment for MIBE with management primarily being supportive. The use of antiviral therapy with ribavirin and interferon has been reported in some cases but overall has been unsuccessful.

9.1.7.4.5 Prognosis

MIBE has a very poor prognosis with over 75% dying from it within a few weeks to ten months (Mustafa et al. 1993). Those that survive usually are left with significant neurological impairment.

9.1.7.4.6 Neuropathophysiology and Neuropathology

The exact mechanism by which the measles virus leads to measles inclusion body encephalitis remains unclear. It is likely similar to SSPE with mutated MeV spreading through the central nervous system (Griffin 2014).

Electron microscopy of brain tissue usually reveals intranuclear and/or intracytoplasmic inclusion bodies within neurons and glial cells comprising of paramyxovirus nucleocapsid (Mustafa et al. 1993; Rima and Paul Duprex 2005). Other histological findings of brain tissue may include gliosis, perivascular cuffing by leukocytes, neuronal necrosis, and demyelination (Jabbour et al. 1969; Mustafa et al. 1993). Defective MeV proteins and RNA can be identified from brain tissue samples (Griffin 2014).

9.1.7.5 Hearing Loss

Prior to widespread immunization against measles, it is reported that 5–10% of bilateral sensorineural hearing loss in the United States was secondary to measles (McKenna 1997). Today, measles-associated hearing loss is uncommon in nations where immunization rates are high. The underlying pathophysiology is unclear. One mechanism may be secondary otitis media with bacterial superinfection (Cohen et al. 2014). On histopathological evaluation of temporal bone, diffuse cochlear destruction is often seen involving the cochlear neurons, organ of Corti, and stria vascularis (McKenna 1997).

9.2 Mumps and Its Neurological Complications

9.2.1 Introduction

Mumps is a contagious, often mild, and self-limiting viral illness that is most commonly clinically associated with parotitis. The annual global incidence of mumps is 100–1000 cases of mumps per 100,000 unvaccinated people (World Health Organization 2007). It affects mostly children between the ages of five and nine. Outbreaks of mumps spike every 2–5 years. People born before 1957 are considered to have natural immunity.

The first mumps vaccination was authorized in the United States in 1967, and the measles, mumps, and rubella (MMR) vaccine was introduced in 1971 (Marlow et al. 2021). With the advent of the MMR vaccine, mumps cases in the United States decreased by more than 99% from 152,209 cases in the year 1968 to just 338 in the year 2000 (CDC 2023; Leung and Marlow 2022). Between 2000 and 2005, the number of cases remained low with annually being around 200–350 until 2006 when there was a peak of 6584 cases reported (CDC 2023; Dayan et al. 2008).

Interestingly, this outbreak occurred among young adults who had already received at least one dose of the MMR vaccine. This outbreak and others continue to occur periodically usually in young adults in prolonged close contact environments like universities (Clemmons et al. 2019). It is unclear why vaccinated individuals may still get mumps but believed to be due to either primary vaccine failure or secondary vaccine failure with waning immunity. Similar to the United States, other nations with high immunization rates like the Netherlands and Sweden have experienced similar outbreaks among vaccinated individuals (Ramanathan et al. 2018).

Despite this, the MMR vaccine is still the most effective preventative measure against mumps. A person's chance of contracting mumps is reduced by roughly 88% after receiving two doses of the MMR vaccine and by 78% after receiving just one (CDC 2021a). Additionally, the symptoms are typically less severe in those with a prior history of vaccination. Vaccination is important in limiting the size of these outbreaks.

In contrast to the United States, the United Kingdom experienced a mumps epidemic predominantly in unvaccinated people between the years 2004 and 2005 with around 56,000 cases being reported in England and Wales (Savage et al. 2006). These occurred mainly in those between the ages of 15–24 years with about 3.3% having received 2 doses of MMR vaccine and 30.1% having received 1 dose of MMR (Savage et al. 2006). The low vaccination rate was due to the mumps vaccine not being introduced in the United Kingdom until 1988 (UK Health Security Agency 2022).

9.2.2 Virology

The mumps disease is caused by the mumps virus (MuV), which like the measles virus is also a single-strand, negative-sense RNA virus of the family Paramyxoviridae, but the genus *Rubulavirus* (Ryan 2022). The mumps genome is comprised of about 15,000 nucleotides helically encapsulated and surrounded by a lipid membrane—similar to the measles virus. The genome encodes for seven main structural proteins—nucleocapsid (NP), phospho (P), matrix (M), fusion (F), small hydrophobic (SH), hemagglutinin–neuraminidase (HN), and large (L) (Li et al. 2009; Maldonado and Shetty 2018b). The P sequence also encodes for 2 nonstructural proteins—V and I (Rubin et al. 2015). This genome is helically encapsulated by nucleocapsid proteins attached to an RNA-dependent RNA polymerase (RdRp) complex derived from L and P proteins (Rubin et al. 2015). This entire structure is surrounded by a lipid membrane with glycoprotein spikes—F and HN proteins.

Like the measles virus, the mumps virus only has humans as known hosts. While there are 12 distinct mumps genotypes identified, only one serotype exists (Maldonado and Shetty 2018b). The SH gene has the greatest genetic variation allowing for heterogeneity in the strains (Jin et al. 1999; McNall et al. 2020).

9.2.3 Pathogenesis

Mumps is a contagious viral disease that is transmitted via the respiratory route when contact occurs with infected oral or respiratory secretions (Cherry and Quinn 2019). The mumps virus (MuV) likely infects the respiratory epithelial cells by hemagglutinin–neuraminidase protein (HN) binding to the host cell's surface sialic acid with the fusion (F) protein allowing for the viral lipid membrane to fuse with the host cell's (Rubin et al. 2015). The M protein facilitates the budding process of new virions from the infected host cell along with HN's neuraminidase activity. The V and SH proteins have been implicated in the evasion of the host's immune system (Xu et al. 2012). The V protein has been shown to interfere with interferon and interleukin-6 signaling, while it appears that the SH protein prevents cell apoptosis (Franz et al. 2017; Rubin et al. 2015; Stinnett et al. 2020; Xu et al. 2012).

Viral replication occurs likely within the respiratory epithelial cells with the virus spreading to lymph nodes with subsequent viremia (Cherry and Quinn 2019). Viremia allows for the spread of the MuV to distant organs such as the central nervous system (CNS), thyroid, pancreas, kidneys, testicles, and ovaries. The viremia itself seems to be transient and limited likely due to the host's humoral response; this may explain why the MuV is rarely detected in blood (Overman and Durham 1958; Rubin et al. 2015).

It is still unknown how the mumps virus enters the CNS. It is possible that it may cross the choroid plexus or use the Trojan horse method entering via an infected leukocyte (Hviid et al. 2008).

9.2.4 Clinical Manifestations

Viral replication occurs in the respiratory epithelial cells with viral shedding into the respiratory tract. Infected individuals are usually most infectious 2 days prior to the onset of parotitis and 5 days following it (Cherry and Quinn 2019).

Mumps has an incubation period of about 14–21 days (Scheid 1961). Classically mumps infection clinically begins with a prodrome period of 1–2 days of usually low-grade fever along with headaches, myalgias, and malaise (Cherry and Quinn 2019; Leung and Marlow 2022; World Health Organization 2007). This is then followed by bilateral parotid swelling (Fig. 9.3) which may be unilateral in up to 20% of people. Acute parotitis may, however, only occur in about 15–20% of infected people (Watson et al. 1998). The parotid swelling lasts for about a week (on average 5 days) (Leung and Marlow 2022). Other salivary glands may be involved in about 10% of infected people including the submaxillary, sublingual, and submandibular glands (Cherry and Quinn 2019; Rubin et al. 2015; World Health Organization 2007). For the majority of people, the symptoms resolve after a week. Approximately 30% of persons can be asymptomatic or only have extremely mild, nonspecific respiratory problems (Rubin et al. 2015; World Health Organization 2007).

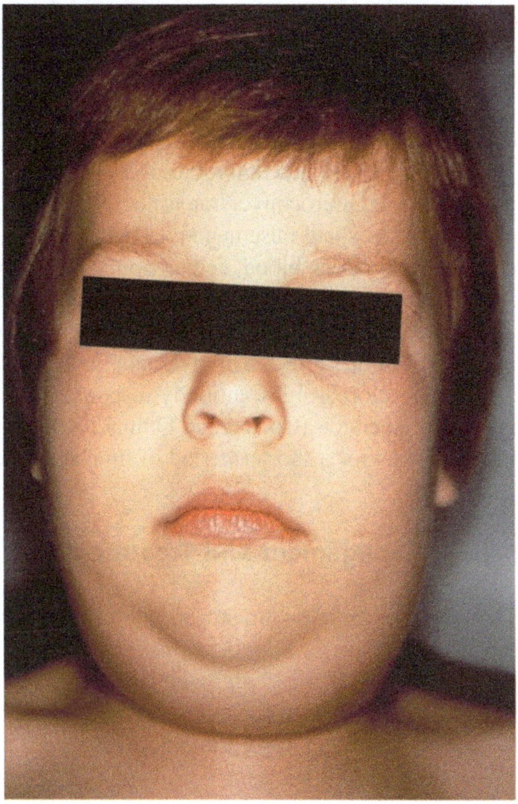

Fig. 9.3 A young boy with bilateral parotitis as well as thyroid swelling secondary to mumps. Image source: Centers for Disease Control and Prevention—Public Health Image Library—ID#1861—(Farmer 1963)—copyright restrictions: none; this image is in the public domain and thus free of any copyright restriction

9.2.5 Diagnosis

The differential diagnosis for parotitis includes other viruses like the Epstein-Barr virus, influenza A virus, cytomegalovirus, human herpesvirus-6, enteroviruses, and human immunodeficiency virus (Cherry and Quinn 2019; Marlow et al. 2021).

A diagnosis of mumps is usually clinically suspected when an individual presents with fever and parotitis, especially during a geographical outbreak of mumps. Diagnosis of mumps can be confirmed virologically through RT-PCR or viral culture or can be done serologically by testing for IgM or IgG antibodies to mumps. The two most common methods for establishing a laboratory diagnosis of mumps infection are testing for anti-mumps virus (anti-MuV) IgM antibodies and using RT-PCR to detect MuV-RNA usually from a buccal or oral swab (Marlow et al. 2021).

The Centers for Disease Control and Prevention (CDC) in the United States advises obtaining a buccal sample for RT-PCR analysis after a 30 s parotid gland massage if parotitis has been present for three or fewer days (CDC 2021b). If it has been more than 3 days since the onset of parotitis or mumps is clinically suspected in the absence of parotitis, the CDC recommends collecting both a buccal specimen for RT-PCR testing and testing serum for anti-MuV IgM antibodies. This is because after 3 days the sensitivity of detecting anti-MuV IgM antibodies increases while by RT-PCR decreases (Rota et al. 2013). Diagnosis may also be confirmed by viral culture from a buccal or urine sample though this can be time-consuming.

Checking serum IgG antibodies to MuV during acute and convalescent phases of infection for IgG seroconversion is not recommended due to the increased possibility of false positives and false negatives (Marlow et al. 2021). Mumps virus can also be cultured from saliva, blood, cerebrospinal fluid, and urine but is not commonly used for establishing diagnosis given the time it takes (Marlow et al. 2021).

Negative virological or serological testing for mumps does not rule out mumps infection. Mumps-infected individuals with a prior history of mumps vaccination may not develop an IgM antibody response or mount a four-fold IgG titer rise for IgG seroconversion (Cherry and Quinn 2019; Marlow et al. 2021). They may also shed the virus for a shorter period of time and thus miss detection by RT-PCR.

9.2.6 Management and Treatment

There is no specific treatment for mumps and for most individuals, it is a short and self-limiting disease. Supportive care targeting symptomatic management such as antipyretics for fever and anti-analgesics for pain may be used. Infected individuals should also self-isolate for at least five days after the onset of parotitis to prevent transmission to other susceptible individuals (CDC 2021c).

9.2.7 Neurological Complications

The most common neurological complications of mumps include meningitis, encephalitis, and hearing loss (Hviid et al. 2008). Other possible but less common neurological complications include acute disseminated encephalomyelitis (ADEM), transverse myelitis, paralysis, cranial nerve palsies, Guillain–Barré syndrome, acquired aqueduct stenosis, and hydrocephalus (Maldonado and Shetty 2018b). Non-neurological complications of mumps include orchitis and epididymitis which can occur in up to 30% of postpubertal men, oophoritis, mastitis, pancreatitis, nephritis, and myocarditis (Marlow et al. 2021; Watson et al. 1998).

9.2.7.1 Meningitis

Meningitis is the most common and usually self-limiting neurological complication of mumps. The incidence of meningitis related to mumps has ranged in past literature from 1% to 30% (Levitt et al. 1970; Russell and Donald 1958; Scheid 1961). It is about 3 times more common in males than in females (Levitt et al. 1970; McLean et al. 1964).

Mumps-associated meningitis usually presents with symptoms of meningeal inflammation—fever, headaches, neck stiffness, and vomiting (Levitt et al. 1970;

McLean et al. 1964). Symptoms begin on average within a few days of parotitis onset but may occur from 6 days prior to 20 days after parotitis (Levitt et al. 1970; McLean et al. 1964). Furthermore, up to 50% may develop meningitis without the involvement of the parotid or any other salivary gland (Levitt et al. 1970; McLean et al. 1964).

Diagnosis can often be made clinically without the need for lumbar puncture or neuroimaging. Cerebrospinal fluid (CSF) analysis, if done, may reveal lymphocytic pleocytosis, mild increase in protein concentrations, and normal glucose levels (Cherry and Lugo 2019; Levitt et al. 1970). MuV may be isolated from CSF for up to 3 days post-onset of symptoms (McLean et al. 1964). Viral culture is usually not done and CSF is tested for anti-mumps virus IgM antibodies and RT-PCR to detect viral RNA.

Complete recovery usually occurs without any complications within 2–3 days with treatment primarily being supportive (McLean et al. 1964; Russell and Donald 1958).

9.2.8 Encephalitis

With the introduction of anti-mumps vaccines, mumps encephalitis today is rare with a reported incidence of around 0.1% (Hviid et al. 2008). It usually occurs about 7–9 days after mumps onset and presents with high fevers, altered mental status, and other focal neurological symptoms like ataxia and hemiplegia (Koskiniemi et al. 1983; Levitt et al. 1970; Russell and Donald 1958). Seizures may also be observed in some with EEG mostly revealing nonspecific changes (Koskiniemi et al. 1983). Parotitis may not occur in some during the clinical course of the mumps infection. The CSF analysis results are comparable to those described in mumps meningitis with some lymphocytic pleocytosis.

In some older literature, mumps encephalitis has been reported to have a high mortality rate with one study reporting it to be around 20% (Russell and Donald 1958). Scheid argued that death was actually uncommon with prior reports of high mortality secondary to encephalitis being incorrectly attributed to mumps infection (Scheid 1961). Koskiniemi et al. reviewed 41 patients with encephalitis from laboratory-confirmed mumps infection in Southern Finland between 1968 and 1980. They reported only one death from acute hydrocephalus occurring in a child with congenital toxoplasma and cytomegalovirus (Koskiniemi et al. 1983). Today, mortality associated with mumps encephalitis is reported to be around 1.5% with neurological long-term complications being rare (Hviid et al. 2008).

The neuropathophysiology of mumps encephalitis is not well understood. Given that the mumps virus can be isolated from CSF, it is clear that it is able to cross the blood–brain barrier (Poggio et al. 2000). Encephalitis may be the consequence of direct viral infection of the brain in some cases or possibly postinfectious presenting as ADEM in others.

Neuropathology findings are difficult to establish given its rarity. However, some have demonstrated perivascular infiltration, hemorrhage, and demyelination (Rubin et al. 2015; Scheid 1961).

9.2.8.1 Hearing Loss

Sensorineural hearing loss is another neurological complication of mumps and can be transient or permanent. It is usually unilateral and reversible occurring within 4–5 days of symptoms onset from mumps (Hall and Richards 1987). The incidence in older literature of permanent hearing loss has been reported to be around 1 per 20,000 cases of mumps (Hall and Richards 1987; Hviid et al. 2008). Atrophied organ of Corti and stria vascularis with endolymphatic hydrops can be seen in histopathological evaluation of temporal bones (McKenna 1997).

References

Adams JM, Baird C, Filloy L (1966) Inclusion bodies in measles encephalitis. JAMA 195:290–298. https://doi.org/10.1001/jama.1966.03100040096026

Aldecoa I, Archilla I, Herrero L, Garcia F, Torres B, Gaig C, Fernández S, Marcos MÁ, Gelpi E (2020) Image of the month: measles inclusion body encephalitis. Clin Neuropathol 39:148–151. https://doi.org/10.5414/NP301283

Anlar B, Saatci I, Kose G, Yalaz K (1996) MRI findings in subacute sclerosing panencephalitis. Neurology 47:1278. https://doi.org/10.1212/WNL.47.5.1278

Baxter R, Lewis E, Goddard K, Fireman B, Bakshi N, DeStefano F, Gee J, Tseng HF, Naleway AL, Klein NP (2016) Acute demyelinating events following vaccines: a case-centered analysis. Clin Infect Dis 63:1456–1462. https://doi.org/10.1093/cid/ciw607

Beaty SM, Lee B (2016) Constraints on the genetic and antigenic variability of measles virus. Viruses 8:109. https://doi.org/10.3390/v8040109

Bellini WJ, Rota JS, Lowe LE, Katz RS, Dyken PR, Zaki SR, Shieh W-J, Rota PA (2005) Subacute sclerosing panencephalitis: more cases of this fatal disease are prevented by measles immunization than was previously recognized. J Infect Dis 192:1686–1693. https://doi.org/10.1086/497169

Bitnun A, Shannon P, Durward A, Rota PA, Bellini WJ, Graham C, Wang E, Ford-Jones EL, Cox P, Becker L, Fearon M, Petric M, Tellier R (1999) Measles inclusion-body encephalitis caused by the vaccine strain of measles virus. Clin Infect Dis 29:855–861. https://doi.org/10.1086/520449

Buchanan R, Bonthius DJ (2012) Measles virus and associated central nervous system sequelae. Semin Pediatr Neurol 19:107–114. https://doi.org/10.1016/j.spen.2012.02.003

Campbell H, Andrews N, Brown KE, Miller E (2007) Review of the effect of measles vaccination on the epidemiology of SSPE. Int J Epidemiol 36:1334–1348. https://doi.org/10.1093/ije/dym207

Carmo RL, do Alves Simão AK, do Amaral LLF, Inada BSY, Silveira CF, Campos CM, Freitas LF, Bonadio V, Marussi VHR (2019) Neuroimaging of emergent and reemergent infections. RadioGraphics 39:1649–1671. https://doi.org/10.1148/rg.2019190020

CDC (2020) Transmission of measles. Centers for Disease Control and Prevention. https://www.cdc.gov/measles/transmission.html. Accessed 11 June 23

CDC (2021a) Mumps vaccination. Centers for Disease Control and Prevention. https://www.cdc.gov/mumps/vaccination.html. Accessed 21 May 23

CDC (2021b) Mumps job-aid template for providers. Centers for Disease Control and Prevention. https://www.cdc.gov/mumps/health-departments/provider-job-aid.html. Accessed 10 June 23

CDC (2021c) Mumps: for healthcare providers. Centers for Disease Control and Prevention. https://www.cdc.gov/mumps/hcp.html. Accessed 10 June 23

CDC (2022a) Genetic analysis of measles virus. Centers for Disease Control and Prevention. https://www.cdc.gov/measles/lab-tools/genetic-analysis.html. Accessed 13 May 23

CDC (2022b) Measles: specimen collection, storage, and shipment. Centers for Disease Control and Prevention. https://www.cdc.gov/measles/lab-tools/rt-pcr.html. Accessed 11 June 23

CDC (2022c) Measles (rubeola). Centers for Disease Control and Prevention. https://www.cdc.gov/measles/hcp/index.html. Accessed 7 June 23

CDC (2023) Mumps cases and outbreaks. Centers for Disease Control and Prevention. https://www.cdc.gov/mumps/outbreaks.html. Accessed 15 April 23

Cherry JD, Lugo D (2019) Measles virus. In: Feigin and Cherry's textbook of pediatric infectious diseases, 8th edn. Elsevier, Amsterdam, pp 1754–1771

Cherry JD, Quinn KK (2019) Mumps virus. In: Feigin and Cherry's textbook of pediatric infectious diseases, 8th edn. Elsevier, Amsterdam, pp 1771–1779

Clemmons NS, Redd SB, Gastañaduy PA, Marin M, Patel M, Fiebelkorn AP (2019) Characteristics of large mumps outbreaks in the United States, July 2010–December 2015. Clin Infect Dis 68: 1684–1690. https://doi.org/10.1093/cid/ciy779

Cohen BE, Durstenfeld A, Roehm PC (2014) Viral causes of hearing loss: a review for hearing health professionals. Trends Hear 18:2331216514541361. https://doi.org/10.1177/2331216514541361

Cole J, Evans E, Mwangi M, Mar S (2019) Acute disseminated encephalomyelitis in children: an updated review based on current diagnostic criteria. Pediatr Neurol 100:26–34. https://doi.org/10.1016/j.pediatrneurol.2019.06.017

Colpak AI, Erdener SE, Ozgen B, Anlar B, Kansu T (2012) Neuro-ophthalmology of subacute sclerosing panencephalitis: two cases and a review of the literature. Curr Opin Ophthalmol 23: 466–471. https://doi.org/10.1097/ICU.0b013e328358b196

Dayan GH, Quinlisk MP, Parker AA, Barskey AE, Harris ML, Schwartz JMH, Hunt K, Finley CG, Leschinsky DP, O'Keefe AL, Clayton J, Kightlinger LK, Dietle EG, Berg J, Kenyon CL, Goldstein ST, Stokley SK, Redd SB, Rota PA, Rota J, Bi D, Roush SW, Bridges CB, Santibanez TA, Parashar U, Bellini WJ, Seward JF (2008) Recent resurgence of mumps in the United States. N Engl J Med 358:1580–1589. https://doi.org/10.1056/NEJMoa0706589

Dyken PR (1985) Subacute sclerosing panencephalitis. Current status. Neurol Clin 3:179–196

Dyken PR (2001) Neuroprogressive disease of post-infectious origin: a review of a resurging subacute sclerosing panencephalitis (SSPE). Ment Retard Dev Disabil Res Rev 7:217–225. https://doi.org/10.1002/mrdd.1030

Dyken P, DuRant R, Shmunes P (1982) Subacute sclerosing panencephalitis surveillance – United States. MMWR Morb Mortal Wkly Rep 31:585–588

Eichenwald HF (1958) ID#3187 - Public Health Image Library (PHIL). CDC. Accessed 10 July 23

Ekmekci O, Karasoy H, Gökçay A, Ulkü A (2005) Atypical EEG findings in subacute sclerosing panencephalitis. Clin Neurophysiol 116:1762–1767. https://doi.org/10.1016/j.clinph.2005.03.026

Farmer CN (1963) ID#1861 - Public Health Image Library (PHIL). CDC. https://phil.cdc.gov/Details.aspx?pid=1861. Accessed 10 July 23

Franz S, Rennert P, Woznik M, Grützke J, Lüdde A, Arriero Pais EM, Finsterbusch T, Geyer H, Mankertz A, Friedrich N (2017) Mumps virus SH protein inhibits NF-κB activation by interacting with tumor necrosis factor receptor 1, interleukin-1 receptor 1, and toll-like receptor 3 complexes. J Virol 91:e01037. https://doi.org/10.1128/JVI.01037-17

Garg RK (2008) Subacute sclerosing panencephalitis. J Neurol 255:1861–1871. https://doi.org/10.1007/s00415-008-0032-6

Gastañaduy P, Goodson J (2023) Rubeola/measles. In: CDC yellow book 2024. Centers for Disease Control and Prevention, Atlanta

Gastañaduy P, Haber P, Rota PA, Patel M (2021) Measles. In: Epidemiology and prevention of vaccine-preventable diseases. Public Health Foundation, Washington, pp 193–206

Gibbs FA, Gibbs EL, Carpenter PR, Spies HW (1959) Electroencephalographic abnormality in "uncomplicated" childhood diseases. J Am Med Assoc 171:1050–1055. https://doi.org/10.1001/jama.1959.03010260006002

Government of India (2005) Measles mortality reduction: India strategic plan, 2005-2010. World Health Organization. https://www.who.int/india/health-topics/measles. Accessed 21 June 23

Griffin DE (2014) Chapter 27 - measles virus and the nervous system. In: Tselis AC, Booss J (eds) Handbook of clinical neurology, neurovirology. Elsevier, Amsterdam, pp 577–590. https://doi.org/10.1016/B978-0-444-53488-0.00027-4

Gutierrez J, Issacson RS, Koppel BS (2010) Subacute sclerosing panencephalitis: an update. Dev Med Child Neurol 52:901–907. https://doi.org/10.1111/j.1469-8749.2010.03717.x

Hall R, Richards H (1987) Hearing loss due to mumps. Arch Dis Child 62:189–191. https://doi.org/10.1136/adc.62.2.189

Hemingway C (2020) Demyelinating disorders of the central nervous system. In: Nelson textbook of pediatrics. Elsevier, Amsterdam, pp 3196–3209

Hughes I, Jenney ME, Newton RW, Morris DJ, Klapper PE (1993) Measles encephalitis during immunosuppressive treatment for acute lymphoblastic leukaemia. Arch Dis Child 68:775–778

Hviid A, Rubin S, Mühlemann K (2008) Mumps. Lancet 371:932–944. https://doi.org/10.1016/S0140-6736(08)60419-5

Jabbour JT, Garcia JH, Lemmi H, Ragland J, Duenas DA, Sever JL (1969) Subacute sclerosing panencephalitis: a multidisciplinary study of eight cases. JAMA 207:2248–2254. https://doi.org/10.1001/jama.1969.03150250078007

Jabbour JT, Duenas DA, Sever JL, Krebs HM, Horta-Barbosa L (1972) Epidemiology of subacute sclerosing panencephalitis (SSPE): a report of the SSPE registry. JAMA 220:959–962. https://doi.org/10.1001/jama.1972.03200070051006

Jin L, Beard S, Brown DWG (1999) Genetic heterogeneity of mumps virus in the United Kingdom: identification of two new genotypes. J Infect Dis 180:829–833. https://doi.org/10.1086/314957

Johnson RT, Griffin DE, Hirsch RL, Wounsky JS, Roedenbeck S, de Soriano IL, Vaisberg A (1984) Measles encephalomyelitis — clinical and immunologic studies. N Engl J Med 310:137–141. https://doi.org/10.1056/NEJM198401193100301

Koskiniemi M, Donner M, Pettay O (1983) Clinical appearance and outcome in mumps encephalitis in children. Acta Paediatr Scand 72:603–609. https://doi.org/10.1111/j.1651-2227.1983.tb09778.x

Krupp LB, Banwell B, Tenembaum S (2007) Consensus definitions proposed for pediatric multiple sclerosis and related disorders. Neurology 68:S7. https://doi.org/10.1212/01.wnl.0000259422.44235.a8

La Bocetta AC, Tornay AS (1964) Measles encephalitis: report of 61 cases. Am J Dis Child 107:247–255. https://doi.org/10.1001/archpedi.1964.02080060249005

Laksono BM, de Vries RD, McQuaid S, Duprex WP, de Swart RL (2016) Measles virus host invasion and pathogenesis. Viruses 8:210. https://doi.org/10.3390/v8080210

Leake JAD, Albani S, Kao AS, Senac MO, Billman GF, Nespeca MP, Paulino AD, Quintela ER, Sawyer MH, Bradley JS (2004) Acute disseminated encephalomyelitis in childhood: epidemiologic, clinical and laboratory features. Pediatr Infect Dis J 23(8):756–764

Lee KY, Cho WH, Kim SH, Kim HD, Kim IO (2003) Acute encephalitis associated with measles: MRI features. Neuroradiology 45:100–106. https://doi.org/10.1007/s00234-002-0884-x

Leung J, Marlow M (2022) Mumps. In: Loscalzo J, Fauci A, Kasper D, Hauser S, Longo D, Jameson JL (eds) Harrison's principles of internal medicine, 21st edn. McGraw-Hill Education, New York

Levitt LP, Rich TA, Kinde SW, Lewis AL, Gates EH, Bond JO (1970) Central nervous system mumps a review of 64 cases. Neurology 20:829–829. https://doi.org/10.1212/WNL.20.8.829

Li M, Schmitt PT, Li Z, McCrory TS, He B, Schmitt AP (2009) Mumps virus matrix, fusion, and nucleocapsid proteins cooperate for efficient production of virus-like particles. J Virol 83:7261–7272. https://doi.org/10.1128/JVI.00421-09

Lytvyn H, Basa N, Stasiv M, Troyanovska O, Dorosh O (2020) Difficulties in diagnosing of measles inclusion body encephalitis in a child with acute lymphoblastic leukemia. IDCases 21:e00877. https://doi.org/10.1016/j.idcr.2020.e00877

Maldonado YA (2008) Rubeola virus (measles and subacute sclerosing panencephalitis). In: Long SS (ed) Principles and practice of pediatric infectious disease, 3rd edn. W.B. Saunders, Edinburgh, pp 1120–1126. https://doi.org/10.1016/B978-0-7020-3468-8.50233-9

Maldonado YA, Shetty AK (2018a) Rubeola virus: measles and subacute sclerosing panencephalitis. In: Long SS, Prober CG, Fischer M (eds) Principles and practice of pediatric infectious diseases, 5th edn. Elsevier, Amsterdam, pp 1169–1176. https://doi.org/10.1016/B978-0-323-40181-4.00227-9

Maldonado YA, Shetty AK (2018b) Mumps virus. In: Long SS, Prober CG, Fischer M (eds) Principles and practice of pediatric infectious diseases, 5th edn. Elsevier, Amsterdam, pp 1157–1162. https://doi.org/10.1016/B978-0-323-40181-4.00224-3

Markand ON, Panszi JG (1975) The electroencephalogram in subacute sclerosing panencephalitis. Arch Neurol 32:719–726. https://doi.org/10.1001/archneur.1975.00490530041002

Marlow M, Haber P, Hickman C, Patel M (2021) Mumps. In: Epidemiology and prevention of vaccine-preventable diseases. Public Health Foundation, Washington, pp 225–238

McKenna MJ (1997) Measles, mumps, and sensorineural hearing loss. Ann N Y Acad Sci 830:291–298. https://doi.org/10.1111/j.1749-6632.1997.tb51899.x

McLean DM, Bach RD, Larke RPB, McNaughton GA (1964) Mumps Meningoencephalitis, Toronto, 1963. Can Med Assoc J 90:458–462

McLean HQ, Fiebelkorn AP, Temte JL, Wallace GS (2013) Prevention of measles, rubella, congenital rubella syndrome, and mumps, 2013. Morb Mortal Wkly Rep 62:1–34

McNall RJ, Wharton AK, Anderson R, Clemmons N, Lopareva EN, Gonzalez C, Espinosa A, Probert WS, Hacker JK, Liu G, Garfin J, Strain AK, Boxrud D, Bryant PW, George KS, Davis T, Griesser RH, Shult P, Bankamp B, Hickman CJ, Wroblewski K, Rota PA (2020) Genetic characterization of mumps viruses associated with the resurgence of mumps in the United States: 2015-2017. Virus Res 281:197935. https://doi.org/10.1016/j.virusres.2020.197935

Measles Photos (2023) Centers for disease control and prevention. Accessed 10 July 23

Menge T, Hemmer B, Nessler S, Wiendl H, Neuhaus O, Hartung H-P, Kieseier BC, Stüve O (2005) Acute disseminated encephalomyelitis: an update. Arch Neurol 62:1673–1680. https://doi.org/10.1001/archneur.62.11.1673

Minta AA, Ferrari M, Antoni S, Portnoy A, Sbarra A, Lambert B, Hauryski S, Hatcher C, Nedelec Y, Datta D, Ho LL, Steulet C, Gacic-Dobo M, Rota PA, Mulders MN, Bose AS, Perea WA, O'Connor P (2022) Progress toward regional measles elimination—worldwide, 2000–202. Morb Mortal Wkly Rep 71:1489–1495

Moss WJ (2017) Measles. Lancet 390:2490–2502. https://doi.org/10.1016/S0140-6736(21)02004-3

Moss WJ, Griffin DE, Feinstone WH (2009) Chapter 30 – measles. In: Vaccines for biodefense and emerging neglected diseases. Academic Press, New York. https://doi.org/10.1016/B978-0-12-369408-9.00030-5

Mühlebach MD, Mateo M, Sinn PL, Prüfer S, Uhlig KM, Leonard VHJ, Navaratnarajah CK, Frenzke M, Wong XX, Sawatsky B, Ramachandran S, McCray PB, Cichutek K, von Messling V, Lopez M, Cattaneo R (2011) Adherens junction protein nectin-4 is the epithelial receptor for measles virus. Nature 480:530–533. https://doi.org/10.1038/nature10639

Mustafa MM, Weitman SD, Winick NJ, Bellini WJ, Timmons CF, Siegel JD (1993) Subacute measles encephalitis in the young immunocompromised host: report of two cases diagnosed by polymerase chain reaction and treated with ribavirin and review of the literature. Clin Infect Dis 16:654–660. https://doi.org/10.1093/clind/16.5.654

NHS (2020) MMR (measles, mumps and rubella) vaccine. United Kingdom National Health Service. https://www.nhs.uk/conditions/vaccinations/mmr-vaccine/. Accessed 20 June 23

Noorbakhsh F, Johnson RT, Emery D, Power C (2008) Acute disseminated encephalomyelitis: clinical and pathogenesis features. Neurol Clin 26:759–780. https://doi.org/10.1016/j.ncl.2008.03.009

Noyce RS, Richardson CD (2012) Nectin 4 is the epithelial cell receptor for measles virus. Trends Microbiol 20:429–439. https://doi.org/10.1016/j.tim.2012.05.006

Overman JR, Durham NC (1958) Viremia in human mumps virus infections. Arch Intern Med 102: 354–356. https://doi.org/10.1001/archinte.1958.00030010354002

Perry RT, Halsey NA (2004) The clinical significance of measles: a review. J Infect Dis 189(1): 4–16. https://doi.org/10.1086/377712

Poggio GP, Rodriguez C, Cisterna D, Freire MC, Cello J (2000) Nested PCR for rapid detection of mumps virus in cerebrospinal fluid from patients with neurological diseases. J Clin Microbiol 38:274–278

Pohl D, Alper G, Van Haren K, Kornberg AJ, Lucchinetti CF, Tenembaum S, Belman AL (2016) Acute disseminated encephalomyelitis: updates on an inflammatory CNS syndrome. Neurology 87:S38–S45. https://doi.org/10.1212/WNL.0000000000002825

Ramanathan R, Voigt EA, Kennedy RB, Poland GA (2018) Knowledge gaps persist and hinder progress in eliminating mumps. Vaccine 36:3721–3726. https://doi.org/10.1016/j.vaccine.2018.05.067

Riedel S, Hobden JA, Miller S, Morse SA, Mietzner TA, Detrick B, Mitchell TG, Sakanari JA, Hotez P, Mejia R (2019) Paramyxoviruses and rubella virus. In: Jawetz, Melnick, & Adelberg's medical microbiology. McGraw-Hill Education, New York

Rima BK, Paul Duprex W (2005) Molecular mechanisms of measles virus persistence. Virus Res 111:132–147. https://doi.org/10.1016/j.virusres.2005.04.005

Rodriguez C, Gouilh MA, Weiss N, Stroer S, Mokhtari K, Seilhean D, Mathon B, Demontant V, N'Debi M, Gricourt G, Woerther P-L, Pawlotsky J-M, Stefic K, Marlet J, Dequin P-F, Guillon A, Pourcher V, Boutolleau D, Vabret A, Burrel S (2020) Fatal measles inclusion-body encephalitis in adult with untreated AIDS, France. Emerg Infect Dis 26:2231–2234. https://doi.org/10.3201/eid2609.200366

Rota JS, Rosen JB, Doll MK, McNall RJ, McGrew M, Williams N, Lopareva EN, Barskey AE, Punsalang A, Rota PA, Oleszko WR, Hickman CJ, Zimmerman CM, Bellini WJ (2013) Comparison of the sensitivity of laboratory diagnostic methods from a well-characterized outbreak of mumps in New York City in 2009. Clin Vaccine Immunol 20:391–396. https://doi.org/10.1128/CVI.00660-12

Rota PA, Moss WJ, Takeda M, de Swart RL, Thompson KM, Goodson JL (2016) Measles. Nat Rev Dis Primers 2:1–16. https://doi.org/10.1038/nrdp.2016.49

Rubin S, Eckhaus M, Rennick LJ, Bamford CG, Duprex WP (2015) Molecular biology, pathogenesis and pathology of mumps virus. J Pathol 235:242–252. https://doi.org/10.1002/path.4445

Russell RR, Donald JC (1958) Neurological complications of mumps. Br Med J 2:27–30

Ryan KJ (2022) Viruses of mumps, measles, rubella, and other childhood exanthems. In: Sherris & Ryan's medical microbiology, 8th edn. McGraw-Hill Education, New York

Saha V, John TJ, Mukundan P, Gnanamuthu C, Prabhakar S, Arjundas G, Sayeed ZA, Kumaresan G, Srinivas K (1990) High incidence of subacute sclerosing panencephalitis in south India. Epidemiol Infect 104:151–156. https://doi.org/10.1017/s0950268800054637

Samia P, Oyieke K, Tunje D, Udwadia-Hegde A, Feemster K, Oncel I, Anlar B (2022) Options in the treatment of subacute sclerosing panencephalitis: implications for low resource areas. Curr Treat Options Neurol 24:99–110. https://doi.org/10.1007/s11940-022-00710-x

Sato Y, Watanabe S, Fukuda Y, Hashiguchi T, Yanagi Y, Ohno S (2018) Cell-to-cell measles virus spread between human neurons is dependent on hemagglutinin and hyperfusogenic fusion protein. J Virol 92:e02166. https://doi.org/10.1128/JVI.02166-17

Savage E, White J, Brown D, Ramsay M (2006) Mumps epidemic–United kingdom, 2004-2005. MMWR Morb Mortal Wkly Rep 55:173–175

Scheid W (1961) Mumps virus and the central nervous system. World Neurol 2:117–133

Singh BK, Li N, Mark AC, Mateo M, Cattaneo R, Sinn PL (2016) Cell-to-cell contact and nectin-4 govern spread of measles virus from primary human myeloid cells to primary human airway epithelial cells. J Virol 90:6808–6817. https://doi.org/10.1128/JVI.00266-16

Stinnett RC, Beck AS, Lopareva EN, McNall RJ, Latner DR, Hickman CJ, Rota PA, Bankamp B (2020) Functional characterization of circulating mumps viruses with stop codon mutations in the small hydrophobic protein. mSphere 5:e00840. https://doi.org/10.1128/mSphere.00840-20

Tenembaum S, Chamoles N, Fejerman N (2002) Acute disseminated encephalomyelitis: a long-term follow-up study of 84 pediatric patients. Neurology 59:1224–1231. https://doi.org/10.1212/WNL.59.8.1224

UK Health Security Agency (2022) MMR for all: general guide. UK Government (GOV.UK). https://www.gov.uk/government/publications/mmr-for-all-general-leaflet/mmr-for-all-general-guide. Accessed 16 June 23

Venkatesan A, Tunkel AR, Bloch KC, Lauring AS, Sejvar J, Bitnun A, Stahl J-P, Mailles A, Drebot M, Rupprecht CE, Yoder J, Cope JR, Wilson MR, Whitley RJ, Sullivan J, Granerod J, Jones C, Eastwood K, Ward KN, Durrheim DN, Solbrig MV, Guo-Dong L, Glaser CA, Sheriff H, Brown D, Farnon E, Messenger S, Paterson B, Soldatos A, Roy S, Visvesvara G, Beach M, Nasci R, Pertowski C, Schmid S, Rascoe L, Montgomery J, Tong S, Breiman R, Franka R, Keuhnert M, Angulo F, Cherry J (2013) Case definitions, diagnostic algorithms, and priorities in encephalitis: consensus statement of the international encephalitis consortium. Clin Infect Dis 57:1114–1128. https://doi.org/10.1093/cid/cit458

Watanabe S, Shirogane Y, Sato Y, Hashiguchi T, Yanagi Y (2019) New insights into measles virus brain infections. Trends Microbiol 27:164–175. https://doi.org/10.1016/j.tim.2018.08.010

Watson JC, Hadler SC, Dykewicz CA, Reef S, Phillips L (1998) Measles, Mumps, and Rubella—vaccine use and strategies for elimination of measles, rubella, and congenital rubella syndrome and control of mumps: recommendations of the advisory committee on immunization practices (ACIP). Morb Mortal Wkly Rep 47:1–57

World Health Organization (2005) Weekly epidemiological record, 2005, vol. 80, 28 (full issue). Wkly Epidemiol Rec 80:241–248

World Health Organization (2007) Mumps virus vaccines. Wkly Epidemiol Rec 82:51–60

World Health Organization (2017) Measles vaccines: WHO position paper – April 2017. Wkly Epidemiol Rec 500(92):205–228

World Health Organization (2022) Measles outbreak toolbox. World Health Organization. https://www.who.int/emergencies/outbreak-toolkit/disease-outbreak-toolboxes/measles-outbreak-toolbox. Accessed 7 June 23

World Health Organization (2023a) History of measles vaccination. World Health Organization. https://www.who.int/news-room/spotlight/history-of-vaccination/history-of-measles-vaccination. Accessed 13 May 23

World Health Organization (2023b) Measles. World Health Organization. https://www.who.int/news-room/fact-sheets/detail/measles. Accessed 15 April 23

Xu P, Luthra P, Li Z, Fuentes S, D'Andrea JA, Wu J, Rubin S, Rota PA, He B (2012) The V protein of mumps virus plays a critical role in pathogenesis. J Virol 86:1768–1776. https://doi.org/10.1128/JVI.06019-11

Zeng S-Z, Zhang B, Zhang Y, Xie L-Y, Xiong J, Yu T, Xie Z, Gao H, Duan Z-J (2016) Identification of 12 cases of acute measles encephalitis without rash. Clin Infect Dis 63:1630–1633. https://doi.org/10.1093/cid/ciw611

Rabies Encephalitis: A Disease Characterized By Complex Neuropathogenic Pathways and Diagnostic Difficulties

10

Sadia Hassaan, Safiya Firoze, Adil Raza, Hiba Sami, and Siraj Ahmad

Abstract

Rabies virus belongs to family Rhabdoviridae and Genus Lyssavirus. It is a lethal disease spread by animal bites that carry the virus in their saliva. Rabies encephalitis has the greatest fatality rate among infectious diseases, with an average time gap of 5 to 7 days for furious or encephalitic rabies and 11 to 14 days for paralytic or dumb rabies from the onset of clinical disease to death. The disease is spread through the bites of dogs and other wild animals. Inhalation in bat-infested caves and laboratory settings are two further methods of transmission. Human-to-human transmission has been described in rare cases due to contaminated corneal transplants. Depression and fever are the first symptoms, followed by agitation, increased salivation, and hydrophobia. The presence of the rabies antigen in nuchal skin biopsies and corneal impression smears is required for antemortem diagnosis using the fluorescent antibody approach. The presence of Negri bodies on histological analysis of the brain confirms the postmortem diagnosis of rabies. The vaccine and rabies immunoglobulin are given as a preventative measure. Till date, the disease has no known treatment.

S. Hassaan (✉) · A. Raza
Department of Microbiology, Jawaharlal Nehru Medical College and Hospital, Aligarh Muslim University, Aligarh, Uttar Pradesh, India

S. Firoze · H. Sami
Department of Microbiology, Jawaharlal Nehru Medical College Hospital, Aligarh, Uttar Pradesh, India

S. Ahmad
Department of Pharmacology, Jawaharlal Nehru Medical College and Hospital, Aligarh Muslim University, Aligarh, Uttar Pradesh, India

H. Sami et al. (eds.), *Viral and Fungal Infections of the Central Nervous System: A Microbiological Perspective*, https://doi.org/10.1007/978-981-99-6445-1_10

Keywords

Rabies · Rhabdoviridae · Encephalitis · Negri bodies

10.1 Introduction

The causative agent of rabies disease is a RNA virus, namely rabies virus (RABV). The characteristics of this RNA virus is that it is single-stranded, negative sense, and non-segmented (Rupprecht 1996). The family Rhabdoviridae has three genera Lyssavirus, Ephemerovirus, and Vesiculovirus (Dietzgen et al. 2017). Rabies virus belongs to the genus Lyssavirus and is the only medically important virus of this genus. Rabies virus causes rapidly progressive acute infection of the CNS in humans and animals. Rabies disease is almost exclusively caused by bites of rabid animals, and prevention may be sought for before exposure and also after exposure to RABV, provided the vaccination is administered on time (Rupprecht 1996). Rabies till now is a major public health problem because it is almost always fatal (Koury and Warrington 2023).

10.2 Rabies Virus Structure

The rabies virus is bullet-shaped. The size of virus is 180 nm in length and 75 nm wide (Willoughby 2012). They have a lipid envelope in which 10 nm long peplomers are embedded. The envelope is lined internally by a layer of matrix protein, and the nucleocapsid has a helical symmetry. The rabies genome encodes five proteins (Fig. 10.1): Nucleoprotein (N), phosphoprotein (P), matrix protein (M), glycoprotein (G), and polymerase (L) (Gérard et al. 2022; King et al. 2012).

10.3 Epidemiology

Rabies has been reported from almost all countries except Australia and Antarctica. Low awareness among people to seek medical advice after a dog bite has claimed lives of more than 55,000 to 60,000 people each year in Asia and Africa (Hampson et al. 2019). India accounts for the most deaths in Asia 65% of human rabies deaths. The age group which is most commonly affected are children less than 15 years and it accounts for about 30–60% of reported rabies cases and deaths in India (Sachdeva et al. 2022). Countries like Middle East, Africa, and Central Asia also have a high burden of the disease.

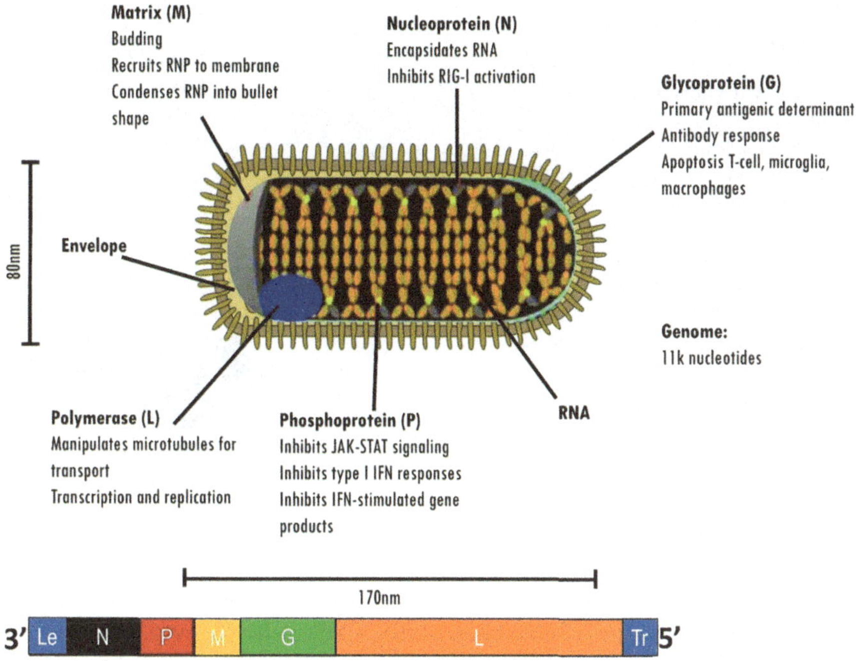

Fig. 10.1 Depiction of structure and genome of rabies virus

10.4 Transmission

The transmission of rabies can be divided into bite and non-bite exposure. In majority of cases, rabies is acquired through bite of dogs and bats. In some case, animals like foxes, raccoons, jackals, and mongooses can also transmit the disease through their bite. Non-bite exposure results from direct saliva contact with open skin or mucous membranes. Other non-bite exposure are transplantation from a donor who was infected with the virus and rarely inhalation of virus-containing aerosols (Hampson et al. 2019; Menezes 2008). Transmission of rabies through fomites or through household contacts has not been reported. During rabies surveillance in the United States, four animals have been identified as reservoir of rabies virus, these are bats, foxes, raccoons, and skunks (Blanton et al. 2012; Zhu et al. 2015).

10.5 Pathogenesis (Fig. 10.2)

Post human transmission via the bite of an infected animal, RABV begins to multiply at the inoculation site. After multiplying, the virus enters local motor and sensory nerves by spreading centripetally (Burrell et al. 2017; Rupprecht 1996). The

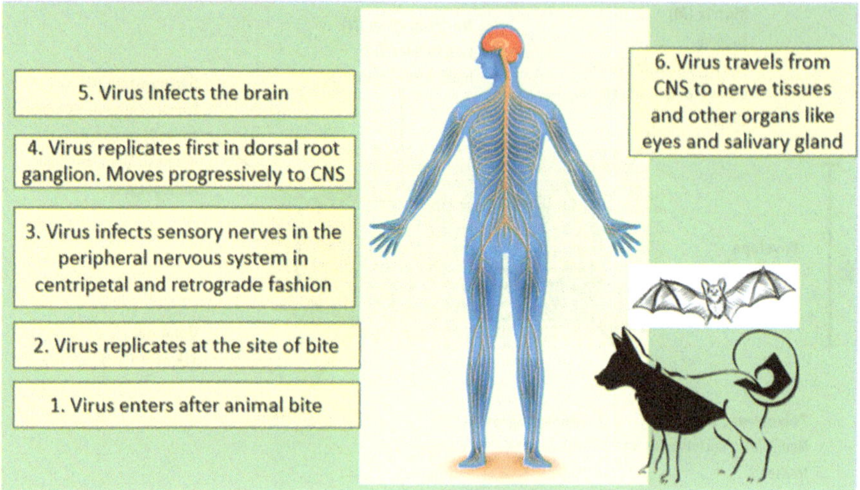

Fig. 10.2 Pathogenesis of rabies

glycoprotein projections on the surface of virus attach to the nicotinic acetylcholine receptors at the neuromuscular junction of the muscle cells. The viruses then enter nerve cells (Lian et al. 2022; Unwin 2013). Lyssaviruses have a predilection for neurons and spread centripetally along motor nerves. Lyssaviruses produce neuronal dysfunction. Malfunctioning of the mitochondria in infected neurons and other cells of the CNS is due to oxidative stress leading to various observed abnormalities (Scott and Nel 2021).

Viruses then travel in a retrograde manner within the axoplasm of nerves at a speed of 50 to 100 mm per day and reaches the dorsal root ganglia of the spinal cord (Tsiang et al. 1989). Rabies virus then reaches up to the spinal cord and the brain, initially infecting the diencephalon, hippocampus, and brainstem (Juntrakul et al. 2005; Mahadevan et al. 2016). Spread of virus occurs centrifugally along the somatic and autonomic nerves resulting in widespread dissemination to salivary glands, cornea, and other organs (Jackson 2011) (Fig. 10.4).

Rabies infection after an exposure is dependent on various factors, including the site of exposure (Rupprecht 1996; Yibrah and Damtie 2015). Exposures of the head or neck area are more likely to result in a productive infection than an exposure to distal part of the body. Host immunity, size of inoculum, and increase virulence of strain are also important factors for developing full blown disease (Grill 2009). In order to evade adaptive or innate immune response and subsequently promote replication and dissemination, rabies virus upregulates a few host pathways (Scott and Nel 2016).

10.6 Immune Evasion

Within the central nervous system, RABV is shown to evade the adaptive immunological response (Fig. 10.3) and also interfere with innate immune response (Fig. 10.4). When RABV infects neurons, it sequesters Toll-like receptor 3 (i.e., TLR3) into Negri bodies, preventing TLR3 from being activated as an innate sensor. Immune sensors (i.e., RIG-I and MDA5) detect RABV-RNA in neuronal cytoplasm, and this recognition of RABV-RNA triggers the expression of innate antiviral type-I interferon (IFN-alpha/beta) through the generation of a complex of proteins that phosphorylate interferon regulatory factors 3 and 7 (IRF3/7). Along with activator protein, AP-1, and nuclear factor B, i.e., NF-B, phosphorylated-IRF3/7 is transported to nuclei in order to stimulate IFN-α/β transcription. Rabies virus phosphoprotein alters IFN-α/β gene transcription by preventing IRF3/7's phosphorylation and nuclear import while leaving activator protein-1 and nuclear factor-B

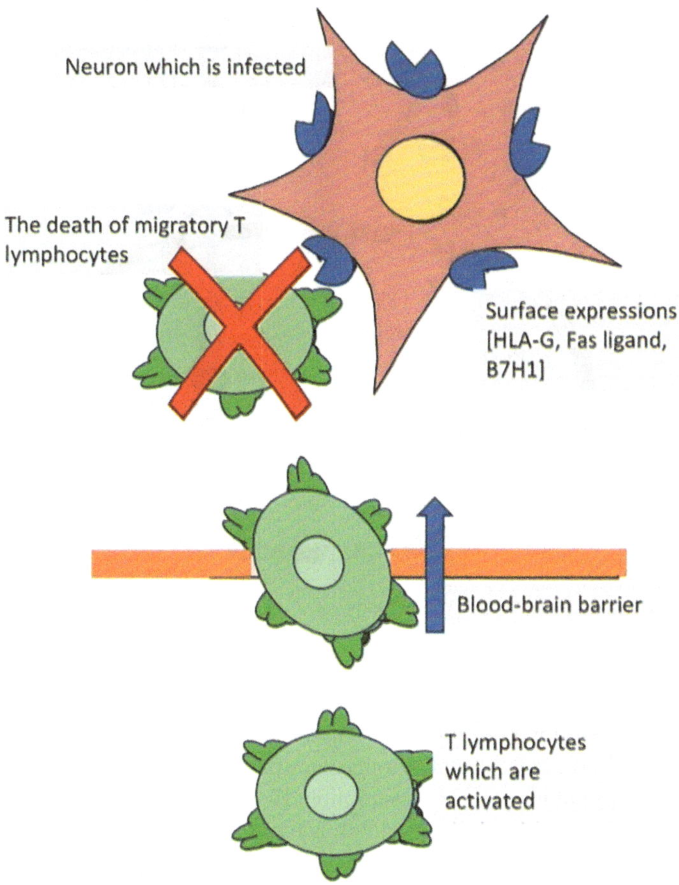

Fig. 10.3 (Adaptive) immune response evasion

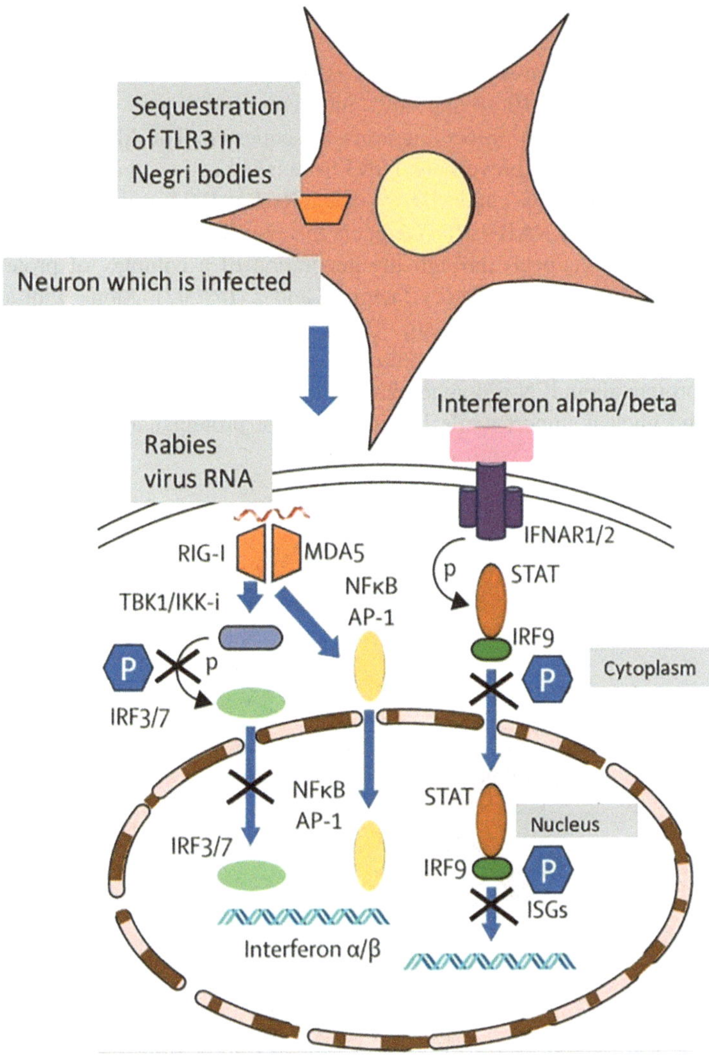

Fig. 10.4 (Innate) immune response inhibition

activation unaffected. The signal transducer and activator of transcription (STAT) signaling pathway is also suppressed by viral phosphoproteins. Viral phosphoprotein inhibits nuclear import of (phosphorylated) STAT, binds to intranuclear STAT-IRF9 complex, and binds to intracytoplasmic (phosphorylated) STAT to suppress interferon stimulated gene (ISG) transcription. Even though the blood–brain barrier is preserved in RABV-infected individuals, activated T lymphocytes and monocytes can penetrate it and enter the central nervous system. RABV-infected neurons that exhibit the immunosubversive components, namely, B7-H1 (B7-homolog 1, also known as programmed death-ligand 1), Fas ligand (FasL), and human leukocyte antigen G (HLA-G) on their surfaces, cause T cells that express the appropriate ligands to bind, which is promptly accompanied by the death of migratory T cells.

10.7 Clinical Presentation

The incubation period of rabies is between 4 weeks to 12 weeks days but can go up to several weeks to many years after an exposure (Rupprecht 1996).

The clinical manifestation has three stages (Table 10.1), which are

- Prodromal phase.
- Acute neurological phase- can either be encephalitic (80% cases) or paralytic (20% cases).
- Coma and death- after the neurological phase the patient goes into coma and eventually dies within 2 weeks. Death is almost certain and recovery and survival is rare.
- The classical symptom of rabies is **hydrophobia**. It occurs in 33 to 50% of patients (Tongavelona et al. 2018). There is discomfort in throat and difficulty in swallowing followed by sudden development of fear for water. The terror of

Table 10.1 Clinical phases of rabies disease (Rupprecht 1996)

Phase	Clinical manifestations
Prodromal phase	Iideally 2 to10 days; marked with nonspecific symptoms: • Fever • Malaise • Anorexia • Nausea and vomiting • Photophobia • Abnormal sensation • Sore throat • Pain at the site of bite
Excitation phase	• Restlessness and tremor • Pharyngeal and laryngeal spasm • Aerophobia • Hydrophobia • Respiratory distress and cardiac arrythmias • Hallucination • Hypertension and autonomic dysfunction • Change in tone and pitch of voice • Death
Paralytic phase	It mostly occurs in persons who are partly vaccinated or infected with bat rabies virus • Flaccid paralysis • Facial paralysis • Paralysis of all the four limbs • Unconsciousness • Death

water leads to involuntary spasms of the muscles of pharynx during attempts to drink (Koury and Warrington 2023; Tongavelona et al. 2018).

- **Aerophobia** is also pathognomonic of rabies (Mahadevan et al. 2016). Painful inspiratory spasms of the diaphragm and accessory inspiratory muscles can lead to aspiration, coughing, choking, vomiting, and hiccups; when severe, these spasms can lead to asphyxiation and respiratory arrest (Kietdumrongwong and Hemachudha 2005).
- **Autonomic hyperactivation** occurs in 25% of cases (Mahadevan et al. 2016). As a result of overactivity excess salivation, lacrimation, sweating, goose flesh, and mydriasis occur. Increased heart rate and cardiac arrhythmias are common and it may be related to myocarditis which results from direct viral injury (Alexander et al. 2021; Park et al. 2019).
- Difficulty in speaking, difficulty in swallowing, and double vision may occur. Dysphagia was reported in approximately half of all cases in one retrospective series (Kietdumrongwong and Hemachudha 2005).
- Aggression and excitation are also common (approximately 50% of patients). During the aggression phase patient is restless, agitated, and disoriented. Hallucination is a common symptom in this phase. After a period marked by agitation, the patient often goes into a period of calmness (Koury and Warrington 2023).
- Respiratory and cardiac complication is the leading cause of death in rabies patients (Alexander et al. 2021).

10.8 Diagnosis

The diagnosis of rabies requires a detailed history taking from the patient and a high level of suspicion. Rabies should be kept in mind as a differential diagnosis in all patients presenting with acute progressive encephalitis irrespective of history of animal bite or exposure (Chacko et al. 2017; Madhusudana and Sukumaran 2008).

There are two methods—direct tests and indirect tests.

Direct tests:

- Antigen detection- sellers staining, direct fluorescence antibody test (DFAT), direct rapid immunohistochemistry test (DRIT), rapid antigen detection test (RADT) (Lembo et al. 2006; Torquato et al. 2020).
- Molecular methods- nucleic acid amplification test (NAAT).
- Virus isolation- rabies tissue culture inoculation tests (RTCIT), mouse inoculation test/biological test (MIT/BT).

Indirect tests:

- Serology—Enzyme linked immunosorbent assay (ELISA), Rapid fluorescent focus inhibition test (RFFIT), indirect fluorescence antibody testing (IFA) (Debnath et al. 2019).

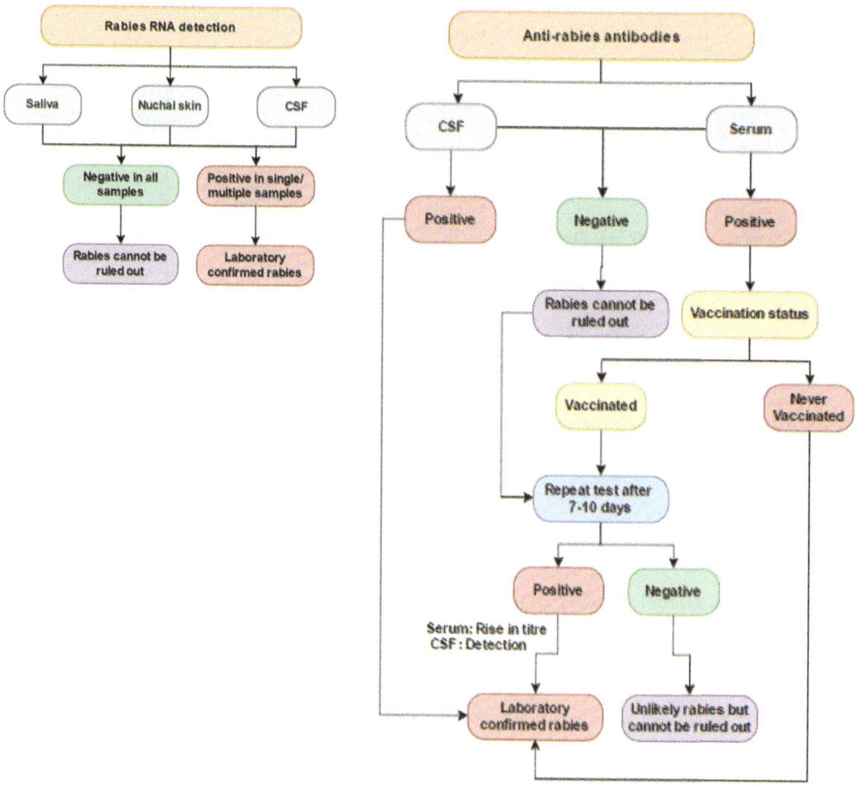

Fig. 10.5 Antemortem laboratory confirmation of rabies

The diagnosis of rabies can be made both antemortem and postmortem. All samples from suspected cases of rabies should be considered as highly infectious and must be properly sealed and labeled in leak-proof containers. Antemortem samples include serum, saliva, CSF, biopsy from skin and corneal impression (Fig. 10.5). Postmortem samples which can be collected are brain, spinal cord, and salivary gland (Fig. 10.6) (Jackson 2011; Madhusudana and Sukumaran 2008).

10.8.1 Sample Collection

10.8.1.1 Saliva

Around 500 μL saliva is collected in a sterile container and sealed securely. No preservative is required. Laboratory tests to be performed include detection of rabies RNA by RT-PCR technique (Madhusudana and Sukumaran 2008; Singh and Ahmad 2018).

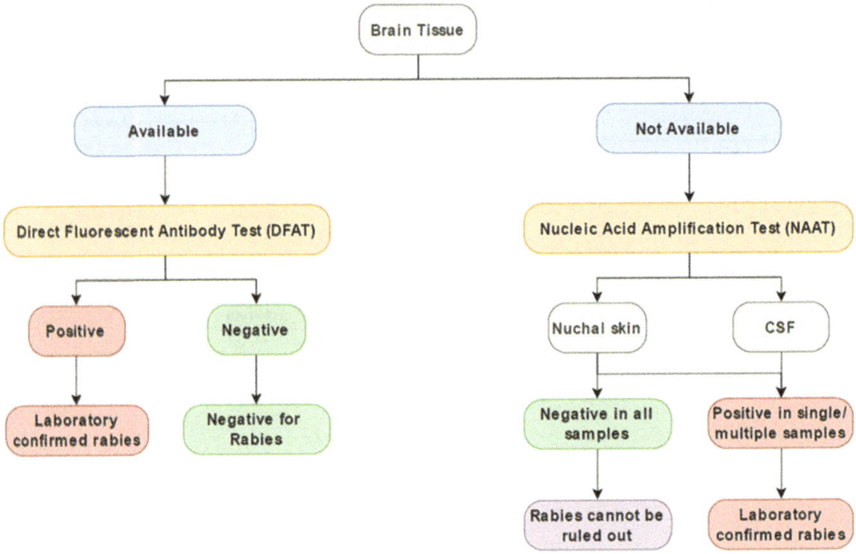

Fig. 10.6 Postmortem laboratory confirmation of rabies

10.8.1.2 Cornea

Eyelids are retracted using thumb and finger. A clean slide is pressed against the cornea. Excess pressure is avoided to prevent any damage to the eye. Corneal impression is air-dried for 10–15 min at room temperature. Then it is treated with chilled acetone and processed further for Direct Fluorescent Antibody Test (DFAT) (Zaidman and Billingsley 1998).

10.8.1.3 Skin Biopsy

From the nape of the neck, 5 to 6 mm of skin tissue sample, with an approximate depth of 5 to 7 mm, is procured. The specimen should contain hair follicles and cutaneous nerves at the base of hair follicles. An excision or punch biopsy may be collected (Madhusudana and Sukumaran 2008; Singh and Ahmad 2018). The skin biopsy is placed on the slide and covered with a piece of sterile moistened gauze to prevent specimen from drying. No fixative is required. The sample can be used for DFAT/DRIT for detecting viral antigens. RT-PCR for detecting viral nucleic acid, and RTCIT for virus isolation.

10.8.1.4 CSF

Two to three millilitre each of serum and CSF is collected in a sterile vial with all aseptic precautions. If vaccine or immunoglobulin has not been given, the presence of antibody to rabies virus in the serum can be diagnosed. Antibody to rabies virus in the CSF suggests a rabies virus infection regardless of the immunization history. Laboratory tests for antibodies include ELISA, indirect immunofluorescence, and virus neutralization. CSF samples can also be processed for RTCIT and molecular technique (NAAT-PCR) (Mani and Madhusudana 2013; Torquato et al. 2020).

10.8.2 Laboratory Testing

10.8.2.1 Negri Bodies

Some viral infections are linked with the production of inclusion bodies in the affected cells. These inclusion bodies are of two types- intranuclear and intracytoplasmic. In Rabies infection, intracytoplasmic inclusion bodies commonly known as Negri bodies which are acidophilic in nature can be demonstrated. Negri body-staining is carried out by the seller's staining test (SST) (Lahaye et al. 2009).

10.8.2.2 DFAT

It is considered as gold standard test. It works on the following principle: Rabies-specific antibodies and antigens combine to form an antigen—antibody complex when mixed and kept under optimum condition The complex formed is not visible to the naked eye. Fluorescent dyes, like fluorescein isothiocyanate (FITC), can be used to visualize this antigen–antibody complex under a fluorescence microscope. To achieve this, the FITC is tagged with an anti-rabies antibody to form the conjugate. As the conjugate directly binds to the antigen, the process is called the direct fluorescent antibody test (DFAT) (Realegeno et al. 2018; Vengatesan et al. 2006).

10.8.2.3 DRIT

Detects rabies antigen in the sample based on specific antigen–antibody reaction followed by detection using a compound microscope. The formaldehyde fixed smears are incubated with polyclonal or monoclonal anti-rabies antibodies which are labeled with biotin moiety. The unbound antibodies are washed away. The biotin moiety has a significant affinity toward streptavidin. The slide is then incubated with streptavidin conjugated with horseradish peroxidase (HRP). After removing unbound reagent, the slide is incubated with chromogen substrate (amino-ethyl carbazole, AEC) in the presence of Hydrogen Peroxide (H_2O_2). The substrate is converted into an insoluble red precipitate which is visible under compound microscope (Debnath et al. 2019).

10.8.2.4 RADT

It is based on immunochromatographic principle in lateral flow format. The assay employs a lateral flow device which has an anti-rabies antibody immobilized on a nitrocellulose membrane. The reagent has a secondary anti-rabies antibody which forms a complex with the virus present in the sample. The secondary antibody is usually tagged with a reporter dye or colloidal gold. Formed complexes in the prepared samples chromatographically move along the lateral flow device till they are captured by immobilized anti-rabies antibodies present at designated bands. This results in formation of a colored band at the site of immobilized anti-rabies antibody in lateral flow device. The band can be observed visibly. A second immobilized antibody captures a control protein from the sample. This control should be clearly visible to interpret the assay results (Wang et al. 2010).

10.8.2.5 NAAT

It is based on the detection of genomic material of the target organism in the sample. Reverse Transcriptase Polymerase Chain Reaction (RT-PCR) can be used for Rabies RNA detection in suspected samples. Therefrom, extracted samples are tested following pan-lyssavirus nested-PCR protocols. Initial reactive processes involve degenerate primers to amplify RNA of all lyssaviruses followed by a second reaction using both pan-lyssavirus degenerate primers and rabies-specific primers. The protocol confirms the presence of rabies and non-rabies lyssavirus RNA in the suspected sample (Biswal et al. 2012).

10.8.2.6 ELISA

It is used to demonstrate rabies antibodies in human and animal (dogs, cats, fox) serum. The assay is a quantitative indirect ELISA which uses WHO's (human samples) and WOAH's (animal samples) traceable standards for quantification of antibodies. The microwells are coated with rabies glycoprotein extracted from inactivated and purified rabies membrane. Therefore, the antibodies detected are specific to rabies glycoprotein (Piza et al. 1999; Servat et al. 2007).

10.8.2.7 RFFIT

It is a type of rabies virus neutralization test. It is carried out in cell culture to demonstrate the rabies virus antibody level in human or animal sera. Immunofluorescence staining is used as an indicator of viral growth (Burgado et al. 2018).

10.8.2.8 RTCIT

It is based on the ability of rabies virus to infect some cell lines in vitro. Murine neuroblastoma 2A (MNA) has been shown to be the most sensitive cell line for this method. However, Baby Hamester Kidney (BHK) cell line is easy to maintain and commonly employed for RTCIT. The growth of the virus in the cell culture can be ascertained by a sensitive technique such as DFAT (Chhabra et al. 2007).

10.8.2.9 MIT/BT

The growth of rabies virus takes place only in living tissues (Rupprecht 1996). These living tissue could be animals, chick embryo, or tissue culture (Koprowski et al. 1954). Rabies virus is pathogenic to all animals when introduced via intracerebral route. Animals for experimental studies are mouse, rat, guinea pigs, hamster, rabbit, and dog (Jackson 2011). RABV-positivity in any sample may be determined by intracerebral inoculation of the specimen into mice, and also by keeping the animal under observation to check for any sickness development or RABV-induced death (Madhusudana and Sukumaran 2008; Rupprecht 1996).

Table 10.2 Differential diagnosis of rabies: different phases of rabies resembles closely to conditions as given in clinical manifestation

Phase	Clinical manifestation
Prodromal phase	• Nonspecific viral illness • Mononucleosis • Meningitis • Bacteremia • Hysteria • Encephalitis
Encephalitis phase	• Herpes simplex virus encephalitis • West Nile virus encephalitis • Central nervous system vasculitis • Toxic or metabolic encephalopathy • Auto immune encephalitis • Delirium tremens • Belladonna poisoning
Paralytic phase	• Acute polyneuritis • Acute transverse myelitis • Poliomyelitis • Guillain Barre syndrome • Neuromuscular junction disorders

10.9 Differential Diagnosis

The nonspecific prodromal phase of rabies may be confused with a wide range of disorders (Table 10.2). In patients with signs and symptoms of encephalitis, more common infection such as herpes simplex virus, west Nile virus and other non-infectious disorders of the CNS should be ruled out. Other causes of muscular rigidity that can be seen with rabies includes: Tetanus, dystonia, and strychnine poisoning.

10.10 Management and Prevention

In rabies endemic countries, there is continuous animal to animal transmission and therefore any animal bite should be considered as bite by rabid animal and postexposure prophylaxis (PEP) should be started as early (Menezes 2008; Radhakrishnan et al. 2020). To maintain uniformity, WHO has classified type of exposure into three classes and has recommended appropriate PEP (Table 10.3) (O'Brien and Nolan 2019).

10.10.1 Wound Toilet

Mostly the virus enters the body through a bite or scratch, so it is important to remove the infected saliva from the wound (Di Quinzio and McCarthy 2008). The wound should be washed with soap and water thoroughly (Scholand et al. 2022). Antiseptic with virucidal activity can be used. Rabies virus multiply locally at the

Table 10.3 WHO guidelines for recommended postexposure treatment against rabies based on categories of exposure

Category	Type of exposure	Recommended postexposure prophylaxis
I	Touching or feeding of animals Licking on unbroken skin Exposure of unbroken skin with secretions/excretions of infected animal/human case	None if reliable case history is available Wash exposed area with water and soap
II	Nibbling of uncovered skin Small scratches or abrasions with no bleeding	Wound cleaning Rabies vaccination
III	One or many transdermal bites or scratches Lick on non-intact skin Staining of mucous membrane with saliva	Wound cleaning Rabies immunoglobulin Rabies vaccination

bitten for a long period and hence wound cleaning must be done even if the patient comes late. Tetanus prophylaxis should be given as per guidelines. Local treatment along with proper instillation of ERIG or HRIG in the wound is also given (Liu et al. 2017; Pounder 2005).

10.10.2 Passive Immunization

10.10.2.1 Equine Rabies Immunoglobulin (ERIG)

Immunoglobulin gives passive immunity in the form of ready-made anti-rabies antibody to overcome the acute phase of the infection (Terryn et al. 2016). RIG binds with the rabies virus, and makes the virus noninfective (Bharti et al. 2017). Immunoglobulin should be started as soon as possible, but can be given up to 7 days.

10.10.2.2 Human Rabies Immunoglobulins (HRIG)

HRIG do not possess the side effects encountered with ERIG. HRIG has a longer half-life, and so their dose is half of ERIG. The Immunoglobulin should be brought to room temperature (20–25 °C) before use (Haradanhalli et al. 2022).

10.10.2.3 Dose of Rabies Immunoglobulins (RIG)

The dose of ERIG is 40 IU per kg body weight and is given after testing of sensitivity, up to a maximum of 3000 IU can be given. (Bharti et al. 2017). The dose of HRIG is 20 IU per kg body weight, maximum of 1500 IU can be given (Madhusudana et al. 2013).

10.10.3 Active Immunization

Administering potent and safe Cell Culture Vaccines (CCVs) bring about active immunization. There is a Rabies immunization schedule (Table 10.4) that has been issued by the World Health Organization (WHO) (O'Brien and Nolan 2019).

Table 10.4 Vaccination routes and their respective schedules for postexposure prophylaxis (PEP) and pre-exposure prophylaxis (PrEP) for rabies virus

Types of prophylaxis	Route of administration*	Dose of vaccine	Day of dose	Injection per visit	No. of visits
Postexposure prophylaxis	ID	0.1 mL per dose	0, 3, 7, and 28	2	4
	IM	1 entire vaccine vial	0, 3, 7, 14, and 28	1	5
Pre-exposure prophylaxis	ID	0.1 mL per dose	0, 7, [plus 21 or 28 in previous regimen]	1	2 [3 in previous regimen]
	IM	1 entire vaccine vial	0, 7 [plus 21 or 28 in previous regimen]	1	2 [3 in previous regimen]
Re-exposure	ID	0.1 mL per dose	0 and 3	1	2
	IM	1 entire vaccine vial	0 and 3	1	2

Immunization with live attenuated vaccines produce humoral and cell mediated immunity within a week and provides protection (Overduin et al. 2019). Immunoglobulin gives instant passive immunity with proven effectiveness over the initial few weeks after exposure in humans (Haradanhalli et al. 2022). The antibodies help in managing the growth of rabies virus infection as they are capable of neutralizing the virus. The induction of a potent cytotoxic T lymphocyte (CTL) response appears to be related to both the effectiveness of postexposure immunization and the long-term benefits of vaccine-induced prophylaxis against rabies (Venkataswamy et al. 2015).

Intramuscular Regime for Rabies Postexposure Prophylaxis—There are three intramuscular schedules for category 2 and 3 exposures:

- **The 5-dose regimen (1–1–1-1-1):**
  Vaccination is given on day 0, 3, 7, 14, and 28.
  Site: Deltoid region for adults and anterolateral aspect of thigh in children.
- **The 2–1-1 regimen (2–0–1-0-1).**
  Vaccination is given on: 2 doses on day 0, one dose on day,7 and one on day 21.
- The 4-dose regimen with RIG in both category 2 and 3.

Intradermal Regime for Rabies Postexposure Prophylaxis:

- 2 Site intradermal method (2–2–2–0–2) given on day 0, 3, 7, 28.
- One dose of vaccine 0.1 mL intradermally at two different sites.

Site of injection:

- **Adult**- deltoid muscle.
- **Infants** and **children**- anterolateral thigh.

References

Alexander B, Lopez-Lopez JP, Saldarriaga C, Ponte-Negretti CI, Lopez-Santi R, Perez GE, del Sueldo M, Lanas F, Liblik K, Baranchuk A (2021) Rabies and the heart. Cardiol Res 12:53–59. https://doi.org/10.14740/cr1216

Bharti OK, Madhusudana SN, Wilde H (2017) Injecting rabies immunoglobulin (RIG) into wounds only: a significant saving of lives and costly RIG. Hum Vaccin Immunother 13:762–765. https://doi.org/10.1080/21645515.2016.1255834

Biswal M, Ratho RK, Mishra B (2012) Role of reverse transcriptase polymerase chain reaction for the diagnosis of human rabies. Indian J Med Res 135:837–842

Blanton JD, Dyer J, McBrayer J, Rupprecht CE (2012) Rabies surveillance in the United States during 2011. J Am Vet Med Assoc 241:712–722. https://doi.org/10.2460/javma.241.6.712

Burgado J, Greenberg L, Niezgoda M, Kumar A, Olson V, Wu X, Satheshkumar PS (2018) A high throughput neutralization test based on GFP expression by recombinant rabies virus. PLoS Negl Trop Dis 12:e0007011. https://doi.org/10.1371/journal.pntd.0007011

Burrell CJ, Howard CR, Murphy FA (2017) Pathogenesis of virus infections. In: Fenner and White's medical virology, pp 77–104. https://doi.org/10.1016/B978-0-12-375156-0.00007-2

Chacko K, Parakadavathu RT, Al-Maslamani M, Nair AP, Chekura AP, Madhavan I (2017) Diagnostic difficulties in human rabies: a case report and review of the literature. Qatar Med J 2016:15. https://doi.org/10.5339/qmj.2016.15

Chhabra M, Mittal V, Jaiswal R, Malik S, Gupta M, Lal S (2007) Development and evaluation of an in vitro isolation of street rabies virus in mouse neuroblastoma cells as compared to conventional tests used for diagnosis of rabies. Indian J Med Microbiol 25:263–266. https://doi.org/10.4103/0255-0857.34772

Debnath A, Pathak DC, Ramamurthy N, Mohd G, Pandey AB, Upmanyu V, Tiwari AK, Saravanan R, Chellappa MM, Dey S (2019) Serological profiling of rabies antibodies by enzyme-linked immunosorbent assay and its comparative analysis with rapid fluorescent focus inhibition test in mouse model. Vet World 12:126–130. https://doi.org/10.14202/vetworld.2019.126-130

Di Quinzio M, McCarthy A (2008) Rabies risk among travellers. CMAJ 178:567. https://doi.org/10.1503/cmaj.071443

Dietzgen RG, Kondo H, Goodin MM, Kurath G, Vasilakis N (2017) The family Rhabdoviridae: mono- and bipartite negative-sense RNA viruses with diverse genome organization and common evolutionary origins. Virus Res 227:158–170. https://doi.org/10.1016/j.virusres.2016.10.010

Gérard FCA, Bourhis J-M, Mas C, Branchard A, Vu DD, Varhoshkova S, Leyrat C, Jamin M (2022) Structure and dynamics of the unassembled nucleoprotein of rabies virus in complex with its phosphoprotein chaperone module. Viruses 14:2813. https://doi.org/10.3390/v14122813

Grill AK (2009) Approach to management of suspected rabies exposures. Can Fam Physician 55:247–251

Hampson K, Ventura F, Steenson R, Mancy R, Trotter C, Cooper L, Abela-Ridder B, Knopf L, Ringenier M, Tenzin T, Ly S, Tarantola A, Moyengar R, Oussiguéré A, Bonfoh B, Narayana DA, Sudarshan MK, Muturi M, Mwatondo A, Wambura G, Andriamandimby SF, Baril L, Edosoa GT, Traoré A, Jayme S, Kotzé J, Gunesekera A, Chitnis N, Hattendorf J, Laager M, Lechenne M, Zinsstag J, Changalucha J, Mtema Z, Lugelo A, Lushasi K, Yurachai O, Metcalf

CJE, Rajeev M, Blanton J, Costa GB, Sreenivasan N, Wallace R, Briggs D, Taylor L, Thumbi SM, Huong NTT (2019) The potential effect of improved provision of rabies post-exposure prophylaxis in Gavi-eligible countries: a modelling study. Lancet Infect Dis 19:102–111. https://doi.org/10.1016/S1473-3099(18)30512-7

Haradanhalli RS, Fotedar N, Kumari N, Narayana DHA (2022) Safety and clinical efficacy of human rabies immunoglobulin in post exposure prophylaxis for category III animal exposures. Hum Vaccin Immunother 18:2081024. https://doi.org/10.1080/21645515.2022.2081024

Jackson AC (2011) Update on rabies. Res Rep Trop Med 2:31–43. https://doi.org/10.2147/RRTM.S16013

Juntrakul S, Ruangvejvorachai P, Shuangshoti S, Wacharapluesadee S, Hemachudha T (2005) Mechanisms of escape phenomenon of spinal cord and brainstem in human rabies. BMC Infect Dis 5:104. https://doi.org/10.1186/1471-2334-5-104

Kietdumrongwong P, Hemachudha T (2005) Pneumomediastinum as initial presentation of paralytic rabies: a case report. BMC Infect Dis 5:92. https://doi.org/10.1186/1471-2334-5-92

King AMQ, Adams MJ, Carstens EB, Lefkowitz EJ (eds) (2012) Family—Rhabdoviridae, in: virus taxonomy. Elsevier, San Diego, CA, pp 686–713. https://doi.org/10.1016/B978-0-12-384684-6.00057-4

Koprowski H, Black J, Nelsen DJ (1954) Studies on chick-embryo-adapted-rabies virus. VI. Further changes in pathogenic properties following prolonged cultivation in the developing chick embryo. J Immunol 72:94–106

Koury R, Warrington SJ (2023) Rabies. In: StatPearls. StatPearls Publishing, Treasure Island (FL)

Lahaye X, Vidy A, Pomier C, Obiang L, Harper F, Gaudin Y, Blondel D (2009) Functional characterization of Negri bodies (NBs) in rabies virus-infected cells: evidence that NBs are sites of viral transcription and replication. J Virol 83:7948–7958. https://doi.org/10.1128/JVI.00554-09

Lembo T, Niezgoda M, Velasco-Villa A, Cleaveland S, Ernest E, Rupprecht CE (2006) Evaluation of a direct, rapid Immunohistochemical test for rabies diagnosis. Emerg Infect Dis 12:310–313. https://doi.org/10.3201/eid1202.050812

Lian M, Hueffer K, Weltzin MM (2022) Interactions between the rabies virus and nicotinic acetylcholine receptors: a potential role in rabies virus induced behavior modifications. Heliyon 8:e10434. https://doi.org/10.1016/j.heliyon.2022.e10434

Liu Q, Wang X, Liu B, Gong Y, Mkandawire N, Li W, Fu W, Li L, Gan Y, Shi J, Shi B, Liu J, Cao S, Lu Z (2017) Improper wound treatment and delay of rabies post-exposure prophylaxis of animal bite victims in China: prevalence and determinants. PLoS Negl Trop Dis 11:e0005663. https://doi.org/10.1371/journal.pntd.0005663

Madhusudana SN, Sukumaran SM (2008) Antemortem diagnosis and prevention of human rabies. Ann Indian Acad Neurol 11:3–12. https://doi.org/10.4103/0972-2327.40219

Madhusudana SN, Ashwin BY, Sudarshan S (2013) Feasibility of reducing rabies immunoglobulin dosage for passive immunization against rabies: results of in vitro and in vivo studies. Hum Vaccin Immunother 9:1914–1917. https://doi.org/10.4161/hv.25431

Mahadevan A, Suja MS, Mani RS, Shankar SK (2016) Perspectives in diagnosis and treatment of rabies viral encephalitis: insights from pathogenesis. Neurotherapeutics 13:477–492. https://doi.org/10.1007/s13311-016-0452-4

Mani RS, Madhusudana SN (2013) Laboratory diagnosis of human rabies: recent advances. ScientificWorldJournal 2013:569712. https://doi.org/10.1155/2013/569712, 1

Menezes R (2008) Rabies in India. CMAJ 178:564–566. https://doi.org/10.1503/cmaj.071488

O'Brien KL, Nolan T (2019) The WHO position on rabies immunization—2018 updates. Vaccine 37:A85–A87. https://doi.org/10.1016/j.vaccine.2018.10.014

Overduin LA, van Dongen JJM, Visser LG (2019) The cellular immune response to rabies vaccination: a systematic review. Vaccines (Basel) 7:110. https://doi.org/10.3390/vaccines7030110

Park SC, Crane IM, Pal K, Cagnina RE (2019) Rabies encephalitis with myocarditis mimicking ST-elevation myocardial infarction. Open Forum Infect Dis 6:ofz260. https://doi.org/10.1093/ofid/ofz260

Piza AS, Santos JL, Chaves LB, Zanetti CR (1999) An ELISA suitable for the detection of rabies virus antibodies in serum samples from human vaccinated with either cell-culture vaccine or suckling-mouse-brain vaccine. Rev Inst Med Trop Sao Paulo 41:39–43. https://doi.org/10.1590/s0036-46651999000100008

Pounder D (2005) Avoiding rabies. BMJ 331:469–470

Radhakrishnan S, Vanak AT, Nouvellet P, Donnelly CA (2020) Rabies as a public health concern in India—a historical perspective. Trop Med Infect Dis 5:162. https://doi.org/10.3390/tropicalmed5040162

Realegeno S, Niezgoda M, Yager PA, Kumar A, Hoque L, Orciari L, Sambhara S, Olson VA, Satheshkumar PS (2018) An ELISA-based method for detection of rabies virus nucleoprotein-specific antibodies in human antemortem samples. PLoS One 13:e0207009. https://doi.org/10.1371/journal.pone.0207009

Rupprecht CE (1996) Rhabdoviruses: rabies virus. In: Baron S (ed) Medical microbiology. University of Texas Medical Branch at Galveston, Galveston (TX)

Sachdeva A, Vinay Kumar T, Yadav G, Tondwal J (2022) Epidemiological profile and management practices of animal bite cases in a tertiary care hospital of Haryana: a cross-sectional study. J Family Med Prim Care 11:7886–7890. https://doi.org/10.4103/jfmpc.jfmpc_549_22

Scholand SJ, Quiambao BP, Rupprecht CE (2022) Time to revise the WHO categories for severe rabies virus exposures–category IV? Viruses 14:1111. https://doi.org/10.3390/v14051111

Scott TP, Nel LH (2016) Subversion of the immune response by rabies virus. Viruses 8:231. https://doi.org/10.3390/v8080231

Scott TP, Nel LH (2021) Lyssaviruses and the fatal encephalitic disease rabies. Front Immunol 12:786953. https://doi.org/10.3389/fimmu.2021.786953

Servat A, Feyssaguet M, Blanchard I, Morize JL, Schereffer JL, Boue F, Cliquet F (2007) A quantitative indirect ELISA to monitor the effectiveness of rabies vaccination in domestic and wild carnivores. J Immunol Methods 318:1–10. https://doi.org/10.1016/j.jim.2006.07.026

Singh CK, Ahmad A (2018) Molecular approach for ante-mortem diagnosis of rabies in dogs. Indian J Med Res 147:513–516. https://doi.org/10.4103/ijmr.IJMR_1705_15

Terryn S, Francart A, Rommelaere H, Stortelers C, Van Gucht S (2016) Post-exposure treatment with anti-rabies VHH and vaccine significantly improves protection of mice from lethal rabies infection. PLoS Negl Trop Dis 10:e0004902. https://doi.org/10.1371/journal.pntd.0004902

Tongavelona JR, Rakotoarivelo RA, Andriamandimby FS (2018) Hydrophobia of human rabies. Clin Case Rep 6:2519–2520. https://doi.org/10.1002/ccr3.1846

Torquato RBC, Iamamoto K, Fernandes ER, Achkar S, Silva SR, Katz ISS, Guedes F (2020) Detection of rabies virus antigen by the indirect rapid immunohistochemistry test in equines and comparisons with other diagnostic techniques. Zoonoses Public Health 67:651–657. https://doi.org/10.1111/zph.12745

Tsiang H, Lycke E, Ceccaldi P-E, Ermine A, Hirardot X (1989) The anterograde transport of rabies virus in rat sensory dorsal root ganglia neurons. J Gen Virol 70:2075–2085. https://doi.org/10.1099/0022-1317-70-8-2075

Unwin N (2013) Nicotinic acetylcholine receptor and the structural basis of neuromuscular transmission: insights from torpedo postsynaptic membranes. Q Rev Biophys 46:283–322. https://doi.org/10.1017/S0033583513000061

Vengatesan D, Raj GD, Raja A, Ramadass P, Gunaseelan L (2006) Detection of rabies virus antigen or antibody using flow cytometry. Cytometry B Clin Cytom 70:335–343. https://doi.org/10.1002/cyto.b.20104

Venkataswamy MM, Madhusudana SN, Sanyal SS, Taj S, Belludi AY, Mani RS, Hazra N (2015) Cellular immune response following pre-exposure and postexposure rabies vaccination by intradermal and intramuscular routes. Clin Exp Vaccine Res 4:68–74. https://doi.org/10.7774/cevr.2015.4.1.68

Wang H, Feng N, Yang S, Wang C, Wang T, Gao Y, Su J, Zheng X, Hou X, Huang H, Yang R, Zou X, Huang G, Xia X (2010) A rapid immunochromatographic test strip for detecting rabies virus antibody. J Virol Methods 170:80–85. https://doi.org/10.1016/j.jviromet.2010.09.002

Willoughby RE (2012) 228—rabies virus. In: Long SS (ed) Principles and practice of pediatric infectious diseases, 4th edn. Elsevier, London, pp 1145–1149.e2. https://doi.org/10.1016/B978-1-4377-2702-9.00230-0

Yibrah M, Damtie D (2015) Incidence of human rabies exposure and associated factors at the Gondar health center, Ethiopia: a three-year retrospective study. Infect Dis Poverty 4:3. https://doi.org/10.1186/2049-9957-4-3

Zaidman GW, Billingsley A (1998) Corneal impression test for the diagnosis of acute rabies encephalitis. Ophthalmology 105:249–251. https://doi.org/10.1016/s0161-6420(98)92860-3

Zhu J, Pan J, Lu Y (2015) A case report on indirect transmission of human rabies. J Zhejiang Univ Sci B 16:969–970. https://doi.org/10.1631/jzus.B1500109

Miscellaneous CNS Viral Infections: Underappreciated Causes of Neurologic Disease

11

Hiba Sami, Safiya Firoze, and Parvez A. Khan

Abstract

Lymphocytic choriomeningitis virus (LCMV) is from the family, Arenaviridae, having single- stranded RNA genome. Adult lymphocytic choriomeningitis viral infection has a biphasic presentation, with a primary phase of fever, malaise, and myalgias. The next phase of neurological disease involves aseptic meningitis, with headaches, fever, photophobia, nuchal rigidity, and vomiting. Congenital infection is generally diagnosed by immunofluorescent antibody testing and enzyme- linked immunoassays. Another important miscellaneous cause of CNS infection is coronavirus. Although coronavirus disease—2019 (COVID-19) principally targets the pulmonary system, it has come to be recognized as a potential neuropathogen. Impaired consciousness/encephalopathy is a widely documented symptom of COVID-19 among the central nervous system (CNS) symptoms. Severely critical cases have been reported of COVID-19-associated encephalitis, in which the patients were critically ill, presenting with new-onset seizures and delirium. We also discuss crucial features of prion-related CNS disorders, as viruses and prions are intimately related. Prion disorders are a category of closely related neurodegenerative illnesses that afflict both animals and humans. Creutzfeldt–Jakob disease (CJD) and Gerstmann–Sträussler–Scheinker (GSS) in humans, bovine spongiform encephalopathy (BSE, or "mad cow disease") in cattle, chronic wasting disease (CWD) in mule deer and elk, and scrapie in sheep are among the diseases covered. All prion illnesses are fatal, and there is now no effective treatment; nevertheless, recent advances in our understanding of their aetiology have led to the prospect of viable therapeutic approaches in the near future.

H. Sami · S. Firoze (✉) · P. A. Khan
Department of Microbiology, Jawaharlal Nehru Medical College Hospital, Aligarh, Uttar Pradesh, India

H. Sami et al. (eds.), *Viral and Fungal Infections of the Central Nervous System: A Microbiological Perspective*, https://doi.org/10.1007/978-981-99-6445-1_11

Keywords

Miscellaneous · Lymphocytic choriomeningitis virus · Neurological disease · Coronavirus · Prion disorders

11.1 Overview and List of Miscellaneous Viral CNS Infections

Infants and people with impaired immune systems are frequently affected by viral infections, which are often linked to central nervous system (CNS) involvement. Encephalitis, meningitis, myelitis, and very infrequent meningoencephalitis are the main clinical manifestations of viral CNS infections (Abdullahi et al. 2020). This comprehensive exploration of miscellaneous CNS viral infections in this chapter, which have not been covered elsewhere, emphasizes their underappreciated role in neurologic diseases across different stages of life. We have also tried to cover prions here. Though prions are not viruses, they share some similarities in terms of their ability to self-propagate and cause infection. By recognizing the potential impacts of these infections and considering them in differential diagnoses, clinicians can improve patient outcomes through early detection, timely interventions, and appropriate management strategies. Future research and collaboration are crucial to further unravel the complexities of these infections and develop effective preventive measures and treatments. Table 11.1 enlists miscellaneous causes of CNS viral infections.

Table 11.1 Miscellaneous viral causes of CNS infections

Infants
Parechovirus A Zika virus
Childhood
Rubella
All ages
Adenoviruses Borna disease virus (BDV) Ebolavirus Henipavirus Hepatitis C virus (HCV) Hepatitis E virus (HEV) John Cunningham virus (JCV) Lymphocytic choriomeningitis virus (LCMV) Severe acute respiratory syndrome coronavirus 2 (SARS-CoV-2) Prion-related CNS disorders (*proteinaceous agents; not viruses, but mentioned here*)

11.2 Parechovirus A

Parechovirus A is a member of the Picornaviridae family's Parechovirus genus that can cause serious sickness in children (Sridhar et al. 2019). The PeV-A virion has a diameter of around 30 nm and is made up of an RNA genome that is protected by an exterior protein capsid that is devoid of a lipid envelope (Sridhar et al. 2019). The 7300 nucleotide (nt) single-stranded positive-sense (+ss) genome has one open reading frame and two untranslated regions (UTRs) (Stanway et al. 1994). The genus Parechovirus consists of four species: Parechovirus A, Parechovirus B (formerly known as Ljungan virus), Parechovirus C (Sebokele virus), and Parechovirus D (ferret parechovirus). Each species represents a distinct group of viruses within the Parechovirus genus. Species PeV-A virus genotypes are capable of infecting humans and causing serious sicknesses such as meningoencephalitis, convulsions, or sepsis-like syndrome (Chiang et al. 2017). There are currently 19 PeV-A types named from PeV-A1 to PeV-A19; PeV-A1 is further split into clusters 1A and 1B (Williams et al. 2009).

Clinical Features Human parechovirus (hPeV) infections have been related to a broad range of clinical symptoms in children, from minor, self-limiting symptoms including rash, diarrhoea, and respiratory sickness to more serious symptoms like sepsis and meningoencephalitis (Felsenstein et al. 2014). Transmission mainly occurs by faeco-oral or respiratory routes, as well as through ambient environmental water sources.

Neonatal central nervous system (CNS) infections caused by HPeV have been linked to white matter abnormalities, leading to long-term consequences, as well as sudden infant death syndrome (Sedmak et al. 2010; Verboon-Maciolek et al. 2008). Neurological disease is generally associated with PeV3 infection. Full- term infants have an onset of disease within the first two weeks, whereas those born premature have an onset between two and three months. Though symptoms from enterovirus (EV) and hPeV are indistinguishable, studies have found that seizures and CNS symptoms are much more common in hPeV-infected children than in EV-infected children (Felsenstein et al. 2014). Even in the most severe cases, the CSF is typically normal, ruling out an infectious cause. Periventricular echogenicity can be seen on ultrasound, but 90% of deep white matter abnormalities can be seen on MRI.

Diagnosis Parechovirus has been identified predominantly by molecular techniques in blood, plasma, throat swabs, nasopharyngeal swabs, bronchoalveolar lavage, faeces, CSF, and tissue biopsies. In cell culture, most HPeV grows ineffectively or not at all. Today, laboratories use molecular methods that are more sensitive than culture and enable type assignment, such as reverse transcriptase polymerase chain reaction. Antibody neutralization was previously used to separate HPeV from EV (Shah and Robinson 2014).

Treatment There are presently no antiviral medications for PeV-A infection that have received clinical approval. Only supportive care and, very rarely, passive immunization with IVIG administration are available as treatments (Wildenbeest et al. 2010).

11.3 Rubella Virus

Rubella, often known as "German" or "3-day" measles, is a single-stranded, positive-sense enveloped RNA virus that belongs to the Togaviridae family. It typically results in an acquired postnatal infection, which is generally benign and self-limiting. This infection is commonly observed in children and is characterized by symptoms such as fever and rash, but the most significant manifestation of rubella infection is when it occurs during early pregnancy. The rubella virus can easily cross the placenta and infect the foetus, which can lead to a number of difficulties. These complications can include spontaneous abortion (miscarriage), stillbirth, or a severe combination of birth defects known as congenital rubella syndrome (CRS) (Frey 2008). Approximately 25 years ago, another central nervous system syndrome associated with the rubella virus was identified. This syndrome is known as progressive rubella panencephalitis, and it is characterized by a progressive and incapacitating encephalopathy. Unfortunately, this condition is uniformly fatal, meaning that it inevitably leads to death.

Prior to the widespread use of effective vaccines, rubella epidemics typically occurred every 6–9 years (Tyor and Harrison 2014). This pattern is still present in some parts of the world where routine immunization is not practised, such as Ethiopia (Getahun et al. 2016). Although the majority of congenital rubella syndrome (CRS)-related symptoms, including rash, fever, and lymphadenopathy, are mild and do not result in long-term consequences, vaccination is the main strategy for preventing these symptoms. The prevalence of rubella has greatly decreased thanks to vaccination campaigns that offer at-risk groups either the MMR or MMRV vaccines against measles, mumps, rubella, and varicella. These shots are normally given between the ages of 12 and 15 months, with a second dose given before to entering kindergarten or first grade. The high seroconversion rate (>95%) achieved through widespread vaccination has greatly contributed to the reduction in rubella cases and CRS (Tyor and Harrison 2014).

Clinical Features The rubella virus (RV) can result in four distinct clinical conditions. Firstly, it can cause an acquired postnatal infection, primarily affecting children, which is generally benign and self-limiting, characterized by fever and rash. Secondly, rubella encephalitis can occur as a complication of the acquired infection but is extremely rare, affecting less than 0.1% of patients. Thirdly, congenital rubella syndrome (CRS) can develop in infants born to mothers who acquired the virus during the first trimester of pregnancy. CRS is characterized by microcephaly, cataracts, and hearing loss. Rubella may also lead to progressive rubella panencephalitis, which is extremely uncommon. This condition, which includes neurological impairment, can appear ten years or more after either the CRS or the acquired postnatal infection.

Exanthem Illness Major symptoms of rubella include a febrile acute infection with the development of a rash, which is erythematous and maculopapular. The severity is less than that of measles (Bale 2014). Most cases of acquired postnatal rubella infection in children are mild or asymptomatic, with studies indicating that up to 40% of children may not show clinical signs of the disease. However, more severe cases tend to occur in adults (Tyor and Harrison 2014). In populations without vaccination, acute rubella infection is commonly observed in school-age children (Bale 2014). Patients with rubella are contagious for two weeks, starting one week before the onset of the rash and continuing for one week afterwards. In the majority of cases, infection or vaccination leads to lifelong immunity. Recurrent infections are rare and typically asymptomatic, without viraemia, and can only be detected through laboratory testing. Lymphadenopathy is a symptom of a clinically obvious rubella infection, and its absence should make other diagnoses more likely. The posterior auricular, posterior cervical, and suboccipital lymph nodes are all involved in the recognizable lymphadenopathy. This lymphadenopathy could last for a few weeks (Bale 2014; Tyor and Harrison 2014).

According to Figueiredo et al. (2008), encephalitis, aseptic meningitis, myelitis, and Guillain–Barre syndrome (GBS) are among the neurologic consequences that might result from a rubella infection. In cases of encephalitis, these sequelae commonly manifest six days after the rash with symptoms like headache, vomiting, somnolence, and convulsions. In rare instances of GBS, symmetric weakness with hyporeflexia may be observed (Connolly et al. 1975). It is important to note that neurologic complications occur in less than 0.1% of rubella cases.

CRS Congenital heart problems, such as patent ductus arteriosus or septal abnormalities, as well as sensorineural hearing loss and cataracts, are among the particular clinical symptoms that define congenital rubella syndrome (CRS). The rubella virus can infect different tissues, resulting in additional manifestations like chorioretinitis, intrauterine growth retardation, microphthalmia, hepatosplenomegaly, hepatitis, the "blueberry muffin" rash (Fig. 11.1), osteopathy, interstitial pneumonitis, or thrombocytopaenia, even though these are considered the hallmarks of CRS (Bale 2014). A study showed that infants infected with rubella before the 11th week of gestation primarily developed congenital heart disease and deafness, while those infected at 13–16 weeks mainly experienced deafness alone. No rubella-related defects were found in children infected after 16 weeks of gestation (Miller et al. 1982). Also at birth, around 25% of infants with congenital rubella syndrome (CRS) exhibit central nervous system (CNS) involvement, which can manifest as symptoms like lethargy, hypotonia (low muscle tone), irritability, or seizures (Desmond et al. 1967).

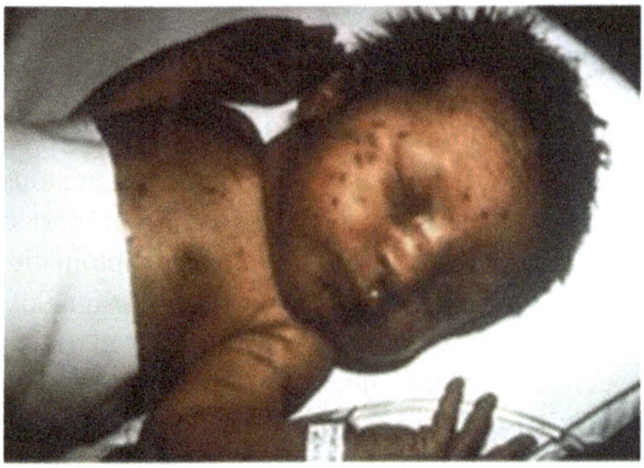

Fig. 11.1 Infant with "blueberry muffin" skin lesions with congenital rubella. Image Source: Centers for Disease Control and Prevention—Public Health Image Library—ID#713—(Lebrun 1978)—Copyright Restrictions: NONE; this image is in the public domain and thus free of any copyright restriction

Progressive Rubella Panencephalitis It is an extremely rare manifestation that can occur lately as a complication of congenital rubella syndrome (CRS) or acute rubella infection (Frey 2008). This neurologic disorder typically presents in the second decade of life and is characterized by symptoms such as spasticity, ataxia, intellectual deterioration, and seizures (Townsend et al. 1975). Compared to subacute sclerosing panencephalitis associated with measles, progressive rubella panencephalitis exhibits more prominent cerebellar symptoms (Rorke and Spiro 1967).

Diagnosis Routine laboratory findings in rubella infection may show leukopaenia (low white blood cell count), atypical lymphocytosis (abnormal increase in certain types of lymphocytes), or thrombocytopaenia (low platelet count). Cerebrospinal fluid (CSF) examination may show a lymphocytic pleocytosis (increased number of lymphocytes) and a higher protein concentration in cases of aseptic meningitis. Rubella-specific IgM antibodies can be used to support the diagnosis of infection, and rubella virus can be found in throat washings or by using the reverse transcription-polymerase chain reaction (RT-PCR) to find rubella virus RNA in CSF, urine, or saliva. The rubella virus' genotype can reveal important epidemiological data (Charlton and Severini 2016).

In cases of congenital rubella syndrome (CRS), CSF analysis typically shows increased protein concentration and lymphocytic pleocytosis (Shukla and Maraqa 2023). Neuroimaging studies such as ultrasonography, CT scans, or MRI scans often reveal findings such as atrophy, calcifications, cystic lesions, or leukomalacia in affected children. Rubella-specific IgM antibodies can be found in serum or CSF to support the diagnosis of CRS, and rubella virus RNA can be found in clinical samples or isolated from urine, CSF, or throat washings to provide confirmation. For several months, infants with CRS may continue to expel the virus in their saliva or urine. Rubella virus in amniotic fluid or rubella-specific IgM antibodies in foetal blood can be used for prenatal diagnosis of CRS.

Treatment Since there is presently no particular antiviral medicine available, supportive care is usually used to treat rubella virus infections, whether acquired or congenital. People who have developed rubella infections and neurologic sequelae typically make a full recovery with supportive care.

On the other hand, children who develop congenital rubella syndrome (CRS), particularly those who contract it during the first trimester of pregnancy, may experience long-term effects like microcephaly, developmental delays, epilepsy, growth impairment, behavioural problems, and permanent deafness. Sensorineural hearing loss is the most prevalent and severe consequence of CRS, affecting 50–60% of infected neonates. Diabetes mellitus, thyroid issues, and the uncommon neurodegenerative condition known as progressive post-rubella panencephalitis—which has certain clinical and EEG characteristics with subacute sclerosing panencephalitis (SSPE)—are other late sequelae of CRS (Townsend et al. 1975).

The measles, mumps, and rubella (MMR) immunization offers protection against both CRS and acquired rubella infections. The MMR vaccine is a useful tool for lowering the prevalence and consequences of rubella virus infections and the complications that go along with them (Bale 2014).

11.4 Hepatitis C Virus and Hepatitis E Virus: Associated Neurological Disorders

The positive-sense, enveloped, ssRNA virus known as hepatitis E virus (HEV) is a common global cause of acute viral hepatitis. HEV is divided into four main genotypes; genotypes 1 and 2 are spread through the consumption of tainted water sources in developing nations, while in both poor and wealthy countries, the ingestion of undercooked pork or game products regularly spreads genotypes 3 and 4 (Kamar et al. 2012). Although the majority of HEV infections are asymptomatic, about 5% of them can result in a range of clinical problems, including hepatitis, acute pancreatitis, kidney damage, neurological disorders, and other immune-mediated symptoms (Pischke et al. 2017). HEV is associated with various neurological syndromes, which include Bell's palsy, encephalitis, myelitis, mononeuritis multiplex, Guillain–Barré syndrome (GBS), neuralgic amyotrophy (NA), and GBS (Dalton et al. 2016). While the relationship between HEV and other neurological diseases has been observed, the causality is not as clear. The frequency of neurological diseases linked to HEV varies between 2.4% and 11% in different cohort studies conducted in the UK, France, Netherlands, and Bangladesh (Wang et al. 2018). The primary neurological manifestations associated with HEV infection are Guillain–Barré syndrome (GBS) and neuralgic amyotrophy (NA), both of which exhibit clinical similarities to other infectious causes. Nerve pain in arm and shoulder is a hallmark of neuralgic amyotrophy, and people frequently test positive for HEV RNA as soon as their symptoms start. On the other hand, HEV RNA is typically not detected in cases of GBS, but serological tests remain positive, indicating an immune-mediated mechanism (Dalton et al. 2016).

Hepatitis C virus, a well-known cause of blood- borne hepatitis, has a long history of links to a variety of neurological conditions, including meningitis, encephalitis, and inflammation of the brain's white matter (leukoencephalitis) (Forton et al. 2006; Zampino et al. 2013). Patients with chronic HCV infection commonly experience symptoms such as fatigue, depression, and cognitive impairment, with fatigue being present in approximately 80% of cases. Numerous investigations have found neurocognitive deficits in people with chronic HCV infection, even in the absence of hepatic encephalopathy. It is yet unknown if viral replication is supported by the central nervous system (CNS). Recent studies have shown that brain microvascular endothelial cells, which are in charge of creating the blood–brain barrier, contain HCV receptors. Interestingly, these endothelial cells are the only cells in the neuronal pool known to possess HCV receptors. This shows that the ability of microvascular endothelial cells to facilitate the entry of HCV into the CNS may be extremely important (Fletcher et al. 2010).

11.5 John Cunningham Virus

The John Cunningham virus (JCV), additionally recognized as human polyomavirus 2 (HPyV2), was previously referred to as a type of papovavirus (Saribas et al. 2010). In around 1965, ZuRhein along with Chou, and Silverman along with Rubinstein, used EM to identify it (Maginnis et al. 2015). Subsequently, culture isolation of the virus in an affected patient with the name of John Cunningham brought about its name. HPyV2-induced progressive multifocal leukoencephalopathy (PML) and other illnesses are seen primarily in association with immune disorders like acquired immunodeficiency syndrome (AIDS) or while taking immunosuppressive medications (such as those undergoing organ transplantation). HPyV 2 is responsible for the disorder affecting the CNS known as PML, which is marked by demyelination with frequent JCV detection in CSF and/or neural tissue (Saribas et al. 2010).

11.5.1 Viral Structure

The John Cunningham virus belongs to the human Betapolyomavirus genus of double-stranded DNA viruses (Boothpur and Brennan 2010). It has a 5.1 kilobase pair, closed, circular genome and is an icosahedral, nonenveloped DNA virus (Ahye et al. 2020). The genome of JCV has a bidirectional variety of expression (Fig. 11.2) (Saribas et al. 2010). T′ proteins (T′): T′135, T′136, and T,165 and antigens (Ag): small T (Sm-t)-Ag and large T (LT)-Ag are all encoded at HPyV2 genome's early coding region (Tyagarajan and Frisque 2006). Structural proteins: viral protein (VP)-1, VP-2, and VP-3 and regulating protein: agnoprotein are encoded in the genome's late coding area (Gasparovic et al. 2006).

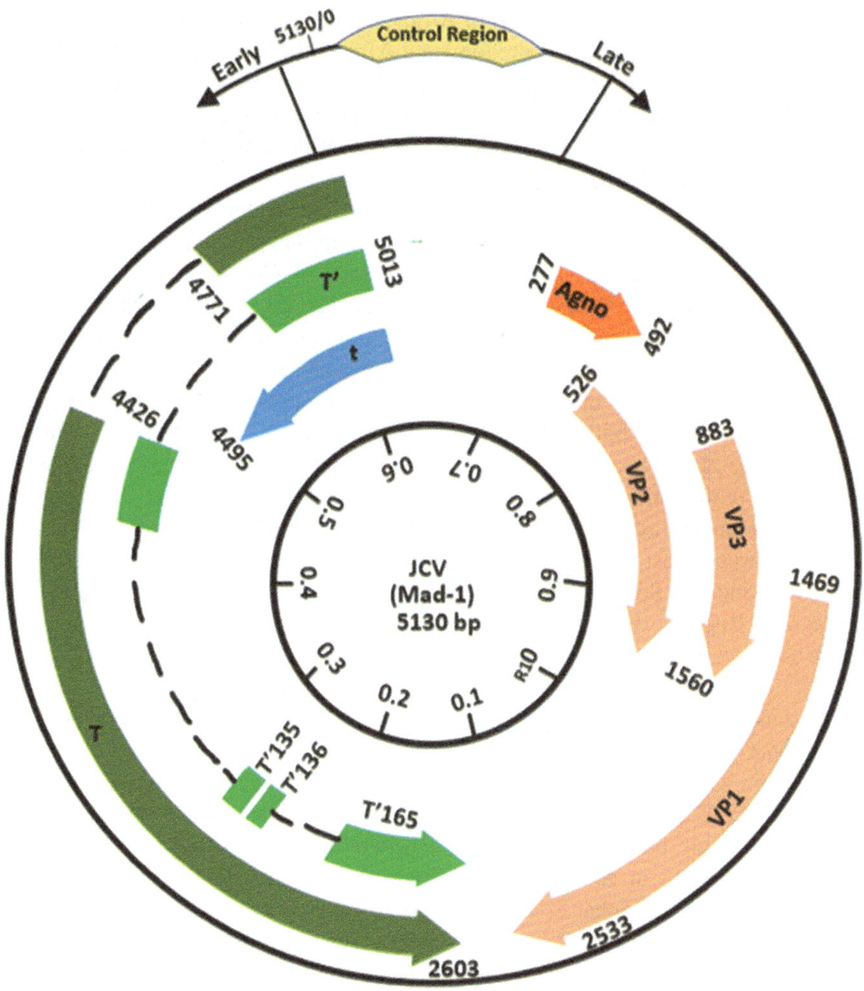

Fig. 11.2 A depiction of human polyomavirus 2's genome

11.5.2 Epidemiological Profile

John Cunningham virus is a relatively prevalent virus that affects between 70% and 90% of the world's population. Individuals typically contract the virus during their early years or as adolescents. Since it is present in urban sewerage in high proportions all around the world, some experts believe that tainted water is a common entry point for the disease (Saribas et al. 2010).

It has proven possible to trace the course of mankind's migrations through the genetic study of HPyV2 specimens since subtle genetic variants are frequently discovered in various parts of the globe (Pavesi 2005). There are 14 identified genotypes or subtypes, each of which is linked to a particular geographic area. In both the western and central regions of Africa, there is an African variety called Af1 (Takasaka et al. 2006). Af2, the predominant African form, is widespread across the African continent, as well as across Western Asia and parts of South Asia (Pavesi 2003; Takasaka et al. 2006). A, B, and C are all prevalent throughout the European continent. B2, CY, B1-a, B1-b, B1-d, SC, and MY are among the acknowledged Asian types (Saruwatari et al. 2002).

Genotypes (1–8) are numbered using a different system that includes extra lettering. In addition to indigenous populations in Japan's northern region, the Northeast Siberian region, and the northern part of Canada, genotypes 1 and 4 (closely related) are also present in European nations (Sugimoto et al. 2002). Africa's sub-Saharan region is linked to type 3 along with type 6. Genotype 3 of HPyV2 has been found in the Republic of Tanzania, South Africa, and Ethiopia. Type 6 has been isolated in Ghana. The Aka and Bantu peoples of Central Africa in particular also exhibit both types (Seyoum et al. 2022). JCV genotype 2 occurs as subtype 2A, 2B, 2D, and others. 2A is mostly present in Japanese and Native American communities, whereas 2B is present in Eurasian populations. 2D is present in Indian populations. Subtype 2E is often present in people from Australia and also from the west coast of the Pacific (Shackelton et al. 2006). Southern Chinese territories and the Southeast Asian regions are linked to subtype 7A. Japan, the Mongolian Republic, and China's northern part all exhibit JCV subtype 7B. Subtype 7C is associated with predominantly north and south Chinese populations (Cui et al. 2004). The south-western Pacific regions and islands are home to subtype 8 (Ryschkewitsch et al. 2000). The worldwide array of HPyV2 types based on genotyping may make it easier for racial tracing.

Infection with human immunodeficiency virus has been shown to be a contributor in 82% of 9675 progressive multifocal leukoencephalopathy cases evaluated in a USA-based study between 1998 and 2005 (Molloy and Calabrese 2009). A low prevalence of PML in Africa and India may be related to diagnostic hurdles and variability in JCV (Shankar et al. 2003). Prior to the development of highly active antiretroviral therapy (HAART), up to 4% of cases of acquired immunodeficiency syndrome included PML. Likewise, the prevalence of PML has declined with the era following HAART implementation (Sidhu and McCutchan 2010). Progressive multifocal leukoencephalopathy happens to be ranked as the third leading contributor to encephalopathy among those with HIV infection, as reported by the Italian Registry Investigative Neuro AIDS (Antinori et al. 2003). From 1996, once HAART gained practical acceptance, PML-linked mortality began to considerably decline in the USA. As there were lesser fatalities among individuals with HIV infection, the mortality rate linked to PML declined from 2.76 lives lost per million (1992–1995) to 0.66 per million (2002–2005) (Christensen et al. 2010).

11.5.3 Aetiopathogenesis

In progressive multifocal leukoencephalopathy, myelin sheathes around neuronal axons become damaged over time; demyelination leads to disrupted nerve-impulse transmissions. Parietal and occipital lobes' subcortical white matter are notably affected (Saji and Gupta 2023). PML causes intra-nuclear inclusions with the destruction of oligodendrocytes. This demyelinating syndrome is comparable to multiple sclerosis but advances considerably more swiftly (Ono et al. 2019). A patient's extent of immunocompromise directly correlates with the rate of myelin degradation (Beltrami and Gordon 2014).

Latest research suggests JCV variants to be sources of additional emerging diseases, despite the fact that human polyomavirus-2 infections are traditionally linked to demyelinating conditions of white matter, as well as PML development (Cortese et al. 2021). In this regard, it was recently established that human polyomavirus-2 infects the cerebellum's granule cell layer, excluding Purkinje fibres, eventually leading to significant cerebellar atrophy (Roux et al. 2011). It is known as JCV GCN, which is short for JCV- associated granule cell layer neuronopathy, distinguished as a viral illness of prolific and lytic nature caused by a VP1 coding mutation (JCV variant) (Dang et al. 2012).

The JC virus variant linked primarily to cortical pyramidal neuron (CPN) infection, as well as affecting astrocytes, is known as JCV-CPN. This explains how HPyV2 seems to have an additional role in mediating encephalopathies (Dang et al. 2012). Evaluation regarding the JCV-CPN variation inferred the presence of a deletion of 143 base pair within the agnogene, which codes a particular 10 amino acid, truncated peptide thought to be responsible for CPN tropism (Sariyer et al. 2008). Furthermore, this virion' subcellular location in axons, cytoplasm, and of course nuclei lead one to believe that JCV-CPN variant can propagate across axons to boost infectiousness (Miskin and Koralnik 2015).

In a study by Miskin and Koralnik, HPyV2 was the sole virus found among the cerebrospinal fluid of a few meningitis patients, suggesting that HPyV2 might additionally serve as a cause of aseptic meningitis. No coding sequence modifications were found in any of the regulatory areas of this JCV meningitis (JCM) variant, when it was subjected to evaluation. The specific molecular processes governing the meningeal tropism by JC virus are still unknown (Miskin and Koralnik 2015).

11.5.4 Diagnosis

Clinical Presentation Contrary with various important opportunistic illnesses, such as focal CNS lesion-causing lymphomas or toxoplasmosis, PML often develops during a span of weeks rather than a matter of hours or days. It also progresses much more quickly in contrast to AIDS–dementia complex (Cinque et al. 2009). The brainstem is far more frequently involved in PML linked to AIDS, as opposed to PML in other conditions. Neurological findings tend to get progressively worse and affect a wider area, in correspondence to concentrical expansion of isolated lesions or extension along white matter segments. By way of example, early symptoms of lower limb weakness may subsequently evolve into haemiparesis (Cinque et al. 2009; Saji and Gupta 2023). Affected patients who have

Table 11.2 Spectrum of clinical symptoms and signs

Main feature	Clinical findings
Classic PML—progressive, mostly demyelinating, multifocal CNS deficits (white matter involvement)	Aphasia, hemianopsia, paresis (hemi-/ quadriparesis), sensory loss, etc.
PML-immune reconstitution inflammatory syndrome	Acute, physical deterioration plus exacerbated signs and symptoms of classic PML
Classic PML with cortical involvement	Seizures plus classic PML findings
Deep grey matter involvement—basal ganglia/thalamus/etc.	Progressing sensory loss and pareses
Behavioural abnormalities	Mental status changes, etc.
JCV-associated meningitis	Aseptic meningitis findings, CSF JCV-DNA (PCR) present
Granule cell neuronopathies	Cerebellar atrophy—ataxia, coordination difficulty, dysarthria, gait impairment
Other findings	Cognitive abnormalities, cranial nerve signs, coma, headache, language impairments, visual impairments (cortical blindness, conjugate gaze dysfunction, hemianopia)

comparatively competent immune systems might encounter a more gradual illness progression compared to individuals who are immunocompromised. Focal CNS features often manifest in the form of aphasia, cortical blindness, dementia, haemiparesis, and hemianopsia, along with an extensive list of other signs and symptoms (Table 11.2). Focal findings frequently pertain to the occipital region and/or other posteriorities of the brain. Clinical signs may sometimes be widespread as opposed to localized/focal (Berger et al. 2013; Saji and Gupta 2023).

Imaging Radiological findings in classic PML are usually found in the form of unifocal or multifocal white matter lesions of non-enhancing variety and typically depict ono mass effect. Lesion(s) can be identified on computed tomography (CT) and MRI scans. Parieto-occipital region is a common location for most associated neuro-pathologies, along with asymmetric infra-tentorial lesions. Less commonly, cortical ribbons may be involved; subcortical matter and spinal cord are rarely affected (Berger et al. 2013). Post-contrast MRI of PML-affected individuals might sometimes reveal an enhanced mass lesion. When assessed against white matter in healthy persons and white matter in HIV-infected, PML-negative patients, the magnetization transfer ratio is found to be mostly low in PML-positive cases (Berger et al. 2013; Iannetta et al. 2013). Hypodense areas can be detected on a computed tomography examination, although they are much less sensitive than scanning with an MRI. T1-weighted (T1W) MRI scans of PML-affected patients often reveal characteristic hypodense lesions, whereas hyperintense areas are appreciated on fluid-attenuated inversion recovery (FLAIR) and T2W MRI (Berger et al. 2013). In cases of inflammatory PML, imaging can reveal unusual signs, such as a widespread impact of associated lesions, with adjacent oedema. These cases might exhibit dramatic contrast enhancement, something that is unusual in classical

PML and usually limited if it does develop. The brain's cerebellum is frequently impacted, although any part comprising white matter is susceptible (Berger et al. 2013; Saji and Gupta 2023).

Lumbar Puncture Although protein amounts in CSF are typically normal, they might have levels a little above normal. Additional aetiologies are ruled out upon assessing normal CSF counts and results. Cerebrospinal fluid pleocytosis is possible on occasion; however, the cell concentration is often below 20 cells per microliter. CSF-JCV culture is typically unhelpful (Berger et al. 2013).

Regarding the identification of JCV in PML cases, PCR of CSF has been demonstrated to be very sensitive and specific, 74–93% and 92–99%, respectively (Lee et al. 2019). False-negative results are occasionally brought about by minimal CSF viral DNA levels, sample volume or quality, and the loss of nucleic acid while sample processing. The rate of false positivity has been estimated as 2%. In individuals on highly active antiretroviral therapy, monitoring CSF-JCV-DNA load acts as an indicator of the illness's progress and may be used in therapeutic studies (Cinque et al. 2009; Landry et al. 2008). This investigation may be able to replace the necessity for an actual brain biopsy, but, in inflammatory PML cases, the likelihood of finding JCV in cerebrospinal fluid gets lower. Subsequently, a repeat CSF study is advised in PML suspects, despite the preliminary report of a negative JCV PCR from CSF (Cinque et al. 2009; Kuhle et al. 2011).

Brain Biopsy In cases with PML, brain biopsy offers high specificity and sensitivity, 92–100% and 74–92%, respectively (Lee et al. 2019). Affected tissue on biopsy reveals minor cortical atrophy. The most effective way to verify the presence of JCV in a biopsied material is through in situ hybridization or by immunohistochemistry-based methods. Numerous demyelinating foci are often visible across cortical grey matter. In more serious cases, demyelinated brainstem and cerebellar and cerebral matter may be visible, with or without confluence of foci. Among the most frequent locations for disease are oligodendrocytes present at grey and white matter junctions. In addition to astrocyte and oligodendrocyte involvement, JCV is able to affect the cerebellum's granule cells (Berger et al. 2013; Lee et al. 2019).

Numerous demyelinate sites are visible under the microscope. Intra-nuclear inclusions, possessing eosinophilic or basophilic natures, amidst oligodendrocytes' enlarged nuclei, often near the edges of lesions, are the condition's morphological hallmark (Saribas et al. 2010; Shishido-Hara et al. 2014). A further distinctive trait is the presence of big, often multi-nucleated astrocytes featuring conspicuous processes (Saribas et al. 2010). Inclusions in JCV infection can be confirmed via means of immunohistochemistry (IHC) and electron microscopy (EM) application (Shishido-Hara et al. 2014). There are pathological findings of perivascular inflammatory infiltrates, although a rarity, cystic, or necrotic changes can also exist (Berger et al. 2013). Sometimes, PML is misinterpreted to be glioma owing to the presence of cellular atypia.

Table 11.3 Clinical, radiology, and virology-based criteria for diagnosing progressive multifocal leukoencephalopathy

Diagnosis	Typical clinical findings	Typical radiological findings	CSF-JCV-DNA (PCR)
Definite PML	Present	Present	Present
Probable PML	Present	Absent	Present
	Absent	Present	Present
Possible PML	Present	Present	Absent/not performed
	Absent	Absent	Present
Not PML	Absent	Absent	Absent
	Present	Absent	Absent
	Absent	Present	Absent

Table 11.4 Histopathology-based criteria for diagnosing progressive multifocal leukoencephalopathy

Diagnosis	Typical histopathologic triad 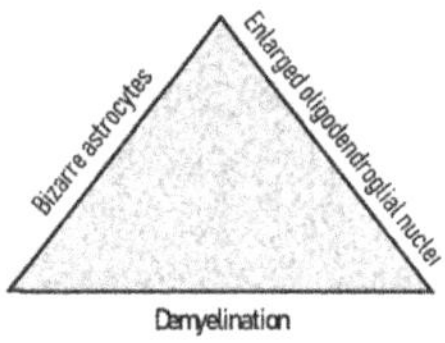	Electron microscopy/ immunohistochemistry proof	Tissue JCV-DNA/ protein (PCR)
Definite PML	Present	Present	Present
	Present	Absent/not performed	Present
	Present	Present	Absent/not performed
Probable PML	Present	Absent	Absent/not performed
Possible PML	Absent	Present	Absent/not performed
Not PML	Absent	Absent	Absent/not performed

Diagnostic Considerations In accordance with the Neuroinfectious Disease Section of the American Academy of Neurology, the criteria for diagnosing PML is based on histopathology evidence, or through clinical, radiology, and virology-based findings (Tables 11.3 and 11.4) (Berger et al. 2013). Progressive multifocal leukoencephalopathy's diagnosis is made based on a characteristic histopathologic triad of widened oligodendroglial nuclei, presence of bizarre astrocytes, and demyelination, as well as proof of John Cunningham virus using PCR, EM, or IHC. Likewise, it is generally accepted that a combination of concordant clinical manifestations and recognized radiologic changes, along with positive PCR results for JCV in cerebrospinal fluid, suffices to conclusively diagnose PML (Berger et al. 2013; Kuhle et al. 2011; Saji and Gupta 2023).

11.5.5 Management, Treatment, and Prevention

The best-available therapy to date for possibly treating PML targets PD 1, which is a programmed cell death protein located on lymphocytic surfaces; pembrolizumab happens to be one such monoclonal antibody. In a preliminary study, 4–6 weekly (total three doses per patient) administration of pembrolizumab to 8 PML-affected subjects was carried out. From the eight patients, five reported betterments in their clinical statuses (Dang et al. 2016; Patnaik et al. 2015). Aside from this novel, investigational medication, the main strategy for treating PML includes the use of antiretroviral medication (Pavlovic et al. 2015). Mefloquine for use in PML patients has also been proposed, as it has shown some degree of activity towards JCV in vitro. Additionally, according to some studies, employing pulsed methylprednisolone for PML treatment in cases of immune reconstitution inflammatory syndrome linked to antiretroviral initiation in HIV infections has proven helpful (Tan et al. 2020). Further assessments are currently being carried out towards building a prospective strategy involving both passive and active immunization against JCV infection and progressive multifocal leukoencephalopathy (Dunham et al. 2020).

11.6 Lymphocytic Choriomeningitis Virus

The lymphocytic choriomeningitis virus (LCMV), which belongs to the Arenaviridae family, is what causes lymphocytic choriomeningitis (LCM). In 1934, American physician, Charles Armstrong came up with the name (Beeman 2007; Zhou et al. 2012). It manifests as meningoencephalitis, encephalitis, or aseptic meningitis, and the viral infection is spread by rodents. Cerebrospinal fluid (CSF), meningeal membranes covering the brain, and the spinal cord can become infected with the virus in LCM (Bonthius 2012). The term was chosen regarding a person's propensity to display very high lymphocyte counts, while they are ill with the infection. A cerebral meningitis referred to as choriomeningitis occurs, which exhibits significant cellular infiltrate throughout the meninges, frequently accompanied by lymphocytic infiltrate into choroid plexuses (Kang and McGavern 2008).

11.6.1 Viral Structure

The lymphocytic choriomeningitis mammarenavirus has a circular envelope and a 50–200 nm ranging diameter (Radoshitzky et al. 2019). Two negative-sense, RNA single-strand segments make up the viral genome in its helical–nucleocapsid (Radoshitzky et al. 2019). The complementary negative- strand RNA to its essential positive mRNA signals that, prior to its translation to the necessary proteins, it needs to be initially transcribed to positive mRNA. An intergenic area divides the L strand's ambisense RNA, which codes for several proteins that bind in opposing orientations. Its size measures to about 7.2 kb and contains the coding for a ring finger motif-containing 11 kDa short polypeptide Z plus a 200 kDa high-molecular-mass protein (L) (Lee et al. 2000). The L protein, which encodes LCMV's RNA -dependent RNA polymerase, has distinctive motifs that are shared among all of the RNA -dependent RNA polymerases. The size of the ambisense S strand is about 3.4 kb (Lee et al. 2000). Nucleo-protein, NP, measuring around 63 kDa, as well as glycoprotein precursor complex, GPC measuring 75 kDa, are both primary structural proteins that are encoded by it (Papageorgiou et al. 2020).

11.6.2 Aetiopathogenesis

Lymphocytic choriomeningitis mammarenavirus involves a variety of strains; however, the two that tend to be frequently utilized include LCMV- Clone 13 and Armstrong. Charles Armstrong discovered the first LCM virus strain in 1934 after isolating it from brain matter, thus the name "Armstrong". It causes strong cytotoxic T lymphocyte responses, which cause the host's immune system to quickly clear it out. The term, acute-Armstrong lymphocytic choriomeningitis is used to describe this (Zhou et al. 2012). The Clone 13 strain is from an Armstrong strain variation that was isolated from a spleen, i.e. receptive towards the internal organs. The variant was initially discovered in mice that had been born with a lifelong LCM (Zhou et al. 2012). LCMV-Clone 13 can eventually remain within its host perpetually because it generates a weaker immune cytotoxic T lymphocyte (CTL) response. It is known as chronic-Clone 13 lymphocytic choriomeningitis (Oldstone et al. 2018).

11.6.3 Diagnosis

The initial diagnosis of lymphocytic choriomeningitis is made on the basis of a suspected history, which is verified by several diagnostic investigations (Bonthius 2012). The most reliable and sensitive diagnostic technique is reverse transcriptase polymerase chain reaction (RT-PCR), which is capable of detecting LCMV in CSF and serum samples (DeBiasi and Tyler 2004). Analysis of affected cerebrospinal fluid usually reveals high opening pressure (with occasional papilledema), lymphocytic pleocytosis ranging from 10 to more than 3000 cells per microliter, increased protein concentrations of 50–300 mg/dl, or sometimes even higher, and less commonly, hypoglycorrhachia (Seregin et al. 2020; Souders et al. 2015).

Initially, in the infection progression, total blood cell counts might demonstrate thrombocytopaenia and leukopaenia. It can be helpful to measure the levels of immunoglobulin M (acute and convalescent) and G from CSF or serum samples. Compared to immunofluorescence-based testing, enzyme-linked immunosorbent assays have higher sensitivities for diagnosing lymphocytic choriomeningitis

(Riera et al. 2005). The usage of complement fixation tests remains quite insensitive; thus, it is of limited application. Apart from reverse transcription PCR of specimens, immunohistochemical staining and viral culture may be somewhat helpful.

11.6.4 Management, Treatment, and Prevention

Research studies for antiviral drugs have not been conducted to treat LCMV infection. For individuals who are immunodeficient, early detection along with supportive care, such as prescribing NSAIDs or replacement of fluids, are crucial. Whenever it is possible, minimizing any further chance of immunosuppression should be aimed for. In the majority of lymphocytic choriomeningitis virus infections, no particular medication is recommended as a treatment. Most individuals experience a spontaneous recovery in 1–3 weeks with little to no long-term effects (Zhou et al. 2012).

Ribavirin has been successfully utilized in recipients of transplants having serious illness because it exhibits in vitro efficacy against LCMV. However, ribavirin for intravenous administration is not offered for commercial use. In accordance with optimal body weight and renal function, its dosage is administered orally. While taking ribavirin, the patients ought to be closely watched for any possible adverse events, notably haemolytic anaemia (Fischer et al. 2006; Mathur et al. 2017).

It has been demonstrated that lymphocytic choriomeningitis virus is inhibited in vitro by favipiravir [T-705], an RNA-dependent RNA polymerase inhibitor. Additionally, it has shown promise in lowering the morbidity of additional arenavirus-associated diseases in experimental animals (Hickerson et al. 2018). To determine whether favipiravir has the potential to be employed for managing arenavirus-linked diseases, such as lymphocytic choriomeningitis in human subjects, more research is required.

LCMV illness is not currently covered by any specific vaccinations. Although multi-epitope vaccination strategies were recently suggested, there are currently no commercially accessible vaccine candidates (Waqas et al. 2023).

The best way to avoid contracting LCMV is to stay away from outdoor mice and use caution whenever caring for domesticated or pet guinea pigs, hamster, or mice (Amman et al. 2007). In the event of a rodent infestation, an individual should refrain from sweeping or vacuuming in such a manner, which may result in stirring up of any debris or dust. A specialist should be called for to clean up the infested locations, including completely soaking the affected region in a solution of bleach to avert aerosol formation (Oldstone et al. 2018).

Those in laboratories who work with mice or and other rodents are more likely to become infected with LCMV. There is no documented way to avoid transmission in these circumstances. While dealing with these animals, it is prudent to wear gloves; particularly in the event, the handler has any cut or abrasion on his/her hands. Consider using respirator masks in circumstances with inevitably high infection rates (Knust et al. 2014). Considering that it is neither specific nor sensitive to determine whether a donor has a pet rodent or not, there are no reliable means to stop the spread of disease through transplantation. Due to the high cost and potential ineffectiveness of PCR of tissue and immunohistochemistry (IHC) evaluation, it cannot be frequently done throughout the organ procurement process (White et al. 2018).

11.7 Severe Acute Respiratory Syndrome Coronavirus 2 (SARS-CoV-2)

The CNS involvement as neurological signs (e.g. the results of hypoxia and thrombosis) in coronavirus disease 2019 (COVID-19) has been shown in several case reports and observational studies (Helms et al. 2020; Lu et al. 2020; Tricco et al. 2018). COVID-19 can exhibit a diverse range of neurological manifestations. While neurological symptoms are more commonly observed in severe cases of the disease, it is crucial to maintain a high level of suspicion in order to diagnose patients who present solely with neurological symptoms, referred to as "neuro-COVID-19 syndrome", at the onset of COVID-19 illness. These neurological presentations may be atypical and require careful evaluation to establish a timely diagnosis and appropriate management. Approximately 25% of COVID-19 cases have reported manifestations of central nervous system (CNS) involvement (Mao et al. 2020). The suggested CNS involvement mechanisms include both direct (neurotropic) and indirect pathways (Fig. 11.3).

The SARS-CoV-2 uses the angiotensin-converting enzyme 2 (ACE2) receptor to enter host cells, but with a far higher affinity than the SARS-CoV-2. High levels of ACE2 receptor expression are seen in adipose tissue, the lung, the heart, the brain, the liver, the vascular endothelium, and the naso-oral mucosa, making these tissues

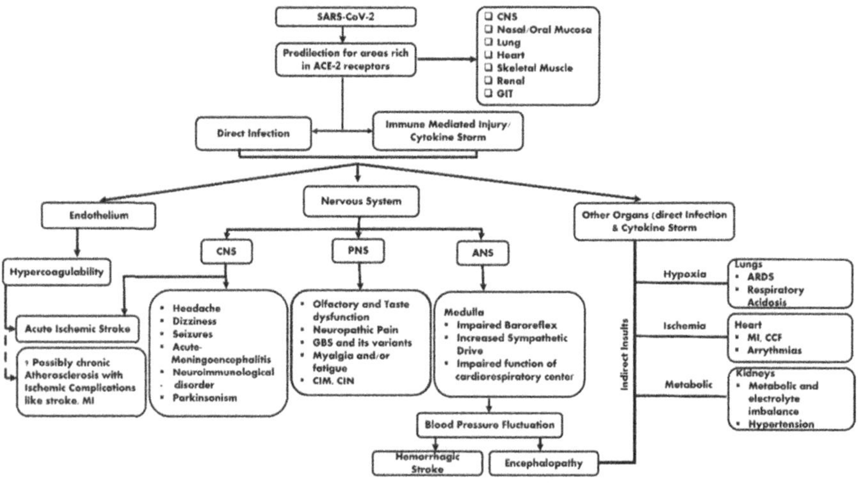

Fig. 11.3 Pathogenesis of CNS involvement by direct and indirect (immune- mediated) mechanism in COVID-19

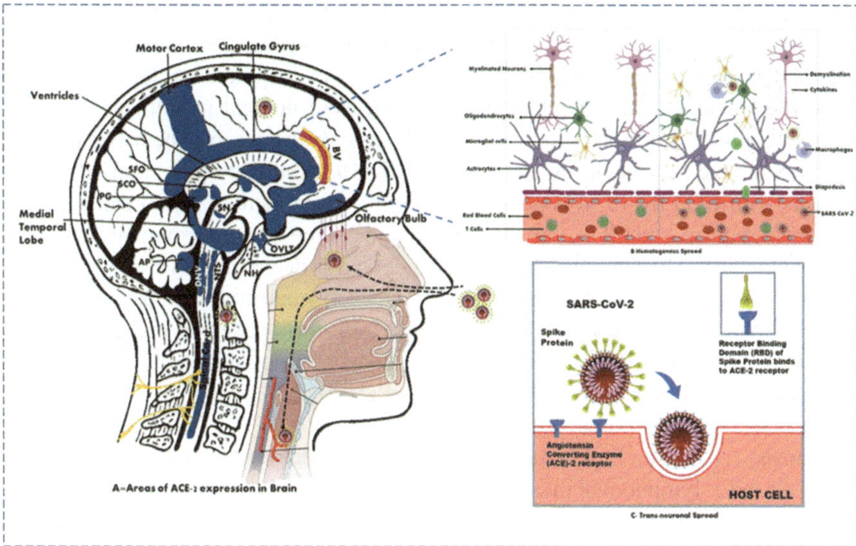

Fig. 11.4 Diagram showing high ACE-2-receptors containing brain regions with possibility of neuro-invasion by SARS-CoV-2

particularly vulnerable to infection (Jin et al. 2020). The olfactory bulb, motor cortex, posterior cingulate cortex, middle temporal gyrus, sympathetic pathways in the brainstem, substantia nigra, ventricles, and circumventricular organs have the highest concentrations of ACE2 receptors in the brain (Fig. 11.4) (Jin et al. 2020; Zubair et al. 2020).

The S1 component of the spike protein is used by the SARS-CoV-2 to bind to the ACE2 receptors in host cells, while the S2 subunit is used for fusion and subsequent endocytosis. Once within the cell, the viral RNA is released, allowing sub-genomic RNA to translate a variety of proteins. The viral particles then assemble, bud, and are expelled from the host cells (Jin et al. 2020; Zubair et al. 2020).

Clinical Features Fever (77–89%), cough (61–81%), shortness of breath (3–26%), sore throat (10%), and gastrointestinal issues (5–9%) are the most common presentation of COVID-19 infection, and Tt is worth noting that neurological symptoms may be present in over 33% of individuals hospitalized with COVID-19 (Galassi and Marchioni 2020; Xu et al. 2020). The neurological symptoms can occur as a result of involvement in either the central nervous system or the peripheral nervous system, either early on or later during the course of the COVID-19 illness. Severe cases of COVID-19, elderly individuals, and patients with pre-existing medical conditions often experience neurological symptoms, which can include muscle pain and/or fatigue, dizziness, headaches, loss of smell (anosmia), loss of taste (ageusia), and changes in mental state. These neurological features are relatively frequent in these specific groups (Metlay et al. 2019; Zubair et al. 2020). Acute encephalopathy has been reported as the most common neurological symptom in COVID-19 patients with CNS involvement in various studies (Chou et al. 2021). The initial signs of moderate neurological symptoms, such as headache, dizziness, hyposmia, hypogeusia (diminished sense of taste), diplopia, and ophthalmoplegia, appeared soon after the outbreak (Wenting et al. 2020). Neurological movement problems such as Guillain–Barré syndrome, Miller– Fisher syndrome, polyneuritis cranialis, and ataxia have also been observed in COVID-19 (Abu-Rumeileh et al. 2021). In patients with COVID-19, it is possible to observe prominent delirium and agitation that may require sedation. On the other hand, some patients may present with encephalopathy, characterized by somnolence and a decreased level of consciousness (Helms et al. 2020; Mao et al. 2020). Corticospinal tract symptoms, such as hyperreflexia and extensor plantar reflexes, are frequently observed in these patients. In people with COVID-19, seizures have also been documented in addition to encephalopathy (Lyons et al. 2020; Somani et al. 2020).

Diagnosis It has been found that an independent predictor of neurological problems is a neutrophil–lymphocyte ratio (NLR) > 3 (Zhang et al. 2020). Regular testing of serum electrolytes and renal and liver functions is recommended to detect multiple organ dysfunction syndrome (MODS) and rule out metabolic encephalopathy in COVID-19 patients (Mao et al. 2020). Abnormalities in the coagulation profile, such as high D-dimer levels, are independent predictors of systemic vascular problems including stroke (Tang et al. 2020; Vonck et al. 2020). Tests for acute -phase reactants including serum ferritin and C-reactive protein (CRP), as well as muscle enzymes like creatine phosphokinase (CPK) and lactate dehydrogenase, are advised. Patients with primarily central nervous system (CNS) symptoms frequently have lower lymphocyte and platelet counts and higher blood urea nitrogen levels, whereas patients with muscle symptoms have significantly higher CPK ($>$200 U/L), neutrophil, CRP, and D-dimer levels along with a lower lymphocyte count (Desai et al. 2021; Mao et al. 2020). Patients with COVID-19 who exhibit unexplained mental instability, agitation, confusion, or localized neurological abnormalities pointing to a recent stroke, meningoencephalitis, or an acute myelitis should have neuroimaging, ideally with MRI, of the relevant region of interest. This is important to further investigate and assess the neurological condition in these individuals.

The presence of albumino-cytological dissociation in cerebrospinal fluid (CSF) can assist in diagnosing Guillain–Barré syndrome (GBS) (Abu-Rumeileh et al. 2021). In COVID-19-related para- or post-infectious neuroimmunological disorders affecting the central or peripheral nervous system, CSF protein levels may be elevated. COVID-19-related meningoencephalitis (Moriguchi et al. 2020; Pilotto et al. 2020) and transverse myelitis (AlKetbi et al. 2020) have been associated with lymphocytic pleocytosis (increased lymphocytes in CSF) and elevated CSF protein levels. However, as of now, no reports have indicated the presence of positive immunological markers in CSF, such as IgG index or oligoclonal bands, in any COVID-19-related cases.

For example, focal or diffuse delta-theta slowing, sporadic epileptiform discharges (symmetrical or asymmetrical), lateralized periodic discharges with overriding fast activity, generalized sharp waves with spikes, triphasic waves, and repetitive focal rhythmic bursts indicating nonconvulsive status epilepticus have all been observed in COVID-19 patients with altered mental status, delirium, or encephalopathy. Ongoing EEG monitoring is suggested for these patients in order to diagnose nonconvulsive status epilepticus. Guillain–Barré syndrome (GBS), myasthenic crisis, inflammatory myopathy, and critical illness neuropathy are examples of rapidly progressing neuromuscular illnesses that are treated with electrophysiological examinations such nerve conduction studies (NCS) and electromyography (EMG) (Helms et al. 2020; Pilato et al. 2022; Vollono et al. 2020). These studies help assess the function and integrity of nerves and muscles in these conditions.

Treatment The majority of COVID-19 cases, over 80%, either exhibit no symptoms or experience mild symptoms that resolve on their own. For symptomatic cases, the primary approach in managing severe cases involves symptomatic management, such as oxygen therapy. In instances of respiratory failure that do not respond to oxygen therapy and for septic shock requiring haemodynamic support, mechanical ventilation is recommended (Galassi and Marchioni 2020; Zubair et al. 2020). There are ongoing evaluations of various antiviral medications that target different stages of the viral replication cycle, as well as convalescent plasma, as potential treatment options for COVID-19. These efforts aim to explore their effectiveness in combating the virus and improving patient outcomes. Due to the limited available data on the management of COVID-19-related neurological manifestations, it is advised to combine supportive management with syndrome-specific therapy that was used prior to the COVID-19 pandemic. This approach involves tailoring treatment to the specific neurological syndrome presented by the patient while providing supportive care. Furthermore, regular monitoring and management of metabolic and electrolyte imbalances are important. Additionally, preventing and treating secondary infections can contribute to improved outcomes in COVID-19 patients with neurological manifestations. By addressing these aspects, healthcare providers aim to optimize patient care and enhance their overall prognosis (Desai et al. 2021).

11.8 Prion-Related CNS Disorders

Prion disorders constitute a sizable collection of neurodegenerative-linked illnesses that impact people and animals (Ritchie and Barria 2021). Mad cow disease, i.e. bovine spongiform encephalopathy (BSE), CJD, short for Creutzfeldt-Jakob disease, Gerstmann Sträussler Scheinker (GSS), scrapie disease of sheep, and chronic wasting disease (CWD) in deer are among the illnesses covered under prion diseases. They have protracted incubation times; however, as soon as clinical symptoms appear, the diseases generally advance rather rapidly. There remains no curative option for any prion illness, which all result in death. Nonetheless, new developments of their aetiopathogenesis and physiology have raised the possibility of possible future medical interventions with promising therapeutic outcomes (Geschwind 2015).

11.8.1 Structure

Addressing the structure that defines prions, a wide range of hypotheses exist, including the notion that they are either composed solely of protein or nucleic acid, that they have neither protein nor nucleic acid, or that they are actually polysaccharides. The protein-only concept is perhaps the most broadly recognized theory. Prusiner coined the name "prion" to denote scrapie's association with the proteinaceous particulate (PrP). PRNP, the PrP gene in humans, is a 253-amino acid protein, which is located on chromosome 20. PrPC is a small cellular glycosylphosphatidylinositol (GPI)-anchored protein and its precise biological purpose is still uncertain. Scientific theories suggest that this protein could play a part in signalling or in cellular adhesions. It is possible that PrP has some function in the metabolism or transportation of copper since its N-terminal comprises a length of five 8 amino acid-sequence repeats (also known as octapeptide repeat regions). The latest research indicates copper imbalance is a significant primary alteration underlying prion disease.

Although PrP exists in the majority of bodily tissues, it is most prominent in the central nervous system, namely within neurons. Additionally, the immune system's cells have high levels of PrP expression. Knockout mice that lack prion protein gene due to genetic engineering do not exhibit any overt pathogenic traits (Steele et al. 2007), but it has been discovered that the circadian patterns, sleep, and synaptic biology of these mice are aberrant (Bouybayoune et al. 2015).

Mice recombinant prion protein and nuclear magnetic resonance were used to initially figure out the secondary molecular structure of cellular prion protein (Baral et al. 2019). Recombinant human and hamster PrP have been employed in recent years to accomplish this (Charco et al. 2017). According to some researches, PrPC contains roughly 40% alpha -helices alongside approximately 3% beta -sheets. PrPSc is extremely aggregated and soluble, which are characteristics, which preclude the utilization of NMR image techniques and other high -resolution structural analyses. PrPSc possesses roughly 45% beta -sheet alongside 30% alpha-helix according to less precise structural techniques like Fourier-transfused infrared spectroscopy (Requena and Wille 2014; Riesner 2003). The elevated beta- sheet concentration is associated with both PrPSc's infectivity and resistance to digestion by enzymes (Sajnani and Requena 2012).

11.8.2 Epidemiological Profile

According to the Creutzfeldt–Jakob disease International Surveillance Network, there are about one to two occurrences of the disease per million people worldwide each year (Watson et al. 2021). Israelis of Libyan descent and some Slovakian groups, where the prevalence for CJD stands sixty to hundred times beyond expected, represent two communities that are unduly impacted by the disease. The widespread prevalence PRNP gene possessing mutation of codon 200 mutation is associated with these localized elevated CJD rates (Chapman et al. 1994; Meiner et al. 1997; Mitrová and Belay 2002).

Diseases associated with prion are extremely fatal and progressive (Geschwind 2015). A sporadic case of CJD typically lasts for 8 months, whereas, with a mean disease duration of 1–1.5 years, the course of variant Creutzfeldt–Jakob disease (vCJD) is lengthier. Also, while Gerstmann–Sträussler–Scheinker disease shows an extended course, lasting roughly 5 years, familial Creutzfeldt–Jakob disease lasts an average of 2 years (Collins et al. 2001; Coulthart and Cashman 2001; Takada and Geschwind 2012).

Persons from every demographic background can get sporadic CJD, which often has comparable symptoms. Certain familial prion conditions, including familial CJD, may display distinctive characteristics among a race. For instance, besides many of the conventional CJD symptoms, familial CJD that affects the Libyan Jews minority (linked to codon 200 mutation) exhibits additional characteristics of peripheral neurological disorder (Sitammagari and Masood 2022). Variant CJD has only been found in Europe, with the majority of cases taking place in England and Wales (Ritchie and Barria 2021).

The average age at which sporadic Creutzfeldt–Jakob disease manifests itself is 60. Its prevalence is approximately one case for every million people, while it is five for every million with people between 60 and 75 years old (Sitammagari and Masood 2022). The age spectrum is wide; reports of cases occurring among individuals who were as young as 17 and those aged at 83 have occurred (Coulthart and Cashman 2001). Having an average onset age of 28 years, vCJD affects younger people. The average onset age for familial Creutzfeldt–Jakob disease, GSS, and fatal familial insomnia (FFI) is between 45 and 49 years old (Cao et al. 2021; Harder et al. 2004; Jeong and Kim 2014; Qi et al. 2020).

In the United Kingdom, 159 variant Creutzfeldt–Jakob disease patients were documented by the beginning of 2006. Towards the late 2005, there were 15 French nationals documented with the illness, three from the country of Ireland, two from the USA, one case from Canada, one from Netherlands, a case in Italy, one case in Portugal, one in Japan, one in Spain, and one documented case in Saudi Arabia. The Canadian, Japanese, American, and one of the Irish patients probably had been infected while residing in Great Britain (Belay et al. 2005; Sánchez-Juan et al. 2007). With the advent of vCJD, the threat of an epidemic of prionoses involving English populations, comparable to the bovine spongiform encephalopathy epidemic in cow populations, has grown (Lee et al. 2013).

11.8.3 Aetiopathogenesis

The neurological pathology associated with prion diseases is a common trait. Grey matter within the CNS is typically affected by these diseases, which result in neuron death, gliosis, and distinctive spongiform alteration. There is a variable degree of vacuolation in the neurons along with vacuolated neuropil (Soto and Satani 2011). Also, plaques are often observed in prion disorders, exhibiting typical amyloid staining characteristics, such as Congo Red stain-associated apple green birefringence with polarizing light (Yakupova et al. 2019). Amyloid can be found in the area of the cerebrum or cerebellum in about 10% of CJD patients (Rodriguez et al. 2015). Cerebellar plaques of multi-centric nature are present in reports of GSS cases (Rossi et al. 2017). Prion-linked amyloid plaques are responsive to prion antibodies, but they are unresponsive with other amyloid proteins (i.e. Alzheimer-linked amyloid-beta) (Rodriguez et al. 2015).

Since prions can spread naturally via ingestion or through transcutaneous pathways, prions' ability to enter the central nervous system is a crucial question. Despite the fact that prion illnesses are neurological disorders, crucial pathogenic processes happen in specific locations outside of the CNS, particularly in lymphoid systems (Cobb and Surewicz 2009; Natale et al. 2011). Initial phases of prion disorders are well known to include lymphoid tissues (Soto and Satani 2011). The lymph nodes and spleen are frequently shown to be among initial locations of PrPSc replication following peripheral route infection and to be considerably altered by intracerebral stimulation. Studies demonstrating the delay of acute symptoms associated with splenectomy or different procedures, which diminish peripheral lymphoid tissues, point to their significance for neurological expropriation following peripheral seeding (Mabbott et al. 2020; O'Connor and Aguzzi 2013).

Investigations that demonstrate that bovine spongiform encephalopathy is transferable among sheep through transfusions of blood indicate haematogenic transmission of prions to central nervous system (Houston et al. 2000). A handful of vCJD cases linked to the use of blood transfusions have likewise been reported. The first case in point was diagnosed with vCJD in the year 2003, roughly 6.5 years after receiving a blood transfusion from a blood donor who had been diagnosed with vCJD just 3.5 years earlier. In another case in 2004, the vCJD patient passed away 5 years after the transfusion, from a non-vCJD -related cause (Dietz et al. 2007; Pincock 2004). Upon autopsy, the patient's splenic tissue and cervical lymph node showed aberrant prion protein; pathological signs of variant CJD could not be seen in brain matter however. Respective prion symptoms appeared in the donor around eighteen months post-donation. A third reported case was of a person confirmed with vCJD in the year 2006, almost eight years after receiving a transfusion; the donor was identified as having the disease 20 months prior to donation (Dietz et al. 2007; Pincock 2004; Ritchie and Barria 2021).

There is evidence that B cells are necessary for haematogenic neurological expropriation (Sisó et al. 2010), but considering that B lymphocytes do not need to express PrPc for neurological expropriation, experts have hypothesized that their primary role is to support and upkeep the follicular dendritic cells (Isaacs et al. 2006). Newer evidence, though, indicates that neurological invasion can occur even without the presence of follicular dendritic cells or B lymphocytes (Sisó et al. 2010). Additional research has linked prion CNS invasion to unique CD11c+ dendritic cells (Aucouturier et al. 2001; Bradford and Mabbott 2012).

Besides haematogenic distribution, parasympathetic vagal route is another means by which prions are transported to the central nervous system (Sisó et al. 2010). As a result, after prions are delivered intraperitoneally, illness may be prevented by sympathectomy or hastened by sympathetic overstimulation of lymphoreticular structures (Glatzel et al. 2001). It is yet unknown which of these two distinct CNS invasion pathways is more significant; this could depend on the particular strain (Solforosi et al. 2013).

11.8.4 Diagnosis

There are numerous varieties of prion illnesses (Table 11.5) (Geschwind 2015). Kuru is among the earliest human prion disease that has been recognized. It is believed that tribal cannibalism is a contributing factor amidst this condition, which affects the Fore folks that reside in the remote foothills of New Guinea. The disease is thought to have started when a person who had sporadic CJD was consumed. Kuru once served as one of the main causes of mortality among Fore

Table 11.5 Various prion-related illnesses

Prion-linked illness	Host	Method of disease contractions
Bovine spongiform encephalopathy (BSE)	Bovines	Contaminated foods
Chronic wasting disease (CWD)	Deer	Uncertain
Creutzfeldt–Jakob disease (CJD)	Humans	
• Familial CJD		PrP gene mutations
• Iatrogenic CJD		Prion-infested material/dura mater, etc.
• Sporadic CJD		PrPc spontaneous conversion to PrPsc/somatic mutation
• vCJD		Consumption of bovine produce contaminated with BSE agent (three known cases via blood transfusion from previously infected/asymptomatic donor)
Exotic ungulate encephalopathy	Antelopes	Contaminated foods
Fatal familial insomnia (FFI)	Humans	Mutation of PrPc into abnormally folded PrPsc
Feline spongiform encephalopathy	Felidae	Contaminated foods
Gerstmann–Straussler–Scheinker disease (GSS)	Humans	PrP gene mutations
Kuru	Humans	Cannibalistic rituals
Scrapie	Goats and sheep	Susceptible lambs
Sporadic fatal insomnia	Humans	Spontaneous conversion of PrPc into PrPsc/somatic mutation

women, but since the termination of tribal cannibalism, this illness is practically gone. Comparable with scrapie, sufferers show medical signs of increasing cerebellar deficit and gait impairment. Demise happens roughly 12 months after the start of clinical symptoms (Liberski et al. 2019).

Kuru shares some degree of similarity in neuropathological findings of other prionoses, including extensive spongiform alteration, astrocytosis, and neuronal death. In kuru, as opposed to Creutzfeldt–Jakob disease, greater intraneuronal vacuolation is seen. Amyloid plaques have been identified in roughly 70% of afflicted cases, with deposits of amyloid constituting a frequent, though not always present, counterpart to the prion diseases (Liberski et al. 2019).

Creutzfeldt–Jakob disease makes up roughly 85% of all cases of human prion disease worldwide (Sitammagari and Masood 2022). In practice, CJD is distinguished by a myoclonic jerk-associated dementia that progresses quickly, along with a variety of cerebellar, pyramidal, and extrapyramidal abnormalities. Electroencephalogram results often reveal noticeable alterations of high-volt, sluggish, 1–2 Hertz, spiky wave complexes over a background of progressively slower and lower voltage (Xu et al. 2021). Creutzfeldt–Jakob disease's annual incidence is around 1 out of million worldwide (Rong et al. 2023). A tenth of these cases present with amyloid plaques, together with gliosis, severe cortical spongiosis, and neuronal death. A familial autosomal dominant type of CJD associated with PrP gene mutations accounts for 10% of cases (Bagyinszky et al. 2018).

Creutzfeldt–Jakob Disease (CJD) Sporadic Creutzfeldt–Jakob (sCJD) disease is defined as a fatal condition of global profound cognitive loss, myoclonic jerks, quickly worsening multifocal neurological impairment, and mortality in around 8 months time. Forty per cent of those who suffer from sporadic CJD have substantially worsening cognitive decline, forty per cent have cerebellar abnormalities, and twenty per cent have both. People exhibiting particular sCJD strains might develop peripheral neuropathy symptoms prior to this. Although PrP(Sc) accumulation is assumed as the cause, the varied peripheral nervous system (PNS) participation raises the possibility of tropism being dependent on the involved strain (Sitammagari and Masood 2022). The diagnostic spectrum quickly broadens to encompass behavioural deviations, cortex defects, including visual cortex function, pyramidal signs, extrapyramidal symptoms, and cerebellar abnormalities (Salehi et al. 2022). The majority of sporadic CJD sufferers experience myoclonic jerks affecting potentially the whole body or a particular limb(s). The jerks are able to be triggered by tactile or auditory stimuli, as well as by unprompted occurrence. Many of those with the disease exhibit a recognizable electroencephalogram pattern featuring periodic (or pseudoperiodic) episodes of sharp waveforms or spiking complexes over a slow-pace backdrop throughout the sporadic CJD progression. The specificity and sensitivity for such paroxysm per EEG are 87% and 67%, accordingly. Serial EEGs, however, are able to reveal abnormal periodic complexes in over 90% of case subjects (Sharma et al. 2009).

A medical professional ought to adhere to recognized case definitions whenever assessing an individual for potential sporadic Creutzfeldt–Jakob disease. Characteristic neuropathology and protease-resistant PrP on a Western blot are required to label the diagnosis as "**definitive CJD**". Progressive dementia together with typical EEG findings and a minimum of two of the following—akinetic mutism, cerebellar manifestations, impairment of vision, and pyramidal or extrapyramidal manifestations—are considered as "**probable CJD**". If a suspected case exhibits progressive dementia, atypical EEG findings or no availability of EEG, and a minimum of two of the following symptoms—akinetic mutism, cerebellar manifestations, impairment of vision, and pyramidal or extrapyramidal manifestations—for a period of under 2 years, the individual is classified as having "possible CJD" (Sitammagari and Masood 2022).

Gerstmann–Sträussler–Scheinker Syndrome (GSS) Individuals having this condition also have dementia in addition to truncal and limb ataxia that progresses gradually. The loss of life follows 3–8 years subsequent to presentation (Zhao et al. 2019). When the patient's brainstem gets heavily involved, indications of olivopontocerebellar degeneration are frequently seen. Autosomal dominant (AD) inheritance is brought on through PrP gene mutations (Prusiner 2013; Riudavets et al. 2013). Aside from the expected gliosis, neuronal death, and spongiform modifications, Gerstmann–Sträussler–Scheinker syndrome's neurological hallmark includes the occurrence of substantial and consistent amyloid deposition. It is relevant to note that substantial neurofibrillary tangle production is observed in a number of GSS Kins (Bruno et al. 2022). PrP-CAA, which is prion protein congophilic angiopathy, or which is also known as prion protein cerebral amyloid angiopathy, is a different type of AD inheritance-linked human prion disorder that is marked by a neurofibrillary tangle component alongside amyloid deposits within cerebral vessels (Jansen et al. 2010). There is also a crucial component known as cerebral amyloid angiopathy (CAA). Both of these prion-related variations strengthen the connection between the pathophysiology of Alzheimer's disease and prion illness (Ghiso et al. 2010).

Fatal Familial Insomnia (FFI) Uncontrollable insomnia, dysautonomic state of tachypnoea, tachycardia, hyperhidrosis, hypertension, plus hyperthermia, along with motor palsy and dementia, are common symptoms in these individuals, but the phenotypic manifestation varies greatly even in cases of identical families (Baldelli and Provini 2019). The approximate onset age might range between 18 and 60 years old (Harder et al. 2004). It is vital that one is cognizant that the earliest stages of FFI can appear with neuro-ophthalmological defects, notably increased saccadic intrusions, despite the fact that manifestations could differ greatly among each person (Mastrangelo et al. 2021). The illness can last anywhere from half a year to 3 years after the manifestations start. Genotype analysis is crucial for a firm diagnosis due to the variety of CNS manifestations of this condition. Due to gliosis and neuronal death, there is significant atrophy occurring in mediodorsal thalamic and anterior ventral nuclei. Spongiform transformation may be minor or not present at all, in contrast with alternative prion-linked conditions (Jankovska et al. 2021). There is a Met located at polymorphic codon 129 and an asparagine replacement of PrP gene's codon 178, which is a missense mutation seen in fatal familial insomnia (Jeong and Kim 2014). D178N mutations on PRNP genes (at chromosome 20) are linked to CJD and FFI. Codon 178's D178N mutation converts aspartate into asparagine. The codon 129 mutation that causes FFI, encodes for methionine. The thalamus region is especially vulnerable when it comes to FFI, but much of the brain's cortex is preserved; as a result, this type of insomnia lies at the extreme tip of the continuum of prion-related illnesses having prevalent neuropsychiatric manifestations (Khan and Bollu 2023; Tan et al. 2020). Among the Basque Autonomous Community of Spain, there appears to be an unusually high prevalence of FFI (Zarranz et al. 2004).

Variant Creutzfeldt–Jakob Disease (vCJD) Within the UK, bovine spongiform encephalopathy has caused the demise of over 160,000 bovines due to unforeseen epidemics of emerging prion diseases (Calza et al. 2001). This illness is believed to be brought on by consumption of bone or meat-based supplementations, which had been contaminated by scrapie-affected sheep and/or BSE-infected cows. A lot of scientific data point to BSE as the cause of variant Creutzfeldt–Jakob disease (Lee et al. 2013). In the year 1995, two British teens were diagnosed with one of the earliest episodes of variant CJD (Diack et al. 2014). Just four occurrences of sporadic CJD in adolescents have ever been documented; the highest occurrence of sporadic CJD emergence occurs in adults between the ages of 60 and 65. These teens stood out due to their early onset and unique CNS pathological findings that featured what are known as florid amyloid plaques that are similar to amyloid PrP plaques of kuru (Qi et al. 2020; Sitammagari and Masood 2022). It is noteworthy that these amyloid plaques likewise serve as a feature of CWD (Sikorska et al. 2009).

Diagnostic Considerations When vCJD first manifests, sensory and psychiatric disturbances are substantially more prevalent than those found in sporadic Creutzfeldt–Jakob disease (Sitammagari and Masood 2022). A fast-deteriorating cognitive health status is revealed by different neuro-psychological assessments. Ataxia and cognitive dysfunction are among the most typical early signs (Caine et al. 2015; Sundaram et al. 2020). There are far less typical variations like association of early cortical blindness (e.g. Heidenhain variant in CJD) (Baiardi et al. 2016). Uncommonly, the initial or predominant feature of CJD might involve unintentional upper limb levitation, which is a symptom of alien limb syndrome (Graff-Radford et al. 2013). Myoclonus represents a significant and almost inevitable clinical manifestation of sporadic Creutzfeldt–Jakob disease, but because it only manifests somewhat late along the progression of the illness, relying solely on this medical evidence might hinder those with the disease from being properly identified and diagnosed with CJD (Collinge 2005). All vCJD -affected individuals show cerebellar abnormalities, whereas only around 40% of sporadic CJD patients exhibit cerebellar impairment symptoms (Will and Ironside 2017).

Differential Diagnosis It is important to screen out any additional possible reasons of dementia, especially those that may be treated, like herpes encephalitis (Geschwind 2015). Increased protein and pleocytosis are present in the cerebrospinal fluid (CSF) in herpes encephalitis, but there are no such findings in CSF analysis of CJD patients. Herpes encephalitis also exhibits typical electroencephalogram and MRI abnormalities, unmatched to prion disease (Coulthart and Cashman 2001). Neurodegenerative conditions like Alzheimer's, familial myoclonic dementia, multisystem atrophy, and Pick's disease are additional disorders to further add into the extensive list of diagnostic differentials. In contrast to Creutzfeldt–Jakob disease, each one of the above diseases has a slower rate of neurodegeneration and associated CNS manifestations (Paterson et al. 2012).

When considering the context of differential diagnosis for CJD, steroid-responsive encephalopathy linked to autoimmune thyroiditis (Hashimoto encephalitis), although rare, should also be taken into account (Seth et al. 2021). Neurological signs and symptoms of Hashimoto encephalitis might be really identical to those associated with possible CJD, including fast -progressing dementia, psychosis, ataxia, and myoclonus, but contrarily, Hashimoto encephalitis is an autoimmune condition and hence responds effectively to immunosuppressive therapy. Detection of raised serum thyroperoxidase autoantibodies are able to aid in diagnosing this condition, along with EEG reporting of abnormal wave complexes and 14-3-3 protein found in cerebrospinal fluid (Payer et al. 2012).

Typical manifestations of CJD seem highly distinct; however, because of a great deal of clinical heterogeneity plus a wide range of clinical features shared by more commonly occurring neurodegenerative illnesses, "CJD-mimics" remain fairly prevalent, making a diagnosis difficult. Hyperthermia, hyponatraemia, and seizures are medical symptoms that point to a "CJD-mimic" condition, instead of CJD (Mead and Rudge 2017). Magnetic resonance image signalling showing hyperintensity beyond the cerebral cortex, thalamus, or striatum is also indicative of CNS pathology besides CJD. Additionally, observing lesions on contrast-enhanced computed tomography, as well as CSF findings of pleocytosis, strongly suggest a "CJD-mimic" (Macfarlane et al. 2007; Zerr and Poser 2002).

Laboratory Evaluations Regulating infectiousness or replication at the very beginning of the illness's evolution has proven difficult by the lack of a clear mechanism for detecting preliminary prion illness. Novel approaches like PMCA, otherwise known as protein misfolding cyclic amplification, and RT-QuIC, which is short for real -time quaking -induced conversion, might nevertheless help us get around this restriction (Barria et al. 2012; Schmitz et al. 2016). RT-QuIC analyses use fluorescent markers to track the formation of misfolded PrP gene aggregates at a real-time basis. The National CJD Research & Surveillance Unit determined that the RT-QuIC test's sensitivity with cerebrospinal fluid specimens was 92% and its specificity was 100% (Green 2019). In individuals receiving screening for Creutzfeldt–Jakob disease, due to any alarming indications, RT-QuIC assays demonstrated to exhibit comparable levels of raised specificity and sensitivity (Green 2019).

Numerous additional CSF indicators, such as non-phosphorylated tau, alpha-synuclein, malate dehydrogenase, and neurogranin have shown diagnostic reliability when applied in electrochemiluminescence -based tests, although some of the above might not yet be used in healthcare settings (Altuna et al. 2022; Ermann et al. 2018). Investigations for dementia ought to be part of the first assessment for neurological disease. Assess erythrocyte sedimentation rate, liver functions, and the usual laboratory workups including complete blood cell tests to exclude off metabolic/toxic encephalopathies. If the presence of bacteria-related illness is believed to exist, take appropriate samples for culture (e.g. blood cultures) (Connor et al. 2019; Rosenbloom and Atri 2011). Run evaluations for neurosyphilis (preferably, fluorescent antibody test) and assess thyroid status, vitamin B-12 levels, folate, and other relevant values. Have HIV tests conducted if there is evidence of any concerning risk factors (Kwon and Kwon 2019). Autoantibodies require being checked whenever any paraneoplastic syndrome is suspected, for instance, if there is a record of neoplasia or an undetected tumour is discovered by neuroimaging (Paterson et al. 2012).

Imaging Magnetic resonance imaging (MRI) serves as a crucial diagnostic tool. Diffusion- weighted or even fluid- attenuated inversion recovery images may portray hyper-intensified basal ganglia, cortical ribbon, and thalamus regions. Pertaining to sporadic CJD, fluid- attenuated inversion recovery imaging and diffusion- weighted imaging were reported as having 95% and 91% specificities and sensitivities, respectively, with a diagnostic accuracy of 94%, as per a current study (Young et al. 2005).

Diffusion-based magnetic resonance imaging has been shown to have diagnostic value, which is in parallel to RT-QuIC when it comes to human prion diseases, with much higher specificity value than CSF chemistries (Foutz et al. 2017). It should also be noted that MRI and other such imaging studies are generally not invasive, and hence of much benefit when compared with some other tests.

There are some radiologically significant findings, such as the hockey stick sign and the pulvinar sign, which are often seen in variant Creutzfeldt–Jakob disease, denoting caudate head nucleus plus putamen involvement, as well as bilateral pulvinar thalamic involvement, respectively (Collie et al. 2003; Van Cauter et al. 2020; Venkatesan and Ramadoss 2015).

PET scan, otherwise known as position emission tomography, done in some affected individuals (but never routinely), has demonstrated local glucosal hypometabolic states corresponding to autopsy findings of associated neuropathology (Renard et al. 2017).

Ataxia, corticobasal syndrome, myoclonus, limb apraxias/dystonias, parkinsonian signs, pyramidal signs, sensory deficits, and visual impairment all come under the umbrella of diverse presentations of CJD (Parmera et al. 2016; Sitammagari and Masood 2022).

Additionally, it may be beneficial to request for abdomen, chest, and pelvis contrast CT imaging, so as to rule off the likelihood of malignancies, which might be causing paraneoplastic conditions (Geschwind 2015).

Electroencephalography (EEG) Most sporadic CJD- affected individuals demonstrate distinct periodic/pseudoperiodic wave paroxysms sometime in the disease progression. These paroxysms are characterized as sharp or associated with spikes with slow backgrounds. They may, however, not be found as often, in cases of variant CJD.

Sleep Analysis Video polysomnography is a helpful technique to detect sleep disorganization and significant drops in total sleep times, which may be experienced in fatal insomnia suspects (Khan and Bollu 2023).

Genetic Analysis Genetic testing to find loci expressing prion-related genetic risk might get more popular in the near future, although, as yet, they are neither widely available, nor recommended (Prusiner 2013).

11.8.5 Management, Treatment, and Prevention

There is no successful or proven therapy for any of the prion disorders, making them all deadly. Today, only symptoms-specific therapy is given to affected individuals (Appleby and Lyketsos 2011). To prevent seizures among certain CJD patients, antiepileptic medications may need to be given, while anti-Parkinson medications must be given to patients who experience extrapyramidal signs and symptoms (Appleby and Lyketsos 2011; Burgyone et al. 2004).

In various experiments, it has been demonstrated that some drugs (i.e. amphotericin B, sulphated polyanions, Congo red analogues, anthracyclines, tetrapyrroles, and pentosan polysulphates) may aid in suppressing prion spread (Barret et al. 2003; Doh-ura et al. 2004; Forloni et al. 2002; Shim et al. 2022; Wiegmans et al. 2019). Chlorpromazine and quinacrine have also been demonstrated to be able to possibly prevent the converting of PrPC into PrPSc, as per some tissue culture trials. Chlorpromazine and quinacrine have also been demonstrated to be able to possibly prevent the converting of PrPC into PrPSc, as per some tissue culture trials (Barret et al. 2003; Shim et al. 2022). For prion-related CNS disorders, and also for different conformational illnesses (e.g. Alzheimer's), substantial research is now being done on techniques for therapy that target the aberrant protein construct linked to the condition (Burchell and Panegyres 2016; Prusiner 2013).

Various other experimental researches on therapeutic approaches for prion illnesses have been carried out, such as chelation therapies (e.g. with copper and D-penicillamine); immunological approaches (e.g. using alpha -helix peptides, vaccines with attenuated strains of CJD, and prion-peptide-based immunization);

use of metformin; and use of antisense oligonucleotides to target PRNP gene; despite such a vast arena of researches on a global scale, till date, an effective treatment for human prion illnesses is yet to be discovered.

Prognosis Prion-related CNS disorders advance quite quickly. Beginning with the moment of diagnoses until death, the average survival period varies between 8 and 60 months. Individuals having sporadic illness typically endure a shorter course compared to people suffering from familial prion illness (Geschwind 2015).

References

Abdullahi AM, Sarmast ST, Jahan N (2020) Viral infections of the central nervous system in children: a systematic review. Cureus 12:e11174

Abu-Rumeileh S, Abdelhak A, Foschi M, Tumani H, Otto M (2021) Guillain-Barré syndrome spectrum associated with COVID-19: an up-to-date systematic review of 73 cases. J Neurol 268: 1133–1170

Ahye N, Bellizzi A, May D, Wollebo HS (2020) The role of the JC virus in central nervous system tumorigenesis. Int J Mol Sci 21:6236

AlKetbi R, AlNuaimi D, AlMulla M, AlTalai N, Samir M, Kumar N, AlBastaki U (2020) Acute myelitis as a neurological complication of Covid-19: a case report and MRI findings. Radiol Case Rep 15:1591–1595

Altuna M, Ruiz I, Zelaya MV, Mendioroz M (2022) Role of biomarkers for the diagnosis of prion diseases: a narrative review. Medicina 58:473

Amman BR, Pavlin BI, Albariño CG, Comer JA, Erickson BR, Oliver JB, Sealy TK, Vincent MJ, Nichol ST, Paddock CD, Tumpey AJ, Wagoner KD, Glauer RD, Smith KA, Winpisinger KA, Parsely MS, Wyrick P, Hannafin CH, Bandy U, Zaki S, Rollin PE, Ksiazek TG (2007) Pet rodents and fatal lymphocytic choriomeningitis in transplant patients. Emerg Infect Dis 13:719–725

Antinori A, Cingolani A, Lorenzini P, Giancola ML, Uccella I, Bossolasco S, Grisetti S, Moretti F, Vigo B, Bongiovanni M, Del Grosso B, Arcidiacono MI, Fibbia GC, Mena M, Finazzi MG, Guaraldi G, Ammassari A, d'Arminio Monforte A, Cinque P, De Luca A, Italian Registry Investigative Neuro AIDS Study Group (2003) Clinical epidemiology and survival of progressive multifocal leukoencephalopathy in the era of highly active antiretroviral therapy: data from the Italian Registry Investigative Neuro AIDS (IRINA). J Neurovirol 9(1):47–53

Appleby BS, Lyketsos CG (2011) Rapidly progressive dementias and the treatment of human prion diseases. Expert Opin Pharmacother 12:1–12

Aucouturier P, Geissmann F, Damotte D, Saborio GP, Meeker HC, Kascsak R, Kascsak R, Carp RI, Wisniewski T (2001) Infected splenic dendritic cells are sufficient for prion transmission to the CNS in mouse scrapie. J Clin Invest 108:703–708

Bagyinszky E, Giau VV, Youn YC, An SSA, Kim S (2018) Characterization of mutations in PRNP (prion) gene and their possible roles in neurodegenerative diseases. Neuropsychiatr Dis Treat 14:2067–2085

Baiardi S, Capellari S, Ladogana A, Strumia S, Santangelo M, Pocchiari M, Parchi P (2016) Revisiting the Heidenhain variant of Creutzfeldt-Jakob disease: evidence for prion type variability influencing clinical course and laboratory findings. J Alzheimers Dis 50:465–476

Baldelli L, Provini F (2019) Fatal familial insomnia and Agrypnia Excitata: autonomic dysfunctions and pathophysiological implications. Auton Neurosci 218:68–86

Bale JF (2014) Measles, mumps, rubella, and human parvovirus B19 infections and neurologic disease. Handb Clin Neurol 121:1345–1353

Baral PK, Yin J, Aguzzi A, James MNG (2019) Transition of the prion protein from a structured cellular form (PrPC) to the infectious scrapie agent (PrPSc). Protein Sci 28:2055–2063

Barret A, Tagliavini F, Forloni G, Bate C, Salmona M, Colombo L, De Luigi A, Limido L, Suardi S, Rossi G, Auvré F, Adjou KT, Salès N, Williams A, Lasmézas C, Deslys JP (2003) Evaluation of quinacrine treatment for prion diseases. J Virol 77:8462–8469

Barria MA, Gonzalez-Romero D, Soto C (2012) Cyclic amplification of prion protein misfolding. Methods Mol Biol 849:199–212

Beeman A (2007) Charles Armstrong, M.D.: a biography, p 183ff

Belay ED, Sejvar JJ, Shieh W-J, Wiersma ST, Zou W-Q, Gambetti P, Hunter S, Maddox RA, Crockett L, Zaki SR, Schonberger LB (2005) Variant Creutzfeldt-Jakob disease death, United States. Emerg Infect Dis 11:1351–1354

Beltrami S, Gordon J (2014) Immune surveillance and response to JC virus infection and PML. J Neurovirol 20:137–149

Berger JR, Aksamit AJ, Clifford DB, Davis L, Koralnik IJ, Sejvar JJ, Bartt R, Major EO, Nath A (2013) PML diagnostic criteria: consensus statement from the AAN neuroinfectious disease section. Neurology 80:1430–1438

Bonthius DJ (2012) Lymphocytic choriomeningitis virus: an under-recognized cause of neurologic disease in the fetus, child, and adult. Semin Pediatr Neurol 19:89–95

Boothpur R, Brennan DC (2010) Human polyoma viruses and disease with emphasis on clinical BK and JC. J Clin Virol 47:306–312

Bouybayoune I, Mantovani S, Gallo FD, Bertani I, Restelli E, Comerio L, Tapella L, Baracchi F, Fernández-Borges N, Mangieri M, Bisighini C, Beznoussenko GV, Paladini A, Balducci C, Micotti E, Forloni G, Castilla J, Fiordaliso F, Tagliavini F, Imeri L, Chiesa R (2015) Transgenic fatal familial insomnia mice indicate prion infectivity-independent mechanisms of pathogenesis and phenotypic expression of disease. PLoS Pathog 11:e1004796

Bradford BM, Mabbott NA (2012) Prion disease and the innate immune system. Viruses 4:3389–3419

Bruno R, Pirisinu L, Riccardi G, D'Agostino C, De Cecco E, Legname G, Cardone F, Gambetti P, Nonno R, Agrimi U, Di Bari MA (2022) Gerstmann–Sträussler–Scheinker disease with F198S mutation induces independent tau and prion protein pathologies in bank voles. Biomolecules 12:1537

Burchell JT, Panegyres PK (2016) Prion diseases: immunotargets and therapy. Immunotargets Ther 5:57–68

Burgyone K, Aduri K, Ananth J, Parameswaran S (2004) The use of antiparkinsonian agents in the management of drug-induced extrapyramidal symptoms. Curr Pharm Des 10:2239–2248

Caine D, Tinelli RJ, Hyare H, De Vita E, Lowe J, Lukic A, Thompson A, Porter M-C, Cipolotti L, Rudge P, Collinge J, Mead S (2015) The cognitive profile of prion disease: a prospective clinical and imaging study. Ann Clin Transl Neurol 2:548–558

Calza L, Manfredi R, Chiodo F (2001) Epidemics of bovine spongiform encephalopathy and new variant of Creutzfeldt-Jakob disease in humans. Most recent findings on prion disease. Recenti Prog Med 92:140–149

Cao L, Feng H, Huang X, Yi J, Zhou Y (2021) Gerstmann-Sträussler-Scheinker syndrome misdiagnosed as cervical spondylotic myelopathy. Medicine 100:e25687

Chapman J, Ben-Israel J, Goldhammer Y, Korczyn AD (1994) The risk of developing Creutzfeldt-Jakob disease in subjects with the PRNP gene codon 200 point mutation. Neurology 44:1683–1686

Charco JM, Eraña H, Venegas V, García-Martínez S, López-Moreno R, González-Miranda E, Pérez-Castro MÁ, Castilla J (2017) Recombinant PrP and its contribution to research on transmissible spongiform encephalopathies. Pathogens 6:67

Charlton CL, Severini A (2016) Dilemmas and pitfalls in rubella laboratory diagnostics in low prevalence or elimination settings. Curr Treat Options Infect Dis 8:329–342

Chiang GPK, Chen Z, Chan MCW, Lee SHM, Kwok AK, Yeung ACM, Nelson EAS, Hon KL, Leung TF, Chan PKS (2017) Clinical features and seasonality of parechovirus infection in an Asian subtropical city, Hong Kong. PLoS One 12:e0184533

Chou SH-Y, Beghi E, Helbok R, Moro E, Sampson J, Altamirano V, Mainali S, Bassetti C, Suarez JI, McNett M, GCS-NeuroCOVID Consortium and ENERGY Consortium (2021) Global incidence of neurological manifestations among patients hospitalized with COVID-19—a report for the GCS-NeuroCOVID consortium and the ENERGY consortium. JAMA Netw Open 4: e2112131

Christensen KLY, Holman RC, Hammett TA, Belay ED, Schonberger LB (2010) Progressive multifocal leukoencephalopathy deaths in the USA, 1979-2005. Neuroepidemiology 35:178–184

Cinque P, Koralnik IJ, Gerevini S, Miro JM, Price RW (2009) Progressive multifocal leukoencephalopathy complicating HIV-1 Infection. Lancet Infect Dis 9:625–636

Cobb NJ, Surewicz WK (2009) Prion diseases and their biochemical mechanisms. Biochemistry 48: 2574–2585

Collie DA, Summers DM, Sellar RJ, Ironside JW, Cooper S, Zeidler M, Knight R, Will RG (2003) Diagnosing variant Creutzfeldt-Jakob disease with the pulvinar sign: MR imaging findings in 86 neuropathologically confirmed cases. AJNR Am J Neuroradiol 24:1560–1569

Collinge J (2005) Molecular neurology of prion disease. J Neurol Neurosurg Psychiatry 76:906–919

Collins S, McLean CA, Masters CL (2001) Gerstmann-Sträussler-Scheinker syndrome, fatal familial insomnia, and kuru: a review of these less common human transmissible spongiform encephalopathies. J Clin Neurosci 8:387–397

Connolly JH, Hutchinson WM, Allen IV, Lyttle JA, Swallow MW, Dermott E, Thomson D (1975) Carotid artery thrombosis, encephalitis, myelitis and optic neuritis associated with rubella virus infections. Brain 98:583–594

Connor A, Wang H, Appleby BS, Rhoads DD (2019) Clinical laboratory tests used to aid in diagnosis of human prion disease. J Clin Microbiol 57:e00769

Cortese I, Reich DS, Nath A (2021) Progressive multifocal leukoencephalopathy and the spectrum of JC virus-related disease. Nat Rev Neurol 17:37–51

Coulthart MB, Cashman NR (2001) Variant Creutzfeldt–Jakob disease: a summary of current scientific knowledge in relation to public health. CMAJ 165:51–58

Cui X, Wang J, Deckhut A, Joseph B, Eberwein P, Cubitt C, Ryschkewitsch C, Agostini H, Stoner G (2004) Chinese strains (type 7) of JC virus are Afro-Asiatic in origin but are phylogenetically distinct from the Mongolian and Indian strains (type 2D) and the Korean and Japanese strains (Type 2A). J Mol Evol 58:568–583

Dalton HR, Kamar N, van Eijk JJJ, Mclean BN, Cintas P, Bendall RP, Jacobs BC (2016) Hepatitis E virus and neurological injury. Nat Rev Neurol 12:77–85

Dang X, Vidal JE, Penalva de Oliveira AC, Simpson DM, Morgello S, Hecht JH, Ngo LH, Koralnik IJ (2012) JC virus granule cell neuronopathy is associated with VP1 C terminus mutants. J Gen Virol 93:175–183

Dang TO, Ogunniyi A, Barbee MS, Drilon A (2016) Pembrolizumab for the treatment of PD-L1 positive advanced or metastatic non-small cell lung cancer. Expert Rev Anticancer Ther 16:13–20

DeBiasi RL, Tyler KL (2004) Molecular methods for diagnosis of viral encephalitis. Clin Microbiol Rev 17:903–925

Desai I, Manchanda R, Kumar N, Tiwari A, Kumar M (2021) Neurological manifestations of coronavirus disease 2019: exploring past to understand present. Neurol Sci 42:773–785

Desmond MM, Wilson GS, Melnick JL, Singer DB, Zion TE, Rudolph AJ, Pineda RG, Ziai M-H, Blattner RJ (1967) Congenital rubella encephalitis: course and early sequelae. J Pediatr 71:311–331

Diack AB, Head MW, McCutcheon S, Boyle A, Knight R, Ironside JW, Manson JC, Will RG (2014) Variant CJD 18 years of research and surveillance. Prion 8:286–295

Dietz K, Raddatz G, Wallis J, Müller N, Zerr I, Duerr H-P, Lefèvre H, Seifried E, Löwer J (2007) Blood transfusion and spread of variant Creutzfeldt-Jakob disease. Emerg Infect Dis 13:89–96

Doh-ura K, Ishikawa K, Murakami-Kubo I, Sasaki K, Mohri S, Race R, Iwaki T (2004) Treatment of transmissible spongiform encephalopathy by intraventricular drug infusion in animal models. J Virol 78:4999–5006

Dunham SR, Schmidt R, Clifford DB (2020) Treatment of progressive multifocal leukoencephalopathy using immune restoration. Neurotherapeutics 17:955–965

Ermann N, Lewczuk P, Schmitz M, Lange P, Knipper T, Goebel S, Kornhuber J, Zerr I, Llorens F (2018) CSF nonphosphorylated Tau as a biomarker for the discrimination of AD from CJD. Ann Clin Transl Neurol 5:883–887

Felsenstein S, Yang S, Eubanks N, Sobrera E, Grimm JP, Aldrovandi G (2014) Human parechovirus central nervous system infections in southern California children. Pediatr Infect Dis J 33:87–91

Figueiredo CA, Klautau GB, Afonso AMS, Castrignano SB, Oliveira MI, Curti SP, Squarcina GG, Narimatsu K, Rasslan Z, Lima CAC, Golin V, Tadeo EF, Spagunolo FJ, Cataldo B, Durigon EL (2008) Isolation and genotype analysis of rubella virus from a case of Guillain-Barré syndrome. J Clin Virol 43:343–345

Fischer SA, Graham MB, Kuehnert MJ, Kotton CN, Srinivasan A, Marty FM, Comer JA, Guarner J, Paddock CD, DeMeo DL, Shieh W-J, Erickson BR, Bandy U, DeMaria A, Davis JP, Delmonico FL, Pavlin B, Likos A, Vincent MJ, Sealy TK, Goldsmith CS, Jernigan DB, Rollin PE, Packard MM, Patel M, Rowland C, Helfand RF, Nichol ST, Fishman JA, Ksiazek T, Zaki SR, LCMV in Transplant Recipients Investigation Team (2006) Transmission of lymphocytic choriomeningitis virus by organ transplantation. N Engl J Med 354:2235–2249

Fletcher NF, Yang JP, Farquhar MJ, Hu K, Davis C, He Q, Dowd K, Ray SC, Krieger SE, Neyts J, Baumert TF, Balfe P, McKeating JA, Wong-Staal F (2010) Hepatitis C virus infection of neuroepithelioma cell lines. Gastroenterology 139:1365–1374

Forloni G, Iussich S, Awan T, Colombo L, Angeretti N, Girola L, Bertani I, Poli G, Caramelli M, Grazia Bruzzone M, Farina L, Limido L, Rossi G, Giaccone G, Ironside JW, Bugiani O, Salmona M, Tagliavini F (2002) Tetracyclines affect prion infectivity. Proc Natl Acad Sci U S A 99:10849–10854

Forton DM, Taylor-Robinson SD, Thomas HC (2006) Central nervous system changes in hepatitis C virus infection. Eur J Gastroenterol Hepatol 18:333–338

Foutz A, Appleby BS, Hamlin C, Liu X, Yang S, Cohen Y, Chen W, Blevins J, Fausett C, Wang H, Gambetti P, Zhang S, Hughson A, Tatsuoka C, Schonberger LB, Cohen ML, Caughey B, Safar JG (2017) Diagnostic and prognostic value of human prion detection in cerebrospinal fluid. Ann Neurol 81:79–92

Frey TK (2008) Neurological aspects of rubella virus infection. Intervirology 40:167–175

Galassi G, Marchioni A (2020) Facing acute neuromuscular diseases during COVID-19 pandemic: focus on Guillain-Barré syndrome. Acta Neurol Belg 120:1067–1075

Gasparovic ML, Gee GV, Atwood WJ (2006) JC virus minor capsid proteins Vp2 and Vp3 are essential for virus propagation. J Virol 80:10858–10861

Geschwind MD (2015) Prion diseases. Continuum 21:1612–1638

Getahun M, Beyene B, Gallagher K, Ademe A, Teshome B, Tefera M, Asha A, Afework A, Assefa E, HaileMariam Y, HaileGiorgis Y, Ketema H, Shiferaw D, Bekele A, Jima D, Kebede A (2016) Epidemiology of rubella virus cases in the pre-vaccination era of Ethiopia, 2009–2015. BMC Public Health 16:1168

Ghiso J, Tomidokoro Y, Revesz T, Frangione B, Rostagno A (2010) Cerebral amyloid angiopathy and Alzheimer's disease. Hirosaki Igaku 61:S111–S124

Glatzel M, Heppner FL, Albers KM, Aguzzi A (2001) Sympathetic innervation of lymphoreticular organs is rate limiting for prion neuroinvasion. Neuron 31:25–34

Graff-Radford J, Rubin MN, Jones DT, Aksamit AJ, Ahlskog JE, Knopman DS, Petersen RC, Boeve BF, Josephs KA (2013) The alien limb phenomenon. J Neurol 260:1880–1888

Green AJE (2019) RT-QuIC: a new test for sporadic CJD. Pract Neurol 19:49–55

Harder A, Gregor A, Wirth T, Kreuz F, Schulz-Schaeffer WJ, Windl O, Plotkin M, Amthauer H, Neukirch K, Kretzschmar HA, Kuhlmann T, Braas R, Hahne HH, Jendroska K (2004) Early age of onset in fatal familial insomnia. Two novel cases and review of the literature. J Neurol 251: 715–724

Helms J, Kremer S, Merdji H, Clere-Jehl R, Schenck M, Kummerlen C, Collange O, Boulay C, Fafi-Kremer S, Ohana M, Anheim M, Meziani F (2020) Neurologic features in severe SARS-CoV-2 infection. N Engl J Med 382:2268–2270

Hickerson BT, Westover JB, Jung K-H, Komeno T, Furuta Y, Gowen BB (2018) Effective treatment of experimental lymphocytic choriomeningitis virus infection: consideration of favipiravir for use with infected organ transplant recipients. J Infect Dis 218:522–527

Houston F, Foster JD, Chong A, Hunter N, Bostock CJ (2000) Transmission of BSE by blood transfusion in sheep. Lancet 356:999–1000

Iannetta M, Bellizzi A, Lo Menzo S, Anzivino E, D'Abramo A, Oliva A, D'Agostino C, d'Ettorre G, Pietropaolo V, Vullo V, Ciardi MR (2013) HIV-associated progressive multifocal leukoencephalopathy: longitudinal study of JC virus non-coding control region rearrangements and host immunity. J Neurovirol 19:274–279

Isaacs JD, Jackson GS, Altmann DM (2006) The role of the cellular prion protein in the immune system. Clin Exp Immunol 146:1–8

Jankovska N, Rusina R, Bruzova M, Parobkova E, Olejar T, Matej R (2021) Human prion disorders: review of the current literature and a twenty-year experience of the national surveillance center in the Czech Republic. Diagnostics 11:1821

Jansen C, Parchi P, Capellari S, Vermeij AJ, Corrado P, Baas F, Strammiello R, van Gool WA, van Swieten JC, Rozemuller AJM (2010) Prion protein amyloidosis with divergent phenotype associated with two novel nonsense mutations in PRNP. Acta Neuropathol 119:189–197

Jeong B-H, Kim Y-S (2014) Genetic studies in human prion diseases. J Korean Med Sci 29:623–632

Jin Y, Yang H, Ji W, Wu W, Chen S, Zhang W, Duan G (2020) Virology, epidemiology, pathogenesis, and control of COVID-19. Viruses 12:372

Kamar N, Bendall R, Legrand-Abravanel F, Xia N-S, Ijaz S, Izopet J, Dalton HR (2012) Hepatitis E. Lancet 379:2477–2488

Kang SS, McGavern DB (2008) Lymphocytic choriomeningitis infection of the central nervous system. Front Biosci 13:4529–4543

Khan Z, Bollu PC (2023) Fatal familial insomnia. In: StatPearls. StatPearls Publishing, Treasure Island

Knust B, Ströher U, Edison L, Albariño CG, Lovejoy J, Armeanu E, House J, Cory D, Horton C, Fowler KL, Austin J, Poe J, Humbaugh KE, Guerrero L, Campbell S, Gibbons A, Reed Z, Cannon D, Manning C, Petersen B, Metcalf D, Marsh B, Nichol ST, Rollin PE (2014) Lymphocytic choriomeningitis virus in employees and mice at multipremises feeder-rodent operation, United States, 2012. Emerg Infect Dis 20:240–247

Kuhle J, Gosert R, Bühler R, Derfuss T, Sutter R, Yaldizli Ö, Radue E-W, Ryschkewitsch C, Major EO, Kappos L, Frank S, Hirsch HH (2011) Management and outcome of CSF-JC virus PCR-negative PML in a natalizumab-treated patient with MS. Neurology 77:2010–2016

Kwon GT, Kwon MS (2019) Diagnostic challenge of rapidly progressing sporadic Creutzfeldt-Jakob disease. BMJ Case Rep 12:e230535

Landry ML, Eid T, Bannykh S, Major E (2008) False negative PCR despite high levels of JC virus DNA in spinal fluid: implications for diagnostic testing. J Clin Virol 43:247–249

Lebrun AJ (1978) ID#713 - Public Health Image Library (PHIL). CDC. Accessed 10 July 23

Lee KJ, Novella IS, Teng MN, Oldstone MBA, de la Torre JC (2000) NP and L proteins of lymphocytic choriomeningitis virus (LCMV) are sufficient for efficient transcription and replication of LCMV genomic RNA analogs. J Virol 74:3470–3477

Lee J, Kim SY, Hwang KJ, Ju YR, Woo H-J (2013) Prion diseases as transmissible zoonotic diseases. Osong Public Health Res Perspect 4:57–66

Lee S, Ko H, Kim S, Lee YS, Son B (2019) Progressive multifocal leukoencephalopathy diagnosed by brain biopsy, not by the DNA test for JC virus. Asian J Neurosurg 14:240–244

Liberski PP, Gajos A, Sikorska B, Lindenbaum S (2019) Kuru, the first human prion disease. Viruses 11:232

Lu L, Xiong W, Liu D, Liu J, Yang D, Li N, Mu J, Guo J, Li W, Wang G, Gao H, Zhang Y, Lin M, Chen L, Shen S, Zhang H, Sander JW, Luo J, Chen S, Zhou D (2020) New onset acute symptomatic seizure and risk factors in coronavirus disease 2019: a retrospective multicenter study. Epilepsia 61:e49–e53

Lyons S, O'Kelly B, Woods S, Rowan C, Brady D, Sheehan G, Smyth S (2020) Seizure with CSF lymphocytosis as a presenting feature of COVID-19 in an otherwise healthy young man. Seizure 80:113–114

Mabbott NA, Bradford BM, Pal R, Young R, Donaldson DS (2020) The effects of immune system modulation on prion disease susceptibility and pathogenesis. Int J Mol Sci 21:7299

Macfarlane RG, Wroe SJ, Collinge J, Yousry TA, Jäger HR (2007) Neuroimaging findings in human prion disease. J Neurol Neurosurg Psychiatry 78:664–670

Maginnis MS, Nelson CDS, Atwood WJ (2015) JC polyomavirus attachment, entry, and trafficking: unlocking the keys to a fatal infection. J Neurovirol 21:601–613

Mao L, Jin H, Wang M, Hu Y, Chen S, He Q, Chang J, Hong C, Zhou Y, Wang D, Miao X, Li Y, Hu B (2020) Neurologic manifestations of hospitalized patients with coronavirus disease 2019 in Wuhan, China. JAMA Neurol 77:683–690

Mastrangelo V, Merli E, Rucker JC, Eggenberger ER, Zee DS, Cortelli P (2021) Neuro-ophthalmological findings in early fatal familial insomnia. Ann Neurol 89:823–827

Mathur G, Yadav K, Ford B, Schafer IJ, Basavaraju SV, Knust B, Shieh W-J, Hill S, Locke GD, Quinlisk P, Brown S, Gibbons A, Cannon D, Kuehnert M, Nichol ST, Rollin PE, Ströher U, Miller R (2017) High clinical suspicion of donor-derived disease leads to timely recognition and early intervention to treat solid organ transplant-transmitted lymphocytic choriomeningitis virus. Transpl Infect Dis 19:4

Mead S, Rudge P (2017) CJD mimics and chameleons. Pract Neurol 17:113–121

Meiner Z, Gabizon R, Prusiner SB (1997) Familial Creutzfeldt-Jakob disease. Codon 200 prion disease in Libyan Jews. Medicine 76:227–237

Metlay JP, Waterer GW, Long AC, Anzueto A, Brozek J, Crothers K, Cooley LA, Dean NC, Fine MJ, Flanders SA, Griffin MR, Metersky ML, Musher DM, Restrepo MI, Whitney CG (2019) Diagnosis and treatment of adults with community-acquired pneumonia. An official clinical practice guideline of the American Thoracic Society and Infectious Diseases Society of America. Am J Respir Crit Care Med 200:e45–e67

Miller E, Cradock-Watson JE, Pollock TM (1982) Consequences of confirmed maternal rubella at successive stages of pregnancy. Lancet 2:781–784

Miskin DP, Koralnik IJ (2015) Novel syndromes associated with JC virus infection of neurons and meningeal cells: no longer a gray area. Curr Opin Neurol 28:288–294

Mitrová E, Belay G (2002) Creutzfeldt-Jakob disease with E200K mutation in Slovakia: characterization and development. Acta Virol 46:31–39

Molloy ES, Calabrese LH (2009) Progressive multifocal leukoencephalopathy: a national estimate of frequency in systemic lupus erythematosus and other rheumatic diseases. Arthritis Rheum 60: 3761–3765

Moriguchi T, Harii N, Goto J, Harada D, Sugawara H, Takamino J, Ueno M, Sakata H, Kondo K, Myose N, Nakao A, Takeda M, Haro H, Inoue O, Suzuki-Inoue K, Kubokawa K, Ogihara S, Sasaki T, Kinouchi H, Kojin H, Ito M, Onishi H, Shimizu T, Sasaki Y, Enomoto N, Ishihara H, Furuya S, Yamamoto T, Shimada S (2020) A first case of meningitis/encephalitis associated with SARS-Coronavirus-2. Int J Infect Dis 94:55–58

Natale G, Ferrucci M, Lazzeri G, Paparelli A, Fornai F (2011) Transmission of prions within the gut and toward the central nervous system. Prion 5:142–149

O'Connor T, Aguzzi A (2013) Prions and lymphoid organs. Prion 7:157–163

Oldstone MBA, Ware BC, Horton LE, Welch MJ, Aiolfi R, Zarpellon A, Ruggeri ZM, Sullivan BM (2018) Lymphocytic choriomeningitis virus Clone 13 infection causes either persistence or acute death dependent on IFN-1, cytotoxic T lymphocytes (CTLs), and host genetics. Proc Natl Acad Sci U S A 115:E7814–E7823

Ono D, Shishido-Hara Y, Mizutani S, Mori Y, Ichinose K, Watanabe M, Tanizawa T, Yokota T, Uchihara T, Fujigasaki H (2019) Development of demyelinating lesions in progressive multifocal leukoencephalopathy (PML): comparison of magnetic resonance images and neuropathology of post-mortem brain. Neuropathology 39:294–306

Papageorgiou N, Spiliopoulou M, Nguyen T-HV, Vaitsopoulou A, Laban EY, Alvarez K, Margiolaki I, Canard B, Ferron F (2020) Brothers in arms: structure, assembly and function of arenaviridae nucleoprotein. Viruses 12:772

Parmera JB, Rodriguez RD, Studart Neto A, Nitrini R, Brucki SMD (2016) Corticobasal syndrome: a diagnostic conundrum. Dement Neuropsychol 10:267–275

Paterson RW, Torres-Chae CC, Kuo AL, Ando T, Nguyen EA, Wong K, DeArmond SJ, Haman A, Garcia P, Johnson DY, Miller BL, Geschwind MD (2012) Differential diagnosis of Jakob-Creutzfeldt disease. Arch Neurol 69:1578–1582

Patnaik A, Kang SP, Rasco D, Papadopoulos KP, Elassaiss-Schaap J, Beeram M, Drengler R, Chen C, Smith L, Espino G, Gergich K, Delgado L, Daud A, Lindia JA, Li XN, Pierce RH, Yearley JH, Wu D, Laterza O, Lehnert M, Iannone R, Tolcher AW (2015) Phase I study of pembrolizumab (MK-3475; anti-PD-1 monoclonal antibody) in patients with advanced solid tumors. Clin Cancer Res 21:4286–4293

Pavesi A (2003) African origin of polyomavirus JC and implications for prehistoric human migrations. J Mol Evol 56:564–572

Pavesi A (2005) Utility of JC polyomavirus in tracing the pattern of human migrations dating to prehistoric times. J Gen Virol 86:1315–1326

Pavlovic D, Patera AC, Nyberg F, Gerber M, Liu M (2015) Progressive multifocal leukoencephalopathy: current treatment options and future perspectives. Ther Adv Neurol Disord 8:255–273

Payer J, Petrovic T, Lisy L, Langer P (2012) Hashimoto encephalopathy: a rare intricate syndrome. Int J Endocrinol Metab 10:506–514

Pilato MS, Urban A, Alkawadri R, Barot NV, Castellano JF, Rajasekaran V, Bagić AI, Fong-Isariyawongse JS (2022) EEG findings in coronavirus disease. J Clin Neurophysiol 39:159–165

Pilotto A, Odolini S, Masciocchi S, Comelli A, Volonghi I, Gazzina S, Nocivelli S, Pezzini A, Focà E, Caruso A, Leonardi M, Pasolini MP, Gasparotti R, Castelli F, Ashton NJ, Blennow K, Zetterberg H, Padovani A (2020) Steroid-responsive encephalitis in coronavirus disease 2019. Ann Neurol 88:423–427

Pincock S (2004) Government confirms second case of vCJD transmitted by blood transfusion. BMJ 329:251

Pischke S, Hartl J, Pas SD, Lohse AW, Jacobs BC, Van der Eijk AA (2017) Hepatitis E virus: infection beyond the liver? J Hepatol 66:1082–1095

Prusiner SB (2013) Biology and genetics of prions causing neurodegeneration. Annu Rev Genet 47: 601–623

Qi C, Zhang J-T, Zhao W, Xing X-W, Yu S-Y (2020) Sporadic Creutzfeldt-Jakob disease: a retrospective analysis of 104 cases. Eur Neurol 83:65–72

Radoshitzky SR, Buchmeier MJ, Charrel RN, Clegg JCS, Gonzalez J-PJ, Günther S, Hepojoki J, Kuhn JH, Lukashevich IS, Romanowski V, Salvato MS, Sironi M, Stenglein MD, de la Torre JC, ICTV Report Consortium (2019) ICTV virus taxonomy profile: arenaviridae. J Gen Virol 100:1200–1201

Renard D, Castelnovo G, Collombier L, Thouvenot E, Boudousq V (2017) FDG-PET in Creutzfeldt-Jakob disease: analysis of clinical-PET correlation. Prion 11:440–453

Requena JR, Wille H (2014) The structure of the infectious prion protein. Prion 8:60–66

Riera L, Castillo E, Del Carmen Saavedra M, Priotto J, Sottosanti J, Polop J, Ambrosio AM (2005) Serological study of the lymphochoriomeningitis virus (LCMV) in an inner city of Argentina. J Med Virol 76:285–289

Riesner D (2003) Biochemistry and structure of PrPC and PrPSc. Br Med Bull 66:21–33

Ritchie DL, Barria MA (2021) Prion diseases: a unique transmissible agent or a model for neurodegenerative diseases? Biomolecules 11:207

Riudavets MA, Sraka MA, Schultz M, Rojas E, Martinetto H, Begué C, de Halac IN, Poleggi A, Equestre M, Pocchiari M, Sevlever G, Taratuto AL (2013) Gerstmann-Sträussler-Scheinker syndrome with variable phenotype in a new kindred with PRNP-P102L mutation. Brain Pathol 24:142–147

Rodriguez FJ, Picken MM, Lee JM (2015) Amyloid deposition in the central nervous system. In: Picken MM, Herrera GA, Dogan A (eds) Amyloid and related disorders: surgical pathology and clinical correlations, current clinical pathology. Springer, Cham, pp 121–131

Rong LL, Lannen NJ, Tank EC, Feistel JL, Therasse CJ, Potluri A, Khan M, Min J (2023) Case report: two clusters of Creutzfeldt–Jakob disease cases within 1 year in West Michigan. Front Neurol 14:1134225

Rorke LB, Spiro AJ (1967) Cerebral lesions in congenital rubella syndrome. J Pediatr 70:243–255

Rosenbloom MH, Atri A (2011) The evaluation of rapidly progressive dementia. Neurologist 17: 67–74

Rossi M, Saverioni D, Di Bari M, Baiardi S, Lemstra AW, Pirisinu L, Capellari S, Rozemuller A, Nonno R, Parchi P (2017) Atypical Creutzfeldt-Jakob disease with PrP-amyloid plaques in white matter: molecular characterization and transmission to bank voles show the M1 strain signature. Acta Neuropathol Commun 5:87

Roux D, Bouldouyre M-A, Mercier-Delarue S, Seilhean D, Zagdanski A-M, Delaugerre C, Simon F, Molina J-M, Legoff J (2011) JC virus variant associated with cerebellar atrophy in a patient with AIDS. J Clin Microbiol 49:2196–2199

Ryschkewitsch C, Friedlaender J, Mgone C, Jobes D, Agostini H, Chima S, Alpers M, Koki G, Yanagihara R, Stoner G (2000) Human polyomavirus JC variants in Papua New Guinea and Guam reflect ancient population settlement and viral evolution. Microbes Infect 2:987–996

Saji AM, Gupta V (2023) Progressive multifocal leukoencephalopathy. In: StatPearls. StatPearls Publishing, Treasure Island

Sajnani G, Requena JR (2012) Prions, proteinase K and infectivity. Prion 6:430–432

Salehi P, Clark M, Pinzon J, Patil A (2022) Sporadic Creutzfeldt-Jakob disease. Am J Emerg Med 52:267

Sánchez-Juan P, Houben MPWA, Hoff JI, Jansen C, Sie MPS, van Rijn MJE, Ironside JW, Will RG, van Duijn CM, Rozemuller A (2007) The first case of variant Creutzfeldt-Jakob disease in the Netherlands. J Neurol 254:958–960

Saribas AS, Ozdemir A, Lam C, Safak M (2010) JC virus-induced progressive multifocal leukoencephalopathy. Futur Virol 5:313–323

Sariyer I, Khalili K, Safak M (2008) Dephosphorylation of JC virus agnoprotein by protein phosphatase 2A: inhibition by small t antigen. Virology 375:464–479

Saruwatari L, Sugimoto C, Kitamura T, Ohno N, Sakai E, Shresta P, Hoa BK, Phi PTP, An HPH, Tuyet NTA, Honjo T, Kobayashi N, Zheng HY, Takasaka T, Yogo Y (2002) Asian domains of four major genotypes of JC virus, Af2, B1-b, CY and SC. Arch Virol 147:1–10

Schmitz M, Cramm M, Llorens F, Müller-Cramm D, Collins S, Atarashi R, Satoh K, Orrù CD, Groveman BR, Zafar S, Schulz-Schaeffer WJ, Caughey B, Zerr I (2016) The real-time quaking-induced conversion assay for detection of human prion disease and study of other protein misfolding diseases. Nat Protoc 11:2233–2242

Sedmak G, Nix WA, Jentzen J, Haupt TE, Davis JP, Bhattacharyya S, Pallansch MA, Oberste MS (2010) Infant deaths associated with human parechovirus infection in Wisconsin. Clin Infect Dis 50:357–361

Seregin A, Yun N, Paessler S (2020) Lymphocytic choriomeningitis, lassa fever, and the South American Hemorrhagic fevers (arenaviruses). In: Mandell, Douglas, and Bennett's principles and practice of infectious diseases. Saunders, Philadelphia, pp 1277–1282

Seth V, Kushwaha S, Verma R, Patel PM, Kiran Gowda R, Bapat P (2021) Autoimmune encephalitis versus Creutzfeldt-Jakob disease in a patient with typical Facio-brachial dystonic seizures: a case report with Diagnostic challenges. Brain Behav Immun Health 13:100236

Seyoum A, Assefa N, Gure T, Seyoum B, Mulu A, Mihret A (2022) Prevalence and genotype distribution of high-risk human papillomavirus infection among Sub-Saharan African women: a systematic review and meta-analysis. Front Public Health 10:890880

Shackelton LA, Rambaut A, Pybus OG, Holmes EC (2006) JC virus evolution and its association with human populations. J Virol 80:9928–9933

Shah G, Robinson JL (2014) The particulars on parechovirus. Can J Infect Dis Med Microbiol 25: 186–188

Shankar SK, Satishchandra P, Mahadevan A, Yasha TC, Nagaraja D, Taly AB, Prabhakar S, Nath A (2003) Low prevalence of progressive multifocal leukoencephalopathy in India and Africa: is there a biological explanation? J Neurovirol 9(1):59–67

Sharma S, Mukherjee M, Kedage V, Muttigi MS, Rao A, Rao S (2009) Sporadic Creutzfeldt-Jakob disease–a review. Int J Neurosci 119:1981–1994

Shim KH, Sharma N, An SSA (2022) Prion therapeutics: lessons from the past. Prion 16:265–294

Shishido-Hara Y, Yazawa T, Nagane M, Higuchi K, Abe-Suzuki S, Kurata M, Kitagawa M, Kamma H, Uchihara T (2014) JC virus inclusions in progressive multifocal leukoencephalopathy: scaffolding promyelocytic leukemia nuclear bodies grow with cell cycle transition through an S-to-G2-like state in enlarging oligodendrocyte nuclei. J Neuropathol Exp Neurol 73:442–453

Shukla S, Maraqa NF (2023) Congenital rubella. In: StatPearls. StatPearls Publishing, Treasure Island

Sidhu N, McCutchan JA (2010) Unmasking of PML by HAART: unusual clinical features and the role of IRIS. J Neuroimmunol 219:100–104

Sikorska B, Liberski PP, Sobów T, Budka H, Ironside JW (2009) Ultrastructural study of florid plaques in variant Creutzfeldt-Jakob disease: a comparison with amyloid plaques in kuru, sporadic Creutzfeldt-Jakob disease and Gerstmann-Sträussler-Scheinker disease. Neuropathol Appl Neurobiol 35:46–59

Sisó S, González L, Jeffrey M (2010) Neuroinvasion in prion diseases: the roles of ascending neural infection and blood dissemination. Interdiscip Perspect Infect Dis 2010:747892

Sitammagari KK, Masood W (2022) Creutzfeldt Jakob disease. In: StatPearls. StatPearls Publishing, Treasure Island

Solforosi L, Milani M, Mancini N, Clementi M, Burioni R (2013) A closer look at prion strains. Prion 7:99–108

Somani S, Pati S, Gaston T, Chitlangia A, Agnihotri S (2020) De novo status epilepticus in patients with COVID-19. Ann Clin Transl Neurol 7:1240–1244

Soto C, Satani N (2011) The intricate mechanisms of neurodegeneration in prion diseases. Trends Mol Med 17:14–24

Souders HT, Byler D, Marupudi N, Patel R, McSherry G (2015) Protracted symptoms in lymphocytic choriomeningitis: a case report. J Child Neurol 30:644–647

Sridhar A, Karelehto E, Brouwer L, Pajkrt D, Wolthers KC (2019) Parechovirus a pathogenesis and the enigma of genotype A-3. Viruses 11:1062

Stanway G, Kalkkinen N, Roivainen M, Ghazi F, Khan M, Smyth M, Meurman O, Hyypiä T (1994) Molecular and biological characteristics of echovirus 22, a representative of a new picornavirus group. J Virol 68:8232–8238

Steele AD, Lindquist S, Aguzzi A (2007) The prion protein knockout mouse. Prion 1:83–93

Sugimoto C, Hasegawa M, Zheng H-Y, Demenev V, Sekino Y, Kojima K, Honjo T, Kida H, Hovi T, Vesikari T, Schalken JA, Tomita K, Mitsunobu Y, Ikegaya H, Kobayashi N, Kitamura T, Yogo Y (2002) JC virus strains indigenous to northeastern Siberians and Canadian

Inuits are unique but evolutionally related to those distributed throughout Europe and Mediterranean areas. J Mol Evol 55:322–335

Sundaram SE, Staffaroni AM, Walker NC, Casaletto KB, Casey M, Golubjatnikov A, Metcalf S, O'Leary K, Wong K, Benisano K, Forner S, Gonzalez Catalan M, Allen IE, Rosen HJ, Kramer JH, Geschwind MD (2020) Baseline neuropsychological profiles in prion disease predict survival time. Ann Clin Transl Neurol 7:1535–1545

Takada LT, Geschwind MD (2012) Prion diseases. In: Long SS (ed) Principles and practice of pediatric infectious diseases, 4th edn. Elsevier, London, pp 333–338

Takasaka T, Kitamura T, Sugimoto C, Guo J, Zheng H-Y, Yogo Y (2006) Phylogenetic analysis of major African genotype (Af2) of JC virus: implications for origin and dispersals of modern Africans. Am J Phys Anthropol 129:465–472

Tan TH-L, Stark RJ, Waterston JA, White O, Thyagarajan D, Monif M (2020) Genetic prion disease: D178N with 129MV disease modifying polymorphism—a clinical phenotype. BMJ Neurol Open 2:e000074

Tang N, Bai H, Chen X, Gong J, Li D, Sun Z (2020) Anticoagulant treatment is associated with decreased mortality in severe coronavirus disease 2019 patients with coagulopathy. J Thromb Haemost 18:1094–1099

Townsend JJ, Baringer JR, Wolinsky JS, Malamud N, Mednick JP, Panitch HS, Scott RA, Oshiro L, Cremer NE (1975) Progressive rubella panencephalitis. Late onset after congenital rubella. N Engl J Med 292:990–993

Tricco AC, Lillie E, Zarin W, O'Brien KK, Colquhoun H, Levac D, Moher D, Peters MDJ, Horsley T, Weeks L, Hempel S, Akl EA, Chang C, McGowan J, Stewart L, Hartling L, Aldcroft A, Wilson MG, Garritty C, Lewin S, Godfrey CM, Macdonald MT, Langlois EV, Soares-Weiser K, Moriarty J, Clifford T, Tunçalp Ö, Straus SE (2018) PRISMA extension for scoping reviews (PRISMA-ScR): checklist and explanation. Ann Intern Med 169:467–473

Tyagarajan SK, Frisque RJ (2006) Stability and function of JC virus large T antigen and T′ proteins are altered by mutation of their phosphorylated threonine 125 residues. J Virol 80:2083–2091

Tyor W, Harrison T (2014) Mumps and rubella. Handb Clin Neurol 123:591–600

Van Cauter S, Severino M, Ammendola R, Van Berkel B, Vavro H, van den Hauwe L, Rumboldt Z (2020) Bilateral lesions of the basal ganglia and thalami (central grey matter)—pictorial review. Neuroradiology 62:1565–1605

Venkatesan EP, Ramadoss K (2015) Pulvinar/hockey stick sign in Creutzfeldt - Jakob disease. Med J Dr DY Patil Univ 8:840

Verboon-Maciolek MA, Groenendaal F, Hahn CD, Hellmann J, van Loon AM, Boivin G, de Vries LS (2008) Human parechovirus causes encephalitis with white matter injury in neonates. Ann Neurol 64:266–273

Vollono C, Rollo E, Romozzi M, Frisullo G, Servidei S, Borghetti A, Calabresi P (2020) Focal status epilepticus as unique clinical feature of COVID-19: a case report. Seizure 78:109–112

Vonck K, Garrez I, De Herdt V, Hemelsoet D, Laureys G, Raedt R, Boon P (2020) Neurological manifestations and neuro-invasive mechanisms of the severe acute respiratory syndrome coronavirus type 2. Eur J Neurol 27:1578–1587

Wang Y, Wang S, Wu J, Jiang Y, Zhang H, Li S, Liu H, Yang C, Tang H, Guo N, Peppelenbosch MP, Wei L, Pan Q, Zhao J (2018) Hepatitis E virus infection in acute non-traumatic neuropathy: a large prospective case-control study in China. EBioMedicine 36:122–130

Waqas M, Aziz S, Bushra A, Halim SA, Ali A, Ullah S, Khalid A, Abdalla AN, Khan A, Al-Harrasi A (2023) Employing an immunoinformatics approach revealed potent multi-epitope based subunit vaccine for lymphocytic choriomeningitis virus. J Infect Public Health 16:214–232

Watson N, Brandel J-P, Green A, Hermann P, Ladogana A, Lindsay T, Mackenzie J, Pocchiari M, Smith C, Zerr I, Pal S (2021) The importance of ongoing international surveillance for Creutzfeldt–Jakob disease. Nat Rev Neurol 17:362–379

Wenting A, Gruters A, van Os Y, Verstraeten S, Valentijn S, Ponds R, de Vugt M (2020) COVID-19 neurological manifestations and underlying mechanisms: a scoping review. Front Psych 11: 860

White SL, Rawlinson W, Boan P, Sheppeard V, Wong G, Waller K, Opdam H, Kaldor J, Fink M, Verran D, Webster A, Wyburn K, Grayson L, Glanville A, Cross N, Irish A, Coates T, Griffin A, Snell G, Alexander SI, Campbell S, Chadban S, Macdonald P, Manley P, Mehakovic E, Ramachandran V, Mitchell A, Ison M (2018) Infectious disease transmission in solid organ transplantation: donor evaluation, recipient risk, and outcomes of transmission. Transplant Direct 5:e416

Wiegmans AP, Saunus JM, Ham S, Lobb R, Kutasovic JR, Dalley AJ, Miranda M, Atkinson C, Foliaki ST, Ferguson K, Niland C, Johnstone CN, Lewis V, Collins SJ, Lakhani SR, Al-Ejeh F, Möller A (2019) Secreted cellular prion protein binds doxorubicin and correlates with anthracycline resistance in breast cancer. JCI Insight 5:e124092

Wildenbeest JG, Harvala H, Pajkrt D, Wolthers KC (2010) The need for treatment against human parechoviruses: how, why and when? Expert Rev Anti-Infect Ther 8:1417–1429

Will RG, Ironside JW (2017) Sporadic and infectious human prion diseases. Cold Spring Harb Perspect Med 7:a024364

Williams ÇH, Panayiotou M, Girling GD, Peard CI, Oikarinen S, Hyöty H, Stanway G (2009) Evolution and conservation in human parechovirus genomes. J Gen Virol 90:1702–1712

Xu X-W, Wu X-X, Jiang X-G, Xu K-J, Ying L-J, Ma C-L, Li S-B, Wang H-Y, Zhang S, Gao H-N, Sheng J-F, Cai H-L, Qiu Y-Q, Li L-J (2020) Clinical findings in a group of patients infected with the 2019 novel coronavirus (SARS-Cov-2) outside of Wuhan, China: retrospective case series. BMJ 368:606

Xu Y-W, Wang J-Q, Zhang W, Xu S-C, Li Y-X (2021) Rarely fast progressive memory loss diagnosed as Creutzfeldt-Jakob disease: a case report. World J Clin Cases 9:10638–10644

Yakupova EI, Bobyleva LG, Vikhlyantsev IM, Bobylev AG (2019) Congo Red and amyloids: history and relationship. Biosci Rep 39:1415

Young GS, Geschwind MD, Fischbein NJ, Martindale JL, Henry RG, Liu S, Lu Y, Wong S, Liu H, Miller BL, Dillon WP (2005) Diffusion-weighted and fluid-attenuated inversion recovery imaging in Creutzfeldt-Jakob disease: high sensitivity and specificity for diagnosis. AJNR Am J Neuroradiol 26:1551–1562

Zampino R, Marrone A, Restivo L, Guerrera B, Sellitto A, Rinaldi L, Romano C, Adinolfi LE (2013) Chronic HCV infection and inflammation: clinical impact on hepatic and extra-hepatic manifestations. World J Hepatol 5:528–540

Zarranz JJ, Digon A, Atarés B, Arteagoitia JM, Carrera N, Fernández-Manchola I, Fernández-Martínez M, Fernández-Maiztegui C, Forcadas I, Galdos L, Ibáñez A, Lezcano E, Martí-Massó JF, Mendibe MM, Urtasun M, Uterga JM, Saracibar N, Velasco F, González de Galdeano L (2004) Familial prion diseases in the Basque Country (Spain). Neuroepidemiology 24:103–109

Zerr I, Poser S (2002) Clinical diagnosis and differential diagnosis of CJD and vCJD. APMIS 110: 88–98

Zhang Y, Xiao M, Zhang S, Xia P, Cao W, Jiang W, Chen H, Ding X, Zhao H, Zhang H, Wang C, Zhao J, Sun X, Tian R, Wu W, Wu D, Ma J, Chen Y, Zhang D, Xie J, Yan X, Zhou X, Liu Z, Wang J, Du B, Qin Y, Gao P, Qin X, Xu Y, Zhang W, Li T, Zhang F, Zhao Y, Li Y, Zhang S (2020) Coagulopathy and antiphospholipid antibodies in patients with Covid-19. N Engl J Med 382:e38

Zhao M-M, Feng L-S, Hou S, Shen P-P, Cui L, Feng J-C (2019) Gerstmann-Sträussler-Scheinker disease: a case report. World J Clin Cases 7:389–395

Zhou X, Ramachandran S, Mann M, Popkin DL (2012) Role of lymphocytic choriomeningitis virus (LCMV) in understanding viral immunology: past, present and future. Viruses 4(11): 2650–2669

Zubair AS, McAlpine LS, Gardin T, Farhadian S, Kuruvilla DE, Spudich S (2020) Neuropathogenesis and neurologic manifestations of the coronaviruses in the age of coronavirus disease 2019: a review. JAMA Neurol 77:1018–1027

Part III

Fungal Pathogens: Pathogenesis, Pathology, Diagnosis and Treatment

Neuro-Infections Caused By Candida Species

12

Nazish Fatima and Shariq Ahmed

Abstract

There has recently been a 6 to 17% rise in candida species infections in the central nervous system. Although *Candida* can cause infections in both immunocompetent and immunosuppressed individuals, disseminated candidiasis is mostly seen in patients with impaired immunity due to AIDS, diabetes, neutropenia, underlying malignancy, and extensive wounds (burns and operations). Patients with disseminated candidiasis or with any neurosurgical procedure like ventricular shunts are observed with central nervous system (CNS) involvement. Another high-risk group for invasive candidiasis consists of newborns and infants, particularly those with low birth weight (<1.5 kg). *Candida* crosses the blood–brain barriers by three proposed mechanisms namely the Trojan horse mechanism, paracellular migration, and transcellular migration. Clinical manifestations of neuro-candidiasis are usually variable; however, meningoencephalitis is the most common. Invasive candidiasis mortality is thought to vary from 10 to 70%, rising to 90% in cases when the central nervous system is affected. Analysis of cerebral spinal fluid plus cultures exhibiting growth of budding, yeast-like cells, is used to make laboratory-based diagnoses. The positive serum beta-D-glucan assay aids in the diagnosis but may be positive in patients with various other invasive fungal infections. Liposomal amphotericin B with better CNS concentrations followed by fluconazole as a step-down therapy is considered an effective treatment strategy.

Keywords

Central nervous system · *Candida albicans* · Meningoencephalitis · Disseminated candidiasis

N. Fatima (✉) · S. Ahmed
Department of Microbiology, JN Medical College, AMU, Aligarh, Uttar Pradesh, India

H. Sami et al. (eds.), *Viral and Fungal Infections of the Central Nervous System: A Microbiological Perspective*, https://doi.org/10.1007/978-981-99-6445-1_12

12.1 Pathogen

Candida neurological infections are almost always due to *C. albicans* (Sánchez et al. 2000). However, the world has seen a gradual rise in the frequency of non-*C. albicans* species over *Candida albicans* over the past decade (Table 12.1). *Candida* infects both immunocompetent and immunosuppressed individuals but disseminated disease mostly occurs in the latter (Cesaro et al. 2017). A remarkable surge from 6 to 17% has been noted in the incidence of CNS infection from *Candida* spp. (Brumble et al. 2017). *Candida albicans* was the cause of 33% of CNS infections among individuals with underlying malignant illnesses or among those with hematopoietic stem cell transplantations. Non-*Candida albicans*, however, accounted for 77% of neuro-infections in such patients; the most common strain found was C. parapsilosis, while isolates of Candida glabrata, guilliermondii, krusei, and tropicalis were also found occasionally. Additionally, multispecies involvements were also reported (Cesaro et al. 2017; McCullers et al. 2000; Murthy and Sundaram 2014).

12.2 Pathogenesis

Although the exact pathogenic mechanisms involved in certain neuro-mycological infections are not entirely known, it is a well-established fact that the fungi must first get past the host defenses in cross the blood–brain barrier (BBB) in order to commence CNS infection. The fungi adhere to the microvasculature of the brain, after which transmigration happens into the brain parenchyma. Trojan horse mechanism, transcellular migration, and paracellular migration are the three mechanisms proposed for crossing the BBB by the fungus (Liu et al. 2011). The building block of BBB is the brain endothelial cells connected with tight junctions; foot process of astrocytes surrounds these cells maintaining its integrity. Fibronectin, vitronectin, and laminin facilitate adhesion of *C. albicans* to the extracellular matrix. Als3 (agglutinin-like sequence 3) and Ssa1 (stress seventy subfamily A 1) are the fungal invasins that mediate invasion of brain endothelial cells by *C. albicans* (Shi and Mody 2016). Als3's affinity for the heat shock protein 96 (gp96), which is expressed on endothelial cells that line the cerebral vessels, promotes the transcytosis of *Candida albicans*. *C. albicans* gets engulfed by the phagocyte in the peripheral blood circulation from where the Trojan horse pathway begins. The fungus inside the phagocyte maneuvers its movement toward the brain, where it adheres on the luminal side of the brain capillaries finally crossing the BBB transcellularly and paracellularly (Koutsouras et al. 2016; Liu et al. 2011; Santiago-Tirado and Doering 2017).

Factors that facilitate the penetration of fungi across the blood–brain barrier are downregulation of immunity by various immunosuppressant conditions which lead to increased permeability of the BBB, direct disruption of the BBB by any neurosurgical procedure, colonization of the foreign devices placed in the ventricles of the brain, or activation of microglia and cytokines: Tumor necrosis factor alpha (TNF-α)

Table 12.1 Overview of clinically relevant Candida species

Order: Saccharomycetales ⇒ Family: Debaryomycetaceae ⇒ Genus: Candida	
Species	Remarks
C. albicans	• The most common *Candida* species identified (50–60%) Pappas et al. (2018) • About 70–80% of the *Candida* species found in candidemia/invasive candidiasis are *C glabrata* and *C albicans* Pappas et al. (2018)
C. auris	• Linked to high mortality rates and outbreaks related to healthcare • Frequently resistant to several antifungal agents • Difficult to distinguish in a routine laboratory • Internationally emerging invasive yeast species Ahmad and Alfouzan (2021) • From the time of initial discovery and February 2019, 587 *C auris* cases were documented in USA • 2377 cases have been documented in 2022 alone Egger et al. (2022)
C. Dubliniensis	• Primarily recovered from patients infected with HIV
C. glabrata	• Formerly recognized as *Torulopsis glabrata* (15–20%) • Due to its rising prevalence around the globe, its link to fluconazole resistance in as many as 20% of clinical samples, and its general decreased sensitivity to various azoles and polyenes, *C. glabrata* has recently gained medical significance Lee et al. (2021)
C. guilliermondii	• (< 5%)
C. haemulonii	• Emerging species • Closely related to *C. auris* • Occasionally been implicated in the development of disseminated/ invasive candida infections Ben-Ami et al. (2017)
C. Kefyr	• (< 5%)
C. krusei	• (1–3%) • Intrinsically resistant to ketoconazole and fluconazole • Less susceptible to various other antifungal drugs such as itraconazole and amphotericin B Badiee and Alborzi (2011)
C. lusitaniae	• (< 5%) • *C. lusitaniae*, albeit less frequent than other species of *Candida*, is significant clinically as it may be inherently resistant to amphotericin B while still being susceptible to the azoles and echinocandins Mendoza-Reyes et al. (2022)
C. parapsilosis	• (10–20%) • Another crucial species to take into account when treating hospitalized individuals • Occurs in infections linked to prosthetic vascular catheters • Echinocandins have a greater in vitro minimum inhibitory concentration against *C parapsilosis* versus other *Candida* species Trofa et al. (2008)
C. tropicalis	• (6–12%) • Repeatedly been cited as a significant contributor to candidemia in malignancies (leukemia) and bone marrow transplant recipients Sendid et al. (2003)

disrupts the tight junctions of the barrier. After the crossing of the BBB, cerebral and subarachnoid space proliferation of the fungus in the brain parenchyma causes inflammation. Invasion of the brain tissue/meninges by the fungal pathogen after surpassing all the effective barriers surrounding the brain facilitated by the immunosuppressive conditions occurs; subsequently, there is release of immune-enhancing and immune-suppressing chemokines and cytokines by the activated nerve cells occurs. The interplay between these cytokines and chemokines determines the pathogenesis of fungal CNS infections (Koutsouras et al. 2016; Sharma et al. 2012).

The rate and severity of neuro-infection are determined by the fungus' pathogenicity and the host's immunological responsiveness (Koutsouras et al. 2016). Host immune defense against the pathogen in the CNS comprises microglia, endothelial cells, astrocytes, and T cells. They inhibit the fungal infection, thereby limiting the infection by interferon-γ, interleukin (IL)1β, IL6, IL12, TNF-α (i.e., chemokines) production, nitric oxide generation, superoxide anion production, major histocompatibility complex class I (MHC I), and MHC II molecular expression (Koutsouras et al. 2016; Sharma et al. 2012). In invasive cerebral candidiasis, microglial cells are the principle effector cells, and it has been reported that on intracerebral administration; it can limit infection and tissue damage (Blasi et al. 1991; Lionakis et al. 2011). Disruption of tight junction of BBB by cytokine such as TNF-α released by microglial activation has been noted The toll-like receptors (TLR) 2, 4, and 9 are able to identify fungal antigens, including *C. albicans*' pseudohyphae. TLR-2, Dectin-1, and CR-3 all recognize mannose and β-glucans, which are present on the surface of *C. albicans*, as well as on Aspergillus fumigatus, respectively (Koutsouras et al. 2016). The ability of microglia to generate pro-inflammatory cytokines when exposed to the pathogen, however, may be diminished by these carbohydrates attenuating TLR-mediated NF-κB activation (Shah et al. 2009). *Candida* spp. produces biofilm which is protective, as it helps to evade the host defense such as phagocytosis, cytokines, and nitric oxide by microglial cells (Koutsouras et al. 2016).

Numerous single-nucleotide polymorphisms, known as SNPs, in immune-associated genes, appear to be linked to an increased vulnerability to invasive candidiasis. Independent risk factor for the development of systemic risk factor was homozygosity for the dysfunctional CX3CR1-M280 allele that resulted in dysfunctional monocyte signaling and survival (Collar et al. 2018; Lionakis et al. 2013). Additional variables that are linked to an increased likelihood of invasive yeast infections include a malfunctioning CXCR1-T276 allele, which impairs degranulation of neutrophils, SNPs inCD58, TAGAP genes, LCE4A-C1orf68 locus, and also SNP mutations in mannose-binding lectin genes. Nevertheless, not all of the aforementioned variables particularly raise the likelihood of candidiasis of the central nervous system (Kumar et al. 2014; Swamydas et al. 2016; van Till et al. 2008). The molecule whose deficiency leads to profound susceptibility to neurocandidiasis is CARD9, a myeloid-expressed signaling adaptor protein (Lionakis et al. 2017). These conclusions are supported by numerous cases that authors have described (Drewniak et al. 2013; Gavino et al. 2014; Glocker et al. 2009; Herbst et al. 2015; Jones et al. 2016).

12.3 Pathology, Risk Factors, and Clinical Manifestations of Neuro-Candidiasis

Primary candidiasis of central nervous system although uncommon is reported to be around 18–52% of disseminated candidiasis (Jain et al. 2007). Susceptibility rate further increases in neonates, where 66% of the newborns with disseminated candidiasis and 80% with *Candida* endocarditis have CNS involvement (Drummond and Lionakis 2018; Góralska et al. 2018). Unrecognized infection of CNS has been estimated in 6% of patients with disseminated candidiasis (Shankar et al. 2007). Risk factors for CNS invasion by *Candida* (Table 12.2) include neonates [LBW/premature] or recent neurosurgery, patients with CNS shunts or catheters, other devices, neutropenia associated with AIDS/diabetes/hematological

Table 12.2 Spectrum of neuro-candidiasis

Neuro-candidiasis	Remarks
Epidemiological distribution	• Human commensal • Worldwide, global
Risk factors/clinical setting	• Abdominal surgery/post-op • Children • Elderly • Hospital setting • Intravenous (IV) catheters/prolonged IV therapy • IV drug abusers • Neutropenic • Preterm neonates • Prolonged broad-spectrum antimicrobial therapy • Total parenteral nutrition
Transmission	• Blood–brain barrier crossing—Transcellular • Direct access—Intracranial devices • Hematogenous—Disseminated candidiasis • Nosocomial—Neurosurgery, neurological devices • Oropharynx/urogenital
Clinical spectrum	• Aseptic meningitis • Chronic meningitis • Diffuse cerebritis along with microabscesses • Fever that is unresponsive to broad-spectrum drugs • Granulomatous vasculitis • Meningoencephalitis • Mental status alterations • Mycotic aneurysms • Ventriculitis
Microbiological analysis	• Budding-yeast-like-cells with or without pseudohyphae • Cerebrospinal fluid culture, analysis
Primary treatment	• Amphotericin B lipid formulation
Secondary treatment	• Amphotericin B deoxycholate • Echinocandins • Fluconazole • Flucytosine • Voriconazole

malignancies/extensive wounds like burns, prolonged use of steroids in high doses, and graft vs. host disease after bone marrow transplantation (Barton et al. 2014).

Candida has predilection for microcirculation particularly in the middle cerebral artery facilitated by its small size results in multiple cerebral microabscesses and leptomeningitis; however, it can also cause aneurysms and thrombosis of microvessels, cerebral vasculitis, and intracerebral or subarachnoid hemorrhage (Sánchez et al. 2000). Primary focus of infection in disseminated candidiasis is mostly lung/respiratory system or gastrointestinal system [GIT] from where it spreads (Kullberg and Arendrup 2015). Dissemination is favored either when impairment of immune system in conditions like AIDS, newborns, organ transplants, hematological malignancies, prolonged corticosteroids therapy, or after interventions like neurosurgical procedures, placement of shunts, catheters surgery (Barton et al. 2014; Cesaro et al. 2017; Katragkou et al. 2017; Kullberg and Arendrup 2015; Neves et al. 2014; O'Brien et al. 2011).

CNS candidiasis often has varied clinical presentations, meningoencephalitis being the most common others being meningitis, brain abscesses, and ventriculitis. Some more infrequent clinical presentation reported are endophthalmitis, intraventricular fungus balls, hydrocephalus, vasculitis, calcifications, cranial neuropathies, numerous cerebral abscesses accompanied by nodular enhancing lesions, or by ring enhancements, and stroke sequelae in a small number of cases (Huang et al. 1998; Katragkou et al. 2017; McCarthy et al. 2017; Pappas et al. 2016; Shankar et al. 2007).

The symptoms of *Candida* meningitis are neck rigidity, fever, and headache-altered mental status which are also similar to acute or chronic bacterial meningitis. However, signs and symptoms may be subtle in neonates and severely neutropenic/immunosuppressed individuals. Neonates particularly preterm neonates are considered a very high-risk group for developing invasive candidiasis and subsequently neuro-candidiasis. It is reported that 25% of newborn with birth weight < 1.5 kg develop systemic candidiasis and CNS invasion is seen in 50% of the neonates with disseminated candidiasis with a fatality rate of 70% (Baradkar et al. 2009; Faix and Chapman 2003; Lewis et al. 2015; Murthy and Sundaram 2014; Sánchez et al. 2000; Yoshida et al. 2022). Overall mortality from neuro-candidiasis varies between 10 and 30% (Sánchez et al. 2000). Disseminated candidiasis risk factors include certain surgical procedures, extended hospital stay, intubation (endotracheal), neonates with Apgar scores of below 5, parental lipids, prolonged catheterization, and total parental feeding (Lewis et al. 2015). Additionally, as a result the commensal *Candida* species' receptive translocation from the GIT into the bloodstream during necrotizing enterocolitis of newborns is a significant determinant of subsequent disseminated candidiasis occurrence (Barton et al. 2014; Huang et al. 1998).

12.4 Diagnosis

Patients experiencing CNS manifestations involving any or all of the signs/symptoms listed below should have neurological candida infection suspected (Sánchez et al. 2000):

- A *Candida* species isolated in the cerebrospinal fluid (CSF).
- *Candida* isolation from a different, typically sterile location among patients with pleocytosis as detected by CSF biochemistries (Table 12.3). Finding *Candida* species in blood cultures is helpful, but candidemia may not be documented in patients with *Candida* meningitis.
- Inadequate response to treatment for suspected mycobacterial or bacterial meningitis.

12.4.1 CSF Analysis

For establishing a confirmatory diagnosis of CNS candidiasis, obtaining CSF for analysis and culture by lumbar puncture is essential. Around 80% of cases reveal positive cultures. Regardless of whether other microorganisms are found, a positive cerebrospinal fluid culture should not be regarded as a contamination, especially in patients with compromised immunity (Sánchez et al. 2000).

The CSF findings are also variable in patients with neurosurgery-related *Candida* meningitis. Some have a neutrophilic pleocytosis similar to that seen in bacterial meningitis (Nguyen and Yu 1995), whereas others have a lymphocytic predominance (Sánchez et al. 2000).

Those with chronic meningitis have a harder time being diagnosed since there is only a minimal pathogen count, and the routine CSF cultures produce poorer results. In patients with chronic *Candida* meningitis, large-volume (10 to 20 mL) spinal taps are frequently necessary to collect enough CSF for culture. The microbiology laboratory should be asked to culture the entire sample or to filter the sample through a Millipore filter and culture the filter on appropriate media.

The cell wall of several fungi, notably *Candida* species, contains 1,3-beta-D-glucan (Fig. 12.1). Several commercial assays are available that detect serum beta-D-glucan. Positivity of the assays indicates toward an invasive fungal infection including invasive candidiasis. Importantly, this test can be positive in the setting of many different fungal infections, and cutoffs to define positivity in CSF are not defined. Test results must therefore be interpreted with caution.

12.4.2 Neuroimaging

In central nervous system candidiasis, neuroradiological abnormalities are frequently nonspecific and are frequently mistaken for meningitis due to Mycobacterial tuberculosis, or are misinterpreted as CNS tumors, or pyogenic abscesses (Jain et al.

Table 12.3 Laboratory techniques to diagnose neuro-candidiasis

Laboratory techniques	Comments
API20C and API32C biochemical assays	–
BactiCard Candida test	–
Cerebrospinal fluid (CSF) biochemistry	• Glucose—May be slightly decreased • Leukocytes—100 s per microliter (mononuclear cell predominance) • Opening pressure—May be increased (>150 mmH_2O) • Protein—Increased (>1000 mg per dL)
CHROMagar	–
Cultures	• CSF culture • Blood culture
Direct microscopy/histopathology	• Potassium hydroxide mount • Periodic acid–Schiff • Grocott methenamine silver stain • Calcofluor staining • Hematoxylin and eosin stain of fixed tissue
Germ tube formation	–
Gram stain, lactophenol cotton blue (LPCB) wet mount, etc.	–
In situ hybridization	• *C. albicans* peptide nucleic acid fluorescence in situ hybridization test
Matrix-associated laser desorption/ionization time-of-flight mass spectrometry	–
Molecular methods	• Polymerase chain reaction • DNA probe-based assays
Murex *C. albicans* test	–
Neuroimaging	–
Sabouraud's dextrose agar, cornmeal agar, etc.	–
Serology	Antibody detection • Agglutination and complement fixation tests • Enzyme-linked immunosorbent assay • Fluorescent antibody tests • Radioimmunoassay Antigen detection • Cytoplasmic antigens • Enolase assay • Glycoproteins • Candida mannan assay • Proteinase • Candida heat labile antigen assay • D-arabinitol assay • (1, 3) β-D-glucan assay
T2 Candida panel	Direct blood test to identify *C. albicans/krusei/glabrata/parapsilosis/tropicalis*

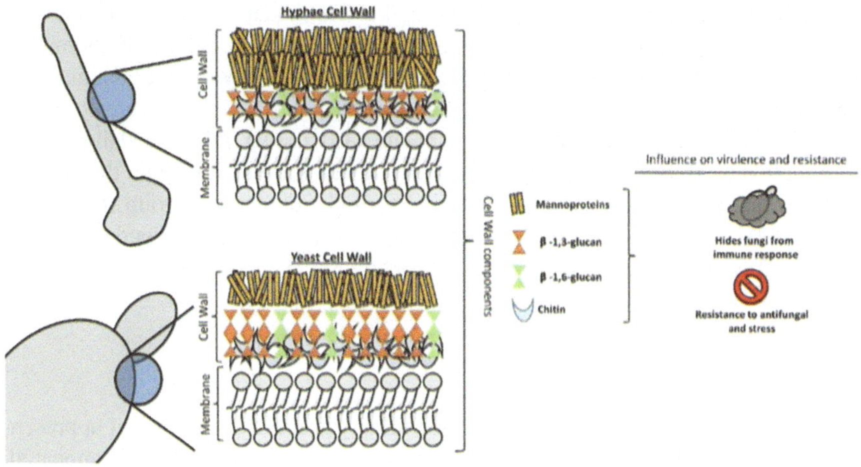

Fig. 12.1 Basic structure of Candida cell wall. (*Image source: Frontiers in Microbiology: Volume 10* (Garcia-Rubio et al. 2020)—*Copyright restrictions: none; this image is from an open-access article distributed under the terms of the Creative Commons Attribution License (CC BY) and thus free of any copyright restriction)*

2007). Microabscesses found at the grey–white junction, cerebellum, and basal ganglia are the most frequent pathologic finding, followed by meningitis and macro-abscesses (Pendlebury et al. 1989; Sánchez et al. 2000). Computed tomography (CT) can detect hydrocephalus, which is particularly common in patients with infected CNS shunts (Chiou et al. 1994; Sánchez-Portocarrero et al. 1994). However, the CT scan is often normal and may not detect microabscesses (Pendlebury et al. 1989; Sánchez et al. 2000). In comparison, magnetic resonance imaging (MRI) can detect microabscesses, which appear as multiple, small, enhanced ring lesions, sometimes with a hemorrhagic component (Lai et al. 1997). MRI scans may be utilized to monitor the effectiveness of antifungal medications whenever these types of lesions are present.

12.5 Treatment

The most thoroughly researched medication for the medical management of neurological candidiasis is amphotericin B deoxycholate; however, optimal treatment evaluation through a randomized controlled trial is still lacking (Geers and Gordon 1999; Nguyen and Yu 1995; Pappas et al. 2016; Sánchez et al. 2000; Voice et al. 1994). In comparison to amphotericin B lipid complex, 5 mg/kg intravenously, daily, liposomal amphotericin B may produce higher CNS concentrations (Groll et al. 2000). After a positive response to lipid amphotericin B is achieved, either with or without flucytosine (5-FC), fluconazole is advised to serve as a step-down regimen. 5-FC is administered 4 times a day to individuals having normalized kidney functions, at an oral dosage of 25 mg/kg. Treatment ought to be maintained for several weeks to months till resolution of CNS manifestations, CSF anomalies, and radiological clearing has taken place. The specialists advise that contaminated ventricular devices be removed (Alothman et al. 2014).

In neonates, amphotericin B deoxycholate (1.0 mg/kg IV once daily) is the formulation used. Neonates tolerate the deoxycholate formulation better than adults, and there is little experience using the lipid formulations in this group (Fernandez et al. 2000; van den Anker et al. 1995). Flucytosine is not recommended because adverse effects are frequent in neonates (Benjamin et al. 2006).

12.5.1 Duration

Antifungal treatment ought to be continued till any abscess(es), if present at presentation, have cleared on MRI, the patient's clinical symptoms/signs have disappeared, and CSF cell counts (pleocytosis), glucose levels, protein concentration, and CSF culture have normalized. The aforementioned steps could take weeks or months to complete, particularly in those suffering from chronic *Candida* meningitis.

There are no data available concerning the frequency of testing, but the following approach seems reasonable:

- Repeat MRI scans for patients with brain abscesses are advised to be done after 2 weeks (or sooner if their condition is deteriorating) and then every month till the abscess clears up.
- In order to confirm that all CSF alterations are reverting to normal as well as that CSF cultures will remain negative, lumbar punctures should be performed weekly during the initial few weeks, in acute meningitis cases.
- For patients with chronic meningitis, lumbar puncture should be repeated as for acute meningitis. The best parameters to follow are the CSF white cell count and protein and glucose concentrations. A negative CSF culture is insufficient to assess treatment response since the microorganism exists in low quantities and routine CSF cultures of CSF provide poor yields (Voice et al. 1994).

Patients who have cerebral abscesses, are markedly immunocompromised, or have chronic *Candida* meningitis need longer courses of therapy, and patients who have infected ventricular shunts or implantable CNS devices must have the device removed to achieve cure for the infection (Chiou et al. 1994; Nguyen and Yu 1995; Pappas et al. 2016; Sánchez-Portocarrero et al. 1994).

If at all possible, infected CNS devices such as chemotherapy delivery biopolymer wafers, prosthetic reconstructive devices, shunts, stimulators, and ventriculostomy drains should be removed. In those with a ventricular device that is impractical to remove, amphotericin B deoxycholate may be injected directly into the ventricle via the device (Pappas et al. 2016).

Adjuvant immune therapies, particularly for CARD9-deficient individuals reported by some authors, are abnormalities in GM-CSF seen in myeloid cells with the p.Y91H CARD9 mutation, corrected by recombinant GM-CSF, and the use of GM-CSF therapy to rectify cytokine (IL-17) production errors in p.Q295X-mutant CARD9-deficient patients (Celmeli et al. 2016; Gavino et al. 2014, 2016).

References

Ahmad S, Alfouzan W (2021) Candida auris: epidemiology, diagnosis, pathogenesis, antifungal susceptibility, and infection control measures to combat the spread of infections in healthcare facilities. Microorganisms 9:807. https://doi.org/10.3390/microorganisms9040807

Alothman AF, Al-Musawi T, Al-Abdely HM, Salman JA, Almaslamani M, Yared N, Butt AA, Raghubir N, Morsi WE, Al Thaqafi AO (2014) Clinical practice guidelines for the management of invasive Candida infections in adults in the middle east region: expert panel recommendations. J Infect Public Health 7:6–19. https://doi.org/10.1016/j.jiph.2013.08.002

Badiee P, Alborzi A (2011) Susceptibility of clinical Candida species isolates to antifungal agents by E-test, southern Iran: a five year study. Iran J Microbiol 3:183–188

Baradkar V, Mathur M, Kumar S (2009) Cerebral CANDIDA abscess in an infant. Indian J Med Microbiol 27:70–72. https://doi.org/10.1016/S0255-0857(21)01762-X

Barton M, O'Brien K, Robinson JL, Davies DH, Simpson K, Asztalos E, Langley JM, Le Saux N, Sauve R, Synnes A, Tan B, de Repentigny L, Rubin E, Hui C, Kovacs L, Richardson SE (2014) Invasive candidiasis in low birth weight preterm infants: risk factors, clinical course and outcome in a prospective multicenter study of cases and their matched controls. BMC Infect Dis 14:327. https://doi.org/10.1186/1471-2334-14-327

Ben-Ami R, Berman J, Novikov A, Bash E, Shachor-Meyouhas Y, Zakin S, Maor Y, Tarabia J, Schechner V, Adler A, Finn T (2017) Multidrug-resistant Candida haemulonii and C. auris, Tel Aviv, Israel. Emerg Infect Dis 23:195–203. https://doi.org/10.3201/eid2302.161486

Benjamin DK, Stoll BJ, Fanaroff AA, McDonald SA, Oh W, Higgins RD, Duara S, Poole K, Laptook A, Goldberg R, National Institute of Child Health and Human Development Neonatal Research Network (2006) Neonatal candidiasis among extremely low birth weight infants: risk factors, mortality rates, and neurodevelopmental outcomes at 18 to 22 months. Pediatrics 117: 84–92. https://doi.org/10.1542/peds.2004-2292

Blasi E, Mazzolla R, Barluzzi R, Mosci P, Bartoli A, Bistoni F (1991) Intracerebral transfer of an in vitro established microglial cell line: local induction of a protective state against lethal challenge with Candida albicans. J Neuroimmunol 32:249–257. https://doi.org/10.1016/0165-5728(91)90195-d

Brumble L, Reza M, Dhakal L, Cruz G, Abu Saleh O, Heckman M (2017) Fungal infections of the central nervous system: clinical, radiographic and laboratory manifestations. J Microbiol Exp 5: 00167. https://doi.org/10.15406/jmen.2017.05.00167

Celmeli F, Oztoprak N, Turkkahraman D, Seyman D, Mutlu E, Frede N, Köksoy S, Grimbacher B (2016) Successful granulocyte colony-stimulating factor treatment of relapsing Candida albicans meningoencephalitis caused by CARD9 deficiency. Pediatr Infect Dis J 35:428–431. https://doi.org/10.1097/INF.0000000000001028

Cesaro S, Tridello G, Castagnola E, Calore E, Carraro F, Mariotti I, Colombini A, Perruccio K, Decembrino N, Russo G, Maximova N, Baretta V, Caselli D (2017) Retrospective study on the incidence and outcome of proven and probable invasive fungal infections in high-risk pediatric onco-hematological patients. Eur J Haematol 99:240–248. https://doi.org/10.1111/ejh.12910

Chiou CC, Wong TT, Lin HH, Hwang B, Tang RB, Wu KG, Lee BH (1994) Fungal infection of ventriculoperitoneal shunts in children. Clin Infect Dis 19:1049–1053. https://doi.org/10.1093/clinids/19.6.1049

Collar AL, Swamydas M, O'Hayre M, Sajib MS, Hoffman KW, Singh SP, Mourad A, Johnson MD, Ferre EMN, Farber JM, Lim JK, Mikelis CM, Gutkind JS, Lionakis MS (2018) The homozygous CX3CR1-M280 mutation impairs human monocyte survival. JCI Insight 3: e95417. https://doi.org/10.1172/jci.insight.95417

Drewniak A, Gazendam RP, Tool ATJ, van Houdt M, Jansen MH, van Hamme JL, van Leeuwen EMM, Roos D, Scalais E, de Beaufort C, Janssen H, van den Berg TK, Kuijpers TW (2013) Invasive fungal infection and impaired neutrophil killing in human CARD9 deficiency. Blood 121:2385–2392. https://doi.org/10.1182/blood-2012-08-450551

Drummond RA, Lionakis MS (2018) Candidiasis of the central nervous system in neonates and children with primary immunodeficiencies. Curr Fungal Infect Rep 12:92–97. https://doi.org/10.1007/s12281-018-0316-y

Egger NB, Kainz K, Schulze A, Bauer MA, Madeo F, Carmona-Gutierrez D (2022) The rise of Candida auris: from unique traits to co-infection potential. Microb Cell 9:141–144. https://doi.org/10.15698/mic2022.08.782

Faix RG, Chapman RL (2003) Central nervous system candidiasis in the high-risk neonate. Semin Perinatol 27:384–392. https://doi.org/10.1016/S0146-0005(03)00065-X

Fernandez M, Moylett EH, Noyola DE, Baker CJ (2000) Candidal meningitis in neonates: a 10-year review. Clin Infect Dis 31:458–463. https://doi.org/10.1086/313973

Garcia-Rubio R, de Oliveira HC, Rivera J, Trevijano-Contador N (2020) The fungal cell wall: Candida, cryptococcus, and aspergillus species. Front Microbiol 10:10

Gavino C, Cotter A, Lichtenstein D, Lejtenyi D, Fortin C, Legault C, Alirezaie N, Majewski J, Sheppard DC, Behr MA, Foulkes WD, Vinh DC (2014) CARD9 deficiency and spontaneous central nervous system candidiasis: complete clinical remission with GM-CSF therapy. Clin Infect Dis 59:81–84. https://doi.org/10.1093/cid/ciu215

Gavino C, Hamel N, Zeng JB, Legault C, Guiot M-C, Chankowsky J, Lejtenyi D, Lemire M, Alarie I, Dufresne S, Boursiquot J-N, McIntosh F, Langelier M, Behr MA, Sheppard DC, Foulkes WD, Vinh DC (2016) Impaired RASGRF1/ERK-mediated GM-CSF response characterizes CARD9 deficiency in French-Canadians. J Allergy Clin Immunol 137:1178–1188.e7. https://doi.org/10.1016/j.jaci.2015.09.016

Geers TA, Gordon SM (1999) Clinical significance of Candida species isolated from cerebrospinal fluid following neurosurgery. Clin Infect Dis 28:1139–1147. https://doi.org/10.1086/514755

Glocker E-O, Hennigs A, Nabavi M, Schäffer AA, Woellner C, Salzer U, Pfeifer D, Veelken H, Warnatz K, Tahami F, Jamal S, Manguiat A, Rezaei N, Amirzargar AA, Plebani A, Hannesschläger N, Gross O, Ruland J, Grimbacher B (2009) A homozygous CARD9 mutation in a family with susceptibility to fungal infections. N Engl J Med 361:1727–1735. https://doi.org/10.1056/NEJMoa0810719

Góralska K, Blaszkowska J, Dzikowiec M (2018) Neuroinfections caused by fungi. Infection 46: 443–459. https://doi.org/10.1007/s15010-018-1152-2

Groll AH, Giri N, Petraitis V, Petraitiene R, Candelario M, Bacher JS, Piscitelli SC, Walsh TJ (2000) Comparative efficacy and distribution of lipid formulations of amphotericin B in experimental Candida albicans infection of the central nervous system. J Infect Dis 182:274–282. https://doi.org/10.1086/315643

Herbst M, Gazendam R, Reimnitz D, Sawalle-Belohradsky J, Groll A, Schlegel P-G, Belohradsky B, Renner E, Klepper J, Grimbacher B, Kuijpers T, Liese J (2015) Chronic Candida albicans meningitis in a 4-year-old girl with a homozygous mutation in the CARD9 gene (Q295X). Pediatr Infect Dis J 34:999–1002. https://doi.org/10.1097/INF.0000000000000736

Huang CC, Chen CY, Yang HB, Wang SM, Chang YC, Liu CC (1998) Central nervous system candidiasis in very low-birth-weight premature neonates and infants: US characteristics and histopathologic and MR imaging correlates in five patients. Radiology 209:49–56. https://doi.org/10.1148/radiology.209.1.9769811

Jain KK, Mittal SK, Kumar S, Gupta RK (2007) Imaging features of central nervous system fungal infections. Neurol India 55:241. https://doi.org/10.4103/0028-3886.35685

Jones N, Garcez T, Newman W, Denning D (2016) Endogenous Candida endophthalmitis and osteomyelitis associated with CARD9 deficiency. BMJ Case Rep 2016:bcr2015214117. https://doi.org/10.1136/bcr-2015-214117

Katragkou A, Fisher BT, Groll AH, Roilides E, Walsh TJ (2017) Diagnostic imaging and invasive fungal diseases in children. J Pediatr Infect Dis Soc 6:S22–S31. https://doi.org/10.1093/jpids/pix055

Koutsouras GW, Ramos RL, Martinez LR (2016) Role of microglia in fungal infections of the central nervous system. Virulence 8:705–718. https://doi.org/10.1080/21505594.2016.1261789

Kullberg BJ, Arendrup MC (2015) Invasive candidiasis. N Engl J Med 373:1445–1456. https://doi.org/10.1056/NEJMra1315399

Kumar V, Cheng S-C, Johnson MD, Smeekens SP, Wojtowicz A, Giamarellos-Bourboulis E, Karjalainen J, Franke L, Withoff S, Plantinga TS, van de Veerdonk FL, van der Meer JWM, Joosten LAB, Bochud P-Y, Marchetti O, Perfect JR, Xavier R, Kullberg BJ, Wijmenga C, Netea MG (2014) Immunochip SNP array identifies novel genetic variants conferring susceptibility to candidemia. Nat Commun 5:4675. https://doi.org/10.1038/ncomms5675

Lai PH, Lin SM, Pan HB, Yang CF (1997) Disseminated miliary cerebral candidiasis. AJNR Am J Neuroradiol 18:1303–1306

Lee Y, Puumala E, Robbins N, Cowen LE (2021) Antifungal drug resistance: molecular mechanisms in Candida albicans and beyond. Chem Rev 121:3390–3411. https://doi.org/10.1021/acs.chemrev.0c00199

Lewis L, Jayashree P, Bhat Y (2015) Multiple fungal brain abscesses in a neonate: successful treatment with liposomal amphotericin B. Manipal J Nurs Health Sci 1:69

Lionakis MS, Lim JK, Lee C-CR, Murphy PM (2011) Organ-specific innate immune responses in a mouse model of invasive candidiasis. J Innate Immun 3:180–199. https://doi.org/10.1159/000321157

Lionakis MS, Swamydas M, Fischer BG, Plantinga TS, Johnson MD, Jaeger M, Green NM, Masedunskas A, Weigert R, Mikelis C, Wan W, Lee C-CR, Lim JK, Rivollier A, Yang JC, Laird GM, Wheeler RT, Alexander BD, Perfect JR, Gao J-L, Kullberg B-J, Netea MG, Murphy PM (2013) CX3CR1-dependent renal macrophage survival promotes Candida control and host survival. J Clin Invest 123:5035–5051. https://doi.org/10.1172/JCI71307

Lionakis MS, Iliev ID, Hohl TM (2017) Immunity against fungi. JCI. Insight 2:e93156. https://doi.org/10.1172/jci.insight.93156

Liu Y, Mittal R, Solis NV, Prasadarao NV, Filler SG (2011) Mechanisms of Candida albicans trafficking to the brain. PLoS Pathog 7:e1002305. https://doi.org/10.1371/journal.ppat.1002305

McCarthy MW, Kalasauskas D, Petraitis V, Petraitiene R, Walsh TJ (2017) Fungal infections of the central nervous system in children. J Pediatr Infect Dis Soc 6:e123–e133. https://doi.org/10.1093/jpids/pix059

McCullers JA, Vargas SL, Flynn PM, Razzouk BI, Shenep JL (2000) Candidal meningitis in children with cancer. Clin Infect Dis 31:451–457. https://doi.org/10.1086/313987

Mendoza-Reyes DF, Gómez-Gaviria M, Mora-Montes HM (2022) Candida lusitaniae: biology, pathogenicity, virulence factors, diagnosis, and treatment. Infect Drug Resist 15:5121–5135. https://doi.org/10.2147/IDR.S383785

Murthy JMK, Sundaram C (2014) Fungal infections of the central nervous system. In: Biller J, Ferro J (eds) Handbook of clinical neurology. Elsevier, New York, NY, pp 1383–1401

Neves N, Santos L, Reis C, Sarmento A (2014) Candida albicans brain abscesses in an injection drug user patient: a case report. BMC Res Notes 7:837. https://doi.org/10.1186/1756-0500-7-837

Nguyen MH, Yu VL (1995) Meningitis caused by Candida species: an emerging problem in neurosurgical patients. Clin Infect Dis 21:323–327. https://doi.org/10.1093/clinids/21.2.323

O'Brien D, Stevens NT, Lim CH, O'Brien DF, Smyth E, Fitzpatrick F, Humphreys H (2011) Candida infection of the central nervous system following neurosurgery: a 12-year review. Acta Neurochir 153:1347–1350. https://doi.org/10.1007/s00701-011-0990-9

Pappas PG, Kauffman CA, Andes DR, Clancy CJ, Marr KA, Ostrosky-Zeichner L, Reboli AC, Schuster MG, Vazquez JA, Walsh TJ, Zaoutis TE, Sobel JD (2016) Clinical practice guideline for the management of candidiasis: 2016 update by the Infectious Diseases Society of America. Clin Infect Dis 62:e1–e50. https://doi.org/10.1093/cid/civ933

Pappas PG, Lionakis MS, Arendrup MC, Ostrosky-Zeichner L, Kullberg BJ (2018) Invasive candidiasis. Nat Rev Dis Primer 4:1–20. https://doi.org/10.1038/nrdp.2018.26

Pendlebury WW, Perl DP, Munoz DG (1989) Multiple microabscesses in the central nervous system: a clinicopathologic study. J Neuropathol Exp Neurol 48:290–300. https://doi.org/10.1097/00005072-198905000-00006

Sánchez PJ, Pérez Cecilia E, Corral O, Romero Vivas J, Picazo JJ (2000) The central nervous system and infection by Candida species. Diagn Microbiol Infect Dis 37:169–179. https://doi.org/10.1016/S0732-8893(00)00140-1

Sánchez-Portocarrero J, Martín-Rabadán P, Saldaña CJ, Pérez-Cecilia E (1994) Candida cerebrospinal fluid shunt infection. Report of two new cases and review of the literature. Diagn Microbiol Infect Dis 20:33–40. https://doi.org/10.1016/0732-8893(94)90016-7

Santiago-Tirado FH, Doering TL (2017) False friends: phagocytes as Trojan horses in microbial brain infections. PLoS Pathog 13:e1006680. https://doi.org/10.1371/journal.ppat.1006680

Sendid B, Caillot D, Baccouch-Humbert B, Klingspor L, Grandjean M, Bonnin A, Poulain D (2003) Contribution of the Platelia Candida-specific antibody and antigen tests to early diagnosis of systemic Candida tropicalis infection in neutropenic adults. J Clin Microbiol 41:4551–4558. https://doi.org/10.1128/JCM.41.10.4551-4558.2003

Shah VB, Williams DL, Keshvara L (2009) β-Glucan attenuates TLR2- and TLR4-mediated cytokine production by microglia. Neurosci Lett 458:111–115. https://doi.org/10.1016/j.neulet.2009.04.039

Shankar SK, Mahadevan A, Sundaram C, Sarkar C, Chacko G, Lanjewar DN, Santosh V, Yasha TC, Radhakrishnan VV (2007) Pathobiology of fungal infections of the central nervous system with special reference to the Indian scenario. Neurol India 55:198. https://doi.org/10.4103/0028-3886.35680

Sharma R, Pawar S, Lad S, Mishra G, Netalkar A, Rege S (2012) Fungal infections of the central nervous system. In: Schmidek and sweet: operative neurosurgical techniques, pp 1691–1732

Shi M, Mody CH (2016) Fungal infection in the brain: what we learned from intravital imaging. Front Immunol 7:7

Swamydas M, Gao J-L, Break TJ, Johnson MD, Jaeger M, Rodriguez CA, Lim JK, Green NM, Collar AL, Fischer BG, Lee C-CR, Perfect JR, Alexander BD, Kullberg B-J, Netea MG, Murphy PM, Lionakis MS (2016) CXCR1-mediated neutrophil degranulation and fungal killing promote Candida clearance and host survival. Sci Transl Med 8:322ra10. https://doi.org/10.1126/scitranslmed.aac7718

Trofa D, Gácser A, Nosanchuk JD (2008) Candida parapsilosis, an emerging fungal pathogen. Clin Microbiol Rev 21:606–625. https://doi.org/10.1128/CMR.00013-08

van den Anker JN, van Popele NM, Sauer PJ (1995) Antifungal agents in neonatal systemic candidiasis. Antimicrob Agents Chemother 39:1391–1397

van Till JWO, Modderman PW, de Boer M, Hart MHL, Beld MGHM, Boermeester MA (2008) Mannose-binding lectin deficiency facilitates abdominal Candida infections in patients with secondary peritonitis. Clin Vaccine Immunol 15:65–70. https://doi.org/10.1128/CVI.00297-07

Voice RA, Bradley SF, Sangeorzan JA, Kauffman CA (1994) Chronic Candidal meningitis: an uncommon manifestation of candidiasis. Clin Infect Dis 19:60–66. https://doi.org/10.1093/clinids/19.1.60

Yoshida K, Matsuda N, Sato T, Watanabe T, Nakamura K, Saito K, Kanai K, Ugawa Y (2022) Candida brain abscesses in a patient with anorexia nervosa receiving total parenteral nutrition. Clin Neurol Neurosurg 212:107058. https://doi.org/10.1016/j.clineuro.2021.107058

Cryptococcosis of the Central Nervous System

13

Nagarathna Siddaiah, Kruthika Perumal, and Shayanki Lahiri

Abstract

Cryptococcosis is the most common opportunistic fungal infection of the central nervous system caused by the genus *Cryptococcus,* a basidiomycete yeast with polysaccharide capsule. *Cryptococcus neoformans* was the predominant species known until *Cryptococcus gattii* was identified as a separate species in 2002. The '*C. neoformans' and 'C. gattii' species complexes* include seven distinct biological and phylogenetic species. *C. neoformans complex* is associated with pigeons and commonly affects the immunocompromised individuals, and meningitis and meningoencephalitis are the common manifestations, whereas *C. gattii species complex* is associated with Eucalyptus trees and commonly affects the immunocompetent hosts and forms granulomatous lesions, which are often misdiagnosed as malignant glioma. Laboratory diagnosis ranges from a simple microscopic examination to whole-genome sequencing to identify the species. Polyenes, flucytosine and azoles are the antifungal agents used in combination and in three-phase therapy. Treatment is challenging because of the limited drug availability and the adverse effects, and WHO has provided new treatment guidelines for CNS cryptococcosis.

N. Siddaiah (✉)
Department of Neuromicrobiology, National Institute of Mental Health and Neurosciences, Bangalore, Karnataka, India

K. Perumal
Department of Microbiology, Sri Ramachandra Medical College and Research Institute, Sri Ramachandra Institute of Higher Education and Research, Chennai, Tamilnadu, India

S. Lahiri
Department of Computational Medicine and Bioinformatics, University of Michigan Medical School, Ann Arbor, MI, USA

H. Sami et al. (eds.), *Viral and Fungal Infections of the Central Nervous System: A Microbiological Perspective*, https://doi.org/10.1007/978-981-99-6445-1_13

Prevention of cryptococcal meningitis is possible on routine screening of HIV seropositive patients followed by pre-emptive fluconazole therapy. New drugs, repurposed drugs and vaccines are in various stages of research. Prevention, early diagnosis and effective treatment will reduce the mortality and morbidity.

Keywords

Cryptococcal meningitis · *Cryptococcus neoformans* · Cryptococcoma · Human immunodeficiency virus · Antifungal therapy

13.1 Introduction

Fungal infections of the central nervous system (FIs-CNS) are gaining importance recently, affecting immunocompetent and predominantly immunocompromised hosts (Bloch and Bailin 2019). Outcome of the infection in terms of mortality and morbidity is decided by the virulence of the fungus and the resistance offered by immune system of the host (Góralska et al. 2018).

There are 1.5 million different species of fungi worldwide, 70,000 of them have received official descriptions. Only 300 of these identified species are known to be pathogenic, and 10 to 15% of them are known to infect the central nervous system (CNS), which includes *Cryptococcus neoformans (C. neoformans), Candida* spp., *Mucor, Rhizopus, Aspergillus* and *Cladophialophora bantiana* (Góralska et al. 2018). The diagnosis of FIs-CNS is difficult as deep-seated fungal granulomas can mimic other disorders; for example, cryptococcomas are frequently mistaken for gliomas, and also, cryptococcal meningitis (CM) and tuberculous meningitis have identical clinical presentations. In order to choose an appropriate therapy for patients with FIs-CNS and lower their mortality, an early diagnosis of fungal infection is required, together with the identification of the etiological agent (Góralska et al. 2018). The best management of FIs-CNS infections necessitates an understanding of their epidemiology, host features, diagnostic criteria and therapeutic choices (Schwartz et al. 2018).

Cryptococcosis is an opportunistic infection caused by the yeast *Cryptococcus* (Maziarz and Perfect 2016). They can disseminate and involve any organ in the body such as skin, prostate, eyes and bone/joints, but have a predisposition for the lungs and the CNS. The incidence of cryptococcosis has increased ever since 1950, with the use of corticosteroids among cancer patients. However, significant rise was observed from 1980 with the surge in AIDS-related cases (Ruschel and Thapa 2022). Approximately 6% of people with AIDS go on to develop cryptococcal infections, and AIDS-related cryptococcosis makes up to 85% of all instances of the disease (Ruschel and Thapa 2022). CNS cryptococcosis happens in people who have HIV infection and have CD4 levels under 100 cells/μL (Li et al. 2020). CM is one of the leading causes of morbidity and mortality among patients with HIV/AIDS (Rajasingham et al. 2017).

Anti-retroviral therapy (ART) access has led to a decrease in the annual occurrence of cryptococcal illness in the developed nations, but the rate is still high among developing nations such as Asia and sub-Saharan Africa (McKenney et al. 2015; Tenforde et al. 2017). Even one-year mortality is high in low-resource settings, ranging from 50% to 100% due to restricted access to antifungals, but it is only 10% to 30% in nations with abundant resources. High death rates are also a result of rising antifungal resistance trends, intolerance and noncompliance (Loyse et al. 2013; Rajasingham et al. 2017).

13.2 Cryptococcus

With more than 50 species that are widely distributed across the environment, the genus *Cryptococcus* belongs to the Basidiomycota phylum (Ruschel and Thapa 2022; Maziarz and Perfect 2016). It is an environmental saprophyte; humans are accidental dead-end host in the life cycle (Bicanic and Harrison 2004). In 1894, for the first time *C. neoformans* was isolated from peach juice, and in the same year, the first case of human disease was reported by Busse. *Cryptococcus* are encapsulated spherical-to-oval yeast cells that are 5–10 μm in diameter. The polysaccharide capsule that may be up to 30 μm thick is a key virulence factor that can be demonstrated as halo around the yeast cells in negative staining and can be detected by cryptococcal antigen detection assays (Page et al. 2013; Bicanic and Harrison 2004).

The name *Cryptococcus*, which means 'hidden sphere' in Greek, refers to a unicellular yeast that is a member of the phylum Basidiomycota, or club fungi, and class Tremellomycetes of the kingdom Mycota. Since its discovery, *C. neoformans* was the sole fungal species responsible for cryptococcosis. In the year 2002, *Cryptococcus neoformans var. gattii* was categorised as a separate species, with advent of newer molecular techniques (Table 13.1). For many years, the five serotypes of cryptococcus were divided into two varieties: serotypes B and C for *C. neoformans var. gattii* and serotypes A, D and AD (now proven to be a hybrid of A and D strains) *C. neoformans var. neoformans* (Kwon-Chung et al. 1982). *C. grubii* was the third variety and was introduced in the year 1999 for serotype A. Later on, the *C. gattii* was given species level in the year 2002. The majority of clinical isolates belongs to serotype A (Bicanic and Harrison 2004).

Several molecular techniques have been used over the past 20 years to demonstrate the significant genetic heterogeneity that exists. Recently, the taxonomy of *C. neoformans/gattii* was revised and seven species and thirteen genotypes were proposed based on PCR fingerprinting, restriction fragment length polymorphism (RFLP), multilocus sequence typing (MLST)/amplified fragment length polymorphism (AFLP) and other molecular techniques such as genome sequencing data. These species comprise two members of the *Cryptococcus neoformans species complex*: (1) *Cryptococcus neoformans* sensu stricto [serotype A; AFLP1/VNI, AFLP1A/VNB/VNII and AFLP1B/VNII], (2) *Cryptococcus deneoformans* [serotype D; AFLP2/VNIV]; six species within *Cryptococcus gattii species complex*:

Table 13.1 Representation of current/proposed nomenclature and classification of *C. neoformans/ C. gattii* species complex

Current Species[a]	Current serotypes[a]	Current nomenclature	Proposed species name[b]	Genotype[c]
Cryptococcus neoformans	Serotype A	Variety: *C. neoformans var grubii*[d]	*Cryptococcus neoformans*	VNI VNII VNBI VNBII VNIII
	Serotype D	Variety: *C. neoformans var neoformans*[d]	*Cryptococcus deneoformans*	VNIV
	Serotype AD	*C. neoformans inter-variety hybrid*	*C. neoformans* X *Cryptococcus deneoformans*	Not proposed
Cryptococcus gattii	Serotype B Serotype C	*Cryptococcus Gattii*	*Cryptococcus gattii*	VGI
			Cryptococcus deuterogattii	VGII
			Cryptococcus bacillisporus	VGIII
			Cryptococcus tetragattii	VGIV
			Cryptococcus decagattii	VGIV/ VGIIIC

[a]Bennett JE et al. Epidemiologic differences among serotypes of *Cryptococcus neoformans.* Am J Epimediol. 1977;105:582–586
[b]Hagen, F., Khayhan, K., Theelen, B., Kolecka, A., Polacheck, I., Sionov, E., Falk, R., Parnmen, S., Lumbsch, H.T. and Boekhout, T., 2015. Recognition of seven species in the Cryptococcus gattii/ Cryptococcus neoformans species complex. *Fungal genet. Boil.*, *78*, pp.16–48
[c]PCR-fingerprinting and RFLP-based genotyping nomenclature as introduced by Meyer et al. (2003)
[d]Introduced by Franzot et al. (1999) for serotype A isolates

(1) *Cryptococcus gattii* sensu stricto [serotype B; AFLP1B/VGI], (2) *Cryptococcus bacillisporus* [serotype B and C; AFLP5/VGIII], (3) *Cryptococcus deuterogattii* [serotype B; AFLP6/VGII], (4) *Cryptococcus tetragattii* [serotype C; AFLP7/ VGIV], (5) *Cryptococcus gattii* [AFLP10/VGIV/VGIIIC] and (6) unnamed. Clinically significant differences in molecular genotypes have not been demonstrated, and the majority of specialists agree that the most helpful clinical differentiation is between the *Cryptococcus neoformans and Cryptococcus gattii species complexes*. However, in settings with limited resources, identifying species may not be possible (Ashton et al. 2019; Hagen et al. 2015; Kwon-Chung et al. 1982).

Apart from the usual, *C. neoformans* and *C. gattii infection*, in rest, *C. laurentii* or *C. albidus* accounts for about 80% of instances of cryptococcal infection. Additionally, infection due to *C. diffluens, C. liquefaciens, C. uniguttulatus, C. adeliensis, C. luteolus* and *C. curvatus* infection have also been reported (Cano et al. 2020).

13.3 Epidemiology

C. neoformans is present globally and typically infects immunosuppressed people, and it is associated with soil that has been contaminated by bird, particularly pigeon excrement. The *C. gattii* species complex has been linked to a number of eucalyptus, fir and oak trees (Ruschel and Thapa 2022). It is present in tropical and subtropical regions as well as temperate areas including British Columbia, the Pacific Northwest of the USA and Southern Australia. Investigations from South America have also demonstrated that both species can be found in the decomposing heartwood of some trees (Tintelnot et al. 2015).

According to reports, CM is the primary cause of morbidity and mortality among people with HIV/AIDS, causing 223,000 illnesses out of which 73% occur in sub-Saharan Africa followed by 19% in Asia and Pacific. Case fatality is reported to be 181,100 deaths annually across the globe in the year 2017. CM causes 19% of AIDS-related deaths globally (Rajasingham et al. 2017).

13.4 Pathogenicity

The infective agent enters the human body in the form of propagules through the respiratory system by inhalation and gets deposited in the pulmonary alveoli (Ruschel and Thapa 2022) (Fig. 13.1). Individuals may experience mild symptoms but largely remain asymptomatic. The *Cryptococci* on reaching the lungs colonise the alveolar spaces. The host immune system, virulence of the organism and the load of the inhaled desiccated fungal cells (either poorly encapsulated yeast cells or basidiospores) determine the pathogenesis. Alveolar macrophages play a pivotal role in tackling the pathogen in the alveoli. To combat the cryptococcal infection, the host cellular and humoral components play important role, and the T cytokines

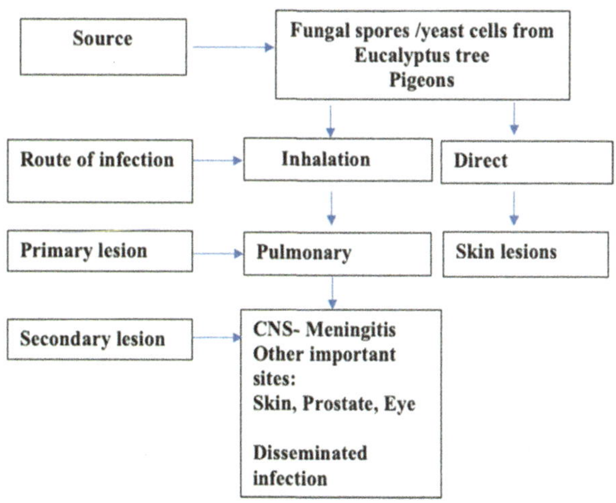

Fig. 13.1 Pathogenesis of Cryptococcus

tumour necrosis factor (TNF), interleukin-2 and interferon-γ, produced by T helper cells, result in the formation of a granuloma (Maziarz and Perfect 2016).

The yeast may stay dormant inside the phagolysosome of the lymph nodes of thorax or as lung granuloma, which could last in an asymptomatic person for years, or it could cause mild symptoms that might go unrecognised (Maziarz and Perfect 2016).

The excellent adaptation to elude the human immune system and boost spreading in the host is particularly impressively exhibited by the morphological transformation of the *C. neoformans/C. gattii species complex* to create larger cells known as titan cells (Dyląg et al. 2020). Titan cells feature a thicker cell wall, a strongly cross-linked capsule and a diameter of more than 12 μm without the capsule (May et al. 2016).

Due to the larger size, titan cells cannot be phagocytosed. It is also resistant to a number of host factors and to some of the antifungals. Titan cell formation is observed only in *C. neoformans/C. gattii species complex* (Dyląg et al. 2020).

Small-sized cryptococcal cells or micro-cells or drop cells are also seen, along with large-sized titan cells. Despite having a thicker cell wall, these micro-cells are just 2–4 μm in size and appear to be specialised for proliferation in macrophages. During sexual reproduction, hyphal or pseudohyphal forms are observed in both laboratory and environment, but these are not seen in vivo, probably because they are effectively removed by host immune response (May et al. 2016).

In the event of immunosuppression, the silent yeasts reactivate, multiply and disseminate to different parts of the body and exhibit neurotropism (May et al. 2016).

13.4.1 The Described Ways Through Which the Organism Crosses the BBB

13.4.1.1 The 'Trojan Horse' Mechanism

It is the most prevailing mode in the dissemination of Cryptococcosis. The yeast cells are taken up by regional phagocytes and cross the epithelial barrier of various tissues of the body, including blood–brain barrier through a process called transcytosis. Yeast cells can move within the phagocytes unharmed by lytic or non-lytic extrusion, by a process called vomocytosis (Alspaugh 2015).

13.4.1.2 Paracellular Passage

The yeast cells can pass between the cells by the paracellular passage. By this process, yeast can move through the blood–brain barrier and the pulmonary endothelium. Paracellular processes are facilitated by urease enzyme production, which triggers the release of IL-33, which causes separation of tight junctions in the epithelium to ease the passage of *Cryptococcus*. Epithelial cells of the blood–brain barrier can be destroyed by urease, other proteolytic enzymes and Plb1 in order to facilitate yeast passage (Alspaugh 2015).

13.4.1.3 Transcellular Passage

Yeast cells pass through the cells by the enhancement of *Cryptococcus* adhesion through the engagement of glucuronoxylomannan (GXM) with the CD14 receptor and binding of yeast palmitic acid to the pulmonary surfactant protein D. Hyaluronic acid binding to CD44 has been linked to a similar process, which likewise promotes cell adhesion and involves metalloproteinase Mpr1 interacting with Annexin A2 to allow yeast to penetrate the blood–brain barrier transcellularly (Alspaugh 2015).

13.4.2 Virulence

The development of melanin pigment, capsule formation and thermotolerance are the three basic and important virulence factors of *C. neoformans*.

13.4.2.1 Capsule

Cryptococcus's capsule contains polysaccharide glucuronoxylomannan and has antiphagocytic properties. The polysaccharides glucuronoxylomannan galactan (GXMGal) and glucuronoxylomannan (GXM) make up the majority of the composition. Depending on the variety of *Cryptococcus* spp., GXM makes up 90 to 95% of the polysaccharide capsule and has a molecular weight between 1700 and 7000 kDa (Israel Diniz-Lima et al. 2022). In addition to cell wall chitin and its derivatives, which suppresses the inflammatory response, it is the capsular polysaccharide, the GXM and galactoxylomannan (GalXM), released abundantly into the surrounding environment, undermine the pro-inflammatory nuclear factor-κB (NF-κB) pathway and weaken the production of pro-inflammatory cytokines such as tumour necrosis factor (TNF) and the inflammatory response (May et al. 2016).

13.4.2.2 Melanin

The presence of the enzyme phenoloxidase in *C. neoformans* protects the yeasts from the oxidative stresses of the host and facilitates the neurotropism towards diphenolic catecholamine-rich areas. Melanin reduces lymphocyte multiplication and tumour necrosis factor synthesis, making the organism resistant to leucocyte attack (Kwon-Chung et al. 2014).

13.4.2.3 Thermotolerance

Growth at high temperature (37 °C) is a fundamental component for virulence. Molecular techniques have found several signalling pathways and enzymes that are required for human pathogenesis (Kwon-Chung et al. 2014).

Phospholipase and urease synthesis and a number of enzymes linked to oxidative stress resistance and survival inside the phagolysosome are additional virulence factors. According to estimates, over genes are crucial for the pathogenesis of yeast in hosts. By altering the phagosome membrane permeability and using nonlytic exocytosis (vomocytosis), the yeast has even evolved complicated escape mechanisms from the intracellular environment. This enables the transfer of the yeast within the phagocytes from one cell to another or within the host compartment (Kwon-Chung et al. 2014; Diniz-Lima et al. 2022).

As the cells age, the changes in the cell wall offer survival benefit to the organism by resisting phagocytosis and antifungals and accumulate in the brain increasing the fungal load. These are called as founder cells. Quorum sensing might have predominant role during cryptococcal pathogenesis, and light-sensing pathways prevent inappropriate filamentation of the cells, which otherwise would initiate a lethal immune response against the organism (May et al. 2016).

13.4.3 Reason for Neurotropism

Demonstration of distinctive neurotropism by *C. neoformans* was first recognised by Versé in the year 1914 and 2 years later by Stoddard and Cutler (1916) (Kwon-Chung et al. 2014). This neurotropism is associated with a numerous cryptococcal-specific factors that promote the blood–brain barrier permeability. These factors include metalloproteinases and ureases enzymes, which cause neuromodulation, and mechanisms like autophagy and high-affinity sugar transporters, which help organisms survive in the nutrient-poor environment of the brain (Ruschel and Thapa 2022). Due to the absence of complements and immunoglobulins, the CSF makes an excellent location for infection. Melanin is produced by an enzyme phenoloxidase of *C. neoformans*. The abundance of substrates that can interact with phenoloxidase in the brain, such as L-dopamine, may explain *C. neoformans'* predilection for the neurological system (Satishchandra et al. 2007). Necrosis and organ destruction may be minimal or absent until a later illness. Organ dysfunction in patients with severe infections may worsen more quickly, mostly as a result of tissue deformation brought on by the expanding fungal burden (Ruschel and Thapa 2022).

13.5 Clinical Manifestations

Majority of the time, the *Cryptococci* enter the CNS hematogenously, mainly from pulmonary foci. It is more common particularly, among immunocompromised (HIV seropositive) populations including transplant recipients, patients with defects in cell-mediated immunity, cancer patients and people who have used antibiotics for an extended period of time. Mortality remains high despite antifungal treatment (Maziarz and Perfect 2016).

Though the reactivation of latent infection is often the reason for the development of the disease, a study cohort reports only 52% of the incidence following solid organ transplantation, suggesting that the traditional view that cryptococcosis is always due to reactivation may not be accurate (Day 2004). The proportion of clinical disease in HIV-positive people that is caused by primary infection as opposed to reactivated latent disease remains undetermined (Day 2004).

The primary route by which *Cryptococcus* enters the body is through the respiratory system.

From asymptomatic colonisation of the airways or a small pulmonary nodule visible on a chest radiograph to potentially fatal pneumonia accompanied by acute respiratory distress syndrome, pulmonary cryptococcosis has a wide spectrum of clinical symptoms. Pulmonary involvement may be evidenced in 10% to 55% of individuals with AIDS-associated cryptococcal meningoencephalitis (Maziarz and Perfect 2016). Meningitis/meningoencephalitis and cryptococcoma are the most common CNS manifestations of cryptococcal infection.

13.5.1 Meningitis/Meningoencephalitis

It is rather meningoencephalitis, because, in patients with immunosuppression or those who are immunocompetent, histopathological evidence (Lee et al. 1996) shows that brain parenchyma is not spared. CM frequently manifests as chronic or subacute meningitis and rarely progresses rapidly (Casadevall and Perfect 1998).

Over a period of weeks, patients typically manifest with fever, malaise, headaches and impaired mental status. Oftentimes, headache is so severe that it mimics a subarachnoid haemorrhage and is accompanied by vomiting and visual problems (Satishchandra et al. 2007). Although the specific cause of such a severe headache is unknown, it could be caused by meningeal involvement, increased intracranial pressure, sino-venous thrombosis or all of these (Satishchandra et al. 2007). A temporary visual disturbance could be caused by increased intracranial pressure (Satishchandra et al. 2007). Only 65% of patients present with fever, while more than 75% of patients present with headache. A bad prognosis is indicated by the development of altered sensorium seen in 20% to 46% of individuals. Approximately 13% of these people may experience vision loss. An estimated 8% to 25% of patients experience seizures (Satishchandra et al. 2007).

Signs of meningism, cranial nerve palsies, papilloedema, various focal neurological deficits, decreased conscious levels and other specific neurological deficits may be seen during an examination. Only 30% of cases have stiff necks. Although focal deficits are rare, patients can show hemiparesis or hemisensory symptoms as a result of arteritis. Isolated occurrences of hydrocephalus or basal arachnoiditis-related cranial neuropathies, particularly of the lower cranial nerves, can damage a specific cranial nerve (II, VII, VIII, IX, X and XII) or several. In immunocompetent people, cryptococcal meningitis is more frequently accompanied by papilloedema, hydrocephalus, focal deficits, seizures and cryptococcomas (Satishchandra et al. 2007).

Due to the lack of particular clinical symptoms, CM should always be considered while determining a differentiating cause of subacute or chronic meningoencephalitis. When CM is suspected, other diagnoses to be taken into account include tuberculous meningitis and carcinomatous/lymphocytic meningitis. One-third of the patients may have lungs, kidney or skin involvement from a disseminated illness. Some patients of suspected CM can benefit from a thorough search for cutaneous cryptococcal lesions (Satishchandra et al. 2007).

Raised intracranial tension without ventricular dilatation may result in severe vision and hearing loss. Less frequently, obstructive hydrocephalus with ventricular dilatation can cause patients to experience cognitive impairment and gait ataxia (Day 2004; Satishchandra et al. 2007).

13.5.2 Cryptococcoma

Apart from the classic meningoencephalitis caused by cryptococcus, CNS cryptococcoma is another important manifestation, which is misdiagnosed most often as glioma. Though cryptococcomas are usually rare, it can be common among immunocompetent individuals and predominant causative agent is *C. gattii*, while classical meningoencephalitis is caused by *C. neoformans* (Chen et al. 2000; Mitchell et al. 1995). Up to 69% of immunocompetent *C. gattii* patients have been observed to have cryptococcomas (Chen et al. 2000; Mitchell et al. 1995; Perfect 2010).

Clinical presentation may be that of space-occupying lesion and may vary depending on the location of the granuloma and immunological status of the patient. There are no classical image findings for cryptococcoma. It mimics other lesions like toxoplasmosis, abscess, lymphoma and malignant glioma (Kovoor et al. 2002). Cryptococcoma will not be in the list of differential diagnosis especially so when the patient is not immunosuppressed. Most of the time the diagnosis is made postoperatively based on microbiological or histopathological evidence when the lesion is biopsied or excised as a part of treatment. CSF examination may appear normal except for cryptococcal antigen in exceptional cases. Antifungal therapy alone may not suffice, resection of the lesion is required for good outcome. In patients with cryptococcomas, the neurological sequelae and morbidity could be high and prolonged antifungal therapy may be required. In regions endemic to *C. gattii*, cryptococcoma is included in the differential diagnosis for tumour and mass lesions of CNS (Ulett et al. 2017).

13.5.3 CNS Cryptococcosis and HIV Seronegative

Globally, CNS cryptococcosis is usually associated with HIV infection (Park et al. 2009). It has been recognised among HIV seronegative individuals also. Patient group that is at risk includes solid organ transplant recipients (kidney transplantation 11.1%), patients with rheumatic diseases, rheumatological diseases, chronic hepatic failure (16%) and those receiving immunosuppressive therapies. It is also seen in otherwise immunocompetent hosts (Beardsley et al. 2019). It is observed that underlying predisposing conditions exist in majority (52.5%–79%) of the cases (Huang et al. 2023; Pappas et al. 2001). In contrast, a study from china has recorded that only 33% of them had underlying predisposing condition (Zhu et al. 2010).

Patients present with classic symptoms of meningitis irrespective of whether HIV seropositive or negative, and prominent characteristics of non-HIV-related cryptococcal meningitis include focal neurological impairments. There could be delay in diagnosis among HIV-negative cases because of insidious onset of the disease and low clinical suspicion (Beardsley et al. 2019).

It is known that *C. gattii* is more commonly associated with HIV-negative cases and pulmonary lesions are more common. Dissemination is probably not easy in an immunocompetent individual, and the disease is localised to the lungs. Various studies report contrasting numbers wherein one study reports that 75% of the infection is localised to lungs and the other study from Australia reports 85% dissemination to the CNS (Galanis et al. 2009). The mortality rate remains the same with or without underlying conditions among cases of CM (Huang et al. 2023). CSF India ink could detect 70% of patients with *C. gattii* as compared to 95% among *C. neoformans* meningitis. The diagnostic sensitivity for *C. neoformans* is lower (30–72%) in HIV-negative patients and increases to 80% among those suffering from AIDS (Diaz and Nguyen 2010; Lalloo et al. 1994).

13.6 Laboratory Diagnosis of Cryptococcal Meningitis

13.6.1 CSF Composition

Raised intracranial pressure is one of the important features of CM, which ranges from 350 mmH_2O to more than 900 mmH_2O. This is because of blockade to free flow of CSF due to *Cryptococci* and accrual of polysaccharide capsule in the arachnoid villi and subarachnoid spaces (Guo et al. 2016).

CSF macroscopically may appear clear/opalescent or slightly hazy depending on the quantity of cryptococci, protein and the cells present in the CSF. In general, there is an increase in CSF protein level (up to 2 g/L), whereas decrease in glucose and chloride levels. Ten to seventeen percent of HIV-positive patients show signs of normal CSF. In our study at NIMHANS by lahari et al., (unpublished data) CSF protein ranged from 23 to 569 mg/dL, and CSF glucose level ranged from 0 to 325 mg/dL.

Typically, there is a CSF pleocytosis. In majority of the cases, there is an increase in CSF WBC count, ranging between 100 and 500 $\times$ 10^6/L with predominant lymphocytes. In a retrospective analysis, only 20% of patients had CSF WBC levels greater than 100 cells/cumm, while 52% of patients had CSF WBC levels less than 20 cells/cumm. Total CSF cell count ranged from nil cells to 1100 cells/cumm. Lymphocyte predominance was observed in 57.6% and nil cells in 38.6% of cases (Satishchandra et al. 2007). The disease status and therapeutic response are predicted by the CSF cell count (Qu et al. 2020).

13.6.2 Microscopic Examination

13.6.2.1 India Ink Stain

Cryptococcus is a capsulated yeast; negative staining methods such as India ink preparation can be employed for the demonstration of capsule (Fig. 13.2). A drop of India ink is mixed with a drop of CSF on a clean glass slide, later covered with a coverslip and examined first under 10X objective followed by 45X. *Cryptococci* are seen as a spherical cell measuring around 2–15 μm showing single or multiple budding forms. Capsule is seen as a clear halo around the yeast. When large number of *Cryptococci* are present, it is seen as starry night under the microscopy. Alternatively, 7% nigrosin can be used. The addition of 2% chromium mercury to the CSF followed by India ink enhances the observation of the cells (Guo et al. 2016; Zerpa et al. 1996).

Direct microscopic observation techniques, especially the India ink preparation, are only preliminary tests and observation/findings need to be confirmed by other gold standard methods. However, there are several advantages for this preparation. In low-resource setting, it is simple to perform, rapid and cost-effective technique, especially in an expert hand, which has good sensitivity (86%) (Boulware et al. 2014) and specificity (Guo et al. 2016).

On several occasions, cryptococcal cells can be easily confused with small lymphocytes, RBCS and air bubbles, and hence, meticulous examination is required. Diagnosis of CM can be established among 90% of AIDS cases and more than 50% of HIV-negative cases. The sensitivity of the test depends upon the number of *Cryptococci* present in the CSF. When the cells are less than 1000 CFU/mL, sensitivity is 42%, and sensitivity can be improved by centrifuging the sample at 500 rpm for about 10 min and examining the pellet (Boulware et al. 2014).

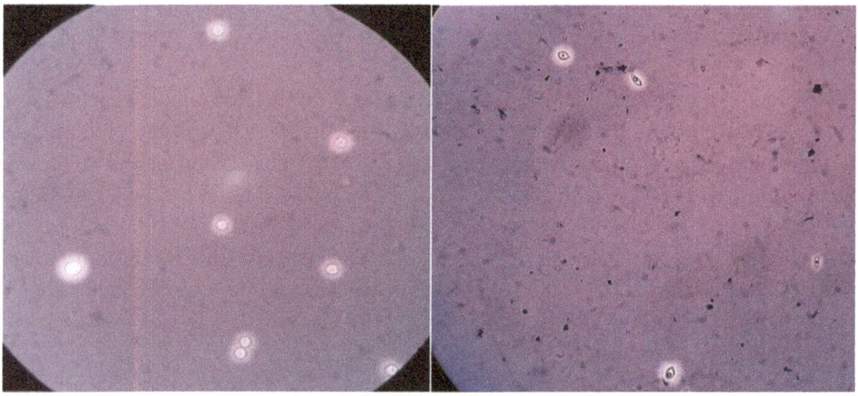

A) *B)*

Fig. 13.2 India ink preparation of the CSF sample showing round budding yeast cells and the capsule is seen as a clear halo around the cells of *Cryptococcus neoformans* (**a**) and slightly oval cells of *Cryptococcus gattii* (**b**)

The microscopic limit of the test is 10^3–10^4 CFU/mL *Cryptococci.* AIDS patients have larger concentrations of yeast cells ranging between 10^5 and 10^7 CFU/mL, hence high sensitivity (Guo et al. 2016).

13.6.2.2 Other Staining Methods

Other staining methods include May–Grunwald–Giemsa (MGG) stain and Alcian blue staining which have better detection rates, however, have low specificity. Gram staining usually reveals poorly stained gram-positive yeast (Guo et al. 2016).

13.6.2.3 The Histopathological Analysis

The histopathological analysis of biopsy specimens is helpful for diagnosis. Yeast forms consistent with *Cryptococcus* can be seen using periodic acid–Schiff stains and Grocott methenamine silver (GMS) stains. Nevertheless, mucicarmine or alcian blue when utilised alongside the Fontana–Masson stain is of higher quality than GMS and dyes like calcofluor -white for recognising cryptococci. Also, they are considerably more specific than these dyes since they accentuate not only just the cell wall (Fontana–Masson), but also the mucin-positive capsule (mucicarmine/alcian blue). Another advantage of the melanin-detecting Fontana–Masson stain is that it can distinguish between nontypical, capsule-negative strains (Perfect 2010; Guo et al. 2016).

13.6.3 Antigen Detection

13.6.3.1 Cryptococcal Antigen (CrAg)

Cryptococcal capsular polysaccharide antigen, an early biomarker that can be detected in CSF, blood and urine samples, can be considered as most reliable method (Guo et al. 2016). It can be detected early in the disease process; test can be performed rapidly and has very good sensitivity. The currently available methods to perform the test are latex agglutination (LA), enzyme immunoassay (EIAs) and newer lateral flow assay (LFA) (Guo et al. 2016; Rajasingham et al. 2012).

13.6.3.2 Latex Agglutination Test (LAT)

It is a simple and semi-quantitative test with sensitivity of 93% to 100% and a specificity of 93% to 98%. Since the test is semi-quantitative, it is of great value in diagnosis and prognosis of the disease. The disadvantages are that the test can be false negatives in some of the patients with autoimmune diseases; since it is not a field test, it requires the basic infrastructure facility for processing the sample and preservation of reagents; the interpretation of the test can go wrong in an inexperienced hand (Rajasingham et al. 2012) (Fig. 13.3).

13.6.3.3 Lateral Flow Assay (LFA)

It is a novel immunochromatographic dipstick assay approved by FDA in 2011.The LFA strips are coated with gold-conjugated anticryptococcal antibody. When specimen containing the cryptococcal antigen is added over the strip, upon conjugation a

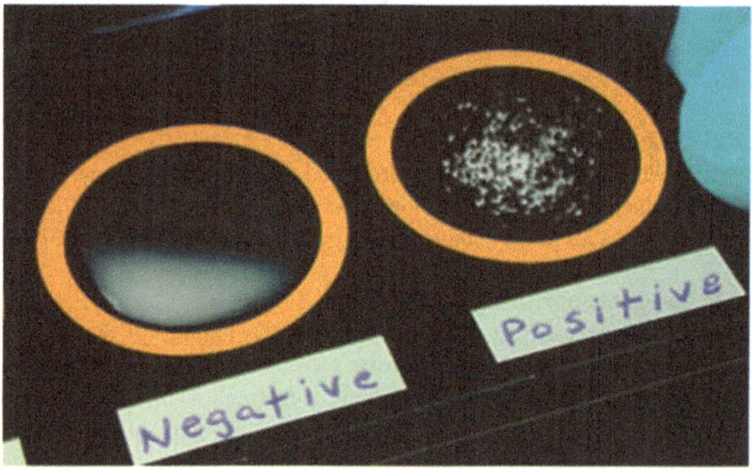

Fig. 13.3 Cryptococcal antigen detection by latex agglutination test. Note: Positive test sample shows clumps of antigen antibody complex with clearing of the rest of the material. Negative test sample shows no clumps

band appears suggestive of positivity. It has many advantages over LAT; it is rapid, results are available within few minutes (10 min), can be used as point of care test (POCT), more economical, obviates the need for basic infrastructure and does not require special skills (Guo et al. 2016; Rajasingham et al. 2019).

For CSF samples, sensitivity and specificity of the test are 99.3% and 99.1%, respectively, which can be more reliable than culture. The test can detect all the serotypes. In a study, LFA identified additional six more cases of CM, which was negative by other methods (Rajasingham et al. 2019; Boulware et al. 2014).

Antigen detection by LAT/LFA in CSF is a semi-quantitative method, and hence, the antigen titre can be considered as an indicator of severity of the infection and to assess the prognosis. However, in few cases, they can continue to remain positive in low titres for a long period of time despite good response to treatment (Rajasingham et al. 2019).

The World Health Organization (WHO) advises routine blood screening for cryptococcal infection in people living with HIV/AIDS (PLWHA) whose CD4+ levels are fewer than 100 cells/μL (Rajasingham et al. 2019).

13.6.3.4 Screening of Blood for Cryptococcal Antigen

The test is particularly useful in situations where lumbar puncture (LP) is contraindicated, especially when patients are presenting with raised ICP, one of the common manifestations of CM. CrAg in blood can be a predictor of meningitis and can be detected weeks to months before the onset of classical symptoms of meningitis among persons suffering from HIV infection. In patients with advanced HIV disease, the prevalence of asymptomatic cryptococcal antigenaemia ranges from 1% to 15%. The probability of development of meningitis is low in whom the serum antigen titre is less than 1 in 80 and increases proportionately with increasing titres. CrAg titres of more than 1:1280 have definite involvement of the CNS. Mortality remains high among asymptomatic patients with positive CrAg titres of more than 1:160 despite pre-emptive fluconazole therapy (Temfack et al. 2021).

Since asymptomatic CrAg positivity is considered to be an independent predictor of meningitis and death, pre-emptive treatment of such cases with high-dose fluconazole may prevent the onset of the disease and mortality. In a study, 28% survival benefit was observed. It has been found that about one-third of asymptomatic patients living with HIV who are positive for cryptococcal antigen in serum had CM, suggesting that routine serum screening for cryptococcal antigen may facilitate the clinician in decision-making about the management. WHO and other national HIV guidelines now recommend routine screening of blood of patients with advanced HIV disease and advice pre-emptive treatment with high-dose fluconazole for positive cases (Temfack et al. 2021).

It demonstrates a positive predictive value of 100%, with 93% detection for cryptococcal meningitis, 100% negative predictive value and 100% agreement with serum or plasma CrAg readings (Temfack et al. 2021).

Instead of whole plasma or serum, finger prick blood sample is enough to test for CrAg in blood. It demonstrates a positive predictive value of 100%, with 93% detection for cryptococcal meningitis, 100% negative predictive value and 100% agreement with serum or plasma CrAg readings (Temfack et al. 2021).

13.6.3.5 Screening of Urine Sample

Screening of urine sample for cryptococcal antigen would be an ideal non-invasive POCT for the diagnosis of cryptococcosis but fails due to reported 50% false positivity (Tenforde et al. 2018).

13.6.4 Culture

CSF culture is the gold standard test in the diagnosis of cryptococcosis (Fig. 13.4). CSF is inoculated using a sterile Pasteur pipette or microtitre pipette onto Sabouraud's dextrose agar medium (SDA) (Guo et al. 2016).

Alternatively, brain–heart infusion agar and potato dextrose agar can be used for the isolation of *Cryptococci*. Two tubes are inoculated, one of the tube is incubated at 37 °C and the another at 25 °C. Minimum of 1 mL of CSF should be inoculated, especially so when very few *Cryptococci* are present in the CSF of the immunocompetent and in patient on antifungals. Colonies appear within 48 to 72 h, they are creamy, mucoid 3–4 mm in size, sometimes the culture may be delayed depending upon the number of *Cryptococci* present in the CSF sample. The disadvantage is that there is a delay in identification and confirmation of the isolate (Rajasingham et al. 2019).

The advantages of culture are that quantitative CSF cultures may predict the rate of clearance of *Cryptococci* from the serial cultures of CSF (Dyal et al. 2016).

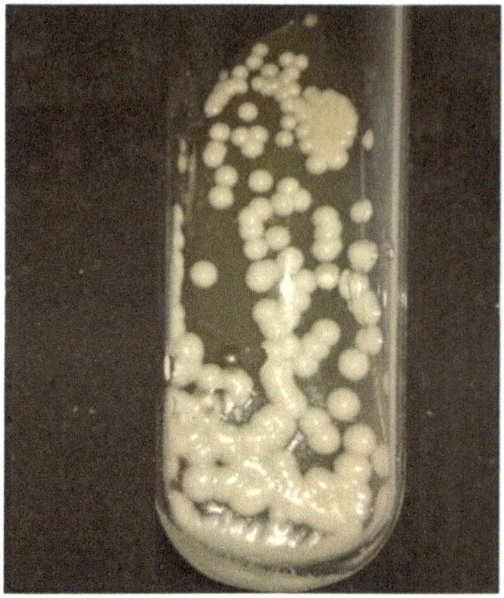

Fig. 13.4 CSF fungal culture on Sabouraud's dextrose agar. Note: Cryptococcal culture on Sabouraud's dextrose agar medium showing creamy white, mucoid colonies measuring 2–3 mm in size

Fungal cultures will also aid in identification of pathogen at the species level with antifungal sensitivity. The advanced molecular and sequencing studies have identified many species of *Cryptococci*, virulence factors and differences in antifungal susceptibility, which has made the species identification rather mandatory. With the advent of automated methods identification and antifungal susceptibility of *Cryptococci* has become easy compared to the conventional methods of growing them on caffeic acid ferric citrate agar (CAFC), detection of urea hydrolysis on urease medium and by using l-canavanine glycine bromothymol blue (CGB) medium to differentiate *C. neoformans* and *C. gattii* (Fig. 13.5) (McTaggart et al. 2011).

13.6.5 MALDI-TOF MS

Matrix-assisted laser desorption/ionization time-of-flight (MALDI-TOF) mass spectrometry (MS) system identifies the isolate within 10 min. The automated VITEK 2 in addition to identification determines the antifungal susceptibility compared with the CLSI and EUCAST BMD reference standards (Siqueira et al. 2019).

13.6.6 Other Tests

Molecular methods like polymerase chain reaction (PCR) are an additional diagnostic test. The meningitis/encephalitis panel (Biomerieux) is a multiplex PCR assay.

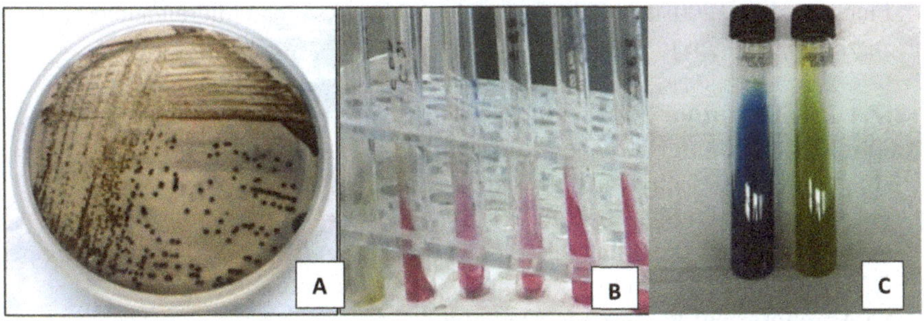

Fig. 13.5 Identification of Cryptococcal species by CAFC test agar, urease medium and CGB medium. (**a**) Caffeic acid ferric citrate test agar (CAFC) medium for melanin production—black-coloured colonies formed by *Cryptococcus* spp. It differentiates *Cryptococcus species* from other yeasts. (**b**) Urease test to demonstrate urea hydrolysis—urease enzyme production by *Cryptococcus species*. Positive test—pink colour indicates cryptococcus spp. Negative test—no change in colour (straw colour)—not cryptococcus spp. (**c**) l-Canavanine glycine bromothymol blue agar medium (CBG). Positive test—turns blue within 3–5 days—indicates *Cryptococcus gattii complex*. Negative test—green colour—indicates *Cryptococcus neoformans complex*

More than 100 CFU/mL is necessary for the detection. Sensitivity is 96% (83%–99%) and specificity is 100%, but it is not cost-effective. A very effective methodology to identify intrathecal IgG production in the CSF is to use isoelectric focusing to identify CSF-specific oligoclonal bands (Satishchandra et al. 2007).

13.7 Treatment of Cryptococcal Meningitis

Conventional treatment protocol includes 2-week amphotericin B deoxycholate-based regimens, which is associated with adverse effects, and the efficacy of fluconazole monotherapy is poor leading to poor outcome, and a 10-week mortality rate of more than 50% is reported (Bicanic and Harrison 2004; Tenforde et al. 2018).

The three antifungal classes, namely the azoles, the polyenes and flucytosine, are frequently combined in a three-part therapy approach to treat disseminated infection: Induction, consolidation and maintenance are effective against *Cryptococcus* spp. (Mourad and Perfect 2018).

13.7.1 WHO Recommendation

WHO Recommendation (adapted from WHO Guidelines for diagnosing, prevention and managing cryptococcal disease among adults, adolescents and Children living with HIV, 2022).

13.7.1.1 Induction Therapy

Liposomal amphotericin B—10 mg/kg—single high dose.

Flucytosine—100 mg/kg per day—4 divided doses daily—2 weeks.

Fluconazole—1200 mg daily for adults—2 weeks; for children/adolescents—12 mg/kg per day, up to maximum of 800 mg daily—2 weeks.

13.7.1.2 Alternative Induction Regimens

If liposomal Amphotericin Is Not Available.

Amphotericin B deoxycholate—1 mg/kg per day, plus.

Fluconazole—100 mg/kg per day, 4 divided doses—1 week, followed by.

Fluconazole—1200 mg daily for adults; for children/adolescents—12 mg/kg per day, up to maximum of 800 mg daily—1 week.

If No Amphotericin Formula is Available:

Fluconazole—1200 mg daily for adults; for children/adolescents—12 mg/kg per day, plus.

Flucytosine—100 mg/kg per day—4 divided doses daily—2 weeks.

If Flucytosine Is Not Available.

Liposomal amphotericin B—3–4 mg/kg per day plus.

Fluconazole—1200 mg daily for adults; for children/adolescents—12 mg/kg per day, up to maximum of 800 mg daily.

If Liposomal Amphotericin B and Flucytosine Are Not Available.

Amphotericin B deoxycholate—1 mg/kg per day—2 weeks and Fluconazole—1200 mg daily for adults; for children/adolescents—12 mg/kg per day, up to maximum of 800 mg daily.

13.7.1.3 Consolidation Therapy (2018 Recommendation)

Fluconazole—800 mg daily for adults; for children/adolescents—6–12 mg/kg/day, up to maximum of 800 mg daily for 8 weeks following the induction phase.

13.7.1.4 Maintenance (2018 Recommendation)

Fluconazole (200 mg daily for adults or 6 mg/kg per day for adolescents and children) is suggested during the maintenance phase until immunological reconstitution (CD4 > 200 mm^3) and reduction in viral loads on ART.

Note

1. The only oral combination treatment that is advised is fluconazole and flucytosine, which has been linked to decreased mortality when compared to fluconazole and amphotericin B deoxycholate.
2. Regimens containing flucytosine are beneficial.
3. It is not advised to treat adults, adolescents or children with HIV-associated cryptococcal meningitis with supplementary corticosteroid medication on a regular basis during the induction period.
4. Due to the risk of increased mortality, immediate ART beginning is not advised for HIV seropositive who have CM in adults, adolescents or children; instead, they should wait 4–6 weeks after the start of antifungal therapy.
5. The most effective induction therapy is hampered by the toxicity of the infusion and the side effects of amphotericin B therapy.

6. When compared to liposomal amphotericin B, toxicities include hypokalaemia, nephrotoxicity and anaemia, which are seen when employing amphotericin B deoxycholate-based regimens. Liposomal amphotericin B as a single high dose (10 mg/kg) followed by amphotericin B deoxycholate on a seventh day is found to be well tolerable than a 14-day amphotericin B deoxycholate regimen.

13.7.2 Managing Raised Intracranial Pressure

Therapeutic lumbar puncture reduces CSF pressure to below 20 cm H_2O by draining or, if the baseline pressure is excessively high, to half that value. The frequency of subsequent therapeutic lumbar punctures should be based on clinical presentation and outcome.

13.7.3 Monitoring Treatment Response

In the initial 2 weeks of induction therapy, the clinical response should be monitored every day. In low- and middle-income nations, for patients who showed excellent response clinically: routine LP post-induction therapy to assess the effectiveness of antifungal therapy (CSF culture for fungi) is not necessitated. Testing for cryptococcal antigen in CSF, serum or plasma is not advised for assessing therapy response in any situation.

13.7.4 Persistent or Recurrent Symptoms

(a) Evaluate the patient's medical history for any indications of treatment failure caused by (1) an insufficient drug regimen, dose or duration; (2) a lack of compliance with fluconazole consolidation and maintenance therapy; or (3) patients who have previously undergone prolonged fluconazole medication have underlying fluconazole drug resistance.
(b) Lumbar puncture to be carried out to assess the opening pressure and to repeat CSF examination for a prolonged fungal culture (incubation time of 2 weeks).
(c) Be mindful about paradoxical cryptococcal immune reconstitution inflammatory syndrome.
(d) Considerations should also be given to other illnesses (including viral, bacterial or tuberculous meningitis) that can exhibit symptoms and signs resembling those of cryptococcal meningitis.
(e) When clinical suspicion exists (culture-positive relapse despite fluconazole adherence), fluconazole susceptibility testing should be carried out whenever practical.

13.7.5 Managing Relapse

1. Initiate or resume induction therapy according to the induction therapy recommendations.
2. Therapeutic lumbar puncture can be used to lower elevated intracranial pressure.
3. Strengthen adherence.
4. If ART has not already begun, it must be started 4–6 weeks after receiving the best antifungal therapy.
5. If available, consider having a fluconazole susceptibility test.

13.7.6 Species-Based Treatment

Though the therapeutic strategies are the same for both *C. gattii* and *C. neoformans* infection, Cryptococcomas/hydrocephalus caused by *C. gattii* requires more radiologic diagnostic attention and follow-up examinations in comparison with *C. neoformans.*

Fluconazole (400 mg daily, orally) is advised for treating pulmonary cryptococcosis (the same as *C. neoformans*), but for the treatment of very large and multiple cryptococcomas, AmB deoxycholate and flucytosine treatment for 4 to 6 weeks, followed by 6 to 18 months of fluconazole, depending on whether any operative measure was undertaken, should be taken into account. Consider surgery if essential structures are compressed, the size of the cryptococcoma does not shrink after 4 weeks of medication, or the patient is not improving.

13.7.7 Treatment in Pregnant Women with Cryptococcosis

Amphotericin B deoxycholate or liposomal amphotericin B (AmBd or LFAmB), with or without flucytosine, for disseminated and CNS illness. Flucytosine is considered as category C medicine for antinatal cases; thus, the benefits and risks of using it must be weighed carefully.

Fluconazole after delivery prevents exposure to fluconazole during the first trimester and, during the final two trimesters, weighs the requirement for ongoing antifungal drug exposure (pregnancy categories) when deciding whether to take fluconazole. In the postpartum period, keep watch out for IRIS.

13.7.8 IRIS

Immune reconstitution inflammatory syndrome (IRIS) among cases of CM is not an unusual manifestation observed following the administration of ART. It is an exaggerated and dysregulated pro-inflammatory immune response. There is CD4+ T-cell recovery accompanied by decreasing peripheral blood HIV viral load (Le and Spudich 2016). It is seen in 25% of HIV patients associated with CM. It occurs within the first 4 months of treatment with ART amounting to 20 +/− 10% mortality (Sereti et al. 2020).

There are two types of IRIS that develop among cases of CM. The first type is 'unmasking' IRIS, seen after the initiation of ART in ART-naïve individuals in whom existence of CM was not known or not diagnosed, sets within 2 to 6 weeks of ART and could be fatal. Patients manifest with clinical features of raised intracranial pressure and inflammation. The second type is 'paradoxical' IRIS, while the patient is on antifungal therapy. It manifests within 6 months following ART showing good response to antifungal and ART. Clinically, there is worsening of symptoms and altered mental status due to raised ICP. The host immunological status, the antigen load and response to ART or antifungals determine the development of IRIS (Shelburne III et al. 2005).

13.7.9 Antifungals

The widespread application of antifungal drugs in health care and agriculture has been attributed for the rise of new fungal species and resistant strains that have developed through time. Antifungal resistance is influenced by hybridization between *Cryptococcus* spp., which also supports a changing population pattern. Under the influence of drugs, hybridization promotes genetic modifications, such as gene duplication and loss of heterozygosity, which hastens the response to antifungal medications. The need for the production of new drugs has increased due to antifungal resistance, although knowledge of the processes underlying drug resistance in *Cryptococcus* spp. is still limited (Bermas and Geddes-McAlister 2020).

13.7.9.1 Polyenes

Amphotericin B (AmB) is a polyene compound, fungicidal, fast acting and efficacious (Klepser 2011). Binding to cell wall ergosterol causes formation of a porin channel and leakage of cellular contents (Gray et al. 2012). In addition to limited availability and parenteral administration, the drug causes nephrotoxicity, anaemia, electrolyte abnormalities and hypomagnesaemia. Liposomal AmB, lipid-based formulations, have lesser side effects; however, the usage is limited by the high cost. Despite AmB being widely used, resistance to AmB is uncommon. Mutations in the enzyme sterol Δ8–7 isomerase, which is involved in the synthesis of ergosterol, are one of the earliest suggested reasons. There are individual case reports of resistance. We have reported 1.85% of clinical isolates of cryptococcus to be resistant to amphotericin B (Lahiri et al. 2023). Following treatment, *C. neoformans* may show altered susceptibility to AMB though acquired resistance is not well known. Poor AmB susceptibility is likely caused by an alteration or suppression of the sterol concentration in fungal cell membranes (Bermas and Geddes-McAlister 2020).

13.7.9.2 Flucytosine

It is a synthetic antifungal agent. This drug cannot be used alone in treatment of cryptococcosis due to threat of emergence of intrinsic resistance, hence used in combination always (Mayers 2009). It exhibits synergistic or addictive effect when combined with AMB, and it is one of the first-line drugs in the treatment of cryptococcosis (Schwarz et al. 2006). Cytosine permease mediates the uptake of 5FC into susceptible fungal cells, where it is transformed into 5-fluorouracil (5FU) via cytosine deaminase (Waldorf and Polak 1983), and protein and DNA synthesis are inhibited. The toxic effects are somewhat similar to AMB and may require hospital admission for administration (Bermas and Geddes-McAlister 2020). However, research on the prevalence of AmB + 5FC resistance in isolates that are 5FC resistant is still lacking and the mechanism of resistance development is not well established (Bermas and Geddes-McAlister 2020).

13.7.9.3 Fluconazole

Early 1990 saw the approval of fluconazole (FLC), which has since become the standard therapy for CM. Lanosterol 14-sterol-demethylase, a fungus-specific cytochrome P450-dependent enzyme, is inhibited in order to target ergosterol manufacture in fungal cells. (Sheehan et al. 1999). The accumulation of sterol precursors brought on by the enzyme's inhibition alters the structure of the plasma membrane, increases cellular permeability, impairs development and causes the leakage of cellular contents (Ghannoum and Rice 1999). In developing countries, fluconazole is the preferred fungicide to treat cryptococcal meningitis. It is a broad-spectrum triazole antifungal drug that is highly effective, less side effects and inexpensively available (Kneale et al. 2016). The other advantages with fluconazole are its good CNS permeability, bioavailability (~100%), less significant drug–drug interactions, tolerance of gastric pH and comprehensive within clinical settings (Brammer et al. 1990). The drawbacks are as follows: It is fungistatic, less effective than AmB, takes longer to sterilise the CSF and produces worse clinical results than AmB-based induction regimens (Saag et al. 1992; Van Der Horst et al. 1997). In clinical settings, extensive usage of FLC has resulted in the development of antifungal resistance among many fungal species. Studies have revealed gradual increase in FLC resistance from 7.3% to 11.7% among *C. neoformans* between 1997 and 2007, and among new cases, relapse cases were 10.6% and 24.1%, respectively (Bermas and Geddes-McAlister 2020).

According to studies, azole resistance is linked to mutations in the ERG11 gene, which codes for the FLC-targeting enzyme lanosterol 14-demethylase. Through heteroresistance, *Cryptococcus* also develops FLC resistance. The establishment of heteroresistance was initially noted in vitro in 1999, and it is believed to be associated with chromosomal aneuploidy (Bermas and Geddes-McAlister 2020). Lahiri et al. (2023) documented 28.5% and 25% FLC resistance in *C. neoformans* and *C. gattii,* respectively.

13.8 Outcome

Outcome is poor among patients with immune ageing. Immunity gets suppressed either due to HIV infection or any other reason, and older individuals of more than 60 years of age are associated with co-morbidities (Fuentes et al. 2017; Qu et al. 2020) A significant sign of poor outcomes for CM patients is altered mentation (Qu et al. 2020). Heavy fungal burden is another important factor in deciding the outcome, augmented by poor inflammatory response (Jarvis et al. 2014; Concha-Velasco et al. 2017). The prognosis could be bad when CSF cell count was less than 40 cells/cmm. CD4/CD8 ratio and CD4+ T-cell percentage is remarkably low in poor outcome and correlates with CSF cell count (Qu et al. 2020). There is inhibition of pro-inflammatory CSF cytokine c (IL-6, IFN- and TNF-2020). Capsular GXM might also have a role in stunting the immune response leading to poor outcome (Retini et al. 1998).

13.9 Repurposing Old Drugs

13.9.1 Fenbendazole

In a mouse model of cryptococcosis, significance of in vitro action against *C. neoformans* and *C. gattii* is shown by the anthelmintic drug fenbendazole. It has growth-inhibitory capabilities, and its inhibition of virulence factors and its reduction in fungal proliferation inside macrophages are the mechanism of action for its effectiveness (Mourad and Perfect 2018).

13.9.2 Macrolides

It inhibits protein synthesis, by weakening this essential component of virulence in *C. gattii*, and the macrolides increase macrophage phagocytosis and the generation of inflammatory cytokines in culture (Mourad and Perfect 2018).

13.9.3 Antimalarial Drug Mefloquine

Mefloquine derivatives have been found to have better in vitro antifungal activity and prevent the development and melanization of *C. neoformans* capsules. Unfortunately, these compounds also showed a slight increase in their in vitro toxicity when tested against human cells compared to mefloquine (Mourad and Perfect 2018).

13.9.4 Sertraline and Tamoxifen

These are two medications that have moved to clinical trials and may be the most notable examples of substances being repurposed to treat cryptococcosis. In a phase III experiment with patients being treated for cryptococcal meningitis, single-drug therapy with sertraline or combination therapy with sertraline, fluconazole and amphotericin B did not unfortunately enhance patient outcomes, despite encouraging results in phase I/II clinical studies. For the treatment of cryptococcal meningitis, a phase II clinical trial (NCT03112031) to evaluate (300 mg daily) tamoxifen's safety/efficacy, when used as an adjunct drug to standard amphotericin B/fluconazole therapy, found no therapeutic benefit (Mourad and Perfect 2018).

13.10 Drug Combination Therapy

Using antifungal drugs that act synergistically and boost efficacy and fungal selectivity can help prevent the development of resistance to antifungal medicines. As a broad-spectrum adjuvant, clofazimine has been found to work in conjunction with caspofungin and fluconazole to combat *C. neoformans* and other fungi.

13.11 Drugs in Pipeline

13.11.1 Monoclonal Antibody 18B7

It targets capsular GXM. It is capable of altering the gene expression and metabolism of *Cryptococci* and also functions as a catalytic antibody that can hydrolyse the *C. neoformans* capsule (Mourad and Perfect 2018).

13.11.2 Benzyl thioureas (BTUs)

They target the integrity of the cell wall through the inhibition of the late post-Golgi secretory pathway (Mourad and Perfect 2018).

13.11.3 Other Drugs

Clofazimine causes cell membrane stress and accelerates the effects of antifungals. APX001 (fosmanogepix), hydrazine3-bromo-N′-(3-bromo-4hydroxybenzylidene) benzohydrazide (B0) and its derivative N′-(3-bromo-4-hydroxybenzylidene)-2-methylbenzohydrazide (BHBM), APX879 and compound 112 (CMPD#112), ibomycin and Hsp90-selective fungi inhibitors are other agents in the pipeline (Mourad and Perfect 2018).

13.12 Vaccines

The development of a vaccine for cryptococcosis is difficult because the disease primarily affects immunocompromised people with reduced CD4 + T cells. It has been discovered that a mutant, live virulent strain of *C. neoformans* (H99), which lacks sterylglucosidase (sgl), can be genetically manipulated to make IFN (Mourad and Perfect 2018). It has also been shown that heat-killed mutant strains activate potent T helper 1 cell responses, offering total defence against the *C. neoformans* assault. As a delivery system and adjuvant for cryptococcal vaccines, glucan particles produced from acapsular mutant strains of *C. neoformans* and *C. gattii* show enormous potential. Mice that were treated with glucan particles specifically carrying cryptococcal antigens developed antigen-specific T-cell responses and maintained their immunity against cryptococcosis (Mourad and Perfect 2018).

13.13 Prevention Strategies: WHO Recommendations

1. When treating people 10 years of age and older who are presenting with severe HIV disease, it is necessary to screen for cryptococcal infection using plasma, serum or whole-blood samples.
2. Before beginning or restarting antiretroviral therapy (ART), it is advised for PLWHA, with CD4+ levels fewer than 100 cells/mm^3 or even levels less than 200 cells/mm^3, to undergo cryptococcal–antigen screening, followed by pre-emptive antifungal medication among CrAg-positive individuals. This is to discourage any possibility of invasive cryptococcal disease development.
3. To rule out cryptococcal meningitis, all HIV-positive individuals with positive results from a cryptococcal antigen test should be assessed for clinical features of meningitis. They should also have their CSF examined for *Cryptococcus*. In cases where antigen screening is unavailable, a negative India ink result must be verified by additional confirmatory testing. Adults and adolescents with HIV are advised to use fluconazole prophylaxis if their CD4 cell count is between 100 and 200 cells/mm^3.

References

Alspaugh JA (2015) Virulence mechanisms and Cryptococcus neoformans pathogenesis. Fungal Genet Biol 78:55–58

Ashton PM, Thanh LT, Trieu PH, Van Anh D, Trinh NM, Beardsley J, Kibengo F, Chierakul W, Dance DAB, Rattanavong S, Davong V (2019) Three phylogenetic groups have driven the recent population expansion of Cryptococcus neoformans. Nat Commun 10(1):2035

Beardsley J, Sorrell TC, Chen A (2019) Central nervous system Cryptococcal infections in non-HIV infected patients. J Fungi 5(3):71. https://doi.org/10.3390/jof5030071

Bennett JE, Kwon-Chung KJ, Howard DH (1977) Epidemiologic differences among serotypes of Cryptococcus neoformans. Am J Epidemiol 105(6):582–586
Bermas A, Geddes-McAlister J (2020) Combatting the evolution of antifungal resistance in Cryptococcus neoformans. Mol Microbiol 114(5):721–734. https://doi.org/10.1111/mmi.14565
Bicanic T, Harrison TS (2004) Cryptococcal meningitis. Br Med bull 72(1):99–118
Bloch KC, Bailin SS (2019) Update on fungal infections of the central nervous system: emerging pathogens and emerging diagnostics. Curr Opin Infect Dis 32(3):277–284
Boulware DR, Rolfes MA, Rajasingham R, von Hohenberg M, Qin Z, Taseera K, Schutz C, Kwizera R, Butler EK, Meintjes G, Muzoora C (2014) Multisite validation of cryptococcal antigen lateral flow assay and quantification by laser thermal contrast. Emerg Infect Dis 20:45
Brammer KW, Farrow PR, Faulkner JK (1990) Pharmacokinetics and tissue penetration of fluconazole in humans. Rev Infect Dis 12(Supplement 3):S318–S326
Cano EJ, Yetmar ZA, Razonable RR (2020) Cryptococcus species other than Cryptococcus neoformans and Cryptococcus gattii: are they clinically significant? Open Forum Infect Dis 7(12):ofaa527. US: Oxford University Press, 7
Casadevall A, Perfect JR (1998) Human cryptococcosis. In: Cryptococcus neoformans, vol 595. ASM Press, Washington, DC, pp 407–456
Chen S, Sorrell T, Nimmo G, Speed B, Currie B, Ellis D, Marriott D, Pfeiffer T, Parr D, Byth K (2000) Epidemiology and host-and variety-dependent characteristics of infection due to Cryptococcus neoformans in Australia and New Zealand. Clin Infect Dis 31(2):499–508
Concha-Velasco F, Gonzalez-Lagos E, Seas C, Bustamante B (2017) Factors associated with early mycological clearance in HIV-associated cryptococcal meningitis. PLoS One 12(3):e0174459
Day JN (2004) Cryptococcal meningitis. Pract Neurol 4(5):274–285
Diaz MR, Nguyen MH (2010) Diagnostic approach based on capsular antigen, capsule detection, β-glucan, and DNA analysis. In: Cryptococcus: from human pathogen to model yeast, pp 547–564
Diniz-Lima I, Fonseca LMD, Silva-Junior EBD, Guimarães-de-Oliveira JC, Freire-de-Lima L, Nascimento DO, Morrot A, Previato JO, Mendonça-Previato L, Decote-Ricardo D, Freire-de-Lima CG (2022) Cryptococcus: history, epidemiology and immune evasion. Appl Sci 12(14): 7086
Dyal J, Akampurira A, Rhein J, Morawski BM, Kiggundu R, Nabeta HW, Musubire AK, Bahr NC, Williams DA, Bicanic T, Larsen RA (2016) Reproducibility of CSF quantitative culture methods for estimating rate of clearance in cryptococcal meningitis. Sabouraudia 54(4): 361–369
Dyląg M, Colon-Reyes RJ, Kozubowski L (2020) Titan cell formation is unique to Cryptococcus species complex. Virulence 11(1):719–729
Franzot SP, Salkin IF, Casadevall A (1999) Cryptococcus neoformans var. grubii: separate varietal status for Cryptococcus neoformans serotype a isolates. J Clin Microbiol 37(3):838–840
Fuentes E, Fuentes M, Alarcón M, Palomo I (2017) Immune system dysfunction in the elderly. An Acad Bras Ciênc 89:285–299
Galanis E, Hoang L, Kibsey P, Morshed M, Phillips P (2009) Clinical presentation, diagnosis and management of Cryptococcus gattii cases: lessons learned from British Columbia. Can J Infect Dis Med Microbiol 20(1):23–28
Ghannoum MA, Rice LB (1999) Antifungal agents: mode of action, mechanisms of resistance, and correlation of these mechanisms with bacterial resistance. Clin Microbiol Rev 12(4):501–517
Góralska K, Blaszkowska J, Dzikowiec M (2018) Neuroinfections caused by fungi. Infection 46: 443–459
Gray KC, Palacios DS, Dailey I, Endo MM, Uno BE, Wilcock BC et al (2012) Amphotericin primarily kills yeast by simply binding ergosterol. Proc Natl Acad Sci U S A 109(7):2234–2239
Guo XS, Bu H, He JY, Zou YL, Zhao Y, Li YY, Cui JZ, Zheng MM, Han WX, Zhao ZY (2016) Current diagnosis and treatment of cryptococcal meningitis without acquired immunodeficiency syndrome. Neuroimmunol Neuroinflamm 3:249–256

Hagen F, Khayhan K, Theelen B, Kolecka A, Polacheck I, Sionov E, Falk R, Parnmen S, Lumbsch HT, Boekhout T (2015) Recognition of seven species in the Cryptococcus gattii/Cryptococcus neoformans species complex. Fungal Genet Boil 78:16–48

Huang Y, Jin X, Wu F, Pan T, Wang X, Chen D, Chen Y (2023) Cryptococcal meningitis in HIV-negative patients: a 12-year single-center experience in China. J Clin Med 12(2):515

Jarvis JN, Bicanic T, Loyse A, Namarika D, Jackson A, Nussbaum JC, Longley N, Muzoora C, Phulusa J, Taseera K, Kanyembe C (2014) Determinants of mortality in a combined cohort of 501 patients with HIV-associated Cryptococcal meningitis: implications for improving outcomes. Clin Infect Dis 58(5):736–745

Klepser M (2011) The value of amphotericin B in the treatment of invasive fungal infections. J Crit Care 26(2):225.e1–225.e10

Kneale M, Bartholomew JS, Davies E, Denning DW (2016) Global access to antifungal therapy and its variable cost. J Antimicrob Chemother 71(12):3599–3606

Kovoor JM, Mahadevan A, Narayan JP, Govindappa SS, Satishchandra P, Taly AV, Shankar SK (2002) Cryptococcal choroid plexitis as a mass lesion: MR imaging and histopathologic correlation. Am J Neuroradiol 23(2):273–276

Kwon-Chung KJ, Polacheck ITZHACK, Bennett JE (1982) Improved diagnostic medium for separation of Cryptococcus neoformans var. neoformans (serotypes a and D) and Cryptococcus neoformans var. gattii (serotypes B and C). J Clin Microbiol 15(3):535–537

Kwon-Chung KJ, Fraser JA, Doering TL, Wang ZA, Janbon G, Idnurm A, Bahn YS (2014) Cryptococcus neoformans and Cryptococcus gattii, the etiologic agents of cryptococcosis. Cold Spring Harb Perspect Med 4(7):a019760

Lahiri S, Maji S, Manjunath N, Bahubali VH, Chandrashekar N (2023) Trends of CNS Cryptococcosis during pre-and post-HIV era: a 38 years' retrospective cohort analysis from South India. J Mycol Med 33:101358

Lalloo D, Fisher D, Naraqi S, Laurenson I, Temu P, Sinha A, Saweri A, Mavo B (1994) Cryptococcal meningitis (C. neoformans var. gattii) leading to blindness in previously healthy Melanesian adults in Papua New Guinea. Q J Med 87(6):343–349

Le LT, Spudich SS (2016) HIV-associated neurologic disorders and central nervous system opportunistic infections in HIV. In Seminars in neurology. 36, 04, 373–381). Thieme Medical Publishers; New York, NY

Lee SC, Dickson DW, Casadevall A (1996) Pathology of cryptococcal meningoencephalitis: analysis of 27 patients with pathogenetic implications. Hum Pathol 27(8):839–847

Li S, Nguyen IP, Urbanczyk K (2020) Common infectious diseases of the central nervous system—clinical features and imaging characteristics. Quant Imaging Med Surg 10(12):2227

Loyse A, Thangaraj H, Easterbrook P, Ford N, Roy M, Chiller T, Govender N, Harrison TS, Bicanic T (2013) Cryptococcal meningitis: improving access to essential antifungal medicines in resource-poor countries. Lancet Infect Dis 13(7):629–637

May RC, Stone NR, Wiesner DL, Bicanic T, Nielsen K (2016) Cryptococcus: from environmental saprophyte to global pathogen. Nat Rev Microbiol 14(2):106–117

Mayers D (2009) Antimicrobial drug resistance handbook, vol 1. Humana Press, Totowa, NJ; https://www.springer.com/gp/book/9781603275927

Maziarz EK, Perfect JR (2016) Cryptococcosis. Infect Dis Clin 30(1):179–206

McKenney J, Bauman S, Neary B, Detels R, French A, Margolick J, Doherty B, Klausner JD (2015) Prevalence, correlates, and outcomes of cryptococcal antigen positivity among patients with AIDS, United States, 1986–2012. Clin Infect Dis 60(6):959–965

McTaggart L, Richardson SE, Seah C, Hoang L, Fothergill A, Zhang SX (2011) Rapid identification of Cryptococcus neoformans var. grubii, C. neoformans var. neoformans, and C. gattii by use of rapid biochemical tests, differential media, and DNA sequencing. J Clin Microbiol 49(7): 2522–2527

Meyer W, Castañeda A, Jackson S, Huynh M, Castañeda E, IberoAmerican Cryptococcal Study Group (2003) Molecular typing of IberoAmerican Cryptococcus neoformans isolates. Emerg Infect Dis 9(2):189

Mitchell DH, Sorrell TC, Allworth AM, Health CH, McGregor AR, Papanaoum MJR, Gottlieb T (1995) Cryptococcal disease of the CNS in immunocompetent hosts: influence of cryptococcal variety on clinical manifestations and outcome. Clin Infect Dis 20:611–616

Mourad A, Perfect JR (2018) Present and future therapy of cryptococcus infections. J fungi (Basel, Switzerland) 4(3):79. https://doi.org/10.3390/jof4030079

Page KR, Chaisson RE, Sande M (2013) Cryptococcosis and other fungal infections (histoplasmosis and coccidioidomycosis) in HIV-infected patients. In: Sande's HIV/AIDS medicine: medical management of AIDS, pp 369–388

Pappas PG, Perfect JR, Cloud GA, Larsen RA, Pankey GA, Lancaster DJ, Henderson H, Kauffman CA, Haas DW, Saccente M, Hamill RJ (2001) Cryptococcosis in human immunodeficiency virus-negative patients in the era of effective azole therapy. Clin Infect Dis 33(5):690–699

Park BJ, Wannemuehler Kathleen A, Marston Barbara J, Nelesh G, Pappas Peter G, Chiller Tom M (2009) Estimation of the current global burden of Cryptococcal meningitis among persons living with HIV/AIDS. AIDS (London, England) 23:525–530

Perfect JR (2010) Cryptococcus neoformans. In: Gerald L, Bennett JE, Dolin R (eds) Mandell, Douglas, and Bennett's principles and practice of infectious diseases, 7th edn, pp 3287–3303

Qu J, Jiang J, Lv X (2020) The utility of cerebrospinal fluid white cell count during the prognostic assessment for cryptococcal meningitis patients: a retrospective study. BMC Infect Dis 20(1): 1–10

Rajasingham R, Meya DB, Boulware DR (2012) Integrating cryptococcal antigen screening and pre-emptive treatment into routine HIV care. J Acquir Immune Defic Syndr 59(5):e85–e91

Rajasingham R, Smith RM, Park BJ, Jarvis JN, Govender NP, Chiller TM, Denning DW, Loyse A, Boulware DR (2017) Global burden of disease of HIV-associated cryptococcal meningitis: an updated analysis. Lancet Infect Dis 17(8):873–881

Rajasingham R, Wake RM, Beyene T, Katende A, Letang E, Boulware DR (2019) Cryptococcal meningitis diagnostics and screening in the era of point-of-care laboratory testing. J Clin Microbiol 57(1):e01238–e01218

Retini C, Vecchiarelli A, Monari C, Bistoni F, Kozel TR (1998) Encapsulation of Cryptococcus neoformans with glucuronoxylomannan inhibits the antigen-presenting capacity of monocytes. Infect Immun 66(2):664–669

Ruschel MAP, Thapa B (2022) Cryptococcal meningitis. In: StatPearls. StatPearls Publishing, Treasure Island, FL

Saag MS, Powderly WG, Cloud GA, Robinson P, Grieco MH, Sharkey PK et al (1992) Comparison of Amphotericin B with fluconazole in the treatment of acute aids-associated cryptococcal meningitis. N Engl J Med 326(2):83–89

Satishchandra P, Mathew T, Gadre G, Nagarathna S, Chandramukhi A, Mahadevan A, Shankar SK (2007) Cryptococcal meningitis: clinical, diagnostic and therapeutic overviews. Neurol India 55(3):226

Schwartz S, Kontoyiannis DP, Harrison T, Ruhnke M (2018) Advances in the diagnosis and treatment of fungal infections of the CNS. Lancet Neurol 17(4):362–372

Schwarz P, Dromer F, Lortholary O, Dannaoui E (2006) Efficacy of amphotericin B in combination with flucytosine against flucytosine-susceptible or flucytosine-resistant isolates of *Cryptococcus neoformans* during disseminated murine cryptococcosis. Antimicrob Agents Chemother 50(1): 113–120

Sereti I, Sheikh V, Shaffer D, Phanuphak N, Gabriel E, Wang J, Nason MC, Roby G, Ngeno H, Kirui F, Pau A (2020) Prospective international study of incidence and predictors of immune reconstitution inflammatory syndrome and death in people living with human immunodeficiency virus and severe lymphopenia. Clin Infect Dis 71(3):652–660

Sheehan DJ, Hitchcock CA, Sibley CM (1999) Current and emerging azole antifungal agents. Clin Microbiol Rev 12(1):40–79

Shelburne SA III, Darcourt J, White AC Jr, Greenberg SB, Hamill RJ, Atmar RL, Visnegarwala F (2005) The role of immune reconstitution inflammatory syndrome in AIDS-related

Cryptococcus neoformans disease in the era of highly active antiretroviral therapy. Clin Infect Dis 40(7):1049–1052

Siqueira LPM, Gimenes VMF, de Freitas RS, Melhem MDSC, Bonfietti LX, da Silva Jr AR, Souza Santos LB, Motta AL, Rossi F, Benard G, de Almeida Jr JN (2019) Evaluation of Vitek MS for differentiation of Cryptococcus neoformans and Cryptococcus gattii genotypes. J Clin Microbiol 57(1):e01282–e01218

Tenforde MW, Shapiro AE, Rouse B, Jarvis JN, Li T, Eshun-Wilson I, Ford N (2018) Treatment for HIV-associated cryptococcal meningitis. Cochrane Database Syst Rev 7(7):CD005647. https://doi.org/10.1002/14651858.CD005647.pub3

Temfack E, Rim JJB, Spijker R, Loyse A, Chiller T, Pappas PG, Perfect J, Sorell TC, Harrison TS, Cohen JF, Lortholary O (2021) Cryptococcal antigen in serum and cerebrospinal fluid for detecting cryptococcal meningitis in adults living with human immunodeficiency virus: systematic review and meta-analysis of diagnostic test accuracy studies. Clin Infect Dis 72(7): 1268–1278

Tenforde MW, Mokomane M, Leeme T, Patel RKK, Lekwape N, Ramodimoosi C, Dube B, Williams EA, Mokobela KO, Tawanana E, Pilatwe T (2017) Advanced HIV disease in Botswana following successful antiretroviral therapy rollout: incidence of and temporal trends in cryptococcal meningitis. Clin Infect Dis 65:779

Tintelnot K, Hagen F, Han CO, Seibold M, Rickerts V, Boekhout T (2015) Pitfalls in serological diagnosis of Cryptococcus gattii infections. Med Mycol J 53(8):874–879

Ulett KB, Cockburn JW, Jeffree R, Woods ML (2017) Cerebral cryptococcoma mimicking glioblastoma. Case Rep 2017:bcr2016218824

Van Der Horst CM, Saag MS, Cloud GA, Hamill RJ, Graybill JR, Sobel JD et al (1997) Treatment of cryptococcal meningitis associated with the acquired immunodeficiency syndrome. N Engl J Med 337(1):15–21

Waldorf AR, Polak A (1983) Mechanisms of action of 5-fluorocytosine. Antimicrob Agents Chemother 23(1):79–85

WHO (2022) WHO Guidelines for diagnosing, preventing and managing cryptococcal disease among adults, adolescents and children living with HIV; https://www.who.int/publications/i/item/9789240052178

Zerpa R, Huicho L, Guillen A (1996) Modified India ink preparation for Cryptococcus neoformans in cerebrospinal fluid specimens. J Clin Microbiol 34(9):2290

Zhu LP, Wu JQ, Xu B, Ou XT, Zhang QQ, Weng XH (2010) Cryptococcal meningitis in non-HIV-infected patients in a Chinese tertiary care hospital, 1997–2007. Med Mycol J 48(4):570–579

Mold Infections of the Central Nervous System

14

Sidhya Choudhary, Richa Thakker, and Arghadip Samaddar

Abstract

Filamentous fungi are often identified as deadly human pathogens. They can cause invasive diseases in immunocompromised hosts, especially in those with hematological malignancies and transplant recipients. *Aspergillus* species are the most common fungi linked to invasive mold disease. Other molds of medical importance are *Fusarium*, *Mucorales*, and *Scedosporium*. This chapter exclusively addresses the most commonly encountered hyaline and dematiaceous molds responsible for central nervous system (CNS) mycoses. The primary insult is almost always associated with the paranasal sinuses and/or the lungs, wherefrom the infection disseminates. Hematogenous spread or direct invasion from adjoining sinuses lead to the involvement of the CNS. *Mucor, Rhizopus, Rhizomucor* and other genera under the order *Mucorales* tend to present as rapidly progressive paranasal sinus or rhino-orbito-cerebral infection, mostly in diabetics and high-dose steroid recipients. The diagnosis of CNS mold infections is challenging. While serological markers, such as galactomannan and (1,3)-β-D-Glucan (BDG) are of value in diagnosing many filamentous mold infections, there are exceptions, such as *Mucorales*, which lack BDG entirely or produce it in very low amounts. Culture-based methods still form the cornerstone for diagnosis of CNS mold diseases. In mucormycosis, the presence of broad, aseptate, ribbon-like hyphae with right angle branching in direct microscopy and histopathological

S. Choudhary
Department of Microbiology, Sawai Man Singh Medical College, Jaipur, Rajasthan, India

R. Thakker
Hospital Infection Control, Healthcare Global Cancer Centre, Mumbai, Maharashtra, India

A. Samaddar (✉)
Department of Neuromicrobiology, National Institute of Mental Health and Neuro Sciences (NIMHANS), Bengaluru, Karnataka, India

H. Sami et al. (eds.), *Viral and Fungal Infections of the Central Nervous System: A Microbiological Perspective*, https://doi.org/10.1007/978-981-99-6445-1_14

examination of necrotic tissue is considered diagnostic. Treatment of CNS mold infections encompasses an early aggressive multidisciplinary approach with surgical debridement, antifungal therapy and correction of underlying disease, if any.

Keywords

Molds · Filamentous fungi · Aspergillus · Invasive mold disease · Fusarium · Mucorales · Scedosporium · Non-dematiaceous molds · Central nervous system

14.1 Introduction

Central nervous system (CNS) mold infections are rare but serious conditions that can cause significant morbidity and mortality. These infections are caused by a variety of molds, most commonly, *Aspergillus, Fusarium,* and *Mucorales.* CNS infection caused by dematiaceous or melanized fungus is known as cerebral phaeohyphomycosis, often referred as black molds or phaeoid fungi. Most of the causative agents of cerebral phaeohyphomycosis belong to the order *Chaetothyriales* and include *Cladophialophora bantiana, Rhinocladiella mackenziei, Exophiala dermatitidis,* and *Fonsecaea monophora.* Other less common causes include *Verruconis gallopava, Acrophialophora fusispora, Neoscytalidium dimidiatum, Curvularia* spp., *Exserohilum* spp., *Chaetomium* spp., and *Nodulisporium* sp. *C. bantiana* is the most common (48%) cause of cerebral phaeohyphomycosis worldwide (Revankar and Sutton 2010).

The incidence of CNS mold infections has increased in recent years, primarily due to the growing population of immunocompromised patients. The risk factors for CNS mold infections include HIV infection, organ transplantation, malignancy, chemotherapy, prolonged neutropenia, long term use of systemic steroids, uncontrolled diabetes, severe renal failure, and intravenous drug usage. Additionally, patients are more likely to develop CNS mold infections if they have a history of prior fungal diseases like aspergillosis or mucormycosis. A high level of suspicion is essential for the diagnosis of CNS mold infections, especially in patients with risk factors.

CNS mold infections are more common in developing countries, where access to healthcare and antifungal medications is limited. In these countries, CNS mold infections are often associated with tuberculosis and other opportunistic infections.

14.2 Etiologic Agents

The causative agents of CNS mold infections include hyaline and dematiaceous fungi. Filamentous fungi encompass moniliaceous molds (light-colored) with septate hyphae, such as *Aspergillus* spp., *Fusarium* spp., and *Mucorales* (*Rhizopus, Mucor, Lichtheimia, Syncephalestrum,* and *Rhizomucor*) (Murthy and Sundaram

2014; Bongomin et al. 2017; McCarthy et al. 2014). Melanized fungi, although less common, are also implicated in some cases, albeit rarely and include *Cladophialophora bantiana* (Asian countries, particularly in India), *Rhinocladiella mackenziei* (Middle East), *Exophiala dermatitidis* (East Asia), *Fonsecaea monophora,* and *Verruconis gallopava* (worldwide). These molds can cause a variety of clinical syndromes, which are classified according to the site of infection and the extent of the disease (Table 14.1). A proper identification of the causative agent is crucial for appropriate management.

14.3 Burden of CNS Fungal Infections

CNS fungal infections impose a significant burden on global healthcare systems and public health. These infections are linked to high morbidity and mortality, with outcomes influenced by factors such as the patient's underlying immune status, the severity and duration of the infection, and the specific fungal pathogen involved. The incidence of CNS fungal infections varies across regions and populations studied. Generally, these infections are more prevalent among immunocompromised individuals, including those with HIV, hematologic malignancies, solid organ and hematopoietic stem cell transplantation, and prolonged neutropenia. The occurrence of CNS fungal infections is also on the rise in terminally diseased patients, such as those receiving mechanical ventilation and prolonged ICU stay.

Cerebral aspergillosis is the most common form of CNS mold infection. The increasing use of immunosuppressive medications and the expanding population of immunocompromised hosts in recent years have contributed to the rise in the incidence of cerebral aspergillosis. Mortality rates for cerebral aspergillosis can

Table 14.1 Clinical manifestations of various fungal

Manifestation	Fungal Species
Brain abscess	*Aspergillus* species *Mucormycetes* Dematiaceous fungi *Penicillium* species
Rhino-cerebral mucormycosis	*Rhizopus arrhizus* *Mucor racemosus* *Lichthemia corymbifera* *Cunninghamella* species *Syncephalastrum racemosum*
Meningitis	*Exserohilum rostrum* *Paecilomyces variotii* (CNS shunt infections)
Skull-base syndromes	*Aspergillus* species
Stroke/infarction	*Mucorales* *Aspergillus* species
Disseminated	*Mucorales* *Aspergillus* species

reach as high as 90% in some individuals, making it a substantial concern in both clinical and public health contexts.

Cerebral mucormycosis and disseminated fusariosis are less common than cerebral aspergillosis, but are associated with a higher mortality rate. Cerebral mucormycosis typically affects patients with uncontrolled diabetes, and its incidence is increasing globally due to the rising prevalence of diabetes mellitus and indiscriminate use of high-dose steroid medications.

The impact of CNS fungal infections extends beyond high morbidity and mortality rates. These infections pose a significant economic burden on healthcare systems due to prolonged hospital stays, increased diagnostic testing, and expensive antifungal treatments. Additionally, CNS fungal infections can have long-term sequelae, including cognitive impairment and neurologic deficits.

14.4 Epidemiology

The epidemiology of CNS mold infections displays notable variations worldwide, encompassing differences in incidence rates, risk factors, and the specific causative agents involved. This section aims to offer a comprehensive overview of the epidemiology of CNS mold infections across various regions, drawing upon relevant studies and references.

North America: Invasive mold infections contribute to increased levels of suffering and death among immunosuppressed patients (Patterson et al. 2016). Cerebral aspergillosis is particularly prevalent among individuals with compromised immune system, including those with hematologic malignancies, solid organ or bone marrow transplantation, and prolonged neutropenia. Furthermore, the use of immunosuppressive medications, such as corticosteroids, presents a significant risk factor for cerebral aspergillosis. In North America, the incidence of cerebral aspergillosis appears to be on the rise, likely due to the use of immunosuppressive therapies and the rising number of immunocompromised hosts.

Cerebral mucormycosis is less common than cerebral aspergillosis in North America, but is associated with similar high mortality rates. Malignancy is currently the main risk factor for mucormycosis in developed nations like the USA, whereas in developing countries, uncontrolled diabetes and overuse of steroids remain the predominant causes. (Ruping et al. 2010; Prakash et al. 2018; Chakrabarti et al. 2009).

Disseminated fusariosis is less common than cerebral aspergillosis in North America. It can affect multiple organs, including the CNS, and is associated with prolonged neutropenia and immunosuppression.

Europe: The epidemiology of CNS mold infections in Europe shares similarities with North America. Cerebral aspergillosis stands out as the predominant CNS mold infection, especially among immunocompromised patients. In Europe, mortality rates due to invasive fungal diseases (IFDs) vary depending on several factors, such as the specific pathogen, geographical location, and underlying patient characteristics. The mortality rates for invasive *Aspergillus* infections vary between 38% and 80% (Lass-Flörl 2009).

Though cerebral mucormycosis is less prevalent than cerebral aspergillosis in Europe, an increasing trend is being observed in recent years. The frequency of cerebral mucormycosis is rising globally as a result of rising cases of diabetes, and is particularly prevalent among individuals with uncontrolled diabetes.

Asia: In Asia, the epidemiology of CNS mold infections differs from that in North America and Europe. While cerebral aspergillosis remains the most common CNS mold infection in Asia, its incidence is comparatively lower than in North America and Europe. In Asia, cerebral aspergillosis is particularly prevalent among immunocompromised patients, including those with hematologic malignancies and individuals who have undergone solid organ or bone marrow transplantation.

Cerebral mucormycosis, on the other hand, is more frequently encountered in Asia compared to North America and Europe. This difference is likely attributed to the higher prevalence of uncontrolled diabetes in the region. In the general population, mucormycosis is considered a rare disease with an estimated incidence of 0.005 to 1.7 cases per million individuals (Jeong et al. 2019). Nevertheless, reports indicate that in India, the prevalence of mucormycosis among diabetic patients is approximately 0.14 per 1000, which is significantly higher (80 times) compared to other regions of the world (Chander et al. 2018). This incidence is also higher than the estimated rate in the general population based on computational modeling (Ray et al. 2022).

Cladophialophora bantiana, a dematiaceous fungus with global distribution, is particularly prevalent in Asian countries such as India. It is highly neurotropic and is a major cause of fungal brain abscess both in immunocompetent and immunocompromised individuals. Other CNS mold infections, including disseminated fusariosis, are less frequently reported from Asia.

Africa: In Africa, the epidemiology of CNS mold infections is poorly understood. However, the existing evidence from case reports and small case series highlights a significant occurrence of cerebral aspergillosis and cerebral mucormycosis among immunocompromised patients in the region, including individuals with HIV and hematologic malignancies.

The epidemiology of CNS mold infections, including aspergillosis, mucormycosis, and fusariosis, varies significantly around the globe.

14.4.1 CNS Aspergillosis

Aspergillosis, the most common CNS mold infection globally, has shown an increasing incidence in recent years. The highest rates of cerebral aspergillosis are observed in immunocompromised patients, including those with HIV, hematologic malignancies, and solid organ or bone marrow transplantation. Furthermore, critically ill patients, particularly those in intensive care units and receiving mechanical ventilation, are at higher risk of developing invasive aspergillosis.

The most common invasive mold infection, especially among individuals with hematopoietic stem cell transplants and hematological malignancies, is invasive aspergillosis (Patterson et al. 2016). Nearly 90% of human infections are attributed to *Aspergillus fumigatus*. However, infections caused by non-fumigatus species, such as *A. niger, A. terreus, A. flavus, A. nidulans, A. oryzae, and A. ustus,* are also being observed. Tropical and subtropical regions like North America, the Middle East, and Southeast Asia have higher incidences of non-fumigatus *Aspergillus* spp. (Zanganeh et al. 2018; Lionakis et al. 2005). In immunosuppressed hosts, *Aspergillus fumigatus* is the primary cause of CNS infections, whereas *Aspergillus flavus* is more prevalent in immunocompetent individuals (Candoni et al. 2019).

14.4.2 Rhino-Orbito-Cerebral Mucormycosis

Rhizopus arrhizus is the most common cause of mucormycosis worldwide, while *Lichtheimia* species is predominant in Europe and *Apophysomyces variabilis* in Asia (Roden et al. 2005; Prakash and Chakrabarti 2019). Rhino-orbital cerebral mucormycosis is the commonest clinical manifestation, with *Rhizopus oryzae* being the most common etiologic agent. Uncontrolled diabetes mellitus is the leading risk factor, accounting for majority of cases (36%), followed by hematologic malignancies (17%) and hematopoietic stem cell or solid organ transplantation (12%) (Roden et al. 2005). Environmental factors like high humidity and temperature, may also contribute to the increased incidence of mucormycosis.

The current survival rate of mucormycosis patients without brain involvement ranges from 50% to 80%, which reduces to 20% when brain is affected. The prognosis and fatality associated with rhino-orbito-cerebral mucormycosis depend on several factors, such as the early identification of underlying disease, aggressive surgical debridement, and the timely administration of appropriate antifungal therapy. (Prakash and Chakrabarti 2019; Tooley et al. 2022).

14.4.3 CNS Fusariosis

Despite being less common than aspergillosis and mucormycosis, disseminated fusariosis, nevertheless, is a serious concern due to high morbidity and mortality. The incidence of fusariosis is highest in patients with prolonged neutropenia, T-cell immunodeficiency, hematologic malignancies, and those undergoing immunosuppressive therapy. The infection has a trimodal distribution in the allogeneic hematopoietic stem cell transplantation (HSCT) population with three peaks. The first peak occurs during neutropenia in the initial post-transplant phase. Patients receiving corticosteroids for acute graft-versus-host disease (GvHD) experience the second peak, and the third peak appears during treatment for chronic severe GvHD more than a year following transplant (Nucci et al. 2004). Notably, the main risk factor for fusariosis in these patients is severe T-cell immunodeficiency, not neutropenia. Fusariosis occurs in about 6 cases for every 1000 HSCTs, with autologous recipients experiencing the lowest incidence (1.5 to 2/1000), matched related and unrelated allogeneic recipients reporting intermediate incidence (2.5 to 5/1000), and mismatched related donor allogeneic recipients having the highest incidence (20/1000) (Nucci et al. 2004).

The epidemiology of CNS mold infections varies significantly around the globe, with incidence rates influenced by various factors, including population demographics, underlying medical conditions, environmental factors, and healthcare infrastructure. To improve the outcomes and alleviate the burden of these infections, it is crucial to raise awareness, facilitate early diagnosis, and ensure appropriate treatment.

14.5 Pathogenesis

CNS mold infections are uncommon, yet highly fatal conditions caused by fungi that invade the brain. The pathogenesis of CNS mold infections involves a complex interplay between the fungal virulence factors, host immune response and the structural and functional characteristics of the CNS.

Fungi can enter the CNS through different routes, including direct extension from a nearby infected site, hematogenous dissemination, or accidental inoculation during neurosurgical procedures or trauma. Once inside the CNS, the fungi can cause a range of pathologic changes, including tissue necrosis, inflammation, hemorrhage, and edema. Some fungi, such as *Aspergillus* and *Rhizopus,* can produce angioinvasive hyphae that penetrate blood vessels and cause thrombosis and infarction, leading to ischemic injury and further tissue damage (Fig. 14.1).

The outcome of CNS mold infections greatly relies on the host immune response. Immunosuppressed patients, such as those with HIV/AIDS, organ transplants, or hematologic malignancies, are particularly vulnerable to fungal infections because of their compromised immune function. Conversely, immunocompetent patients possess the ability to mount an effective immune response, which helps contain the infection and facilitate fungal clearance. The immune response to fungi involves innate, as well as adaptive mechanisms, which includes phagocytic cell-activation (neutrophils and macrophages), in addition to the cytokine production and chemokines that recruit and stimulate immune cells.

The pathogenesis of CNS mold infections is complex and involves several steps. These include:

- Direct inoculation or inhalation of fungal spores: The fungi enter the body either through inhalation or by accidental inoculation during surgery or trauma. In certain instances, the infection may originate from a primary site of infection located elsewhere in the body, which then spreads to other areas. There may be contiguous spread from nearby sites such as sinuses, mastoid or orbit. Other sources, such as intravenous drug use, or contaminated medical supplies have been reported in patients infected with *Exophiala dermatitidis* and *Exserohilum rostratum.*

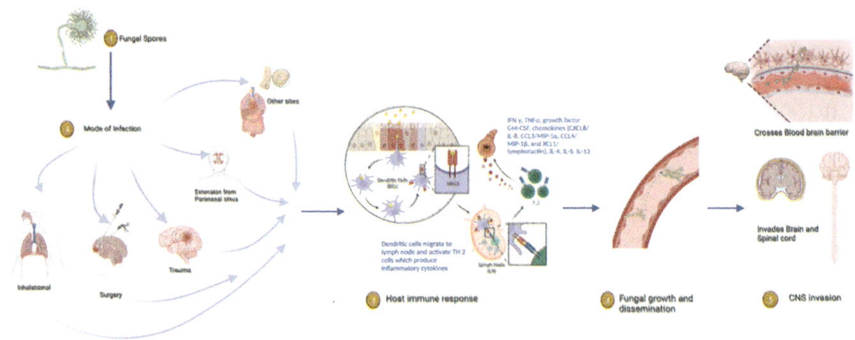

Fig. 14.1 The pathogenesis of CNS mold infection involves a series of events, as depicted in the diagram (1) the portal of entry, typically the respiratory or in few cases extra-pulmonary site, extension from the ear or paranasal sinuses, or direct inoculation through cranial trauma or neurosurgery. (2) The host immune response involving secretion of cytokines and chemokines to clear the fungal infection. (3) Fungal elements multiply and disseminate via the bloodstream. (4) The fungi invade the central nervous system by disrupting the blood-brain barrier. Mycotoxins may contribute to this process, leading to damage to neurons and astrocytes

- Adherence and invasion: The fungal cells adhere to the host tissue and form a biofilm. The biofilm allows the fungi to evade host defenses and invade the host tissue, and also acts as permeability barrier to antifungal drugs.
- Host immune response: The effective containment of fungal infection depends substantially on the host immunological response. However, in immunocompromised patients, the immune response is often inadequate allowing the fungi to multiply and disseminate rapidly.
- Fungal growth and dissemination: Fungal growth can cause tissue damage and may lead to the dissemination of the fungi to other organs, including the CNS.
- CNS invasion: In CNS mold infections, the fungi invade the brain, spinal cord, or meninges. The invasion can lead to inflammation, necrosis, and hemorrhage.

Immunocompromised state causes the BBB to become more permeable, which makes it easier for fungi to invade the brain. Penetration of the fungal pathogen is facilitated but increased permeability of the blood brain barrier and the breach of the barrier occurs by one of the following mechanisms:

Trans-cellular migration,
Para-cellular migration, and
Trojan Horse Mechanism (inside infected phagocytes)

Once inside the brain parenchyma, the pathogens begin to multiply and cause inflammation. Fungal invasion is typically linked to immunocompromised states because the pathogens must breach the effective defenses encircling the brain. Once the BBB, cerebral and subarachnoid spaces are crossed, the CNS is involved by the invading fungus. Hence, the development of CNS fungal infections is influenced by the interaction between fungal cells and nerve cells, as well as the production of immune-suppressing and immune-enhancing chemokines and cytokines (Sharma et al. 2012). The blood-brain barrier, subarachnoid space, and brain parenchyma are affected by fungal invasion (Koutsouras et al. 2017). Various factors, such as surgical intervention, trauma, cytokine release, and microglial activation can disrupt blood-brain barrier, facilitating this process.

14.6 Pathogenesis by Specific Fungi

14.6.1 Aspergillus species

CNS infections due to *Aspergillus* species are intricate and exhibit variations depending on the specific species involved. Risk factors, such as prolonged neutropenia and corticosteroid usage increase the likelihood of *Aspergillus* brain infection. Moreover, individuals with cytomegalovirus infection or cerebral trauma, immunocompromised patients, hematological malignancies and organ transplant recipients are at increased risk of neuroaspergillosis. CARD-9 deficiencies are associated with impaired neutrophil accumulation and predispose to neuroaspergillosis.

CNS infection may occur either through hematogenous dissemination of the fungus from a primary pulmonary focus or through contiguous spread from adjoining regions, such as ear, paranasal sinuses or mastoids (Candoni et al. 2019). Immunocompromised individuals are more prone to develop CNS infection through hematogenous route, whereas direct extension from sinuses, following trauma and neurosurgical procedures are more likely to occur in the immunocompetent (Candoni et al. 2019; Pasqualotto and Denning 2006; Jensen et al. 2010; Gonzales Zamora et al. 2018). The exact mechanism by which *Aspergillus* species cross the blood-brain barrier and enter the CNS is unclear. Studies have demonstrated that mycotoxins produced by *Aspergillus* species, such as gliotoxins and aflatoxins, prevent phagocytosis and decrease opsonization of conidia during invasion (Lewis et al. 2005a, b; Eichner et al. 1986; Murayama et al. 1996). These mycotoxins possess the ability to disrupt the integrity of the blood-brain barrier, leading to neuronal, astrocytic and microglial damage or apoptosis (Patel et al. 2018). *Aspergillus* angioinvasion in the brain can cause hemorrhage, cerebral infarction, meningitis, and mycotic aneurysms in patients with severe immunosuppression, with mortality rates as high as 90% (Economides et al. 2017). *Aspergillus* spp. is the most common cause of intracranial mycotic aneurysms, with the internal carotid artery being the most common site involved. Aneurysm formation is proposed to occur by (a) direct invasion of the arterial wall from within the lumen, (b) invasion of the arterial wall from adjacent structures (sinus, bone, and meninges), (c) immune complex deposition leading to vascular injury, and (d) embolic occlusion of the vasa vasorum. In immunocompetent individuals, granulomas, brain abscess and meningitis may develop with mortality rates ranging from 40–80% (Nadkarni and Goel 2005).

14.6.2 Mucorales

Fungi belonging to the order Mucorales can cause various infections in humans, including CNS infections. The pathogenesis of rhino-orbito-cerebral mucormycosis is complex and is greatly influenced by underlying host factors. Majority (~80%) of the patients have uncontrolled diabetes. Other risk factors include hematological malignancy (13-22%), solid-organ transplantation (7-14%), and chronic kidney disease (~10%) (Sundaram et al. 2005; Abdollahi et al. 2016; Vironneau et al. 2014). During the recent Covid-19 pandemic, India encountered several cases of rhino-orbito-cerebral mucormycosis, predominantly associated with indiscriminate use of corticosteroids, poor glycemic control, COVID-19-associated immunosuppression and blanket use of broad-spectrum antibiotics.

Individuals with uncontrolled diabetes exhibit a sluggish immune response, impaired functioning of neutrophils, monocytes and macrophages, and suppressed inflammatory response. In addition, elevated serum levels of free iron in the background of diabetic ketoacidosis favors fungal growth, leading to angioinvasion and ischaemic tissue necrosis (Koutsouras et al. 2017; Sundaram et al. 2005; Gen et al. 2013).

Mucorales reach the brain through the cerebral vasculature (Bannykh et al. 2018). The internal elastic lamina of the vessel wall is invaded by fungal hyphae, leading to obliteration of vessel lumen, intravascular thrombosis and intimal hyperplasia (Bannykh et al. 2018; Higo et al. 2015). Hemorrhagic necrosis and cerebral infarction are caused by vascular occlusion. In advanced stages, hyphal invasion of the brain parenchyma occurs, often as a preterminal event (Malik et al. 2014). This is attributed to larger morphology of hyphal elements, which prevents access to the meningeal microcirculation.

The outcome of rhinocerebral mucormycosis depends significantly on the host immune status. In immunocompromised patients, the immune response is often insufficient to contain fungal invasion, allowing rapid fungal proliferation, tissue damage and dissemination to the CNS. Giant cell infiltration and the development of granulomas suggest a substantially unaltered immune response, and these features are linked to improved outcomes (Economides et al. 2017).

14.6.3 Fusarium species

Invasive infections caused by *Fusarium* spp. have become more common in immunocompromised individuals in recent years. *Fusarium* is the second leading etiologic agent of invasive mold infections in immunocompromised patients after *Aspergillus* spp. (Peterson et al. 2014; Garcia et al. 2015; Kleinschmidt-Demasters 2009). In patients with prolonged neutropenia, recipients of stem cell transplants, and those with hematological malignancies, *Fusarium solani* is the predominant species responsible for invasive infections (Dignani and Anaissie 2004; Koutsouras et al. 2017; Nucci et al. 2015). Like *Aspergillus* spp., *Fusarium* possesses the ability to produce mycotoxins, such as fumonisins and trichothecenes, which primarily impede cellular and humoral immunity. Fumonisin B1, the predominant mycotoxin produced by Fusarium spp., has been linked to cerebral invasion, resulting in impaired mitochondrial function and axonal degeneration (Bertero et al. 2018).

Cerebral fusariosis is usually a manifestation of disseminated infection (Góralska et al. 2018). Patients may present with single or multiple cerebral abscesses, cutaneous nodules, endophthalmitis, chorioretinitis, meningitis, and fungemia (McCarthy et al. 2014). Among the medically relevant Fusarium species, *F. oxysporum, F. solani,* and *F. moliniforme* complexes are most commonly encountered in clinical settings (Peterson et al. 2014). As for other invasive mold diseases, a definitive diagnosis of CNS fusariosis requires direct microscopy and culture of CSF and brain biopsies to identify the etiologic agent. In some cases, positive blood cultures can aid in the diagnosis, as disseminated fusariosis is often associated with fungemia (Nucci et al. 2015). *Fusarium* spp. can often be isolated from blood cultures in up to 40% of invasive fusariosis cases, with faster detection of growth in fungal blood culture bottles compared with standard aerobic bottles.

14.6.4 Scedosporium species

Lomentospora prolificans (formerly, *Scedosporium prolificans*) is an emerging fungal pathogen that is commonly found in soil, polluted water, and plant sources worldwide. Individuals at a higher risk include those who are immunocompromised or have hematological malignancies (Cortez et al. 2008; Guarro et al. 2006). There is a particular risk of CNS infections with *Scedosporium apiospermum (Pseudallescheria boydii)* in individuals who have experienced near-drowning incidents in water contaminated with fungal propagules (Guarro et al. 2006). The Scedosporium genus consists of three medically significant species: *Scedosporium apiospermum, Scedosporium boydii* (formerly known as *Pseudoallescheria boydii),* and *Scedosporium aurantiacum* (Lackner et al. 2014).

Scedosporium species are among the pathogenic fungi capable of causing infections in both immunocompetent and immunocompromised individuals, acting as primary or opportunistic pathogens (Lackner et al. 2014; Richardson and Lass-Flörl 2008). Immunocompromised patients have been reported to develop CNS infection by *L. prolificans*, which commonly results in rapid hematogenous dissemination with high mortality rate.

Colonization by *Scedosporium* species can progress to invasive and disseminated infections involving the CNS, especially following organ transplantation, resulting in poor outcomes (Husain et al. 2005). The outcome of CNS infections caused by *Scedosporium* species is significantly influenced by the host immune status. In immunocompromised individuals, the immunologic response is often inadequate, allowing rapid fungal multiplication and dissemination to the CNS.

The infection is acquired by inhalation of airborne conidia. In healthy individuals, the inhaled conidia are effectively cleared by mucociliary action of the airways and phagocytic action of alveolar macrophages. However, individuals with one or more defects in the terminal airways and innate immunity, fail to expel the inhaled conidia, leading to colonization and infection. The conidia germinate into hyphal morphotypes, the tissue-invading form of the fungus, which penetrate through the endothelial layer into the vascular lumen, leading to angioinvasion and subsequent hematogenous dissemination to the CNS (Ortoneda et al. 2002; Riddell 4th et al. 2004).

Other potential mediators of pathogenesis are:

1. Extracellular **protease peptidases** produced by *Scedosporium apiospermum and Pseudoallescheria boydii* have been identified as metallo-proteases that depend on zinc and exhibit their highest activity in an acidic pH of 5.5. This specific pH environment may function as a mechanism to evade host effector cells, such as fibronectin-activated monocytes and macrophages. For instance, *S. apiospermum* secretes a 33-kDa extracellular serine protease (Larcher et al. 1996), while *P. boydii* releases two extracellular peptidases, one of 28 kDa and another of 35 kDa (Silva et al. 2006). These peptidases are capable of degrading human fibrinogen, contributing to tissue damage and inflammation. Moreover, the degradation of matrix proteins by these peptidases may facilitate the invasion of fungal cells into adjacent tissues and dissemination throughout the body.
2. **Melanin** is thought to play an important role in the virulence of fungi and the ability to defend against host immune response. It allows fungi to tolerate environmental stress and evade host defenses (Rosas et al. 2002). One of the functions of melanin is to scavenge nitrogen and oxygen free radicals, which are generated by phagocytic cells in the course of the oxidative burst. Additionally, it imparts heat resistance by securing host defence proteins, cross-linking, and protecting the components of the cell wall from hydrolytic enzymes (Dixon et al. 1987; Sutton et al. 1998). These mechanisms help fungi, such as *Scedosporium prolificans*, to evade host immune responses and contribute to the high level of fungemia observed in patients with these infections. Moreover, the ability of the fungus to undergo in vivo adventitious sporulation may further enhance its survival and dissemination within the host.
3. **Siderophore activity:** *S. apiospermum* and *L. prolificans* have been found to exhibit siderophore activity, indicating their dependence on iron. This iron dependency may contribute to their neurotropism, as the central nervous system (CNS) contains more free iron compared to the bloodstream (de Hoog et al. 1994). Multiple pathways allow *S. apiospermum* to enter the central nervous system (CNS), which include:
   (a) direct inoculation resulting from trauma (Pérez et al. 1988),
   (b) hematogenous spread from a primary pulmonary source (such as, inhalation of spores following near-drowning) (Fisher et al. 1982),

(c) introduction of intravenous catheter (Yoo et al. 1985), and,
(d) direct spread from infected paranasal sinuses (Bryan et al. 1980).

14.6.5 Cladophialophora bantiana

The pathogenesis of CNS infections caused by the neurotropic dematiaceous fungus *Cladophialophora bantiana* is not fully understood. However, several factors have been identified that contribute to the pathogenesis of these infections. *C. bantiana* is the most common etiological agent of cerebral phaeohyphomycosis, accounting for nearly 48% of the cases (Góralska et al. 2018). About half of the patients with cerebral phaeohyphomycosis have no underlying risk factors. The infection is more common in men (male : female ratio for *C. bantiana* infection is 3:1 worldwide, whereas India presents a much higher ratio of 14:1).

The virulence of *C. bantiana* involves the ability of the fungus to produce melanin, which can contribute to tissue damage and evasion of host defenses. Melanin is a pigment that protects the fungal cells from phagocytosis, oxidative stress, ultraviolet radiation, and antifungal agents. Additionally, *C. bantiana* secretes extracellular enzymes, such as carbohydrate-active enzymes, proteases and phospholipases, which contribute to tissue damage and facilitate CNS invasion. The host immune status is also critical in determining the outcome of the infection, as immunocompromised patients are more likely to have invasive disease and poor prognosis.

14.6.6 Verruconis gallopava

Verruconis gallopava (formerly, *Ochroconis gallopava*) are another group of neurotropic dematiaceous fungi widely found in the environment and are capable of causing various infections in humans, including pulmonary and CNS diseases. It primarily infects the respiratory system and specifically targets the CNS in solid organ transplant recipients and those with underlying immunodeficiency, such as chronic granulomatous disease (Shoham et al. 2008; Meriden et al. 2012).

14.7 Clinical Presentation

The manifestations of CNS mold infections vary depending on the particular mold species, the location and extent of the lesion and the presence of underlying conditions. However, some common clinical features include:

1. Headache: One of the most commonly observed signs of CNS mold infections, which may be accompanied by fever and other symptoms of a systemic illness.

2. Seizures: Seizures are a common presentation of CNS mold infections and can occur even in the absence of other neurological symptoms.
3. Focal neurologic deficits: Patients may experience focal neurologic deficits, in the form of weakness or paresthesia, visual disturbances, unsteady gait or dysphasia, depending on the location and severity of the infection.
4. Altered mental status: Patients with CNS mold infections may experience confusion, disorientation and altered mental status.
5. Meningismus: It refers to the triad of fever, headache, and nuchal rigidity, which can be present in some cases of CNS mold infections.
6. Cranial nerve involvement: Certain types of CNS mold infections, such as aspergillosis, may involve the cranial nerves, leading to symptoms such as diplopia or facial palsy.

Children with CNS invasive mold infection often present with non-specific symptoms like fever and failure to respond to antibiotic treatment. They may present with cerebritis or brain abscess (Luckowitsch et al. 2021).

The clinical symptoms of CNS mold infections are non-specific and may overlap with other neurological disorders. Therefore, maintaining a high level of suspicion and performing appropriate diagnostic investigations are vital for achieving an accurate diagnosis. The clinical presentation of some important mold infections of the CNS are as follows:

14.7.1 CNS Aspergillosis

Fungal infections of the CNS are often missed as they lack classic remarkable clinical features, and are often erroneously diagnosed as brain tumors, pyogenic abscesses, or tuberculous meningitis. The clinical presentation of cerebral aspergillosis includes recurring headaches, altered mental status, localized neurologic impairments, and loss of consciousness. Immunosuppressed patients are more likely to get CNS infections due to granulocytopenia and defective cellular and humoral immune responses. The most common *Aspergillus* species infecting humans is *Aspergillus fumigatus*. The lungs and maxillary sinuses are the most common sites for primary *Aspergillus* infection. The infection may reach the brain by contiguous spread from the adjecent paranasal sinuses or through the hematogenous route from a primary focus in the lungs. A characteristic feature of neuroaspergillosis is the presence of one or more abscesses, with evidence of angioinvasion and thrombosis on histopathologic examination of brain biopsies. Neuro aspergillosis should be considered in the differential diagnosis for immunocompromised patients who develop new onset focal neurologic deficits and focal seizures, especially those presenting with intracranial space-occupying lesions and characteristic pulmonary infiltrates (McCarthy et al. 2014).

14.7.2 Rhino-Orbito-Cerebral Mucormycosis

The clinical manifestations of mucormycosis are categorized on the basis of site of involvement. The most common clinical presentation is rhino-orbito-cerebral mucormycosis (ROCM) that characteristically involves the nose, paranasal sinuses, orbit, and brain. The characteristic "Red flag" signs of ROCM include:

- Altered sensorium.
- Epistaxis.
- Eyelid/periocular/facial—edema and/or discoloration.
- Facial pain and/or palsy.
- Facial paresthesia.
- Fever.
- Focal seizures.
- Foul smell.
- Nasal discharge—mucoid/purulent/bloody.
- Nasal mucosal erythema/ inflammation/purple or blue discoloration/ulceration/ ischemia/eschar.
- Nasal stuffiness.
- Ptosis, proptosis.
- Regional pain—orbit, paranasal sinus, or dental pain.
- Restriction of ocular motility, diplopia.
- Sudden loss of vision.
- Worsening headache.

Based on the anatomical progression of the disease, ROCM is categorized into four stages:

Stage I: Involvement of nasal mucosa
Stage II: Involvement of paranasal sinuses
Stage III: Involvement of the orbit
Stage IV: Involvement of central nervous system

14.7.3 CNS Fusariosis

The clinical presentation of fusariosis varies depending on the site of infection, and the extent and duration of immune suppression. In comparison to aspergillosis, fusariosis is characterized by a notably high rate of positive blood cultures, primarily observed in cases of disseminated disease. The most prevalent and fatal clinical form of fusariosis in immunocompromised patients is disseminated disease, accounting for over 70% of cases in individuals with protracted neutropenia, acute leukemia, and hematopoietic stem cell transplant recipients (Nucci and Anaissie 2007; Vadhan et al. 2022).

Other mold infections of the central nervous system are mentioned in Table 14.2.

Table 14.2 Other mold infections of the central nervous system (McCarthy et al. 2014)

Mold	Risk factors	Clinical presentation	Imaging findings
Scedosporium apiospermum	Drowning, trauma, neutropenia, stem cell transplant	Focal neurologic deficits, seizures	Cerebral infarction and abscess
Cladophialophora bantiana	Immunocompetent host, trauma	Fever, headache, focal neurologic deficits, seizures	A brain tumor-like growth in isolation and an abscess with a satellite lesion
Verruconis gallopava	Neutropenia, hematopoietic stem cell and solid organ transplant	Focal neurologic deficits	Cerebral infarct, parenchymal hemorrhage, brain abscess
Exserohilum rostratum	Injection of contaminated methylprednisolone	Focal neurologic deficits, meningitis, arachnoiditis	Brain abscess, epidural abscess, mycotic aneurysm, arachnoiditis, and cerebral infarction

14.8 Diagnosis

Diagnosis of CNS mold infections can be challenging due to the non-specific and diverse clinical presentation and limitation of diagnostic tests. Definitive diagnosis requires a combination of clinical, radiological, microbiological, and histopathological findings.

The clinical samples obtained from patients need to be processed promptly, and if immediate processing is not possible, they should be stored at 4–5 °C. It is important to keep biopsy specimens moist and avoid placing them in formalin. Fungal growth obtained in culture should be identified to the species level, as significant variations in antifungal susceptibility exist between different species of fungi. Antifungal susceptibility testing is recommended. Culture is gold-standard but lacks sensitivity and a negative culture report does not exclude fungal infection.

Histopathological analysis of the affected brain tissue demonstrates the presence of hyphal angioinvasion, characterized by the penetration of fungal hyphae into small and large blood vessels, leading to thrombosis. This invasive process is accompanied by hemorrhagic infarction, coagulative necrosis, vasculitis, and the formation of granulomas. Moreover, the susceptibility to angioinvasion increases the likelihood of developing mycotic aneurysms (McCarthy et al. 2014).

In pediatric cases of invasive mold infections, the MRI findings exhibit non-specific characteristics, usually appearing hyperintense with varying enhancement on T2-weighted and FLAIR imaging (Fig. 14.2). It might be challenging to differentiate perilesional edema cerebritis set on by an abscess from hemorrhage in both adults and children (Fig. 14.3). It is extremely unusual for children to develop isolated cerebritis without CNS hemorrhage or an abscess that shows little to no

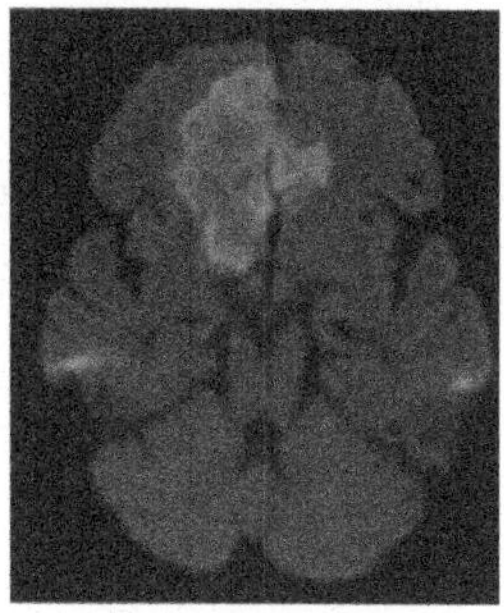

Fig. 14.2 Diffusion-weighted image (DWI) showing restricted diffusion in the region of frontal cerebritis (Luckowitsch et al. 2021)

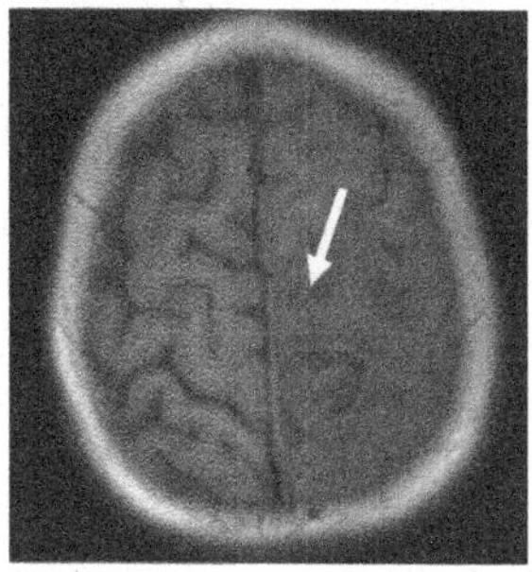

Fig. 14.3 Axial T2-weighted MRI showing a mass with low signal intensity in the left central region, surrounded by hyperintense perilesional edema and a well-defined ring-like hypointense area (arrow) (Luckowitsch et al. 2021)

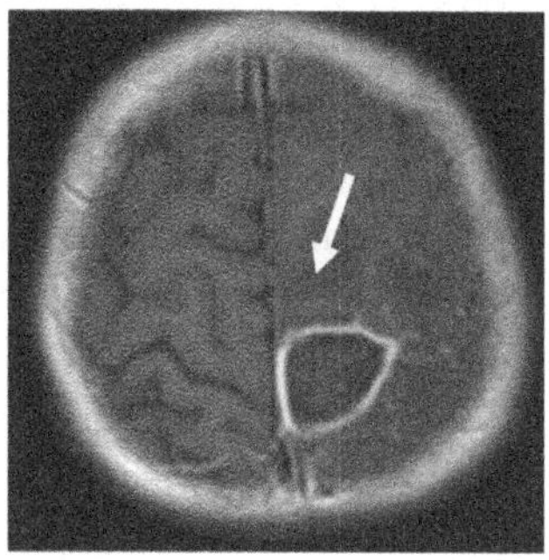

Fig. 14.4 T1-weighted image showing a robust ring enhancement (arrow) (Luckowitsch et al. 2021)

enhancement. Differentiating fungal abscesses from pyogenic and tubercular abscesses based on MRI features is difficult as they exhibit hypointense ring enhancement and restricted diffusion in T2-weighted images.

Fungal abscesses, unlike bacterial abscesses, tend to be multiple, affecting deep gray matter nuclei and are situated close to the grey-white matter junction, resulting from the hematogenous spread of the fungal pathogen (Figs. 14.4 and 14.5) (Luckowitsch et al. 2021). The diagnostic approaches are summarized in Table 14.3.

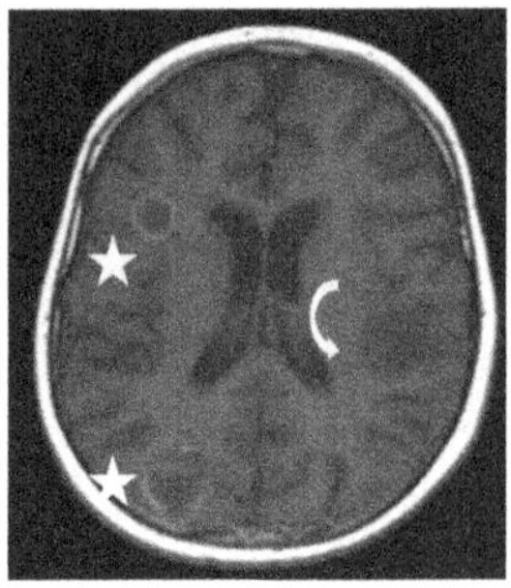

Fig. 14.5 Enhanced T1-weighted image showing multiple ring-enhancing fungal abscesses on the right (indicated by star), while no enhancement is observed on the left side (Luckowitsch et al. 2021)

14.8.1 CNS Aspergillosis

Diagnosis of CNS aspergillosis is challenging and is often a complex task. Aspergillosis is diagnosed on direct examination by Calcofluor-potassium hydroxide stain demonstrating thin, hyaline, septate hyphae, along with fungal growth in culture. The diagnostic modalities include:

Culture: The gold standard for diagnosis of CNS aspergillosis is growth of Aspergillus species in CSF culture. However, it has low sensitivity and takes a long time to provide results.

Histopathology: Histopathology of brain tissue can be diagnostic, but it requires invasive procedures, which can be risky in some patients.

Imaging studies: Magnetic resonance imaging (MRI) and computed tomography (CT) scan of brain, with or without intravenous contrast, can be used for detecting the size, location and extent of CNS lesion(s) (Fig. 14.6), but they are non-specific and should always be correlated with direct microscopy, histopathology and culture results.

PCR assay: The utilization of Aspergillus PCR testing on blood samples has the potential to assist in diagnosing or ruling out invasive aspergillosis (IA). For neutropenic patients who are at a high risk of developing IA and are not getting mold-active prophylaxis, this test is particularly valuable as a screening tool, given its high negative predictive value. (Egger et al. 2020).

Detection of biomarkers: Detection of *Aspergillus*-specific biomarkers, such as galactomannan antigen and (1,3)-β-D-Glucan in CSF or serum can provide early and specific diagnosis of CNS aspergillosis.

14.8.2 Rhino-Orbito-Cerebral Mucormycosis

Direct Microscopy: Direct microscopy is of utmost importance for an evidence-based diagnosis and also, for deciding the significance of a positive culture. The sensitivity of direct microscopy can be enhanced by employing optical brighteners like Calcofluor white. The presence of hyaline, broad (6–16 μm), non-septate

Table 14.3 Diagnosis of CNS mold infections

		Molds				
S. No.	Investigations	Aspergillus	Mucorales	Fusarium	Scedosporium	Cladophialophora
1	Direct microscopy	Hyaline, septate hyphae, dichotomous branching at acute angles	Hyaline, broad, aseptate, ribbon-like hyphae, branching at right angles	Hyaline multicellular macroconidia that are hyaline and banana-shaped and contain a foot cell at their base	Irregular branching hyphal patterns, may contain terminal or intercalary chlamydospores, which can sometimes be mistaken for yeasts	The yeast-like forms of Scedosporium are typically lightly pigmented and can appear as short chains of spores and hyphae
2	Culture	Gold standard, low sensitivity	Gold standard, low sensitivity, positive only in 50% of direct microscopy positive cases	Do not require the presence of cycloheximide. Blood cultures often yield growth and are easily identified from the sickle or canoe shaped macroconidia.	CSF culture for *Scedosporium* infection might exhibit delayed growth or yield negative results. Optimal growth temperature is 25–35 °C, while some strains can grow at 42 °C High rate of positive blood cultures	The colonies exhibit an olive-gray to brown appearance with a slightly folded olivaceous-black underside and grow moderately. They thrive at 40°C. the conidia are pale olivaceous, ellipsoidal to spindle-shaped, forming strongly coherent chains with infrequent branching from undifferentiated hyphae
3	Histopathology	Diagnostic, but risky	High accuracy with immunohistochemical staining	Adventitious sporulation in tissue may indicate	Brain abscesses may involve angioinvasion and	Long, fragile, branched chains of hydrophobic conidia

(continued)

Table 14.3 (continued)

S. No.	Investigations	Molds				
		Aspergillus	Mucorales	Fusarium	Scedosporium	Cladophialophora
				fusariosis in individuals at high risk, particularly when hyphae and yeast-like structures are detected together	tissue invasion. Meningitis may present with purulent CSF, increased white blood cell count (predominantly neutrophils), and multinucleate giant cells	are present, and yeast cells are absent
4	Imaging	Nonspecific; discrete ring-enhancing lesions on CT scan	Thrombosis of the cavernous sinus, brain infarction, and occlusion of the internal carotid artery	Cerebral cortex is most frequently involved	Multiple ring enhancing lesions with perilesional edema and regions of cerebritis.	Nonspecific; unilateral, circumscribed mass lesions in the cerebral cortex are common, may manifest as purulent meningitis and brain abscess
5	PCR assay	High negative predictive value (may be positive in CSF of patients)	Primers specific to Mucorales, such as CotH gives accurate diagnosis	CSF PCR can provide early and specific diagnosis	Not an alternative to clinical suspicion and traditional diagnostic procedures.	NA
6	Fungal markers	Galactomannan antigen and (1,3)-β-D-Glucan in CSF or serum	NA	Negative galactomannan and positive (1,3)-β-D-Glucan test	NA	NA

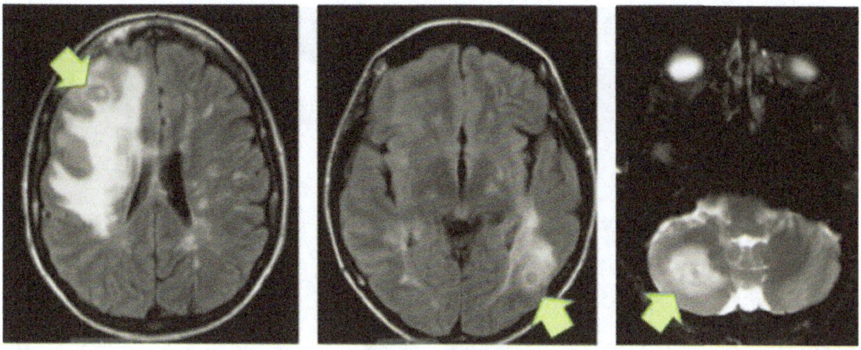

Fig. 14.6 MRI brain showing hyperintense T2 and FLAIR signal in the cerebral white matter and deep gray nuclei, with discrete ring enhancing lesions and perilesional edema in the left occipital lobe and right cerebellum (arrows). FLAIR, Fluid attenuated inversion recovery (Miceli 2019)

(or pauci-septate) haphae, having branching at right angles is suggestive of *Mucorales* infection.

Culture: Culture of Mucorales from biopsies is the gold standard for diagnosis and, in addition, enables fungal speciation and antifungal susceptibility testing. However, culture lacks sensitivity and is positive only in 50% of the microscopy positive cases.

Histopathology: When mucormycosis is suspected, it is important to avoid homogenization of tissue specimens as nonseptate *Mucorales* hyphae have a high risk of being damaged by shear stress. When compared to other fungi, *Mucorales* may require longer staining durations to be detected in histopathological investigation. Grocott Gomori methenamine silver (GMS) and periodic acid Schiff (PAS) stains are quite effective in delineating fungal hyphae and can aid in the detection of *Mucorales*. Using immunohistochemical staining on tissue samples has demonstrated high levels of accuracy in detecting *Mucorales* species, which improves the ability to distinguish between mucormycosis and aspergillosis based on morphology.

Due to the potential for neurological complications associated with brain biopsy, CNS mucormycosis diagnosis is frequently made by determining the primary focus in the lungs or sinuses. Patients who have symptoms suggestive of rhino-orbito-cerebral mucormycosis should be assessed by an experienced ENT surgeon. Fiberoptic endoscopy of the nasal cavity and paranasal sinuses can help in detection of necrotic or ischemic lesions that may require biopsy for further evaluation.

Imaging studies: Imaging modalities, such as MRI and CT can be employed to identify brain lesions (Fig. 14.7), which may exhibit nonspecific nodular mucosal thickening attributing to small blood vessel occlusion, leading to mucosal ischemia.

Cavernous sinus thrombosis, internal carotid artery blockage, and brain infarction are the most frequent imaging findings in intracranial mucormycosis. Brain imaging on T2-weighted sequences typically reveals involvement of the lower regions of the

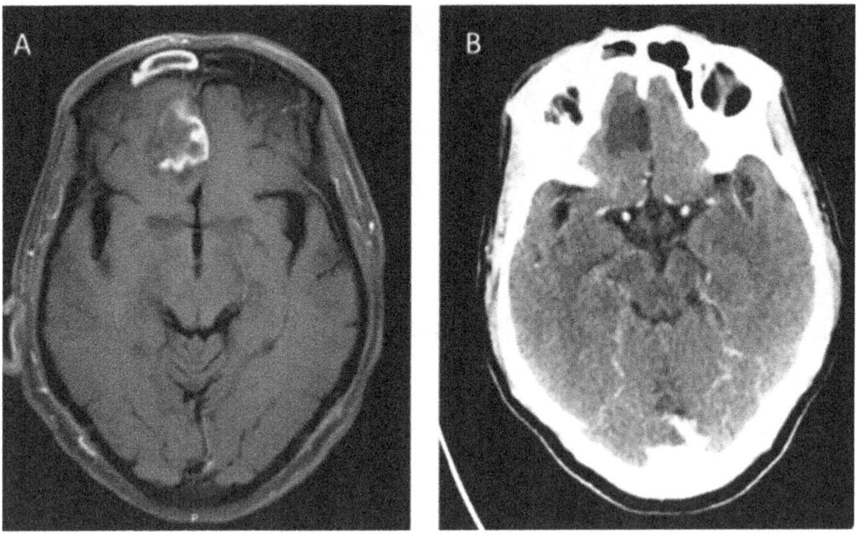

Fig. 14.7 T1-weighted MRI (**a**) and CT scan (**b**) of brain showing mucormycosis with cerebral extension, involving the right frontal lobe (Chikley et al. 2019)

frontal lobes, with the lesions appearing either hyperintense or hypointense. Diffusion-weighted imaging (DWI) demonstrates notable restriction of diffusion. The potential utility of magnetic resonance spectroscopy in differentiating bacterial causes of cerebritis from CNS mucormycosis has been proposed. However, more research is necessary to confirm or validate this finding. Cavernous sinus syndrome can also be caused by metastatic cancer or Tolosa-Hunt syndrome, and therefore, these conditions should be considered as potential differential diagnosis.

PCR assay: Early and specific diagnosis of CNS mucormycosis can be achieved by detecting fungal DNA in CSF through PCR. PCR-based tests utilizing universal fungal primers or primers specific to *Mucorales*, such as CotH, offer promising results in accurately detecting the infection. Unfortunately, there are presently no biomarkers for screening or diagnosing mucormycosis available for clinical use (Dadwal and Kontoyiannis 2018).

14.8.3 CNS Fusariosis

CNS fusariosis can be diagnosed using different methods, including culture, histopathology, imaging studies, and PCR-based detection of fungal DNA. When a hyaline mold is identified from the blood culture of a patient with prolonged neutropenia, it strongly indicates the possibility of *Fusarium* infection.

Culture: The gold standard for diagnosis is CSF culture, but it has low sensitivity. *Fusarium* species tend to grow rapidlyin a broad range of media and typically do not require the presence of cycloheximide. The fungus can be recognized by its

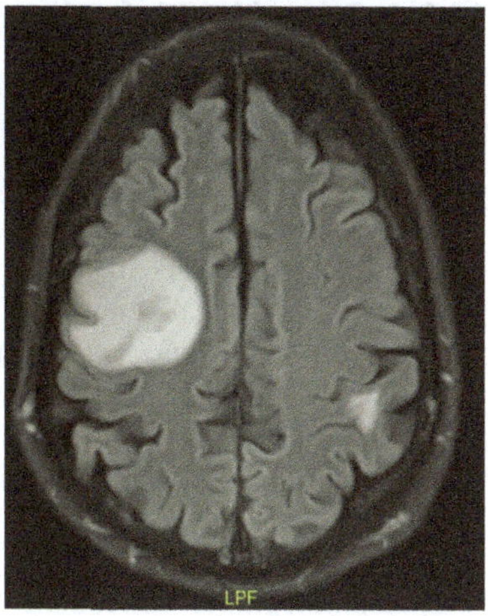

Fig. 14.8 MRI brain showing enhancing intraparenchymal lesion in the right frontoparietal region with mild diffusion restriction and surrounding hyperintense FLAIR, suggesting vasogenic edema (Vadhan et al. 2022)

ability to generate multicellular macroconidia that are hyaline and banana-shaped and contain a foot cell at their base. Severely immunocompromised patients with disseminated fusariosis may exhibit skin lesions, such as cellulitis or metastatic lesions, along with positive blood cultures for the mold. Unlike aspergillosis, fusariosis frequently results in positive blood cultures. This happens because Fusarium species create yeast-like structures which help them spread and thrive in the blood, a process known as adventitious sporulation (Liu et al. 1998).

Histopathology: Histopathology of brain tissue may help in establishing a confirmatory diagnosis of fusariosis. In the absence of fungal growth, it may be difficult to differentiate fusariosis from other types of hyalohyphomycosis. In such cases, the diagnosis of fusariosis can be achieved by in situ hybridization in paraffin-embedded tissue specimens. In tissue, the hyphae of *Fusarium* species resemble those of *Aspergillus*, with hyaline septate filaments having dichotomous acute or right angle branching. However, the presence of adventitious sporulation within tissue and the simultaneous presence of both hyphae and yeast-like structures are highly suggestive of fusariosis in high-risk individuals.

Imaging studies: CT scan and MRI are employed to identify brain lesions, although their findings are not specific to a particular condition. The cerebral cortex is the most commonly affected region within the brain (Fig. 14.8), but intracranial fusariosis can involve any part of the brain, without predilection to any particular region.

Detection of fungal DNA: PCR-based detection of fungal DNA in CSF and brain tissues enables early and precise diagnosis of *Fusarium* infection.

Detection of biomarkers: Among at-risk individuals with mold infections, positive (1,3)-β-D-Glucan and negative galactomannan in CSF is strongly suggestive of CNS fusariosis. However, the (1,3)-β-D-Glucan test cannot distinguish between *Fusarium* and other fungal infections. Hence, biomarker tests should be correlated with clinical, radiological and culture results.

14.8.4 Other Mold Infections

- **Scedosporiosis**: The diagnosis of CNS scedosporiosis is routinely done through a combination of imaging studies, CSF analysis, and fungal cultures. Intracranial lesions revealed through imaging studies like CT or MRI can suggest invasive scedosporiosis, while CSF analysis can show elevated protein levels, low glucose levels, and fungal elements. The presence of *Scedosporium* species can be confirmed by fungal culture of CSF or tissue samples. Unlike *S. apiospermum*, *Lomentospora prolificans* (formerly, *S. prolificans*) may be recovered from blood cultures. Molecular methods like PCR or sequencing can provide rapid and accurate identification. Diagnosis can also be achieved by directly observing the fungal elements in histological samples or isolating it in cultures. Though real-time PCR-based assays can detect Scedosporiosis from clinical samples quickly and accurately, culture remains the most useful tool for diagnosis.
- ***Cladophialophora* infection:** Diagnosing CNS infections caused by *Cladophialophora bantiana* can be difficult since these infections are uncommon and may show nonspecific symptoms. Imaging techniques such as MRI or CT may show intracranial space-occupying lesions, mimicking tuberculosis or malignancy. Fungal culture of CSF or brain biopsies can confirm the presence of *C. bantiana*. However, culture can be time-consuming and pose safety risks for laboratory workers. Molecular techniques, such as PCR or sequencing can provide rapid and accurate identification of *C. bantiana*. While culture is still considered as the gold standard for diagnosing CNS fungal infections, in recent years, advanced techniques, such as real-time PCR, amplified fragment length polymorphism analysis, matrix-assisted laser desorption/ionization-time of flight, and rolling circle amplification have emerged as alternative methods.

14.9 Management

There are four critical components in the management of CNS mold infections: prompt diagnosis, treatment with antifungal medication, assessment and intervention by healthcare providers, and addressing underlying immunodeficiency. Failure to address these components adequately can result in unfavorable outcomes, particularly in immunocompromised individuals who may exhibit solitary brain abscesses resembling CNS tumors (McCarthy et al. 2014). The approach to management may vary depending on the specific mold involved.

The traditional approach for the treatment of CNS mold infections includes:

14.9.1 Antifungal Therapy

Antifungal therapy is the cornerstone of treatment for CNS mold infections. The choice of antifungal drug depends on the type of mold causing the infection, the patient's medical history and overall health status. Voriconazole is the recommended first-line treatment for CNS aspergillosis due to its extensive distribution in the cerebrospinal fluid (Walsh et al. 2008; Ullmann et al. 2018). Therapeutic drug monitoring is crucial for optimizing its efficacy and safety. Voriconazole, fluconazole, and flucytosine have good CNS penetration, but fluconazole and flucytosine have a limited spectrum of activity. Voriconazole is the preferred treatment option for CNS aspergillosis and may also be considered for CNS infections caused by other fungi that demonstrate susceptibility, including *Lomentospora* and *Scedosporium* species.

Isavuconazole is a viable option for treating invasive aspergillosis, including cerebral aspergillosis, as it exhibits excellent CNS penetration and distributes effectively in infected brain tissue. Clinical evidence indicates that isavuconazole demonstrates satisfactory efficacy against invasive aspergillosis and disseminated mucormycosis, particularly when they affect the CNS. Itraconazole and posaconazole have limited CNS penetration, but can be considered for patients who do not tolerate or respond to voriconazole. Echinocandins are not advised for CNS aspergillosis due to limited distribution, although there have been cases where caspofungin and micafungin have been promising.

Amphotericin B liposomal formulations are better tolerated and show promising results in CNS aspergillosis and mucormycosis, while conventional amphotericin B has limited CNS penetration and associated toxicities.

Echinocandins have a higher molecular mass (1140 to 1292 daltons) and therefore, have poor penetration of blood–CSF and blood–brain barrier. In comparison, voriconazole is a relatively small (349-dalton), moderately lipophilic molecule with a CSF : plasma concentration ratio of 1:2. For CNS aspergillosis, voriconazole is recommended as primary therapy, with a serum trough concentration of 2–5 μg/ml, while liposomal amphotericin B (L-AmB) is reserved for intolerant or refractory cases. Studies have shown that use of voriconazole improves survival rates to about 35%. Isavuconazole has been proved in recent trials to be non-inferior to voriconazole. For patients with severe disease, addition of echinocandin is recommended. For CNS fusariosis, pending susceptibility testing results, voriconazole plus a lipid formulation of amphotericin B is advised.

14.9.2 Surgery

Surgical intervention may be necessary to remove infected tissues and relieve pressure on the brain caused by swelling or abscesses. The decision to perform surgery is made on a case-by-case basis and depends on the severity and location of the infection. Whenever feasible, it is recommended to consider surgical excision of lesions and infected adjoining structures. Several studies have indicated that surgical debridement combined with early effective antifungal therapy may improve survival rates.

Table 14.4 Treatment of CNS mold infections

Mold	Antifungal drug therapy		Surgery	Comments
	Adults	Children		
Aspergillus	Voriconazole: A maintenance dose of 4 mg/kg IV every 12 h after a loading dose of 6 mg/kg given intravenously every 12 h. Alternatively, for patients intolerant or resistant to voriconazole, liposomal amphotericin B is administered at a dose of 5–7.5 mg/kg/day IV	Voriconazole: A maintenance dose of 8 mg/kg IV is given after a loading dose of 9 mg/kg IV every 12 h Alternatively, for patients intolerant or resistant to voriconazole, liposomal amphotericin B is administered at a dose of 5–7.5 mg/kg/day IV	It is also an option when systemic therapy is ineffective against the suspected organism. Surgical debridement enhances antifungal penetration in Aspergillus brain abscess. Complete surgical debridement can be curative. Stereotactic aspiration is preferred for most abscesses, aiding diagnosis, relieving intracranial pressure, and improving antifungal drug efficacy. Nadkarni and Goel (2005)	In adults, treatment is continued until clinical improvement and radiological resolution of abnormalities, while in pediatric patients, treatment duration is determined on an individual basis. Gradual reduction of steroid dosage is likely to improve the outcome. The decision to undergo neurosurgical intervention should be made after careful consideration of each individual case
Mucorales	The recommended intravenous doses of liposomal amphotericin B range from 5 to 7.5 mg/kg/day, while amphotericin B lipid complex is given at 5 mg/kg/day	For liposomal amphotericin B, the intravenous dosage is 5–7.5 mg/kg/day, while for amphotericin B lipid complex, it is 5 mg/kg/day	Early surgical debridement alongside antifungal therapy improves clinical outcomes in CNS mucormycosis. Endoscopic sinus debridement is recommended for early-stage infections outside the brain, while open surgery is preferred for advanced cases.	The duration of antifungal treatment should be personalized, typically lasting for at least 6–8 weeks (28 days) in adults. In children, the duration is determined on a case-by-case basis. The need for neurosurgical intervention should be assessed

			Neurosurgery may be necessary in CNS infections involving increased intracranial pressure, obstructive hydrocephalus, or spinal cord compression, such as hemispheric stroke. Chikley et al. (2019)	individually. Hyperbaric oxygen therapy may be considered for ROCM
***Fusarium* spp.**	The first dose of voriconazole is 6 mg/kg IV every 12 h, and the maintenance dose is 4 mg/kg IV every 12 h. Amphotericin B lipid complex is administered at a dose of 5 mg/kg/day IV and liposomal amphotericin B at a dose of 5 mg/kg/day IV	A loading dose of 9 mg/kg IV of voriconazole is given every 12 h, followed by a maintenance dose of 8 mg/kg IV every 12 h. Liposomal amphotericin B is given at a dosage of 5–7.5 mg/kg/day IV, and amphotericin B lipid complex at 5 mg/kg/day IV	Surgical debridement of infected tissue can be considered in cases of focal disease, such as *fusarium* brain abscess	For neutropenic patients, growth factors such as G-CSF or GM-CSF, as well as granulocyte transfusions may be employed. If neutrophil counts are adequate, gamma interferon and/or GM-CSF can be considered as treatment options. Nucci and Anaissie (2007)
***Scedosporium*spp.**	For voriconazole, the initial dose is 6 mg/kg IV every 12 h on the first day, followed by a maintenance dose of 4 mg/kg IV every 12 h, until symptoms resolve	In adolescents aged 13–18 years, voriconazole is administered at a dosage of 4 mg/kg IV every 12 h. In children aged 2–12 years, the dosage is 8 mg/kg IV every 12 h	Surgical resection is preferred for focal diseases like brain abscess due to its resistance to currently available antifungal agents	Treatment for CNS infections caused by Scedosporium species lacks a universally accepted protocol.
Dematiaceous mold (*Cladophialophora bantiana, Verruconis gallopava, Exserohilum rostratum*)	Start with an IV loading dosage of 6 mg/kg of voriconazole every 12 h, then administer a maintenance dose of 4 mg/kg every 12 h	Voriconazole is administered IV at a loading dose of 9 mg/kg every 12 h, followed by a maintenance dose of 8 mg/kg IV every 12 h.	Antifungal therapy alone has not been proved effective. Surgical debridement along with combined antifungal therapy is crucial for	There are no established interpretive breakpoints for in vitro antifungal susceptibility tests for CNS infections due to *Verruconis*

(continued)

Table 14.4 (continued)

Mold	Antifungal drug therapy		Surgery	Comments
	Adults	Children		
	In cases of severe or refractory infection, supplement with liposomal amphotericin B at a dosage of 5 mg/kg/day IV	For severe or refractory infections, add liposomal amphotericin B in a dose of 5 mg/kg/day IV	successful outcomes, particularly in cases involving *Cladophialophora bantiana*	*gallopava*. Spinal epidural abscess decompression can be considered as an additional therapy in case of *Exserohilum rostratum*

14.9.3 Adjunctive Therapies

Reversal of neutropenia and underlying immunodeficiency, glycemic control and discontinuing glucocorticoids should be employed.

14.9.4 Supportive Care

Patients with CNS mold infections may require supportive care to manage symptoms such as fever, headache, and seizures. This may include medications to control pain and reduce inflammation, as well as anticonvulsants to prevent seizures.

14.9.5 Monitoring

Patients with CNS mold infections require close monitoring to track their response to treatment and to watch for any complications. This may include regular imaging studies, such as CT or MRI scans to evaluate the progression of the infection and the effectiveness of treatment.

It is important to note that the management of CNS mold infections is complex and requires a multidisciplinary approach involving infectious disease specialists, neurologists, neurosurgeons, and critical care physicians. The treatment plan should be tailored based-on the patient's needs and should be regularly re-evaluated to assess the response to therapy. Fungal neuroinfections exhibit a dismal prognosis and higher mortality rate compared to viral, bacterial and parasitic infections. An early diagnosis and aggressive treatment are essential for improving the outcomes. The recommended treatment guidelines for CNS mold infections are stated in Table 14.4 (Góralska et al. 2018; McCarthy et al. 2014; Miceli 2019).

14.10 Conclusion

CNS mold infections present significant challenges in diagnosis and treatment, with high morbidity and mortality rates. CNS mold infections exhibit variable presentations, which make them difficult to differentiate for other common space occupying lesions, such as malignancy or tuberculoma, causing delay in appropriate treatment. Fungal culture is specific but not sensitive, and a negative result does not exclude fungal infection. Fungal biomarkers have the potential to be an effective diagnostic tool; however, standardization is necessary. The prognosis is favorable in patients when surgical debridement is combined with at least one antifungal medication. For neuroaspergillosis, voriconazole is the preferred medication. It is also the preferred drug for infections caused by dematiaceous molds, as it offers broad-spectrum activity and has excellent CNS penetration. Moreover, as immunodeficiency is the predominant risk factor, the management of these conditions should have a strong emphasis on reversal of immune suppression. An increased awareness with a high index of suspicion may help in establishing a timely diagnosis and initiating appropriate treatment, which can improve the survival in such cases.

References

Abdollahi A, Shokohi T, Amirrajab N et al (2016) Clinical features, diagnosis, and outcomes of rhino-orbito-cerebral mucormycosis—a retrospective analysis. Curr Med Mycol 2:15–23

Bannykh SI, Hunt B, Moser F (2018) Intra-arterial spread of Mucormycetes mediates early ischemic necrosis of brain and suggests new venues for prophylactic therapy. Neuropathology 38:539–541

Bertero A, Spicer LJ, Caloni F (2018) Fusarium mycotoxins and in vitro species-specific approach with porcine intestinal and brain in vitro barriers: a review. Food Chem Toxicol 121:666–675

Bongomin F, Gago S, Oladele RO, Denning DW (2017) Global and multi-national prevalence of fungal diseases—estimate precision. J Fungi 3:57

Bryan CS, DiSalvo AF, Kaufman L, Kaplan W, Brill AH, Abbott DC (1980) Petriellidium boydii infection of the sphenoid sinus. Am J Clin Pathol 74:846–851

Candoni A, Klimko N, Busca A et al (2019) Fungal infections of the central nervous system and paranasal sinuses in onco-haematologic patients. Epidemiological study reporting the diagnostic-therapeutic approach and outcome in 89 cases. Mycoses 62:252–260

Chakrabarti A, Chatterjee SS, Das A et al (2009) Invasive zygomycosis in India: experience in a tertiary care hospital. Postgrad Med J 85:573–581

Chander J, Kaur M, Singla N et al (2018) Mucormycosis: battle with the deadly enemy over a five-year period in India. J Fungi 4:46

Chikley A, Ben-Ami R, Kontoyiannis DP (2019) Mucormycosis of the central nervous system. J Fungi 5:59

Cortez KJ, Roilides E, Quiroz-Telles F et al (2008) Infections caused by Scedosporium spp. Clin Microbiol Rev 21:157–197

Dadwal SS, Kontoyiannis DP (2018) Recent advances in the molecular diagnosis of mucormycosis. Expert Rev Mol Diagn 18:845–854

de Hoog GS, Marvin-Sikkema FD, Lahpoor GA, Gottschall JC, Prins RA, Guého E (1994) Ecology and physiology of the emerging opportunistic fungi Pseudallescheria boydii and Scedosporium prolificans. Mycoses 37:71–78

Dignani MC, Anaissie E (2004) Human fusariosis. Clin Microbiol Infect 10:67–75

Dixon DM, Polak A, Szaniszlo PJ (1987) Pathogenicity and virulence of wild-type and melanin-deficient Wangiella dermatitidis. J Med Vet Mycol 25:97–106

Economides MP, Ballester LY, Kumar VA et al (2017) Invasive mold infections of the central nervous system in patients with hematologic cancer or stem cell transplantation (2000–2016): uncommon, with improved survival but still deadly often. J Infect 75:572–580

Egger M, Jenks JD, Hoenigl M, Prattes J (2020) Blood aspergillus PCR: the good, the bad, and the ugly. J Fungi 6:18

Eichner RD, Al Salami M, Wood PR, Müllbacher A (1986) The effect of gliotoxin upon macrophage function. Int J Immunopharmacol 8:789–797

Fisher JF, Shadomy S, Teabeaut JR, Woodward J, Michaels GE, Newman MA, White E, Cook P, Seagraves A, Yaghmai F, Rissing JP (1982) Near-drowning complicated by brain abscess due to Petriellidium boydii. Arch Neurol 39:511–513

Garcia RR, Min Z, Narasimhan S, Bhanot N (2015) Fusarium brain abscess: case report and literature review. Mycoses 58:22–26

Gen R, Horasan ES, Vaytsoglu Y, Arpaci RB, Esoz G, Ozcan C (2013) Rhino-orbito-cerebral mucormycosis in patients with diabetic ketoacidosis. J Craniofac Surg 24:e144–e147

Gonzales Zamora JA, Henry Z, Gultekin SH (2018) Central nervous system aspergillosis: an unexpected complication following neurosurgery. Diseases 6:46

Góralska K, Blaszkowska J, Dzikowiec M (2018) Neuroinfections caused by fungi. Infection 46: 443–459
Guarro J, Kantarcioglu AS, Horré R et al (2006) Scedosporium apiospermum: changing clinical spectrum of a therapy-refractory opportunist. Med Mycol 44:295–327
Higo T, Kobayashi T, Yamazaki S et al (2015) Cerebral embolism through hematogenous dissemination of pulmonary mucormycosis complicating relapsed leukemia. Int J Clin Exp Pathol 8: 13639–13642
Husain S, Muñoz P, Forrest G et al (2005) Infections due to Scedosporium apiospermum and Scedosporium prolificans in transplant recipients: clinical characteristics and impact of antifungal agent therapy on outcome. Clin Infect Dis 40:89–99
Jensen J, Guinea J, Torres-Narbona M, Muñoz P, Peláez T, Bouza E (2010) Post-surgical invasive aspergillosis: an uncommon and under-appreciated entity. J Infect 60:162–167
Jeong W, Keighley C, Wolfe R et al (2019) The epidemiology and clinical manifestations of mucormycosis: a systematic review and meta-analysis of case reports. Clin Microbiol Infect 25: 26–34
Kleinschmidt-Demasters BK (2009) Disseminated fusarium infection with brain abscesses in a lung transplant recipient. Clin Neuropathol 28:417–421
Koutsouras GW, Ramos RL, Martinez LR (2017) Role of microglia in fungal infections of the central nervous system. Virulence 8:705–718
Lackner M, de Hoog GS, Yang L et al (2014) Proposed nomenclature for *Pseudallescheria, Scedosporium* and related genera. Fungal Divers 67:1–10
Larcher G, Cimon B, Symoens F, Tronchin G, Chabasse D, Bouchara JP (1996) A 33 kDa serine proteinase from Scedosporium apiospermum. Biochem J 315:119–126
Lass-Flörl C (2009) The changing face of epidemiology of invasive fungal disease in Europe. Mycoses 52:197–205
Lewis RE, Wiederhold NP, Chi J et al (2005a) Detection of gliotoxin in experimental and human aspergillosis. Infect Immun 73:635–637
Lewis RE, Wiederhold NP, Lionakis MS, Prince RA, Kontoyiannis DP (2005b) Frequency and species distribution of gliotoxin-producing aspergillus isolates recovered from patients at a tertiary-care cancer center. J Clin Microbiol 43:6120–6122
Lionakis MS, Lewis RE, Torres HA, Albert ND, Raad II, Kontoyiannis DP (2005) Increased frequency of non-fumigatus aspergillus species in amphotericin B- or triazole-pre-exposed cancer patients with positive cultures for aspergilli. Diagn Microbiol Infect Dis 52:15–20
Liu K, Howell DN, Perfect JR, Schell WA (1998) Morphologic criteria for the preliminary identification of fusarium, Paecilomyces, and Acremonium species by histopathology. Am J Clin Pathol 109:45–54
Luckowitsch M, Rudolph H, Bochennek K, Porto L, Lehrnbecher T (2021) Central nervous system mold infections in children with hematological malignancies: advances in diagnosis and treatment. J Fungi 7:168
Malik AN, Bi WL, McCray B, Abedalthagafi M, Vaitkevicius H, Dunn IF (2014) Isolated cerebral mucormycosis of the basal ganglia. Clin Neurol Neurosurg 124:102–105
McCarthy M, Rosengart A, Schuetz AN, Kontoyiannis DP, Walsh TJ (2014) Mold infections of the central nervous system. N Engl J Med 371:150–160
Meriden Z, Marr KA, Lederman HM et al (2012) Ochroconis gallopava infection in a patient with chronic granulomatous disease: case report and review of the literature. Med Mycol 50:883–889
Miceli MH (2019) Central nervous system infections due to aspergillus and other hyaline molds. J Fungi 5:79
Murayama T, Amitani R, Ikegami Y, Nawada R, Lee WJ, Kuze F (1996) Suppressive effects of aspergillus fumigatus culture filtrates on human alveolar macrophages and polymorphonuclear leucocytes. Eur Respir J 9:293–300
Murthy JMK, Sundaram C (2014) Fungal infections of the central nervous system. In: Biller J, Ferro JM (eds) Handbook of clinical neurology. Elsevier, New York, NY, pp 1383–1401
Nadkarni T, Goel A (2005) Aspergilloma of the brain: an overview. J Postgrad Med 51:S37–S41

Nucci M, Anaissie E (2007) Fusarium infections in immunocompromised patients. Clin Microbiol Rev 20(4):695–704. https://doi.org/10.1128/CMR.00014-07; PMID: 17934079; PMCID: PMC2176050
Nucci M, Marr KA, Queiroz-Telles F et al (2004) Fusarium infection in hematopoietic stem cell transplant recipients. Clin Infect Dis 38:1237–1242
Nucci F, Nouér SA, Capone D, Anaissie E, Nucci M (2015) Fusariosis. Sem Res Crit Care Med 36: 706–714
Ortoneda M, Pastor FJ, Mayayo E, Guarro J (2002) Comparison of the virulence of Scedosporium prolificans strains from different origins in a murine model. J Med Microbiol 51:924–928
Pasqualotto AC, Denning DW (2006) Post-operative aspergillosis. Clin Microbiol Infect 12:1060–1076
Patel R, Hossain MA, German N, Al-Ahmad AJ (2018) Gliotoxin penetrates and impairs the integrity of the human blood-brain barrier in vitro. Mycotoxin Res 34:257–268
Patterson TF, Thompson GR 3rd, Denning DW et al (2016) Practice guidelines for the diagnosis and management of Aspergillosis: 2016 update by the Infectious Diseases Society of America. Clin Infect Dis 63:e1–e60
Pérez RE, Smith M, McClendon J, Kim J, Eugenio N (1988) Pseudallescheria boydii brain abscess. Complication of an intravenous catheter. Am J Med 84:359–362
Peterson A, Pham MH, Lee B et al (2014) Intracranial fusarium fungal abscess in an immunocompetent patient: case report and review of the literature. J Neurol Surg Rep 75:e241–e245
Prakash H, Chakrabarti A (2019) Global epidemiology of mucormycosis. J Fungi 5:26
Prakash H, Ghosh AK, Rudramurthy SM et al (2018) A prospective multicenter study on mucormycosis in India: epidemiology, diagnosis, and treatment. Med Mycol 57:395–402
Ray A, Aayilliath KA, Banerjee S, Chakrabarti A, Denning DW (2022) Burden of serious fungal infections in India. Open Forum Infect Dis 9:ofac603
Revankar SG, Sutton DA (2010) Melanized fungi in human disease. Clin Microbiol Rev 23:884–928
Richardson M, Lass-Flörl C (2008) Changing epidemiology of systemic fungal infections. Clin Microbiol Infect 14(Suppl 4):5–24
Riddell J 4th, Chenoweth CE, Kauffman CA (2004) Disseminated Scedosporium apiospermum infection in a previously healthy woman with HELLP syndrome. Mycoses 47:442–446
Roden MM, Zaoutis TE, Buchanan WL et al (2005) Epidemiology and outcome of zygomycosis: a review of 929 reported cases. Clin Infect Dis 41:634–653
Rosas AL, MacGill RS, Nosanchuk JD, Kozel TR, Casadevall A (2002) Activation of the alternative complement pathway by fungal melanins. Clin Diagn Lab Immunol 9:144–148
Ruping MJ, Heinz WJ, Kindo AJ et al (2010) Forty-one recent cases of invasive zygomycosis from a global clinical registry. J Antimicrob Chemother 65:296–302
Sharma RR, Pawar SJ, Lad SD, Mishra GP, Netalkar AS, Rege S (2012) Fungal infections of the central nervous system. In: Schmidek and sweet: operative neurosurgical techniques, pp 1691–1732
Shoham S, Pic-Aluas L, Taylor J et al (2008) Transplant-associated Ochroconis gallopava infections. Transpl Infect Dis 10:442–448
Silva BA, Pinto MR, Soares RM, Barreto-Bergter E, Santos AL (2006) Pseudallescheria boydii releases metallopeptidases capable of cleaving several proteinaceous compounds. Res Microbiol 157:425–432
Sundaram C, Mahadevan A, Laxmi V et al (2005) Cerebral zygomycosis. Mycoses 48:396–407
Sutton DA, Fothergill AW, Rinaldi MG (1998) Guide to clinically significant fungi, 1st edn. Williams & Wilkins, Baltimore, MD
Tooley AA, Bradley EA, Woog JJ (2022) Rhino-orbital-cerebral Mucormycosis-another deadly complication of COVID-19 infection. JAMA Ophthalmol 140:73–74
Ullmann AJ, Aguado JM, Arikan-Akdagli S et al (2018) Diagnosis and management of aspergillus diseases: executive summary of the 2017 ESCMID-ECMM-ERS guideline. Clin Microbiol Infect 24:e1–e38

Vadhan JD, Melo AJ, Shogan JC, Singh V, Carrillo M (2022) Fast and Fusariosis: a systematic review and case report of a rapidly fatal central nervous system infection. J Emerg Crit Care Med 6:24

Vironneau P, Kania R, Morizot G, Elie C, Garcia-Hermoso D, Herman P et al (2014) Local control of rhino-orbito-cerebral mucor- mycosis dramatically impacts survival. Clin Microbiol Infect 20:O336–O339

Walsh TJ, Anaissie EJ, Denning DW et al (2008) Treatment of aspergillosis: clinical practice guidelines of the Infectious Diseases Society of America. Clin Infect Dis 46:327–360

Yoo D, Lee WH, Kwon-Chung KJ (1985) Brain abscesses due to Pseudallescheria boydii associated with primary non-Hodgkin's lymphoma of the central nervous system: a case report and literature review. Rev Infect Dis 7:272–277

Zanganeh E, Zarrinfar H, Rezaeetalab F et al (2018) Predominance of non-fumigatus aspergillus species among patients suspected to pulmonary aspergillosis in a tropical and subtropical region of the middle east. Microb Pathog 116:296–300

Role of Dimorphic Fungi in CNS Infections 15

Shariq Ahmed, Nazish Fatima, and Uzma Tayyaba

Abstract

Dimorphic fungi, which exist in yeast form at 37 °C and hyphal at 22–25 °C, primarily cause pulmonary infections except for *Sporothrix schenkii* which infects the skin. Central Nervous system infections occur when there is dissemination of the disease mainly in immunosuppressed individuals. These organisms are generally found in warm and temperate regions of the world and mostly cause Endemic mycosis restricted to respective geographical areas, exception to this is *Sporothrix schenkii* species, which is distributed worldwide. This chapter describes CNS infections caused by *Blastomyces dermatitidis*, *Histoplasma capsulatum, Paracoccidioides brasiliensis/lutzii*, *Coccidioides immitis/posadasii, Talaromyces marneffei*, and the *Sporothrix* species complex. Clinical manifestations of CNS infection vary among different dimorphic fungi. Apart from common presentation of meningitis (acute and chronic), *Histoplasma* may present with encephalitis, myelopathy, solitary cerebral or spinal cord mass lesions, hydrocephalous, and stroke. Blastomycosis CNS manifestations are isolated chronic meningitis, epidural or brain abscess, or space-occupying lesions. CNS dissemination of *Coccidioides* presents with coccidioidal meningitis, which is more common in immunocompromised individuals. However, *Paracoccidioides* commonly causes intracerebral mass-like lesions, than meningitis. *Talaromyces marneffei* (previously known as *Penicillium marneffei*) rarely cause CNS infection. Most reported cases are part of widespread dissemination seen in advanced HIV infection. CNS Sporotrichosis presents either as isolated chronic meningitis, where no other organ is involved, seen primarily in immunocompetent individuals, or as a part of widespread dissemination in immunocompromised individuals.

S. Ahmed (✉) · N. Fatima · U. Tayyaba
Department of Microbiology, JN Medical College, AMU, Aligarh, Uttar Pradesh, India

H. Sami et al. (eds.), *Viral and Fungal Infections of the Central Nervous System: A Microbiological Perspective*, https://doi.org/10.1007/978-981-99-6445-1_15

Keywords

Dimorphic fungi · Histoplasmosis · Blastomycosis · Coccidioidomycosis · Paracoccidioidomycosis · Talaromycosis · Sporotrichosis

15.1 Dimorphic Fungal Infection

CNS fungal infections have risen globally over the past few decades. The etiologic factors of neuroinfections are yeasts [*Cryptococcus neoformans, Candida* spp., *Trichosporon* spp.], moniliaceous moulds [*Aspergillus* spp., *Fusarium* spp.], Mucoromycetes [*Mucor* spp., *Rhizopus* spp.], dimorphic fungi [*Blastomyces dermatitidis, Coccidioides* spp., *Histoplasma capsulatum*], and dematiaceous fungi [*Cladophialophora bantiana, Exophiala dermatitidis*] (Góralska et al. 2018).

Dimorphic fungi are organisms that are present in a hyphal form in the environment, but are able to survive in the human host by changing their morphology in which temperature is the main host signal (Sil and Andrianopoulos 2015). The conversion is between hyphal form at 22–25 °C and yeast form at 37 °C (Gauthier 2017). The three major phyla of fungi: the Zygomycota, Basidomycota, and Ascomycota, all have dimorphic fungi, However, few of them are major human pathogens, namely *Blastomyces dermatiditis*, *Coccidioides immitis/posadasii, Histoplasma capsulatum*, *Paracoccidioides brasiliensis/lutzii*, *Sporothrix schenckii*, and *Talaromyces marneffei*. Among these (*Blastomyces, Histoplasma, Coccidioides*, and *Paracoccidioides*) are generally geographically restricted to specific endemic areas and significant contributors to CNS infections by dimorphic fungi (Góralska et al. 2018).

These thermally dimorphic fungi can infect both immunosuppressed and immunocompetent individuals. The morphological switch to yeast form inside human body is essential for virulence. The host immune defence mechanism is better averted by dimorphic fungi in yeast form. Although temperature is the predominant stimulus that influences the phase transition—hyphae at 22–25 °C and yeast at 37 °C, additional stimuli that impact the dimorphic switch include carbon dioxide (CO_2) tension, exogenous cysteine, and estradiol.

15.1.1 Dimorphic Fungi Involving the CNS

Dimorphic fungi (*Blastomyces, Histoplasma, Coccidioides, Paracoccidioides, Sporothrix schenkii*) primarily cause pulmonary disease, but when there is dissemination to other organ via hematogenous or other routes it infects the Central Nervous System. Since they occur in yeast form in human body due to their smaller size, they enter capillaries and subarachnoid spaces causing meningitis and subpial ischaemic lesions (Shankar et al. 2007).

These organisms are generally found in warm, temperate regions across the globe and are mostly restricted to these areas. Mode of infection is ingestion or inhalation of microconidia where they penetrate the oral nasal and pharyngeal mucosa giving rise to local lesions and then spreading to various other organs like lungs, reticulo-endothelial organs, and rarely to CNS. Lungs are primarily infected by inhalation of microconidia which is engulfed by alveolar macrophages. The primary lesion found in lymph node or lung is usually self-limiting and the CNS involvement is apparently seen after 1–3 months of primary infection (Shankar et al. 2007).

15.2 Histoplasmosis

In 1906, a pathologist in Panama named Samuel Taylor Darling first described Histoplasmosis. He believed it to be a protozoal disease caused by an encapsulated *Plasmodium* (Darling 1906). To honour the discoverer, it was also called *Darling's disease*. Caver's and Miner's disease, Bat disease, Cave disease, Tingo Maria fever, Ohio Valley disease, cytomycosis, and reticuloendotheliosis are some of the other names this disease is addressed locally.

The causative agent of the disease is *Histoplasma capsulatum. Histoplasma* can infect both immunocompromised and immunocompetent individuals, making it a true fungal pathogen; however, severe and disseminated disease is more common in patients having impaired immunity. Most of the incidence is reported from Ohio, Mississippi, and St. Lawrence rivers of USA and Latin America. Histoplasmosis is not uncommon in India also; cases are reported endemically from Assam, Delhi, Haryana, West Bengal and Uttar Pradesh (Agrawal et al. 2020; De and Nath 2015).

There are three varieties of *Histoplasma capsulatum,* namely *Histoplasma capsulatum* var. *capsulatum, Histoplasma capsulatum* var. *duboisii, and Histoplasma capsulatum* var. *farciminosum*. The disease is widespread throughout Europe, North Africa, India, and South Asia. Geographical distribution and clinical manifestations of these varieties also differ; *Histoplasma capsulatum* var. *capsulatum* causes classic histoplasmosis predominantly found in central and south-eastern parts of United States including some parts of Latin America, *Histoplasma capsulatum* var. *duboisii* causing African histoplasmosis and is endemic in the tropical areas of Africa, and *Histoplasma. capsulatum* var. *farciminosum* causing epizootic histoplasmosis/epizootic lymphangitis, i.e. lesions of the skin or inflammation of lymphatic system in horses and mules (Kasuga et al. 1999).

Soil is the source of infection in human beings. Histoplasma has been isolated from moist soils having high nitrogen content from droppings of bat and birds usually in caves or river banks area. Birds and bats themselves don't get infected due to higher body temperature than humans; however, Histoplasma persist in soil for years, and on drying, reaches human on inhalation of these dried soil particles or bird/bat droppings. After inhalation of aerosolized mycelia of the Histoplasma, it reaches the lung and converts into yeast form which is phagocytosed by alveolar macrophages. From here, it is subsequently disseminated intracellularly to other body organs including CNS (Wheat et al. 2018). Histoplasmosis has been one of the common AIDS-defining illnesses. When CD4 count is under 200 per mm^3, more so when it reaches below 50 per mm^3, disseminated histoplasmosis is seen (Nacher et al. 2014). An estimated 5–10% of cases of progressive disseminated histoplasmosis will have CNS infection (Wheat et al. 1990).

15.2.1 Pathogenesis

Clinical disease of the *Histoplasma* infections is due to the complex interplay of the CD8$^+$ T lymphocytes and different subpopulations of CD4$^+$ T lymphocytes (Th1, Th2, Th17, and Treg) mediated by the virulence factors of the intracellular yeast forms. Cytokines produced by respective T lymphocytes can be protective and pro-inflammatory response mediated by Th1 lymphocytes, responsible for the production of pro-inflammatory cytokines, such as: IL-2, IL-12, IL-18, IFN-γ, TNF-α, CCL2, CCL3, CCL5, and CXCL8 and non-protective immune response mediated by Th2 lymphocytes that produce the anti-inflammatory cytokines IL-4, IL-5, IL-6, IL-10, IL-13, and TGF-β can antagonize the pro-inflammatory response. Disruption of balance between these two types of cytokines leads to disease progression due to ineffective host defence. If the pathogen clearance is hampered, yeast cells survive and proliferate inside the macrophages constantly that give rise to uncontrolled stimulation of host immune response causing more tissue damage than protection. Granuloma formations primarily play a protective role by inhibiting the spread of the disease, but it may weaken the host defence thereby preventing the optimal elimination of the pathogen. Pathogenesis of CNS histoplasmosis is not well defined and only few studies are available that report that histoplasma causes damage to the meninges in the superficial area, in brain tissue it presents as small brain infarct or focal lesions (DuBois et al. 2016; Kauffman 2007; Nguyen et al. 2013; Ramírez et al. 2023).

15.2.2 Clinical Presentation

Risk Factors—Endemic areas like Ohio and Mississippi River Valleys of United states (Bahr et al. 2015). Other endemic areas, namely Argentina (Mochi and Edwards 1952), Brazil (Mochi and Edwards 1952), India (De and Nath 2015), Western and Central Africa (Gugnani and Muotoe-Okafor 1997), and Europe (Antinori et al. 2021). Histoplasma can infect both immunocompetent and immunosuppressed hosts (Wheat et al. 2018); however, risk of complications, disseminated disease, and CNS involvement is more in HIV-AIDS, transplant recipients, and patients on immunosuppressive drugs like corticosteroids (Cuellar-Rodriguez et al. 2009; Hajjeh et al. 2001; McKinsey and McKinsey 2011; Smith and Kauffman 2009).

Signs and Symptoms—Histoplasmosis presents as Acute pulmonary histoplasmosis, chronic pulmonary histoplasmosis, cutaneous mucocutaneous histoplasmosis, and disseminated histoplasmosis. CNS manifestations are in disseminated histoplasmosis accounting to about 5–10% of the total cases (Wheat et al. 2018).

Reported rate of CNS involvement is quite variable, from no involvement of CNS in 27 disseminated histoplasmosis cases and 12.7% from a Venezuelan study to 62.5% CNS involvement in 40 infants (Mata-Essayag et al. 2008; McLeod et al. 2011; Odio et al. 1999). A recent review of 83 paediatric histoplasmosis has shown 10.4% CNS involvement: 5% in immunocompromised and 13% in immunocompetent (MacInnes and Warris 2021).

The clinical findings of CNS histoplasmosis are often subtle, and not necessarily characteristic of an infectious process. It is often misdiagnosed due to the wide spectrum of neurological presentation; such as meningitis (acute and chronic) which is the most common presentation, encephalitis, myelopathy, solitary cerebral or spinal cord mass lesions, hydrocephalous, and stroke (Gonzalez et al. 2019; Khalaf et al. 2022; Nguyen et al. 2013; Wheat et al. 2005).

Reported symptoms in CNS histoplasmosis are fever, headache, weakness, confusion, lethargy, focal neurological deficits, altered mental status, seizure, and hydrocephalous (Riddell and Wheat 2019). However, there is slight difference in presenting symptoms among immunocompetent and immunocompromised patients, while headache, confusion, and ataxia were more common in former; these are less common in the latter (Schestatsky et al. 2006; Wheat et al. 1990, 2018). Common signs that can be elicited in CNS histoplasmosis are altered mental status focal deficit, neck rigidity, visual impairment, and seizure (Wheat et al. 2018). An unusual presentation is hemichorea (Estrada-Bellmann et al. 2016). In a case series, it was reported to be misdiagnosed as Creutzfeldt-Jakob disease, but found to be histoplasmosis of CNS in autopsy (Batra et al. 2016). Wheat et al. depicted that duration of CNS symptoms is also quite variable, about 35% have it for weeks or less and around 12% have it last more than 26 weeks (Wheat et al. 2018).

15.2.3 Diagnosis

The diagnosis of CNS histoplasmosis is often delayed and challenging. The main cause of such delay is due to diversity of physical manifestations and failure to suspect fungal aetiology, especially Histoplasma infection. Other factors responsible are low sensitivity of gold standard tests, false positives of serological tests, and cross-reactivity with other fungal diseases (Wheat et al. 2018). The delay is usually more than 4 weeks and sometimes it is only found on autopsy that the patient was suffering from CNS histoplasmosis (Myint et al. 2014; Wheat et al. 2018). A multicentric retrospective study has reported that only 28% of diagnosis was made within 2 weeks of presentation and it took more than 26 weeks to establish the diagnosis in 27% patients; also, it was more [50%] in immunocompetent than immunocompromised individuals (Wheat et al. 2018). CNS Histoplasmosis is often misdiagnosed as similarly presenting diseases like toxoplasmosis, tuberculosis, neurosarcoidosis, malignancy, multiple sclerosis, and Behcet's syndrome.

Multimodal approach including radiodiagnosis, histopathological, mycological culture and serological, and other laboratory tests is required to establish a confirmatory diagnosis. Presence of multiple ring enhancing lesions throughout brain and spinal cord is the primary radiographic finding on magnetic resonance imaging (Gonzalez et al. 2019; Kauffman 2007). James Riddle IV et al. reported mass lesion [20%], followed by ventricular enlargement [9%], meningeal enhancement [8%], T2 flare signal [7%], infarcts [3%], other findings in 8%, and multiple abnormalities in 17% of patients (Riddell and Wheat 2019).

Laboratory diagnosis—Laboratory diagnostic modalities for meningitis in Histoplasmosis or other CNS manifestations include culture of CSF of about 10 mL, incubated for 35 days, 3 blood cultures, and serological testing including various histoplasma antigen and antibody. If the test reveals an unequivocal result, then repeat testing at least once is recommended. Biopsy of meninges or focal lesions for culture and histopathology being invasive is used as last resort (Wheat et al. 2005). It has been reported by some authors where the final diagnosis was established only on autopsy (Batra et al. 2016). Chronic meningitis due to *Histoplasma capsulatum* can be a diagnostic dilemma, therefore a delay of 8–24 weeks in diagnosis is not infrequent (Schestatsky et al. 2006; Wheat et al. 2018). Delays ranging in years (4–10) have also been reported (Berger and Greenberg 2010; Hamada and Tsuji 2009; Rivera et al. 1992).

CSF/brain tissue culture has been the gold standard confirmatory diagnostic test owing to its very high specificity and positive predictive value, but its major limitations being a very low sensitivity. Also taking 2–6 weeks for growth to occur delays the diagnosis and hampers the effective treatment selection (Bloch et al. 2018). Reported sensitivity of CSF culture is from 27% to 65%; however, in the most recent large case series, it was only 19% (Riddell and Wheat 2019; Wheat et al. 2005, 2018).

Other CSF findings are increased white blood cell count, low glucose level, and elevated proteins. However, these findings are not always present resulting in failure to suspect CNS histoplasmosis as shown by a case series by Wheat et al. They reported low glucose (<40 mg/mL) in 53%, leucocytosis (>5 cell/μL) in 66%, protein (>50 mg/mL) in 77% patients only; 17% of the patients had normal CSF cell count, glucose, and protein (Wheat et al. 2018).

The clinical sample like CSF/brain tissue is inoculated on 2 sets of Sabouraud's dextrose agar (SDA) media with or without antibiotics and incubated at 25 and 37 °C. Growth takes about 6–12 weeks; at 25 °C, initially the colonies are smooth, which become filamentous, cottony, and turn brownish with time. Hyaline septate hyphae with macro and micro-conidias are seen under microscope. Colonies at 37 °C are smooth white to brown in colour which on microscopic examination are small, oval yeasts of size around 2–4 μm. However, to confirm the diagnosis of *H. capsulatum*, conversion from mycelial to yeast form is necessary. Enriched media like Brain-heart infusion agar (BHI) is used to achieve conversion incubated at 35–37 °C. Some special media like BHI with 6% blood, Kelley's agar, and Pine's media can be used to demonstrate better conversion (Eissenberg and Goldman 1991; Guimarães et al. 2006) (Figs. 15.1, 15.2, and 15.3).

In severely ill patients, histopathological examination of the biopsied brain lesions needs to be done as repeated CSF evaluation is time-consuming and the former provides definitive diagnosis more quickly; other indications for

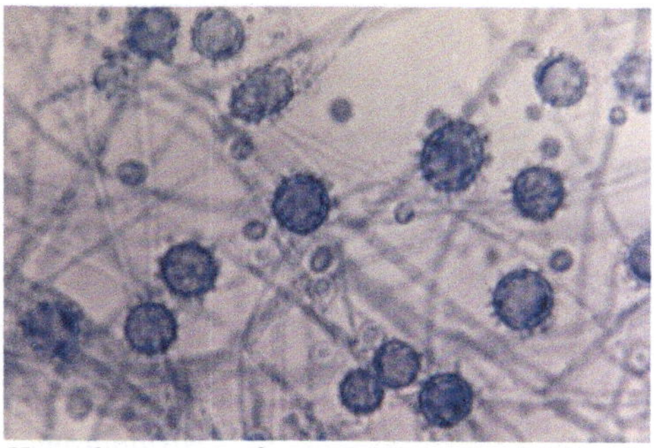

Fig. 15.1 *Histoplasma capsulatum*; LCB mount showing numerous mycelial filaments, tuberculated, thick-walled, round macroconidia and tiny, ellipsoidal microconidia (Magnification ×1200). (***Image Source*** *– Centers for Disease Control and Prevention – Public Health Image Library – ID#22300* (Georg 1968); *ID#22970* (Hardin 1965a, b) *– Copyright Restrictions: NONE; these images are in the public domain and thus free of any copyright restriction*)

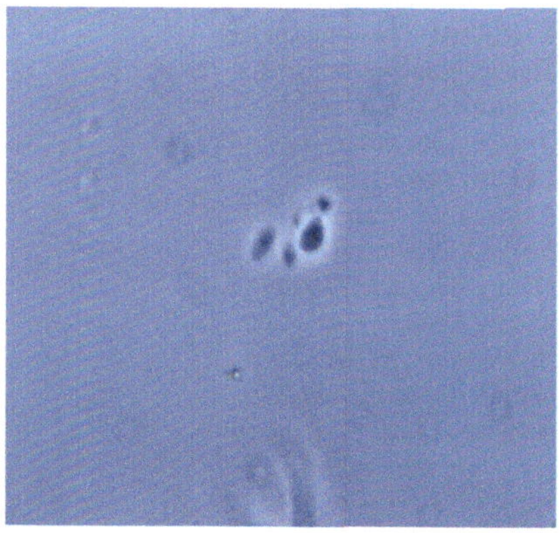

Fig. 15.2 *Histoplasma capsulatum;* budding yeast form cells (Magnification ×900). (***Image Source*** *– Centers for Disease Control and Prevention – Public Health Image Library – ID#22300* (Georg 1968); *ID#22970* (Hardin 1965a, b) *– Copyright Restrictions: NONE; these images are in the public domain and thus free of any copyright restriction*)

histopathology are when obtaining CSF is not possible and all the other diagnostic tests have not provided conclusive results (Wheat et al. 2018). Tissue sections should be stained using PAS (Periodic Acid Schiff), Grocott's methenamine silver (GMS), or Gram stain. A study reported that 8 of 10 patients with diagnosed CNS histoplasmosis had positive CSF Giemsa staining (Mata-Essayag et al. 2008) (Fig. 15.4).

Non-culture-based diagnosis of histoplasmosis—Non-culture-based methods of diagnosis include antigen, antibody testing, and PCR. Detecting circulating anti-*Histoplasma* antibodies and antigen in the CSF provides a rapid diagnosis with good

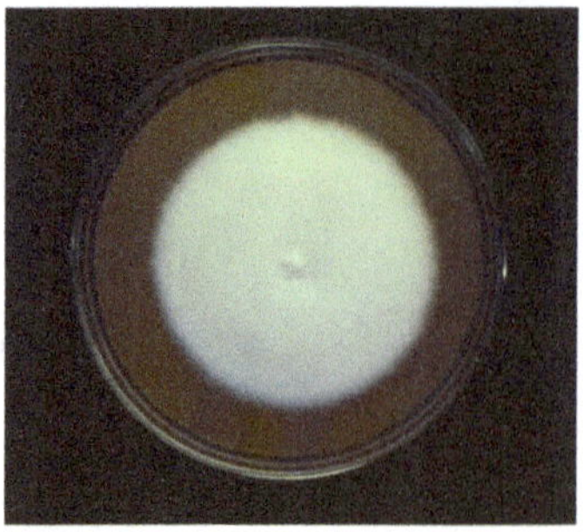

Fig. 15.3 *Histoplasma capsulatum*; growth on Sabouraud dextrose agar showing large white, smoothly textured colony. (***Image Source*** *– Centers for Disease Control and Prevention – Public Health Image Library – ID#22054* (Georg 1964a, b); *ID#21832* (Georg 1968) *– Copyright Restrictions: NONE; these images are in the public domain and thus free of any copyright restriction*)

Histoplasma capsulatum; Growth on Sabouraud dextrose agar showing large white, smoothly-textured colony.

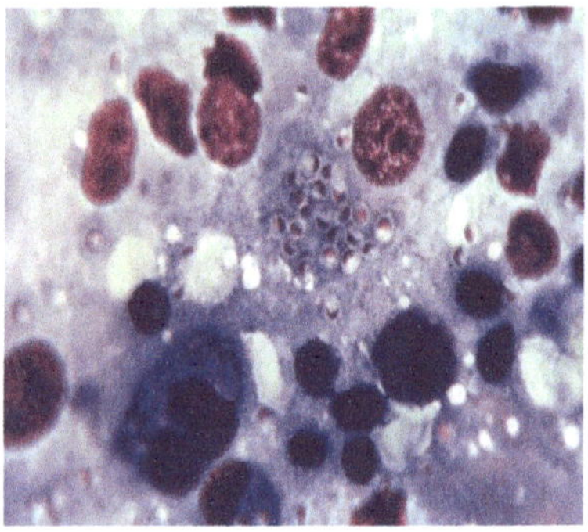

Fig. 15.4 *Histoplasma capsulatum*; yeast-stage in liver tissue specimen of disseminated histoplasmosis. These organisms were of a Tennessee strain, 106-D. (***Image Source*** *– Centers for Disease Control and Prevention – Public Health Image Library – ID#22054* (Georg 1964a, b); *ID#21832* (Georg 1968) *– Copyright Restrictions: NONE; these images are in the public domain and thus free of any copyright restriction*)

Histoplasma capsulatum; yeast-stage in liver tissue specimen of disseminated histoplasmosis. These organisms, were of a Tennessee strain, 106-D.

sensitivities, although not as specific as culture (Bloch et al. 2018). The false positive results decrease the specificity of the antibody detection serological tests, reason being the cross-reactivity of antibodies that recognize similar or shared antigens of other pathogens (Wheat et al. 2018). The antibodies appear 4–6 weeks after onset of primary acute disease and decline after 2–5 years.

Antibody detection—Ease of availability and rapid turnaround time make the antibody detection tests the go to diagnostic tests in suspected histoplasmosis (Guimarães et al. 2006). Complement fixation test and immunodiffusion test are the routinely available methods for antibody detection. The popularity of these assays is due to convenience and satisfactory accuracy, but is not recommended for disseminated histoplasmosis. Also cross reactions are also seen in other dimorphic fungal infections like blastomycosis, coccidioidomycosis, paracoccidioidomycosis, and other fungal infections like aspergillosis, candidiasis, and cryptococossis (Negroni et al. 1976; Wheat et al. 1986). Histoplasmin is the most important antigen in detecting the antibody response. It is antigenic extract obtained from *H. capsulatum* mycelial culture. Main components of histoplasmin are C [galactomannan], M [catalase], and H [b-glucosidase] antigens. Identification of anti-H and anti-M antibodies has diagnostic value (Azuma et al. 1974; Deepe and Durose 1995; Hamilton et al. 1990). Precipitating antibodies (H and M precipitin lines or bands) are measured by the immunodiffusion test which is highly specific but of low sensitivity; especially in acute disease as antibodies appear 4–6 week after infection (Johnson et al. 1984). Presence of both H and M precipitin band is diagnostic for histoplasmosis. Complement fixation test is more sensitive, but less specific than immunodiffusion assay (Bauman and Smith 1976). Limitations of this test are cross-reactivity with antibodies formed in blastomycosis, candidiasis, coccidioidomycosis, and paracoccidioidomycosis, and false negative results due to formation of blocking antibodies or rheumatoid factors and cold agglutinins formed during histoplasmosis infection (Johnson and Roberts 1976). Complement fixing antibodies appear around 3–6 weeks and a titre of 1:32 or more and a fourfold rise in titre is diagnostic of active histoplasmosis (Guimarães et al. 2006; Wheat et al. 2005). Due to the limitations of the tests mentioned above, latex agglutination tests have been developed, but they can also give false positive results in certain diseases like tuberculosis (DiSalvo and Corbett 1976). Immunoenzymatic assays such as ELISA and Western blot are also available. Western blot helps in epidemiological evaluation of the disease as well as diagnosis of the disease using *Histoplasma* antigens of 91, 83, 70, and 38 kDa reacting with serum antibodies. However, a more widely used method is ELISA using diverse antigen preparations (Richardson and Warnock 1983; Torres et al. 1993). Patients having immunosuppressive conditions like HIV-AIDS have diminished immune response; histoplasmosis in such patients can be missed if relied only on antibody detection assays. Bloch et al. depicted sensitivities of Enzyme-immunoassay for IgG and IgM anti-Histoplasma (82%) > Combination of ID and CF (51%) > complement fixation (CF) (50%) > Immunodiffusion (44%). It was seen that antigen testing was more positive in immunocompetent patients than the immunodiffusion or complement fixation or enzyme immune assay (Bloch et al. 2018).

Antigen detection—*H. capsulatum* antigens can be detected in samples like blood, CSF, bronchoalveolar lavage, urine, and pleural fluid. It is considered better than antibody detection, particularly in acute disease and immunosuppressed patients with disseminated histoplasmosis (Wheat et al. 1991, 2002, 2005). High titres of histoplasma antigenuria have been reported in disseminated histoplasmosis cases and monitoring of these titres can help corelating the prognosis of the patient (Wheat et al. 2002). Both urine and serum should be tested in case of disseminated disease. CSF should be tested if CNS histoplasmosis is suspected as it has been found that CNS antigen levels are higher than serum antigen levels as there is active production of antigen in the CNS accompanying the diffusion of antigen across the blood brain barrier (Wheat et al. 2005). The sensitivity of antigen detection in CSF, urine, and serum is 38–67%, 71%, respectively, and 38%; on the other hand, sensitivity of antibody detection in CSF is 80–89% (Wheat et al. 2005). However, in case of chronic disease due to low burden of antigen, these antigen detection tests can be negative but antibody detection can give positive results. Therefore, it is always emphasized to take a multimodal approach in establishing the diagnosis (Joseph Wheat 2003). Authors have reported a sensitivity of 98% and specificity of 90.8% by newer methods and assays for antigen and anti-histoplasma antibodies detection in CSF (Gonzalez et al. 2019).

Histoplasmin skin test—It is an intradermal skin test in which histoplasmin antigen in injected and individual response is evaluated. It is based on delayed type of hypersensitivity response and is analogous to tuberculin test; only useful for epidemiological purpose and has no diagnostic value (Torres-Rodríguez et al. 2000).

Molecular-based methods—To confirm the diagnosis of Histoplasmosis, depiction of Mycelial to yeast form or production of exoantigen is required which takes at least 3–4 weeks, as differentiation from other dimorphic fungi or saprophytic fungi is difficult on culture. Molecular test like PCR can decrease this time and provide specific, sensitive, and accurate result. Various molecular PCR have been developed targeting several different genes of histoplasma capsulatum like (18S) rRNA gene, specific primers targeting the gene coding for a specific 100 kDa protein, M gene, etc. (Guimarães et al. 2006). Molecular assays have shown variable performances according to the stage and clinical form of disseminated histoplasmosis. However, nested PCR has given quite encouraging results. Hcp 100 was found to be most commonly used target gene used in this molecular method. A meta-analysis has shown a sensitivity and specificity of 95% and 99% overall which is quite similar to antigen-based assays (Caceres et al. 2019; da Silva Vasconcellos et al. 2019).

Other tests—Some authors have also studied the test like detection of β-D-glucan in CSF. The Fungitell® immunochromatographic assay was used and a 50% positivity was seen in cases of CNS histoplasmosis and 20% among controls (Myint et al. 2018). However, owing to its poor sensitivity, it is not routinely used in the diagnosis of Histoplasma meningitis.

Diagnosis of Disseminated Histoplasmosis is comparatively easy in comparison to isolated CNS involvement since histoplasma can be identified from various organs; however, in case of latter, positive results may only be obtained from CSF, meninges, or brain tissue (Wheat et al. 2005). Lastly, although non-culture-based methods like antigen and antibody detection assay provide a rapid diagnostic support, there is always a possibility of cross reaction and false positivity; therefore, careful clinical correlation should be made and repeat testing is always recommended to increase the reliability of the initial test result.

15.2.4 Treatment and Subsequent Diagnostic Testing

Early diagnosis with optimal treatment initiated and the immunocompetent status of the patient provides the best prognosis in CNS histoplasmosis. It is quite unfortunate that response to antifungals in CNS histoplasmosis is comparatively poor than other forms of histoplasmosis. The three antifungals most commonly used in the treatment of CNS histoplasmosis are amphotericin B, itraconazole, and fluconazole. The IDSA treatment guideline recommends "initial treatment with Liposomal amphotericin B in doses of 5.0 mg/kg for a total of 175 mg/kg for 4–6 weeks followed by itraconazole 200 mg 2 or 3 times daily with blood levels drawn to ensure therapeutic dosing for at least 1 year" (Wheat et al. 2007). With this treatment regimen, a multicentric study reported 74% survival rate and 6% recurrence rate (Wheat et al. 2018). Quantifiable concentration in brain tissue is achieved by all formulations of amphotericin B; on the other hand, none reaches quantifiable concentration in CSF. Liposomal amphotericin B is the drug of choice due to its superior pharmacodynamics and can be given in higher doses as it is less nephrotoxic and also it achieves higher concentration on brain tissue; however, none of the formulation of amphotericin B has been superior or more efficacious than another for CNS histoplasmosis (Wheat et al. 2007). Deoxycholate and Liposomal Amphotericin B reduced the fungal burden in the brain more completely than amphotericin B lipid complex or amphotericin B colloidal dispersion. Although fluconazole attains a higher CSF concentration level (>70% of serum level) and it crosses blood brain barrier freely, it has reduced activity against *H. capsulatum*. On the other hand, itraconazole attains only 1%, but has better activity against *H. capsulatum*, therefore recommended by majority of authors as maintenance therapy in CNS Histoplasmosis (Como and Dismukes 1994; Mata-Essayag et al. 2008; McKinsey et al. 1996; Wheat et al. 1995, 1997). Mortality rate was quite high (20–40%) 2–3 decades before with 50% relapse rate, which now has reduced with early aggressive treatment and use of azole as maintenance therapy (Saccente et al. 2003; Schestatsky et al. 2006; Wheat et al. 1990).

Some authors recommend 5 mg/kg/day of liposomal amphotericin B for a total of 100–150 mg/kg over 6–12 weeks, followed by a year of maintenance therapy with either fluconazole (600–800 mg daily) or 200 mg itraconazole twice or three times daily for histoplasma meningitis; and for the immunocompromised patients in which restoration of the healthy immune status is not possible, lifelong maintenance therapy is recommended (Saccente et al. 2003; Schestatsky et al. 2006). Successful treatment of CNS histoplasmosis by newer azoles like Posaconazole and voriconazole has been reported by individual case reports; still more extensive studies will be required for them to be recommended (Hott et al. 2003; Restrepo et al. 2007).

15.3 Blastomycosis

Blastomyces dermatitidis, a thermally dimorphic fungus, is the causative organism of the disease Blastomycosis, belonging to the phylum Ascomycota and the family Agellomycetaceae, with asexual state known as *Ajellomyces dermatitidis.* Genus *Blastomyces* is now known to have several species due to advanced taxonomical techniques, most important of which are *Blastomyces dermatitidis* and *Blastomyces gilchristii* which are very closely related and cause identical clinical disease (Brown et al. 2013a, b). Being a thermally dimorphic fungus, phase variation is seen; at body temperature [37 °C], it is seen in yeast form, whereas at room temperature [25 °C] it occurs in mycelial form. Colony of mould form is white and fluffy, whereas colony of yeast form is brown in colour. Intracellularly, it is seen as large (8–10 μm) broad-based budding yeast cell (Saccente and Woods 2010).

15.3.1 Epidemiology

Blastomycosis is an endemic mycosis mostly confined to United States of America and Canada. Soil in the regions around the borders of the Great Lakes and the St. Lawrence River, as well as in the Mississippi River and Ohio River basins, is considered the habitat of the fungus. Yearly incidence rate ranges from 1 to 2/100,000 population with highest incidence reported from Wisconsin state ranging from 10 to 40 cases/100,000 population (Baumgardner et al. 1992; "Blastomycosis -- Wisconsin, 1986-1995," n.d.; Carlos et al. 2010; Herrmann et al. 2011). Authors have also reported cases from Africa, particularly South Africa, Zimbabwe, and also from India (Baily et al. 1991; Frean et al. 1989; Kumar et al. 2014). Although blastomycosis affects all age groups and sexes, adults and male are more affected than females and paediatric population owing to recreational and occupational exposure (Bradsher 1992). *Blastomyces* causes invasive fungal disease even in immunocompetent patients; that's why it is considered as true fungal pathogen just like *Histoplasma,* but there is a constant rise of blastomycosis among immunocompromised patients also, particularly patients with AIDS, solid organ transplant, and hematopoietic cell transplant recipients, on tumour necrosis factor-α (TNF-α) inhibitors. They suffer from more severe and disseminated disease (Castillo et al. 2016; Gauthier et al. 2007; Pappas et al. 1992; Smith and Kauffman 2009).

15.3.2 Pathogenesis

Infections in humans are due to inhalation of conidia containing the *Blastomyces dermatitidis*. The conidia reach the alveoli where it is phagocytised by alveolar macrophages, neutrophils, and monocytes and thus eliminated, resulting in asymptomatic infection in some individuals (Drutz and Frey 1985). However, if the natural defence mechanism is not adequate, these conidia transform into the pathogenic yeast form due to thermal stimulation, which is a key factor in the pathogenesis.

These yeast forms possess a thick capsule that hampers their phagocytosis by the polymorphonuclear leucocytes, leading to proliferation in alveoli (Saccente and Woods 2010; Sugar and Picard 1991). Also, they are relatively resistant to reactive oxygen species produced by macrophages and neutrophils (Rocco et al. 2011). The yeast cells then get disseminated from lungs by lymphatic/haematogenous route with the help of the immune-modulating glycoprotein BAD-1 that facilitates its binding to macrophages (Brandhorst et al. 2013). Extrapulmonary infection accounting approximately 25–30% of patients involve skin (most common) followed by bones, male genitourinary system, and CNS (Chapman et al. 1997; Sarosi and Davies 1979). Pyogranulomatous lesions are seen in the affected organs. Innate immunity prevents the establishment of infection, cell-mediated immunity hampers the progression of the diseases, whereas humoral immunity does not play any significant role (Bradsher 1992; Klein et al. 1990). Cell-mediated immunity lasting for about 2 years is developed in the host following recovery from blastomycosis (Klein et al. 1990).

15.3.3 Clinical Features

Mode of infection in humans is by inhalation of airborne conidia; less common mode are inoculation in skin directly, accidental laboratory mishap, or dog bite. Human to human or animal to human transmission is not reported. About 50% of the infections are asymptomatic/subclinical, rest develop a mild flue like illness with fever, cough, myalgia, arthralgia, and pleurisy, which can further progress to acute or chronic pneumonia, ARDS, or disseminated disease (Sarosi et al. 1974). After inhalation of conidia or mycelial fragments, it takes from about 3 weeks to 3 months for symptoms to develop (Klein et al. 1987). Dissemination to extrapulmonary sites like cutaneous, osseous, genitourinary, and CNS is seen in 25–40% of symptomatic patients (Chapman et al. 2008). Five to ten percent of disseminated Blastomycosis have CNS manifestations presenting as meningitis or epidural abscess or brain abscess (Kravitz et al. 1981; Roos et al. 1987). CNS involvement can occur either by haematogenous seeding from the primary lesion or direct invasion through untreated skull-based osteomyelitis (Bariola et al. 2010). Signs and symptoms of neuroblastomycosis are headache (most common) followed by altered mental status, confusion, focal neurological deficit, blurring of vision, and seizures (Bariola et al. 2010). Complications of CNS involvement reported are hydrocephalous, cerebral edema, infarction, cerebral herniation, loss of concentration at school, and panhypopituitarism (Bariola et al. 2010).

Blastomycosis is mostly misdiagnosed as tuberculosis. Sometimes it mimics malignancy and shows features of atypical pneumonia. Among the endemic mycosis, it is mostly confused with paracoccidiodomycosis. In context of CNS blastomycosis, it usually impersonates malignancy and meningiomas.

15.3.4 Diagnosis

Radiodiagnosis—No radiographic abnormalities are absolutely diagnostic of blastomycosis; however, Magnetic resonance imaging (MRI) is the most sensitive imaging technique. Involved area of the brain is seen as ring-enhancing lesions which may be single or multiple, and meningitis can be seen as meningeal enhancement (Kauffman 2019).

Laboratory diagnosis—Laboratory diagnosis depends on the demonstration of organisms in tissues or exudates, growth on culture media, and detection of antigen or antibody in serum.

Direct Examination—Clinical sample shows characteristic double contoured, thick-walled, multinucleated round to oval, 8–15 μm in diameter yeast forms with single broad-based budding daughter cells; however, CSF has very low sensitivity. Histopathological sections from brain tissue/meninges are stained by haematoxylin and eosin, periodic acid-Schiff stain (PAS), and Gomori Methamine Silver (GMS), which provide excellent contrast. Yeast forms may be found inside or outside the infected cells (Saccente and Woods 2010).

Culture—Specimen like CSF/brain tissue should be inoculated on Sabouraud dextrose agar (SDA) at 25 and 37 °C for 2–4 weeks and incubation could be extended till 6 weeks if no growth occurs. Other media that can be used are potato dextrose agar, brain heart infusion agar (BHI), potato flake agar, blood glucose cysteine agar, or inhibitory mould agar. BHI supplemented with blood is recommended for tissue sections and for better demonstration of mould to yeast conversion. Other media used to demonstrate mycelial to yeast conversion are Kelly's agar, cysteine heart agar, and haemoglobin agar. Grown at 25–30 °C is yeast like initially that are off-white and waxy which then become downy and fluffy white within 14 days as the ariel hyphae develop lastly becoming grey to brown or tan coloured with time. Microscopic examination by Lactophenol Cotton Blue (LCB) mount shows fine branched septate hyphae bearing pyriform or oval single cell conidia measuring 2–4 μm in diameter over short or long conidiophore. However, these features are not diagnostic as *Pseudallescheria boydii* and *Chrysosporium* species have a similar appearance, but can be differentiated by the fact that *Pseudallescheria boydii* and *Chrysosporium* spp. can be inhibited by using cycloheximide containing media; also *Chrysosporium* spp. does not grow at 37 °C whereas *B. dermatitidis* does (Saccente and Woods 2010). Colonies on SDA at 37 °C are moist and pasty macroscopically. Microscopically, yeast cell of size 8–15 μm, spherical with single broad-based budding, is seen giving rise to a characteristic *figure of eight* appearance (Figs. 15.5 and 15.6).

Immunodiagnosis—The serological tests used for diagnosing blastomycosis are complement fixation test, immunodiffusion, precipitation test, enzyme immunoassays, and radioimmunoassay, utilizing antigen A. The two antigens preparations most commonly used are cell free culture filtrate of mycelial phase [Blastomycin] and inactivated whole yeast phase cells. Earlier the most commonly used test was complement fixation test which utilized yeast phase antigen, but it had a unsatisfactory sensitivity of 57% and specificity of 30% due to cross-reactivity of

Blastomyces dermatitidis: medical illustration

Fig. 15.5 *Blastomyces dermatitidis*: medical illustration. (***Image Source*** *– Centers for Disease Control and Prevention – Public Health Image Library – ID#24210* (Gandhi and Rossow 2020); *ID#22960* (Hardin 1965a, b) *– Copyright Restrictions: NONE; these images are in the public domain and thus free of any copyright restriction*)

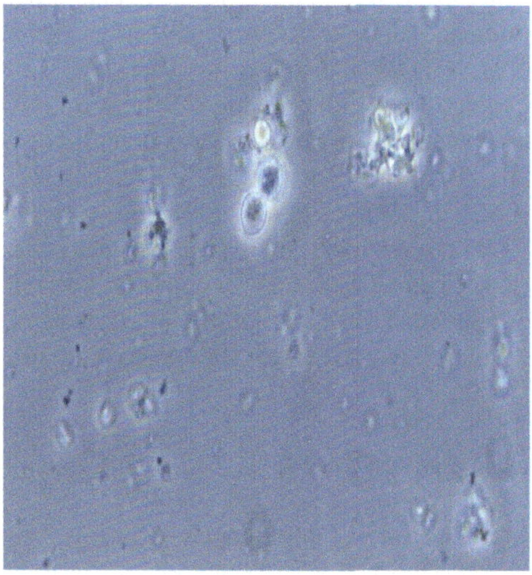

Blastomyces dermatitidis; lactophenol cotton blue (LPCB) stain showing a yeast form (Magnification 400X; phase contrast illumination).

Fig. 15.6 *Blastomyces dermatitidis*; lactophenol cotton blue (LPCB) stain showing a yeast form (Magnification ×400; phase contrast illumination). (***Image Source*** *– Centers for Disease Control and Prevention – Public Health Image Library – ID#24210* (Gandhi and Rossow 2020); *ID#22960* (Hardin 1965a, b) *– Copyright Restrictions: NONE; these images are in the public domain and thus free of any copyright restriction*)

antibodies formed against histoplasmin and blastomycin (Turner et al. 1986). Use of purified *B. dermatitidis* A antigen in immunodiffusion test improved the sensitivity to 65–80% (Turner et al. 1986). Enzyme immunoassay had shown better sensitivity than the complement fixation and immunodiffusion test, but specificity was slightly less; however, it can be used in an outbreak setting as an epidemiological tool to identify acute *B. dermatitidis* infection (Klein et al. 1987).

B. dermatitidis antigen detection tests are also available; urine is the specimen of choice as it is most extensively studied but CSF, serum, other sterile fluids, and bronchoalveolar lavage fluid can also be used. A study has reported positivity rate of 92.9% in proven blastomycosis cases and 96.3% positivity in disseminated cases, but specificity was very low as cross reactions were seen in urine samples of histoplasmosis, paracoccidioidomycosis, and penicilliosis (Durkin et al. 2004). Supportive evidence can be made detecting (1 → 3)-β-D-glucan in serum which is a marker of invasive fungal infection. BAD1 protein is specific for *Blastomyces* and a newer enzyme immunoassay has shown quite good results with sensitivity (87%) and specificity (94–99%). Antibodies against BAD1 does not cross react with *Histoplasma* antigen (Richer et al. 2014). An EIA manufactured by MiraVista Diagnostics, Indianapolis, IN, USA, that detects galactomannans in urine, BAL, CSF, and serum and is commercially available in USA, has shown a sensitivity of 76–90% in various studies (Bariola et al. 2011; Baumgardner 2018; Connolly et al. 2012; Durkin et al. 2004; Frost and Novicki 2015).

Nucleic Acid Amplification tests—They are still not standardized or commercially available and therefore used quite infrequently for regular diagnosis. Only few reference laboratories currently utilize these in-house developed assays (Babady et al. 2011).

Focusing on diagnosing CNS blastomycosis, culture of CSF/brain tissue is the gold standard; however, negative culture does not exclude the disease (Walkty et al. 2018). Antigen or antibody detecting assay lacks adequate sensitivities and specificities (Ryan et al. 2019). However, a presumptive diagnosis can be made by CSF analysis and these antigen detection tests while awaiting culture result (Kauffman 2019). CSF analysis in patient of *B. dermatitidis* meningitis shows lymphocytic or neutrophilic pleocytosis with elevated protein and normal or decreased glucose level (Bariola et al. 2010).

15.3.5 Treatment

Mortality rate of blastomycosis ranges between 4 and 22%; however, for CNS histoplasmosis it is still unknown ("Blastomycosis -- Wisconsin, 1986-1995," n.d.; Cano et al. 2003; Dworkin et al. 2005). According to IDSA guidelines "Liposomal Amphotericin B, at a dosage of 5 mg/kg per day over 4–6 weeks followed by an oral azole [fluconazole, 800 mg per day, itraconazole, 200 mg 2 or 3 times per day, or voriconazole, 200–400 mg twice per day] for at least 12 months and until resolution of CSF abnormalities is recommended" (Chapman et al. 2008). However, voriconazole is preferred over fluconazole and itraconazole as it attains a desirable CNS concentration and antifungal activity. Nephrotoxicity is the major concern while using Amphotericin B formulations; Amphotericin B deoxycholate is more nephrotoxic than lipid amphoteric preparations like liposomal amphotericin B, amphotericin B colloidal dispersion, and amphotericin B lipid complex; therefore, the latter is preferred; moreover, it has best penetration of blood brain barrier, making it the drug of choice for CNS blastomycosis. Other methods by which nephrotoxicity can be reduced are by infusing 0.9% normal saline before and after the dose, avoiding other neurotoxic drugs and diuretics, and frequently monitoring creatinine and electrolytes (Chapman et al. 2008). Newer antifungals from echinocandin group like micafungin and anidulafungin have only an intermediate activity against *B. dermatitidis* in vitro and therefore not recommended for the treatment of any kind of blastomycosis (Espinel-Ingroff 1998).

15.4 Coccidioidomycosis

Coccidiodes immitis is another important common dimorphic fungus causing CNS infection after *Histoplasma capsulatum* (Zarrin and Mahmoudabadi 2010). Fungus *Coccidiodes* is principally fungus of western hemisphere. It is endemic in South-western United States and part of Mexico and Central and South America (Shehab and Shehab 2013). Two species of *Coccidiodes* known to infect humans are *C. immitis* and *C. posadasii*. *C. posadasii* is more frequently found in Texas and Central and South America. However, *C. immitis* is peculiarly common in California, especially the southern San Joaquin Valley of California, thus giving the name "Valley fever" to the infection caused by *C. immitis* (Galgiani et al. 2016). But there is no difference in clinical manifestation of the two species. In endemic regions, coccidiodomycosis is reported to be associated with around 30% mortality, and more than 15% risk of dissemination (Johnson and Einstein 2006).

However, newer epidemiological studies suggest increase in incidence as well as distribution of infection (Mazi et al. 2023), around 10% of adult infections (Brown et al. 2013a, b; Centers for Disease Control and Prevention (CDC) 2013) and 40% of hospitalized paediatric patients with coccidiodomycosis were reported from non-endemic regions (Chu et al. 2006).

15.4.1 Clinical Features

Coccidiodomycosis is a systemic infection; however, most of the times it is asymptomatic or self-limiting. Even though most commonly it presents with pulmonary symptoms only, distal dissemination may occur in virtually any extra-pulmonary sites, including skin and soft tissue, bone, joints and CNS, via lymphatic and/or hematogenous spread. Around 30–50% of disseminated infections present with coccidioidal meningitis (Johnson and Einstein 2006). In a study of paediatric infections, 37% of disseminated disease were reported to be presented with CNS involvement (Dimitrova and Ross 2016). In most of the cases, dissemination occurs within few months of primary infection. In few cases, dissemination has been reported even without primary respiratory illness (Galgiani 2009).

Even though immunocompetent individuals also get infection, immunocompromised individuals, such as People living with HIV/AIDS, transplant recipients, diabetics, patients on chronic steroid therapy, and pregnant females are at much higher risk of dissemination (Dupont 2003). Coccidioidal meningitis is one of the AIDS defining illnesses.

Clinical presentation of coccidioidal meningitis, the most dreadly form of dissemination, is erratic. Common symptoms include headache, altered mental status, nausea, vomiting, and neck rigidity (Sondermeyer et al. 2016; Vincent et al. 1993). Additional findings are fever, personality changes, visual loss, and signs of focal neurological deficit. However, in half of cases only meningismus is seen, without any sign of inflammation of meninges (Johnson and Einstein 2006).

Most frequent complication of coccidioidal meningitis is Hydrocephalus, which is seen in around 30–50% of cases (Murthy and Sundaram 2014). Moreover, in about 15–20% of cases, cerebral and vasculitic infarctions are observed (Williams 2001), which may be a presenting feature itself or may occur due to long-term effect of treatment. Rare complications of infection include mass lesion or abscess formation in the cerebrum (Bañuelos et al. 1996; Kleinschmidt-DeMasters et al. 2000). Spinal arachnoiditis may occur as a complication to intrathecal therapy, especially to amphotericin B deoxycholate (Johnson and Einstein 2006; Winn 1964).

15.4.2 Diagnosis

CSF analysis shows raised cell count, majority of which are lymphocytes. Eosinophilic pleocytosis has high diagnostic significance, although presence of eosinophils is not common (Ragland et al. 1993). Additional findings are elevated CSF proteins (mostly $\geqslant$150 mg/dL), reduced glucose level, and high opening pressure.

The definitive diagnosis of coccidioidal meningitis needs histopathological findings, a positive CSF culture, or serological tests (Johnson and Einstein 2006). Although histopathological examination is not appropriate for routine testing in coccidioidal meningitis, any respiratory or other focus of dissemination, if present, may show endosporulating spherules on histopathological examination. Similarly, isolation of *Coccidiodes* species from CSF sample is diagnostic, but uncommon . On rare occasion, fungal elements can be seen on direct microscopy of the sample, but that will indicate high burden of *Coccidiodes* in the CSF. Furthermore, presence of arthroconidia or mycelial forms in the CSF is more troublesome, as most of these cases are seen in patients with ventriculoperitoneal or other shunts (Kleinschmidt-DeMasters et al. 2000) (Figs. 15.7 and 15.8).

CSF serological tests are the most crucial tools for diagnosis, as well as for patient follow-up. Presence of antibodies against *Coccidiodes* in CSF has substantially low sensitivity and high specificity. Demonstration of IgG antibody in CSF confirms the diagnosis of coccidioidal meningitis (Michelson et al. 1983). However, late positivity after clinical features and low seropositivity, especially in immunocompromised individuals, are disadvantages of serological tests (Blair and Logan 2001).

Various formats of serological tests available are immunodiffusion (ID), enzyme immunoassay (EIA), and complement fixation (CFT). The two antigens primarily

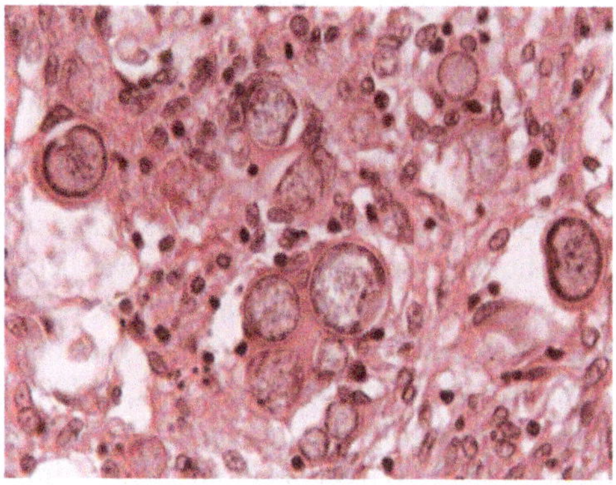

Fig. 15.7 *Coccidioides immitis*; histopathological detail showing a number of spherules were visible in various stages of development (Magnification ×500). (***Image Source** – Centers for Disease Control and Prevention – Public Health Image Library – ID#20515* (Brodsky 1966); *ID#20513* (Georg 1966) – *Copyright Restrictions: NONE; these images are in the public domain and thus free of any copyright restriction*)

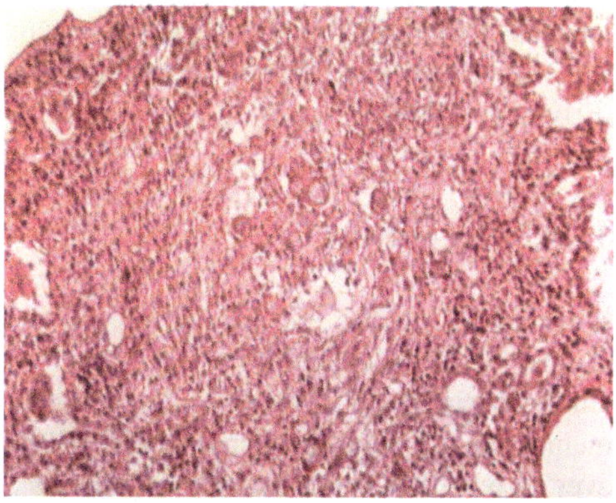

Fig. 15.8 Same photomicrograph at magnification ×125. (***Image Source** – Centers for Disease Control and Prevention – Public Health Image Library – ID#20515* (Brodsky 1966); *ID#20513* (Georg 1966) – *Copyright Restrictions: NONE; these images are in the public domain and thus free of any copyright restriction*)

used in different tests are tube-precipitin antigen and complement-fixing antigen. Tube-precipitin antigen is used to detect IgM antibodies, which appear early in the course of illness. However, CF antigens are used to detect IgG antibodies, which appear later and is used to monitor the course of illness (Blair et al. 2006; Galgiani 1997). CFT titre of 1:32 or more is reported to be associated with dissemination (Shehab and Shehab 2013). One study demonstrated that although immunocompromised individuals have low seropositivity rates for each type of tests when compared with immunocompetent individuals, combining all three types of test results will significantly increase the sensitivity of serologic tests in immunosuppressed individuals (Blair et al. 2006).

Coccidioides antigen (CAg) detection is another useful test developed recently for the early diagnosis of coccidioidomycosis (Kassis et al. 2015). Sensitivity of the test varies in different studies and different samples. Antigenemia reported ranges from 70 to 84% of cases (Durkin et al. 2009; Kassis et al. 2015), whereas presence of CAg in urine was demonstrated in around 71% of coccidioidomycosis patients in different studies (Durkin et al. 2008; Kassis et al. 2015). Providentially, in a study of coccidioidal meningitis cases, CAg was detected in CSF samples with high sensitivity of 93% and specificity of 100% (Kassis et al. 2015). However, cross-reactivity seen in around 11–22% of patients with other endemic mycosis, such as histoplasmosis and blastomycosis (Durkin et al. 2009; Kuberski et al. 2007), is the disadvantage of the test. Moreover, combining testing of serum and urine reported to have higher sensitivity (Kassis et al. 2015). Similarly, combined testing of antigen and antibody in the CSF will be conducive for diagnosis of coccidioidal meningitis.

Another promising diagnostic tool is PCR (polymerase chain reaction), but its application and utility are still being evaluated.

Neuro-imaging—CT (Computed Tomography) scan of brain is useful in detecting Hydrocephalus (Shetter et al. 1985), most common complication of coccidioidal meningitis. However, MRI (Magnetic Resonance Imaging) is better to evaluate meningeal involvement and extent of disease. Moreover, complications such as vasculitic infarctions and spinal arachnoiditis are better detected by MRI (Erly et al. 1999).

15.4.3 Treatment

Polyenes and azoles are the two groups of antifungal agents which are effective against CNS coccidioidomyosis. Previously in the last century, intrathecal amphotericin B deoxycholate was being used for the treatment of coccidioidal meningitis and was considered as the "gold standard" (Einstein et al. 1961). However, in recent years the Infectious Diseases Society of America (IDSA) has revised its guidelines and recommends "Fluconazole 400–1200 mg orally once daily or Itraconazole 200 mg 2–4 times in a day" (Galgiani et al. 2016). As a result, Fluconazole has now become the most commonly used oral agent for the treatment of coccidioidal meningitis; however, Itraconazole, Voriconazole, and Posaconazole are also used by few (Mathisen et al. 2010).

15.5 Paracoccidioidomycosis

Paracoccidioidomycosis is caused more frequently by *Paracoccidioides brasiliensis*, and less frequently by *Paracoccidioides lutzii*, later being recently described species (Teixeira et al. 2014). Paracoccidioidomycosis is endemic to Latin America, which includes Brazil, Colombia, Argentina, and other south American countries (Teixeira et al. 2014). Infection has been reported both in immunocompetent and immunosuppressed individuals, including HIV patients and solid organ transplant recipients (de Almeida et al. 2018).

Paracoccidioidomycosis clinically manifests in two forms: more commonly as chronic form (80%) seen in adults, and acute or subacute or "juvenile" form seen in children and young adults. Chronic form usually involves lung and upper airways, skin, and oral mucosa. Juvenile form usually involves skin, lymph nodes, liver, and spleen (Bellissimo-Rodrigues et al. 2013). In immunosuppressed individuals, it manifests as rapidly progressive disease and with high morbidity (de Almeida et al. 2018).

CNS dissemination (Neuroparacoccidioidomycosis or NPCM) occurs in 10–15% of Paracoccidioidomycosis and is seen in chronic form (de Almeida et al. 2004; Pedroso et al. 2012). It clinically manifests as space-occupying lesions mainly, most commonly as intra-cerebral lesions, and rarely as meningitis. Patient may present with headache, focal deficits, and seizures. Space-occupying lesions can also be seen in cerebellum (present with gait abnormality) and spinal cord. Intracranial pressure may also be raised (Kauffman 2019).

Diagnosis—The definitive diagnosis of Paracoccidioidomycosis relies on growth and identification of *Paracoccidioides* on culture. But since culture growth takes time, early diagnosis can be made by direct microscopic demonstration of fungi from specimens, stained by silver or PAS stained or by calcofluor white. *Paracoccidioides* appears as yeast cells with thick wall and narrow base budding (Kauffman 2019; Shikanai-Yasuda et al. 2017). However, this fungus is rarely seen or grown from CSF, thus making it difficult to diagnose NPCM. Thus, presumptive diagnosis of NPCM can be made when patient with disseminated Paracoccidioidomycosis presents with features of CNS involvement or with suggestive neuro-imaging findings (Reis et al. 2013). Diagnosis can be confirmed on histopathological examination of CNS lesion biopsy, which shows granulomatous lesion and necrosis (Figs. 15.9 and 15.10).

Various serological tests available for antibody detection include enzyme immunoassays, double agar gel immunodiffusion (DID),

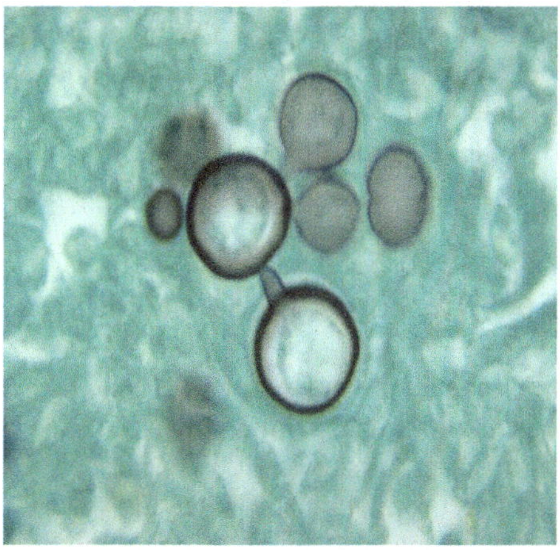

Paracoccidioides brasiliensis; methenamine silver stained tissue sample showing

Fig. 15.9 *Paracoccidioides brasiliensis*; methenamine silver-stained tissue sample showing multiple buds projecting from one of the yeast form cells. (***Image Source*** *– Centers for Disease Control and Prevention – Public Health Image Library – ID#525; ID#520* (Georg 1963) *– Copyright Restrictions: NONE; these images are in the public domain and thus free of any copyright restriction*)

Fig. 15.10 *Paracoccidioides brasiliensis*; methenamine silver-stained liver tissue sample showing minute buds projecting from several yeast form cells. (***Image Source*** *– Centers for Disease Control and Prevention – Public Health Image Library – ID#525; ID#520* (Georg 1963) *– Copyright Restrictions: NONE; these images are in the public domain and thus free of any copyright restriction*)

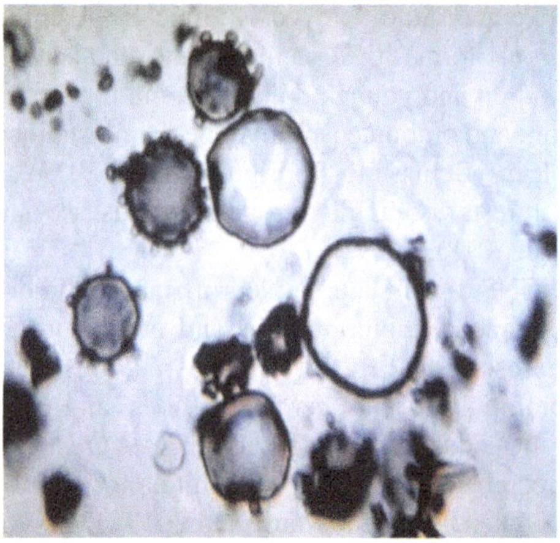

Paracoccidioides brasiliensis; methenamine silver stained liver tissue sample showing

counterimmunoelectrophoresis (CIE), and immunoblot techniques (de Camargo 2008; Shikanai-Yasuda et al. 2017). All these different tests are reported to have varying sensitivities for the detection of antibodies from serum, but none has proven to be useful for CSF (de Camargo 2008). Detection of circulating glycoprotein antigen (gp43 and gp70) in the serum or CSF has been recently described as a promising diagnostic tool (da Silva et al. 2005). Recently, application of nested PCR for the diagnosis was evaluated by molecular detection for Paracoccidioides gp43 membrane protein (Gaviria et al. 2015).

Treatment—For the treatment of Paracoccidioidomycosis, Brazilian guidelines recommend use of Itraconazole 200 mg daily for 6–12 months (Shikanai-Yasuda et al. 2017). Besides Itraconazole, Voriconazole has shown to be more effective in the management of Neuro-paracoccidioidomycosis (NPCM), as it attains much higher concentration in CSF (Telles et al. 2007). Another effective treatment option documented is trimethoprim-sulfamethoxazole (Shikanai-Yasuda et al. 2017). Amphoteracin B can be used in severe Paracoccidioidomycosis (Peçanha et al. 2016). Neurosurgical procedures may be indicated in some cases such as intracranial hypertension and spinal cord compression (Elias et al. 2005).

15.6 Talaromyces Marneffei

Earlier popularly known as *Penicillium marneffei*, it is the only species in the genus *Penicillium* which is a dimorphic fungi and causes systemic disease in humans. Once a rare illness, systemic disease caused by *Talaromyces marneffei* is now third commonest opportunistic infection in HIV (Human immunodeficiency virus) infected people and immunocompromised individuals of Southeast Asia, China, and North-eastern India (Ranjana et al. 2002; Vanittanakom et al. 2006). The only two opportunistic infections preceding this are tuberculosis and cryptococcosis (Chang et al. 1995; Deng and Connor 1985).

Talaromyces marneffei is a facultative, intracellular pathogen, just like *Histoplasma capsulatum*. It characteristically produces soluble red pigment, which diffuses into the agar medium. Mannoprotein present in its cell wall is antigenic.

Infections have been reported both in immune-compromised (>80%) and immune-competent individuals (Duong 1996). Common clinical features are fever, weight loss, anaemia, and characteristic umbilicated skin lesions, similar to molluscum contagiosum. Dissemination, especially in immunosuppressed patients, can occur to reticulo-endothelial system, skin, respiratory system, gastrointestinal system, and in rare cases to CNS. In cases of disseminated penicillosis, fungi were isolated most commonly from skin, blood, bone marrow, and lymph nodes (Supparatpinyo et al. 1994). Dissemination to central nervous system has been reported in only few cases (Le et al. 2010; Noritomi et al. 2005). Clinical symptoms in these cases were non-specific and included fever, altered mental status, agitation, confusion, and depressed consciousness (Le et al. 2010). Multiple brain abscesses were also reported in few (Noritomi et al. 2005).

Diagnosis—The diagnosis can be made by isolation of fungi from clinical specimens, such as blood, bone marrow aspirate, skin biopsy, or CSF. Isolation of the organism from blood and CSF suggests CNS involvement. *T. marneffei* have been isolated from CSF samples (Le et al. 2010; Noritomi et al. 2005; Supparatpinyo et al. 1994). Since culture growth takes time, presumptive early diagnosis can be made by direct microscopic examination of the Wright's or Giemsa-stained smears from bone-marrow aspirate and skin biopsies, which shows intracellular basophilic yeast-like cells. Central septation in the yeast cells is the characteristic feature of *T. marneffei*. However, no budding is seen, which distinguishes *T. marneffei* from *Histoplasma capsulatum* (Supparatpinyo et al. 1994) (Figs. 15.11 and 15.12). However, direct visualization of *T. marneffei* yeast cells in CSF is difficult.

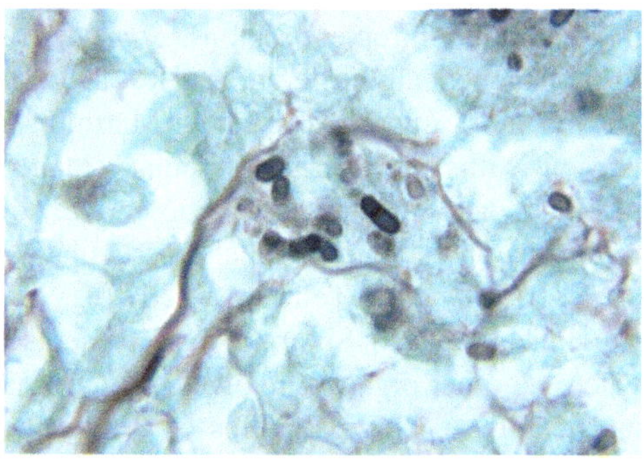

Talaromyces marneffei (formerly known as Penicillium marneffei) methenamine silver stain of tissue section showing globe-shaped yeast cells undergoing multiplication through fission.

Fig. 15.11 *Talaromyces marneffei* (*formerly known as Penicillium marneffei*) methenamine silver stain of tissue section showing globe-shaped yeast cells undergoing multiplication through fission. (***Image Source*** *– Centers for Disease Control and Prevention – Public Health Image Library – ID#4235* (Ajello 1972a, b); *ID#4241* (Ajello 1972a, b) *– Copyright Restrictions: NONE; these images are in the public domain and thus free of any copyright restriction*)

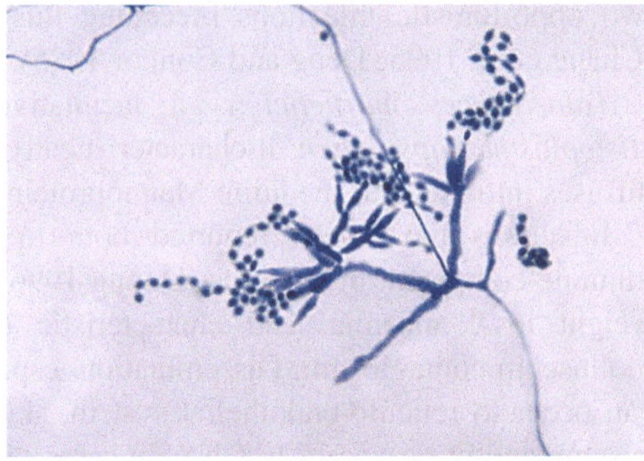

Talaromyces marneffei, figure showing chains of single-celled, teardrop-shaped conidia, each emanating from its respective, flask-shaped phialide.

Fig. 15.12 *Talaromyces marneffei*, figure showing chains of single-celled, teardrop-shaped conidia, each emanating from its respective, flask-shaped phialide. (***Image Source*** *– Centers for Disease Control and Prevention – Public Health Image Library – ID#4235* (Ajello 1972a, b); *ID#4241* (Ajello 1972a, b) *– Copyright Restrictions: NONE; these images are in the public domain and thus free of any copyright restriction*)

Recently, developed serological tests, such as detection of *T. marneffei*-specific mannoprotein (Mp1p) antigen in patient's serum (Wang et al. 2015) and Mp1p-specific antibody (Wang et al. 2013), are also available to aid in the diagnosis.

Treatment—Antifungal agents proved effective against *Talaromyces marneffei* infection and are amphotericin B as primary treatment and Itraconazole as secondary prophylaxis (Supparatpinyo et al. 1998).

15.7 Sporotrichosis

Sporotrichosis is caused by another thermally dimorphic fungus *Sporothrix schenckii*. But unlike other dimorphic fungi, *Sporothrix schenckii* is not confined to any geographical area, but distributed worldwide, though more common in tropical and subtropical regions. *Sporothrix schenckii* species complex is a group of phylogenetically different species, among which *S. schenckii* and *S. brasiliensis* are the most important species (Marimon et al. 2006). Sporotrichosis primarily present with characteristic nodular cutaneous lesions, which later on become ulcerated and painful. Although rare, dissemination can occur, mostly in immunosuppressed individuals, particularly in advanced HIV infection (Moreira et al. 2015; Queiroz-Telles et al. 2019).

Clinically, CNS Sporotrichosis presents either as isolated chronic meningitis, where no other organ is involved, seen primarily in immunocompetent individuals, and manifests with symptoms of chronic headache, confusion, fever, and ataxia. This presentation resembles tubercular meningitis, is associated with poor prognosis, and is difficult to diagnose (Klein et al. 1966). Other type of CNS Sporotrichosis, which is primarily seen in immunocompromised individuals, presents as a part of widespread dissemination, with features of acute/subacute meningoencephalitis, with acute onset headache, confusion, and signs of focal deficits or seizures

(Donabedian et al. 1994; Freitas et al. 2015). Hydrocephalus may also occur as a complication.

Diagnosis—The diagnosis of CNS Sporotrichosis is established by the growth of fungi in culture. However, growth of *Sporothrix* species from CSF sample in cases of isolated chronic meningitis is difficult. But *Sporothrix* species are frequently isolated from CSF samples in disseminated sporotrichosis, as there is high fungal burden in case of dissemination (Moreira et al. 2015; Queiroz-Telles et al. 2019). Fungus grows easily on saborauds dextrose agar usually in 3–5 days. It produces black or brown pigmentation on cornmeal or oatmeal agar after prolonged incubation. *Sporothrix* appears characteristically as oval to cigar-shaped yeast cells on tissues stained by methenamine silver stain. Serological tests for the diagnosis of CNS Sporotrichosis lack standardization. Different formats available are immunodiffusion, enzyme immunoassays, and latex agglutination, with variable sensitivities reported (Rudramurthy and Chakrabarti 2017; Scott et al. 1987; Scott and Muchmore 1989).

Treatment—As per the guidelines by Infectious Disease Society of America (IDSA), CNS Sporotrichosis is initially treated with liposomal amphotericin B, 5 mg/kg daily for the period of 4–6 weeks, followed by oral Itraconazole 200 mg twice daily for 1 year (Kauffman et al. 2000).

References

Agrawal J, Bansal N, Arora A (2020) Disseminated histoplasmosis in India presenting as addisonian crisis with epiglottis involvement. IDCases 21:e00844

Ajello L (1972a) ID#4235 - Public Health Image Library (PHIL). CDC

Ajello L (1972b) ID#4241 - Public Health Image Library (PHIL). CDC

Antinori S, Giacomelli A, Corbellino M, Torre A, Schiuma M, Casalini G, Parravicini C, Milazzo L, Gervasoni C, Ridolfo AL (2021) Histoplasmosis diagnosed in Europe and Israel: a case report and systematic review of the literature from 2005 to 2020. J Fungi 7:481. https://doi.org/10.3390/jof7060481

Azuma I, Kanetsuna F, Tanaka Y, Yamamura Y, Carbonell LM (1974) Chemical and immunological properties of galactomannans obtained from Histoplasma duboisii, Histoplasma capsulatum, Paracoccidioides brasiliensis and Blasomyces dermatitidis. Mycopathol Mycol Appl 54:111–125. https://doi.org/10.1007/BF02055979

Babady NE, Buckwalter SP, Hall L, Le Febre KM, Binnicker MJ, Wengenack NL (2011) Detection of Blastomyces dermatitidis and Histoplasma capsulatum from culture isolates and clinical specimens by use of real-time PCR. J Clin Microbiol 49:3204–3208. https://doi.org/10.1128/JCM.00673-11

Bahr NC, Antinori S, Wheat LJ, Sarosi GA (2015) Histoplasmosis infections worldwide: thinking outside of the Ohio River valley. Curr Trop Med Rep 2:70–80. https://doi.org/10.1007/s40475-015-0044-0

Baily GG, Robertson VJ, Neill P, Garrido P, Levy LF (1991) Blastomycosis in Africa: clinical features, diagnosis, and treatment. Rev Infect Dis 13:1005–1008. https://doi.org/10.1093/clinids/13.5.1005

Bañuelos AF, Williams PL, Johnson RH, Bibi S, Fredricks DN, Gilroy SA, Bhatti SU, Aguet J, Stevens DA (1996) Central nervous system abscesses due to Coccidioides species. Clin Infect Dis 22:240–250. https://doi.org/10.1093/clinids/22.2.240

Bariola JR, Perry P, Pappas PG, Proia L, Shealey W, Wright PW, Sizemore JM, Robinson M, Bradsher RW (2010) Blastomycosis of the central nervous system: a multicenter review of diagnosis and treatment in the modern era. Clin Infect Dis 50:797–804. https://doi.org/10.1086/650579

Bariola JR, Hage CA, Durkin M, Bensadoun E, Gubbins PO, Wheat LJ, Bradsher RW (2011) Detection of Blastomyces dermatitidis antigen in patients with newly diagnosed blastomycosis. Diagn Microbiol Infect Dis 69:187–191. https://doi.org/10.1016/j.diagmicrobio.2010.09.015

Batra V, Khararjian A, Wheat J, Zhang SX, Crain B, Baras A (2016) From suspected Creutzfeldt-Jakob disease to confirmed histoplasma meningitis. BMJ Case Rep 2016:bcr2016214937. https://doi.org/10.1136/bcr-2016-214937

Bauman DS, Smith CD (1976) Comparison of immunodiffusion and complement fixation tests in the diagnosis of histoplasmosis. J Clin Microbiol 2:77–80

Baumgardner DJ (2018) Use of urine antigen testing for blastomyces in an integrated health system. J Patient-Centered Res Rev 5:176–182. https://doi.org/10.17294/2330-0698.1452

Baumgardner DJ, Buggy BP, Mattson BJ, Burdick JS, Ludwig D (1992) Epidemiology of blastomycosis in a region of high endemicity in north central Wisconsin. Clin Infect Dis 15:629–635. https://doi.org/10.1093/clind/15.4.629

Bellissimo-Rodrigues F, Bollela VR, Da Fonseca BAL, Martinez R (2013) Endemic paracoccidioidomycosis: relationship between clinical presentation and patients' demographic features. Med Mycol 51:313–318. https://doi.org/10.3109/13693786.2012.714529

Berger JR, Greenberg RN (2010) Isolated central nervous system histoplasmosis in an immunocompetent patient: 53-month hiatus to diagnosis and treatment. J Neurovirol 16:472–474. https://doi.org/10.1007/BF03210853

Blair JE, Logan JL (2001) Coccidioidomycosis in solid organ transplantation. Clin Infect Dis 33:1536–1544. https://doi.org/10.1086/323463

Blair JE, Coakley B, Santelli AC, Hentz JG, Wengenack NL (2006) Serologic testing for symptomatic coccidioidomycosis in immunocompetent and immunosuppressed hosts. Mycopathologia 162:317–324. https://doi.org/10.1007/s11046-006-0062-5

Blastomycosis -- Wisconsin, 1986-1995 (n.d.) www.document. https://www.cdc.gov/mmwr/preview/mmwrhtml/00043101.htm. Accessed 18 Jun 2023

Bloch KC, Myint T, Raymond-Guillen L, Hage CA, Davis TE, Wright PW, Chow FC, Woc-Colburn L, Khairy RN, Street AC, Yamamoto T, Albers A, Wheat LJ (2018) Improvement in diagnosis of histoplasma meningitis by combined testing for histoplasma antigen and immunoglobulin G and immunoglobulin M anti-histoplasma antibody in cerebrospinal fluid. Clin Infect Dis 66:89–94. https://doi.org/10.1093/cid/cix706

Bradsher RW (1992) Blastomycosis. Clin Infect Dis 14(Suppl 1):S82–S90. https://doi.org/10.1093/clinids/14.supplement_1.s82

Brandhorst TT, Roy R, Wüthrich M, Nanjappa S, Filutowicz H, Galles K, Tonelli M, McCaslin DR, Satyshur K, Klein B (2013) Structure and function of a fungal adhesin that binds heparin and mimics thrombospondin-1 by blocking T cell activation and effector function. PLoS Pathog 9:e1003464

Brodsky (1966) ID#20515 - Public Health Image Library (PHIL). CDC

Brown EM, McTaggart LR, Zhang SX, Low DE, Stevens DA, Richardson SE (2013a) Phylogenetic analysis reveals a cryptic species blastomyces gilchristii, sp. nov. within the human pathogenic fungus blastomyces dermatitidis. PLoS One 8:e59237. https://doi.org/10.1371/journal.pone.0059237

Brown J, Benedict K, Park BJ, Thompson GR (2013b) Coccidioidomycosis: epidemiology. Clin Epidemiol 5:185–197. https://doi.org/10.2147/CLEP.S34434

Caceres DH, Knuth M, Derado G, Lindsley MD (2019) Diagnosis of progressive disseminated histoplasmosis in advanced HIV: a meta-analysis of assay analytical performance. J Fungi 5:76. https://doi.org/10.3390/jof5030076

Cano MV, Ponce-de-Leon GF, Tippen S, Lindsley MD, Warwick M, Hajjeh RA (2003) Blastomycosis in Missouri: epidemiology and risk factors for endemic disease. Epidemiol Infect 131: 907–914. https://doi.org/10.1017/s0950268803008987

Carlos WG, Rose AS, Wheat LJ, Norris S, Sarosi GA, Knox KS, Hage CA (2010) Blastomycosis in Indiana: digging up more cases. Chest 138:1377–1382. https://doi.org/10.1378/chest.10-0627

Castillo CG, Kauffman CA, Miceli MH (2016) Blastomycosis. Infect Dis Clin North Am 30:247–264. https://doi.org/10.1016/j.idc.2015.10.002

Centers for Disease Control and Prevention (CDC) (2013) Increase in reported coccidioidomycosis--United States, 1998-2011. MMWR Morb Mortal Wkly Rep 62:217–221

Chang CC, Liao ST, Huang WS, Liu JD, Shih LS (1995) Disseminated Penicillium marneffei infection in a patient with acquired immunodeficiency syndrome. J Formos Med Assoc 94:572–575

Chapman SW, Lin AC, Hendricks KA, Nolan RL, Currier MM, Morris KR, Turner HR (1997) Endemic blastomycosis in Mississippi: epidemiological and clinical studies. Semin Respir Infect 12:219–228

Chapman SW, Dismukes WE, Proia LA, Bradsher RW, Pappas PG, Threlkeld MG, Kauffman CA, Infectious Diseases Society of America (2008) Clinical practice guidelines for the management of blastomycosis: 2008 update by the Infectious Diseases Society of America. Clin Infect Dis 46:1801–1812. https://doi.org/10.1086/588300

Chu JH, Feudtner C, Heydon K, Walsh TJ, Zaoutis TE (2006) Hospitalizations for endemic mycoses: a population-based national study. Clin Infect Dis 42:822–825. https://doi.org/10.1086/500405

Como JA, Dismukes WE (1994) Oral azole drugs as systemic antifungal therapy. N Engl J Med 330:263–272. https://doi.org/10.1056/NEJM199401273300407

Connolly P, Hage CA, Bariola JR, Bensadoun E, Rodgers M, Bradsher RW, Wheat LJ (2012) Blastomyces dermatitidis antigen detection by quantitative enzyme immunoassay. Clin Vaccine Immunol 19:53–56. https://doi.org/10.1128/CVI.05248-11

Cuellar-Rodriguez J, Avery RK, Lard M, Budev M, Gordon SM, Shrestha NK, van Duin D, Oethinger M, Mawhorter SD (2009) Histoplasmosis in solid organ transplant recipients: 10 years of experience at a large transplant center in an endemic area. Clin Infect Dis 49:710–716. https://doi.org/10.1086/604712

da Silva SHM, Colombo AL, Blotta MHSL, Queiroz-Telles F, Lopes JD, de Camargo ZP (2005) Diagnosis of neuroparacoccidioidomycosis by detection of circulating antigen and antibody in cerebrospinal fluid. J Clin Microbiol 43:4680. https://doi.org/10.1128/jcm.43.9.4680-4683.2005

da Silva Vasconcellos IC, Dalla Lana DF, Pasqualotto AC (2019) The role of molecular tests in the diagnosis of disseminated histoplasmosis. J Fungi 6:1. https://doi.org/10.3390/jof6010001

Darling ST (1906) A protozoön general infection producing pseudotubercles in the lungs and focal necroses in the liver, spleen and lymphnodes. J Am Med Assoc XLVI:1283–1285. https://doi.org/10.1001/jama.1906.62510440037003

de Almeida SM, Queiroz-Telles F, Teive HAG, Ribeiro CEL, Werneck LC (2004) Central nervous system paracoccidioidomycosis: clinical features and laboratorial findings. J Infect 48:193–198. https://doi.org/10.1016/j.jinf.2003.08.012

de Almeida JN, Peçanha-Pietrobom PM, Colombo AL (2018) Paracoccidioidomycosis in immunocompromised patients: a literature review. J. Fungi 5:2. https://doi.org/10.3390/jof5010002

de Camargo ZP (2008) Serology of paracoccidioidomycosis. Mycopathologia 165:289–302. https://doi.org/10.1007/s11046-007-9060-5

De D, Nath UK (2015) Disseminated histoplasmosis in immunocompetent individuals—not a so rare entity, in India. Mediterr J Hematol Infect Dis 7:e2015028. https://doi.org/10.4084/MJHID.2015.028

Deepe GS, Durose GG (1995) Immunobiological activity of recombinant H antigen from Histoplasma capsulatum. Infect Immun 63:3151–3157. https://doi.org/10.1128/iai.63.8.3151-3157.1995

Deng ZL, Connor DH (1985) Progressive disseminated penicilliosis caused by Penicillium marneffei. Report of eight cases and differentiation of the causative organism from Histoplasma capsulatum. Am J Clin Pathol 84:323–327. https://doi.org/10.1093/ajcp/84.3.323

Dimitrova D, Ross L (2016) Coccidioidomycosis: experience from a Children's Hospital in an Area of Endemicity. J Pediatr Infect Dis Soc 5:89–92. https://doi.org/10.1093/jpids/piu091

DiSalvo AF, Corbett DS (1976) Apparent false positive histoplasmin latex agglutination tests in patients with tuberculosis. J Clin Microbiol 3:306–308

Donabedian H, O'Donnell E, Olszewski C, MacArthur RD, Budd N (1994) Disseminated cutaneous and meningeal sporotrichosis in an AIDS patient. Diagn Microbiol Infect Dis 18:111–115. https://doi.org/10.1016/0732-8893(94)90075-2

Drutz DJ, Frey CL (1985) Intracellular and extracellular defenses of human phagocytes against blastomyces dermatitidis conidia and yeasts. J Lab Clin Med 105:737–750

DuBois JC, Pasula R, Dade JE, Smulian AG (2016) Yeast transcriptome and in vivo hypoxia detection reveals histoplasma capsulatum response to low oxygen tension. Med Mycol 54:40–58. https://doi.org/10.1093/mmy/myv073

Duong TA (1996) Infection due to Penicillium marneffei, an emerging pathogen: review of 155 reported cases. Clin Infect Dis 23:125–130. https://doi.org/10.1093/clinids/23.1.125

Dupont B (2003) Fungal infections of the central nervous system. In: Anaissie E, McGinnis M, Pfaller M (eds) Clinical mycology. Churchill Livinstone, New York, pp 539–553

Durkin M, Witt J, Lemonte A, Wheat B, Connolly P (2004) Antigen assay with the potential to aid in diagnosis of blastomycosis. J Clin Microbiol 42:4873–4875. https://doi.org/10.1128/JCM.42.10.4873-4875.2004

Durkin M, Connolly P, Kuberski T, Myers R, Kubak BM, Bruckner D, Pegues D, Wheat LJ (2008) Diagnosis of coccidioidomycosis with use of the Coccidioides antigen enzyme immunoassay. Clin Infect Dis 47:e69–e73. https://doi.org/10.1086/592073

Durkin M, Estok L, Hospenthal D, Crum-Cianflone N, Swartzentruber S, Hackett E, Wheat LJ (2009) Detection of Coccidioides antigenemia following dissociation of immune complexes. Clin Vaccine Immunol 16:1453–1456. https://doi.org/10.1128/CVI.00227-09

Dworkin MS, Duckro AN, Proia L, Semel JD, Huhn G (2005) The epidemiology of blastomycosis in Illinois and factors associated with death. Clin Infect Dis 41:e107–e111. https://doi.org/10.1086/498152

Einstein HE, Holeman C, Sandidge L, Holden D (1961) Coccidioidal meningitis—the use of amphotericin B in treatment. Calif Med 94:339

Eissenberg LG, Goldman WE (1991) Histoplasma variation and adaptive strategies for parasitism: new perspectives on histoplasmosis. Clin Microbiol Rev 4:411–421. https://doi.org/10.1128/CMR.4.4.411

Elias J, dos Santos AC, Carlotti CG, Colli BO, Canheu A, Matias C, Furlanetti L, Martinez R, Takayanagui OM, Sakamoto AC, Serafini LN, Chimelli L (2005) Central nervous system paracoccidioidomycosis: diagnosis and treatment. Surg Neurol 63:S13–S21. https://doi.org/10.1016/j.surneu.2004.09.019

Erly WK, Bellon RJ, Seeger JF, Carmody RF (1999) MR imaging of acute coccidioidal meningitis. AJNR Am J Neuroradiol 20:509–514

Espinel-Ingroff A (1998) Comparison of in vitro activities of the new triazole SCH56592 and the echinocandins MK-0991 (L-743,872) and LY303366 against opportunistic filamentous and dimorphic fungi and yeasts. J Clin Microbiol 36:2950–2956. https://doi.org/10.1128/JCM.36.10.2950-2956.1998

Estrada-Bellmann I, Camara-Lemarroy CR, Flores-Cantu H, Calderon-Hernandez HJ, Villareal-Velazquez HJ (2016) Hemichorea in a patient with HIV-associated central nervous system histoplasmosis. Int J STD AIDS 27:75–77. https://doi.org/10.1177/0956462414564608

Frean JA, Carman WF, Crewe-Brown HH, Culligan GA, Young CN (1989) Blastomyces dermatitidis infections in the RSA. South Afr Med J 76:13–16

Freitas DFS, Lima MA, de Almeida-Paes R, Lamas CC, do Valle ACF, Oliveira MME, Zancopé-Oliveira RM, Gutierrez-Galhardo MC (2015) Sporotrichosis in the central nervous system caused by Sporothrix brasiliensis. Clin Infect Dis 61:663–664. https://doi.org/10.1093/cid/civ361

Frost HM, Novicki TJ (2015) Blastomyces antigen detection for diagnosis and management of blastomycosis. J Clin Microbiol 53:3660–3662. https://doi.org/10.1128/JCM.02352-15

Galgiani JN (1997) Coccidioidomycosis. Curr Clin Top Infect Dis 17:188–204

Galgiani JN (2009) Coccidioides species. In: Mandell G, Bennett J, Dolin R (eds) Principles and practice of infectious diseases. Elsevier Churchill Livingstone, Philadelphia, pp 3333–3344

Galgiani JN, Ampel NM, Blair JE, Catanzaro A, Geertsma F, Hoover SE, Johnson RH, Kusne S, Lisse J, MacDonald JD, Meyerson SL, Raksin PB, Siever J, Stevens DA, Sunenshine R, Theodore N (2016) Executive summary: 2016 Infectious Diseases Society of America (IDSA) clinical practice guideline for the treatment of Coccidioidomycosis. Clin Infect Dis 63:717–722. https://doi.org/10.1093/cid/ciw538

Gandhi P, Rossow S (2020) ID#24210 - Public Health Image Library (PHIL). CDC

Gauthier GM (2017) Fungal dimorphism and virulence: molecular mechanisms for temperature adaptation, immune evasion, and in vivo survival. Mediat Inflamm 2017:8491383. https://doi.org/10.1155/2017/8491383

Gauthier GM, Safdar N, Klein BS, Andes DR (2007) Blastomycosis in solid organ transplant recipients. Transpl Infect Dis 9:310–317. https://doi.org/10.1111/j.1399-3062.2007.00227.x

Gaviria M, Rivera V, Muñoz-Cadavid C, Cano LE, Naranjo TW (2015) Validation and clinical application of a nested PCR for paracoccidioidomycosis diagnosis in clinical samples from Colombian patients. Braz J Infect Dis 19:376–383. https://doi.org/10.1016/j.bjid.2015.04.008

Georg LK (1963) ID#520 – Public Health Image Library (PHIL). CDC

Georg LK (1964a) ID#22054 - Public Health Image Library (PHIL). CDC

Georg LK (1964b) ID#22300 - Public Health Image Library (PHIL). CDC

Georg LK (1966) ID#20513 - Public Health Image Library (PHIL). CDC

Georg LK (1968) ID#21832 - Public Health Image Library (PHIL). CDC

Gonzalez HH, Rane M, Cioci A, Goodman S, Espinosa PS, Gonzalez HH, Rane M, Cioci A, Goodman S, Espinosa PS (2019) Disseminated central nervous system histoplasmosis: a case report. Cureus 11:e4238. https://doi.org/10.7759/cureus.4238

Góralska K, Blaszkowska J, Dzikowiec M (2018) Neuroinfections caused by fungi. Infection 46: 443–459. https://doi.org/10.1007/s15010-018-1152-2

Gugnani HC, Muotoe-Okafor F (1997) African histoplasmosis: a review. Rev Iberoam Micol 14: 155–159

Guimarães AJ, Nosanchuk JD, Zancopé-Oliveira RM (2006) Diagnosis of histoplasmosis. Braz J Microbiol 37:1–13. https://doi.org/10.1590/S1517-83822006000100001

Hajjeh RA, Pappas PG, Henderson H, Lancaster D, Bamberger DM, Skahan KJ, Phelan MA, Cloud G, Holloway M, Kauffman CA, Wheat LJ (2001) Multicenter case-control study of risk factors for histoplasmosis in human immunodeficiency virus-infected persons. Clin Infect Dis 32:1215–1220. https://doi.org/10.1086/319756

Hamada M, Tsuji S (2009) [Central nervous system histoplasmosis]. Brain Nerve Shinkei Kenkyu No Shinpo 61:129–134

Hamilton AJ, Bartholomew MA, Figueroa J, Fenelon LE, Hay RJ (1990) Evidence that the M antigen of Histoplasma capsulatum var. capsulatum is a catalase which exhibits cross-reactivity with other dimorphic fungi. J Med Vet Mycol 28:479–485

Hardin (1965a) ID#22960 - Public Health Image Library (PHIL). CDC

Hardin (1965b) ID#22970 - Public Health Image Library (PHIL). CDC

Herrmann JA, Kostiuk SL, Dworkin MS, Johnson YJ (2011) Temporal and spatial distribution of blastomycosis cases among humans and dogs in Illinois (2001-2007). J Am Vet Med Assoc 239: 335–343. https://doi.org/10.2460/javma.239.3.335

Hott JS, Horn E, Sonntag VKH, Coons SW, Shetter A (2003) Intramedullary histoplasmosis spinal cord abscess in a nonendemic region: case report and review of the literature. J Spinal Disord Tech 16:212–215. https://doi.org/10.1097/00024720-200304000-00016

Johnson RH, Einstein HE (2006) Coccidioidal meningitis. Clin Infect Dis 42:103–107. https://doi.org/10.1086/497596

Johnson JE, Roberts GD (1976) Blocking effect of rheumatoid factor and cold agglutinins on complement fixation tests for histoplasmosis. J Clin Microbiol 3:157–160. https://doi.org/10.1128/jcm.3.2.157-160.1976

Johnson JE, Jeffery B, Huppert M (1984) Evaluation of five commercially available immunodiffusion kits for detection of Coccidioides immitis and Histoplasma capsulatum antibodies. J Clin Microbiol 20:530–532

Joseph Wheat L (2003) Current diagnosis of histoplasmosis. Trends Microbiol 11:488–494. https://doi.org/10.1016/j.tim.2003.08.007

Kassis C, Zaidi S, Kuberski T, Moran A, Gonzalez O, Hussain S, Hartmann-Manrique C, Al-Jashaami L, Chebbo A, Myers RA, Wheat LJ (2015) Role of coccidioides antigen testing in the cerebrospinal fluid for the diagnosis of coccidioidal meningitis. Clin Infect Dis 61:1521–1526. https://doi.org/10.1093/cid/civ585

Kasuga T, Taylor JW, White TJ (1999) Phylogenetic relationships of varieties and geographical groups of the human pathogenic fungus histoplasma capsulatum Darling. J Clin Microbiol 37:653–663

Kauffman CA (2007) Histoplasmosis: a clinical and laboratory update. Clin Microbiol Rev 20:115–132. https://doi.org/10.1128/CMR.00027-06

Kauffman CA (2019) Central nervous system infection with other endemic mycoses: rare manifestation of blastomycosis, paracoccidioidomycosis, talaromycosis, and sporotrichosis. J Fungi 5:64. https://doi.org/10.3390/jof5030064

Kauffman CA, Hajjeh R, Chapman SW, Mycoses Study Group (2000) Practice guidelines for the management of patients with sporotrichosis. Clin Infect Dis 30:684–687. https://doi.org/10.1086/313751

Khalaf SA, Patel P, Caruso CR, Parrett T, Bran A (2022) CNS histoplasmosis as a gliosarcoma mimicker: the diagnostic dilemma of solitary brain lesions. IDCases 27:e01364. https://doi.org/10.1016/j.idcr.2021.e01364

Klein RC, Ivens MS, Seabury JH, Dascomb HE (1966) Meningitis due to Sporotrichum schenckii. Arch Intern Med 118:145–149. https://doi.org/10.1001/archinte.1966.00290140049010

Klein BS, Vergeront JM, Kaufman L, Bradsher RW, Kumar UN, Mathai G, Varkey B, Davis JP (1987) Serological tests for blastomycosis: assessments during a large point-source outbreak in Wisconsin. J Infect Dis 155:262–268. https://doi.org/10.1093/infdis/155.2.262

Klein BS, Bradsher RW, Vergeront JM, Davis JP (1990) Development of long-term specific cellular immunity after acute Blastomyces dermatitidis infection: assessments following a large point-source outbreak in Wisconsin. J Infect Dis 161:97–101. https://doi.org/10.1093/infdis/161.1.97

Kleinschmidt-DeMasters BK, Mazowiecki M, Bonds LA, Cohn DL, Wilson ML (2000) Coccidioidomycosis meningitis with massive dural and cerebral venous thrombosis and tissue arthroconidia. Arch Pathol Lab Med 124:310–314. https://doi.org/10.5858/2000-124-0310-CMWMDA

Kravitz GR, Davies SF, Eckman MR, Sarosi GA (1981) Chronic blastomycotic meningitis. Am J Med 71:501–505. https://doi.org/10.1016/0002-9343(81)90189-3

Kuberski T, Myers R, Wheat LJ, Durkin M, Connolly P, Kubak BM, Bruckner D, Pegues D (2007) Diagnosis of coccidioidomycosis by antigen detection using cross-reaction with a histoplasma antigen. Clin Infect Dis 44:e50–e54. https://doi.org/10.1086/511684

Kumar A, Sreehari S, Velayudhan K, Biswas L, Babu R, Ahmed S, Sharma N, Kurupath VP, Jojo A, Dinesh KR, Karim S, Biswas R (2014) Autochthonous blastomycosis of the adrenal: first case report from Asia. Am J Trop Med Hyg 90:735–739. https://doi.org/10.4269/ajtmh.13-0444

Le T, Huu Chi N, Kim Cuc NT, Sieu TPM, Shikuma CM, Farrar J, Day JN (2010) AIDS-associated Penicillium marneffei infection of the central nervous system. Clin Infect Dis 51:1458–1462. https://doi.org/10.1086/657400

MacInnes R, Warris A (2021) Paediatric histoplasmosis 2000–2019: a review of 83 cases. J Fungi 7:448. https://doi.org/10.3390/jof7060448

Marimon R, Gené J, Cano J, Trilles L, Dos Santos Lazéra M, Guarro J (2006) Molecular phylogeny of Sporothrix schenckii. J Clin Microbiol 44:3251–3256. https://doi.org/10.1128/JCM.00081-06

Mata-Essayag S, Colella MT, Roselló A, de Capriles CH, Landaeta ME, de Salazar CP, Magaldi S, Olaizola C, Calatroni MI, Garrido L (2008) Histoplasmosis: a study of 158 cases in Venezuela, 2000-2005. Medicine (Baltimore) 87:193–202. https://doi.org/10.1097/MD.0b013e31817fa2a8

Mathisen G, Shelub A, Truong J, Wigen C (2010) Coccidioidal meningitis: clinical presentation and management in the fluconazole era. Medicine (Baltimore) 89:251–284. https://doi.org/10.1097/MD.0b013e3181f378a8

Mazi PB, Sahrmann JM, Olsen MA, Coler-Reilly A, Rauseo AM, Pullen M, Zuniga-Moya JC, Powderly WG, Spec A (2023) The geographic distribution of dimorphic mycoses in the United States for the modern era. Clin Infect Dis 76:1295–1301. https://doi.org/10.1093/cid/ciac882

McKinsey DS, McKinsey JP (2011) Pulmonary histoplasmosis. Semin Respir Crit Care Med 32: 735–744. https://doi.org/10.1055/s-0031-1295721

McKinsey DS, Kauffman CA, Pappas PG, Cloud GA, Girard WM, Sharkey PK, Hamill RJ, Thomas CJ, Dismukes WE (1996) Fluconazole therapy for histoplasmosis. The National Institute of Allergy and Infectious Diseases Mycoses Study Group. Clin Infect Dis 23:996–1001. https://doi.org/10.1093/clinids/23.5.996

McLeod DSA, Mortimer RH, Perry-Keene DA, Allworth A, Woods ML, Perry-Keene J, McBride WJH, Coulter C, Robson JMB (2011) Histoplasmosis in Australia: report of 16 cases and literature review. Medicine (Baltimore) 90:61–68. https://doi.org/10.1097/MD.0b013e318206e499

Michelson JB, Belmont JB, Higginbottom P (1983) Juxtapapillary choroiditis associated with chronic meningitis due to Coccidioides immitis. Ann Ophthalmol 15:666–668

Mochi A, Edwards PQ (1952) Geographical distribution of histoplasmosis and histoplasmin sensitivity. Bull World Health Organ 5:259–291

Moreira JAS, Freitas DFS, Lamas CC (2015) The impact of sporotrichosis in HIV-infected patients: a systematic review. Infection 43:267–276. https://doi.org/10.1007/s15010-015-0746-1

Murthy JMK, Sundaram C (2014) Fungal infections of the central nervous system. Handb Clin Neurol 121:1383–1401. https://doi.org/10.1016/B978-0-7020-4088-7.00095-X

Myint T, Anderson AM, Sanchez A, Farabi A, Hage C, Baddley JW, Jhaveri M, Greenberg RN, Bamberger DM, Rodgers M, Crawford TN, Wheat LJ (2014) Histoplasmosis in patients with human immunodeficiency virus/acquired immunodeficiency syndrome (HIV/AIDS). Medicine (Baltimore) 93:11–18. https://doi.org/10.1097/MD.0000000000000016

Myint T, Chow FC, Bloch KC, Raymond-Guillen L, Davis TE, Wright PW, Woc-Colburn L, Khairy RN, Street AC, Yamamoto T, Albers A, Wheat LJ, Hage CA (2018) Detection of (1,3)-β-d-glucan in cerebrospinal fluid in histoplasma meningitis. J Clin Microbiol 56: e00663-18. https://doi.org/10.1128/JCM.00663-18

Nacher M, Adenis A, Blanchet D, Vantilcke V, Demar M, Basurko C, Gaubert-Maréchal E, Dufour J, Aznar C, Carme B, Couppié P (2014) Risk factors for disseminated histoplasmosis in a cohort of HIV-infected patients in French Guiana. PLoS Negl Trop Dis 8:e2638. https://doi.org/10.1371/journal.pntd.0002638

Negroni R, De Flores CI, Robles AM (1976) [Study of serologic cross reactions between the antigens of Paracoccidioides brasiliensis and Histoplasma capsulatum]. Rev Asoc Argent Microbiol 8:68–73

Nguyen FN, Kar JK, Zakaria A, Schiess MC (2013) Isolated central nervous system histoplasmosis presenting with ischemic pontine stroke and meningitis in an immune-competent patient. JAMA Neurol 70:638–641. https://doi.org/10.1001/jamaneurol.2013.1043

Noritomi DT, Bub GL, Beer I, da Silva ASF, de Cleva R, Gama-Rodrigues JJ (2005) Multiple brain abscesses due to Penicillium spp infection. Rev Inst Med Trop Sao Paulo 47:167–170. https://doi.org/10.1590/s0036-46652005000300010

Odio CM, Navarrete M, Carrillo JM, Mora L, Carranza A (1999) Disseminated histoplasmosis in infants. Pediatr Infect Dis J 18:1065–1068. https://doi.org/10.1097/00006454-199912000-00007

Pappas PG, Pottage JC, Powderly WG, Fraser VJ, Stratton CW, McKenzie S, Tapper ML, Chmel H, Bonebrake FC, Blum R (1992) Blastomycosis in patients with the acquired immunodeficiency syndrome. Ann Intern Med 116:847–853. https://doi.org/10.7326/0003-4819-116-10-847

Peçanha PM, de Souza S, Falqueto A, Grão-Veloso TR, Lírio LV, Ferreira CUG, Santos AR, Costa HG, de Souza LRM, Tuon FF (2016) Amphotericin B lipid complex in the treatment of severe paracoccidioidomycosis: a case series. Int J Antimicrob Agents 48:428–430. https://doi.org/10.1016/j.ijantimicag.2016.06.011

Pedroso VSP, Lyon AC, Araújo SA, Veloso JMR, Pedroso ERP, Teixeira AL (2012) Paracoccidioidomycosis case series with and without central nervous system involvement. Rev Soc Bras Med Trop 45:586–590. https://doi.org/10.1590/S0037-86822012000500009

Queiroz-Telles F, Buccheri R, Benard G (2019) Sporotrichosis in immunocompromised hosts. J Fungi 5:8. https://doi.org/10.3390/jof5010008

Ragland AS, Arsura E, Ismail Y, Johnson R (1993) Eosinophilic pleocytosis in coccidioidal meningitis: frequency and significance. Am J Med 95:254–257. https://doi.org/10.1016/0002-9343(93)90276-u

Ramírez JA, del Rocío Reyes-Montes M, Rodríguez-Arellanes G, Pérez-Torres A, Taylor ML (2023) Central nervous system histoplasmosis: an updated insight. Pathogens 12:681. https://doi.org/10.3390/pathogens12050681

Ranjana KH, Priyokumar K, Singh TJ, Gupta CC, Sharmila L, Singh PN, Chakrabarti A (2002) Disseminated Penicillium marneffei infection among HIV-infected patients in Manipur state, India. J Infect 45:268–271. https://doi.org/10.1053/jinf.2002.1062

Reis F, Collier PP, Souza TF, Lopes GP, Bronzatto E, Silva Junior NA, Pereira RM, Appenzeller S (2013) Neuroparacoccidioidomycosis (NPCM): magnetic resonance imaging (MRI) findings. Mycopathologia 175:181–186. https://doi.org/10.1007/s11046-012-9607-y

Restrepo A, Tobón A, Clark B, Graham DR, Corcoran G, Bradsher RW, Goldman M, Pankey G, Moore T, Negroni R, Graybill JR (2007) Salvage treatment of histoplasmosis with posaconazole. J Infect 54:319–327. https://doi.org/10.1016/j.jinf.2006.05.006

Richardson MD, Warnock DW (1983) Enzyme-linked immunosorbent assay and its application to the serological diagnosis of fungal infection. Sabouraudia 21:1–14. https://doi.org/10.1080/00362178385380031

Richer SM, Smedema ML, Durkin MM, Brandhorst TT, Hage CA, Connolly PA, Leland DS, Davis TE, Klein BS, Wheat LJ (2014) Development of a highly sensitive and specific blastomycosis antibody enzyme immunoassay using blastomyces dermatitidis surface protein BAD-1. Clin Vaccine Immunol 21:143–146. https://doi.org/10.1128/CVI.00597-13

Riddell J, Wheat LJ (2019) Central nervous system infection with histoplasma capsulatum. J Fungi 5:70. https://doi.org/10.3390/jof5030070

Rivera IV, Curless RG, Indacochea FJ, Scott GB (1992) Chronic progressive CNS histoplasmosis presenting in childhood: response to fluconazole therapy. Pediatr Neurol 8:151–153. https://doi.org/10.1016/0887-8994(92)90040-6

Rocco NM, Carmen JC, Klein BS (2011) Blastomyces dermatitidis yeast cells inhibit nitric oxide production by alveolar macrophage inducible nitric oxide synthase. Infect Immun 79:2385–2395. https://doi.org/10.1128/IAI.01249-10

Roos KL, Bryan JP, Maggio WW, Jane JA, Scheld WM (1987) Intracranial blastomycoma. Medicine (Baltimore) 66:224–235. https://doi.org/10.1097/00005792-198705000-00006

Rudramurthy SM, Chakrabarti A (2017) Sporotrichosis: update on diagnostic techniques. Curr Fungal Infect Rep 11:134–140. https://doi.org/10.1007/s12281-017-0283-8

Ryan D, Lippert WC, Fenwick AJ, Bhatt M, Jones CR (2019) Central nervous system blastomycosis with multiple brain abscesses presenting as right upper extremity weakness. Neurohospitalist 9:230–234. https://doi.org/10.1177/1941874419829276

Saccente M, Woods GL (2010) Clinical and laboratory update on blastomycosis. Clin Microbiol Rev 23:367–381. https://doi.org/10.1128/CMR.00056-09

Saccente M, McDonnell RW, Baddour LM, Mathis MJ, Bradsher RW (2003) Cerebral histoplasmosis in the azole era: report of four cases and review. South Med J 96:410–416. https://doi.org/10.1097/01.SMJ.0000051734.53654.D0

Sarosi GA, Davies SF (1979) Blastomycosis. Am Rev Respir Dis 120:911–938. https://doi.org/10.1164/arrd.1979.120.4.911

Sarosi GA, Hammerman KJ, Tosh FE, Kronenberg RS (1974) Clinical features of acute pulmonary blastomycosis. N Engl J Med 290:540–543. https://doi.org/10.1056/NEJM197403072901004

Schestatsky P, Chedid MF, Amaral OB, Unis G, Oliveira FM, Severo LC (2006) Isolated central nervous system histoplasmosis in immunocompetent hosts: a series of 11 cases. Scand J Infect Dis 38:43–48. https://doi.org/10.1080/00365540500372895

Scott EN, Muchmore HG (1989) Immunoblot analysis of antibody responses to Sporothrix schenckii. J Clin Microbiol 27:300–304. https://doi.org/10.1128/jcm.27.2.300-304.1989

Scott EN, Kaufman L, Brown AC, Muchmore HG (1987) Serologic studies in the diagnosis and management of meningitis due to Sporothrix schenckii. N Engl J Med 317:935–940. https://doi.org/10.1056/NEJM198710083171505

Shankar SK, Mahadevan A, Sundaram C, Sarkar C, Chacko G, Lanjewar DN, Santosh V, Yasha TC, Radhakrishnan VV (2007) Pathobiology of fungal infections of the central nervous system with special reference to the Indian scenario. Neurol India 55:198. https://doi.org/10.4103/0028-3886.35680

Shehab Z, Shehab K (2013) Coccidioidomycosis. In: Feigin and Cherry's textbook of pediatric infectious diseases. W. B. Saunders, Philadelphia, pp 2769–2779

Shetter AG, Fischer DW, Flom RA (1985) Computed tomography in cases of coccidioidal meningitis, with clinical correlation. West J Med 142:782–786

Shikanai-Yasuda MA, Mendes RP, Colombo AL, de Queiroz-Telles F, Kono ASG, Paniago AMM, Nathan A, Valle ACF, Bagagli E, Benard G, Ferreira MS, de Melo Teixeira M, Silva-Vergara ML, Pereira RM, de Souza Cavalcante R, Hahn R, Durlacher RR, Khoury Z, de Camargo ZP, Moretti ML, Martinez R (2017) Brazilian guidelines for the clinical management of paracoccidioidomycosis. Rev Soc Bras Med Trop 50:715–740. https://doi.org/10.1590/0037-8682-0230-2017

Sil A, Andrianopoulos A (2015) Thermally dimorphic human fungal pathogens—polyphyletic pathogens with a convergent pathogenicity trait. Cold Spring Harb Perspect Med 5:a019794. https://doi.org/10.1101/cshperspect.a019794

Smith JA, Kauffman CA (2009) Endemic fungal infections in patients receiving tumour necrosis factor-alpha inhibitor therapy. Drugs 69:1403–1415. https://doi.org/10.2165/00003495-200969110-00002

Sondermeyer GL, Lee LA, Gilliss D, McCarty JM, Vugia DJ (2016) Epidemiology of pediatric coccidioidomycosis in California, 2000-2012. Pediatr Infect Dis J 35:166–171. https://doi.org/10.1097/INF.0000000000000952

Sugar AM, Picard M (1991) Macrophage- and oxidant-mediated inhibition of the ability of live Blastomyces dermatitidis conidia to transform to the pathogenic yeast phase: implications for the pathogenesis of dimorphic fungal infections. J Infect Dis 163:371–375. https://doi.org/10.1093/infdis/163.2.371

Supparatpinyo K, Khamwan C, Baosoung V, Nelson KE, Sirisanthana T (1994) Disseminated Penicillium marneffei infection in southeast Asia. Lancet 344:110–113. https://doi.org/10.1016/s0140-6736(94)91287-4

Supparatpinyo K, Perriens J, Nelson KE, Sirisanthana T (1998) A controlled trial of itraconazole to prevent relapse of Penicillium marneffei infection in patients infected with the human immunodeficiency virus. N Engl J Med 339:1739–1743. https://doi.org/10.1056/NEJM199812103392403

Teixeira MM, Theodoro RC, Nino-Vega G, Bagagli E, Felipe MSS (2014) Paracoccidioides species complex: ecology, phylogeny, sexual reproduction, and virulence. PLoS Pathog 10:e1004397. https://doi.org/10.1371/journal.ppat.1004397

Telles FQ, Goldani LZ, Schlamm HT, Goodrich JM, Ingroff AE, Yasuda MAS (2007) An open-label comparative pilot study of oral voriconazole and itraconazole for long-term treatment of paracoccidioidomycosis. Clin Infect Dis 45:1462–1469. https://doi.org/10.1086/522973

Torres M, Díaz H, Herrera T, Sada E (1993) Evaluation of enzyme linked immunosorbent-assay and western blot for diagnosis of histoplasmosis. Rev Investig Clin Organo Hosp Enfermedades Nutr 45:155–160

Torres-Rodríguez JM, Ribas-Forcadell E, Gascón J, López-Jodra O, Espasa M (2000) [Diagnostic usefulness of the intradermal test with histoplamin in non-endemic areas of histoplasmosis]. Rev Iberoam Micol 17:97–101

Turner S, Kaufman L, Jalbert M (1986) Diagnostic assessment of an enzyme-linked immunosorbent assay for human and canine blastomycosis. J Clin Microbiol 23:294–297

Vanittanakom N, Cooper CR, Fisher MC, Sirisanthana T (2006) Penicillium marneffei infection and recent advances in the epidemiology and molecular biology aspects. Clin Microbiol Rev 19: 95–110. https://doi.org/10.1128/CMR.19.1.95-110.2006

Vincent T, Galgiani JN, Huppert M, Salkin D (1993) The natural history of coccidioidal meningitis: VA—Armed Forces cooperative studies, 1955–1958. Clin Infect Dis 16:247–254. https://doi.org/10.1093/clind/16.2.247

Walkty A, Keynan Y, Karlowsky J, Dhaliwal P, Embil J (2018) Central nervous system blastomycosis diagnosed using the MVista® blastomyces quantitative antigen enzyme immunoassay test on cerebrospinal fluid: a case report and review of the literature. Diagn Microbiol Infect Dis 90: 102–104. https://doi.org/10.1016/j.diagmicrobio.2017.10.015

Wang Y, Zeng L, Chen X, Hao W, Yang M, Cai J, Wang Y, Yuan G, Che X (2013) [A double-antigen sandwich ELISA for detecting Penicillium marneffei Mp1p-specific antibody]. Nan Fang Yi Ke Da Xue Xue Bao 33:439–443

Wang Y-F, Xu H-F, Han Z-G, Zeng L, Liang C-Y, Chen X-J, Chen Y-J, Cai J-P, Hao W, Chan JF-W, Wang M, Fu N, Che X-Y (2015) Serological surveillance for Penicillium marneffei infection in HIV-infected patients during 2004–2011 in Guangzhou, China. Clin Microbiol Infect 21:484–489. https://doi.org/10.1016/j.cmi.2014.12.014

Wheat J, French ML, Kamel S, Tewari RP (1986) Evaluation of cross-reactions in Histoplasma capsulatum serologic tests. J Clin Microbiol 23:493–499. https://doi.org/10.1128/jcm.23.3.493-499.1986

Wheat LJ, Batteiger BE, Sathapatayavongs B (1990) Histoplasma capsulatum infections of the central nervous system. A clinical review. Medicine (Baltimore) 69:244–260. https://doi.org/10.1097/00005792-199007000-00006

Wheat LJ, Connolly-Stringfield P, Blair R, Connolly K, Garringer T, Katz BP (1991) Histoplasmosis relapse in patients with AIDS: detection using Histoplasma capsulatum variety capsulatum antigen levels. Ann Intern Med 115:936–941. https://doi.org/10.7326/0003-4819-115-12-936

Wheat J, Hafner R, Korzun AH, Limjoco MT, Spencer P, Larsen RA, Hecht FM, Powderly W (1995) Itraconazole treatment of disseminated histoplasmosis in patients with the acquired immunodeficiency syndrome. AIDS Clinical Trial Group. Am J Med 98:336–342. https://doi.org/10.1016/s0002-9343(99)80311-8

Wheat J, MaWhinney S, Hafner R, McKinsey D, Chen D, Korzun A, Shakan KJ, Johnson P, Hamill R, Bamberger D, Pappas P, Stansell J, Koletar S, Squires K, Larsen RA, Cheung T, Hyslop N, Lai KK, Schneider D, Kauffman C, Saag M, Dismukes W, Powderly W (1997) Treatment of histoplasmosis with fluconazole in patients with acquired immunodeficiency syndrome. National Institute of Allergy and Infectious Diseases Acquired Immunodeficiency Syndrome Clinical Trials Group and Mycoses Study Group. Am J Med 103:223–232. https://doi.org/10.1016/s0002-9343(97)00151-4

Wheat LJ, Garringer T, Brizendine E, Connolly P (2002) Diagnosis of histoplasmosis by antigen detection based upon experience at the histoplasmosis reference laboratory. Diagn Microbiol Infect Dis 43:29–37. https://doi.org/10.1016/s0732-8893(02)00367-x

Wheat LJ, Musial CE, Jenny-Avital E (2005) Diagnosis and management of central nervous system histoplasmosis. Clin Infect Dis 40:844–852. https://doi.org/10.1086/427880

Wheat LJ, Freifeld AG, Kleiman MB, Baddley JW, McKinsey DS, Loyd JE, Kauffman CA, Infectious Diseases Society of America (2007) Clinical practice guidelines for the management of patients with histoplasmosis: 2007 update by the Infectious Diseases Society of America. Clin Infect Dis 45:807–825. https://doi.org/10.1086/521259

Wheat J, Myint T, Guo Y, Kemmer P, Hage C, Terry C, Azar MM, Riddell J, Ender P, Chen S, Shehab K, Cleveland K, Esguerra E, Johnson J, Wright P, Douglas V, Vergidis P, Ooi W, Baddley J, Bamberger D, Khairy R, Vikram H, Jenny-Avital E, Sivasubramanian G, Bowlware K, Pahud B, Sarria J, Tsai T, Assi M, Mocherla S, Prakash V, Allen D, Passaretti C, Huprikar S, Anderson A (2018) Central nervous system histoplasmosis. Medicine (Baltimore) 97:e0245. https://doi.org/10.1097/MD.0000000000010245

Williams PL (2001) Vasculitic complications associated with coccidioidal meningitis. Semin Respir Infect 16:270–279. https://doi.org/10.1053/srin.2001.29319

Winn W (1964) The treatment of coccidioidal meningitis—the use of amphotericin B in a group of 25 patients. Calif Med 101:78

Zarrin M, Mahmoudabadi AZ (2010) Central nervous system fungal infections; a review article. Jundishapur J Microbiol 3(2):41–47

Melanized Fungi in CNS Infections 16

Adesh Shrivastava, Rakesh Mishra, and Karuna Tadepalli

Abstract

There are increasing reports of central nervous system infections due to melanized fungi worldwide. These brown pigment-containing fungi are often grouped under dematiaceous mycosis or phaeohyphomycosis. They are predominantly responsible for cerebral infections and brain abscesses, *Cladophialophora bantiana, Exophiala dermatitidis, and Phialophora.* Fungi under the *Bipolaris genus are found in* immunocompetent individuals who present with a chronic sinusitis which aggravates into secondary cerebral infection. The vehicle of transmission of phaeohyphomycosis is usually of exogenous nature, e.g., thorn/ wood pricks. The melanin-like pigment found in their cell walls is responsible for immune evasion mechanisms. These golden-brown pigments can assist in making a presumptive diagnosis based on their microscopic visualization when examining histopathology samples. There are also cases of hematogenous spread leading to central nervous system involvement, due to penetrating head injuries and direct spread. *Curvularia* is a melanized mold found ubiquitously in soil, which although extremely rare, has the potential to cause life-threatening brainstem infection.

A. Shrivastava (✉)
Department of Neurosurgery, All India Institute of Medical Sciences, Bhopal, Madhya Pradesh, India

R. Mishra
Department of Neurosurgery, IMS Banaras Hindu University, Varanasi, Uttar Pradesh, India

K. Tadepalli
Department of Microbiology, All India Institute of Medical Sciences, Bhopal, Madhya Pradesh, India
e-mail: karuna.microbiology@aiimsbhopal.edu.in

H. Sami et al. (eds.), *Viral and Fungal Infections of the Central Nervous System: A Microbiological Perspective*, https://doi.org/10.1007/978-981-99-6445-1_16

Keywords

Central nervous system infections · Melanized fungi · Phaeohyphomycosis · *Cladophialophora* · *Exophiala* · *Phialophora* · *Bipolaris* · *Curvularia*

16.1 Introduction

The central nervous system (CNS) is more privileged than other organ systems from infections because of anatomical and physiological barriers like blood, brain, and cerebrospinal fluid (CSF). Even though CNS infections are not as common as lung or kidney infections, there has been an increase in the rate of CNS infections in the past few decades owing to several reasons for decreasing immunity status. Though CNS infections are uncommon, they result in unprecedented mortality and morbidity. They significantly impact the quality of life of the affected individual as a residual disability, even if nonfatal. Therefore, there is a merit in understanding of and detailed accounts of CNS infections.

Usually, CNS infections are encountered as opportunistic infections and primary infections primarily by bacteria or viruses. Tuberculosis is a significant burden in poor socioeconomic status and is mainly endemic in regions of southeast Asia and Africa and rarely seen in the west. Further rare are the fungal infections of the CNS, and CNS mycoses often result in decreased systemic immunity. Unlike other infections, several classes of fungus can infect the CNS, and the site, severity, underlying disease, and clinical course differ according to the fungal species (Perfect and Durack 1997). The present chapter focuses on melanized fungi in CNS infections where melanization of the fungi is the primary feature either as phenotypic appearance or in the histology.

As we recognize that the field will continue to develop and broaden as a result of our growing understanding of and experience with these clinically significant fungi, our goal is not to review every publication on melanized fungi but rather to provide a comprehensive but in-depth overview of the field as it stands at the moment. We will be restricting the present chapter to CNS melanized fungi clinical aspects as the details on the ecology, classification, taxonomy, nomenclature, and identification shall be covered in other chapters.

16.2 Melanized Fungi

Melanin is a highly stable molecule resistant to damage by the physicochemical processes owing to its molecular structure. Its role in the pathogenesis of several fungi has been explained in detail in several studies (Bloomfield and Alexander 1967; Casadevall et al. 2000; Gómez and Nosanchuk 2003; Jacobson 2000; Liu and Nizet 2009; Nosanchuk and Casadevall 2003). Melanocytes produce melanin by oxidation of the amino acid tyrosine, followed by its polymerization. There are several mechanisms by which melanin provides unique protective features to fungal agents. For example, melanin may protect the fungi from harmful ionizing radiations, heat and cold shock, and chelating metal ions that may be toxic to the fungal cells (Hong and Simon 2007; Pacelli et al. 2017; Cordero et al. 2018; Khajo et al. 2011; Wang and Casadevall 1994). There have been changes in the terms used to denote melanized fungi. *Sporothrix schenckii* was one of the oldest melanized fungi, and later, all the fungi with similar characteristics were termed "Sporotrichoid." However, the gradually more useful term "Phaeooid" represents melanized fungi. Phaeo is a Greek term that means dark. For example, phaeohyphomycoses means infection by dark-walled fungi. Some people also use the word "dematiaceous" for melanized fungi; however, others mention that "dematiaceous" is a misnomer as the Greek phrase "deme" means bundle (Pappagianis and Ajello 1994).

There is confusion about grouping a fungal agent as melanized fungi. This is because melanin has been found to some extent in almost all clinically significant fungi like *Histoplasma capsulatum*, *Paracoccidioides brasiliensis*, *Aspergillus* spp., and even *Candida albicans* as there is no reliable method to quantify the amount of the melanin in the fungus. Therefore, some people advocate that the term melanized fungi should be reserved for the fungus with brown-pigmented hyphae in the tissue seen without any staining. Melanized fungi are commonly present in the environment; still, the infection tends to be rare. Most cases of infections by melanized fungi occur as opportunistic infections; however, there are reports of fatal infections in individuals with normal immunity (Revankar et al. 2002, 2004).

The clinical syndromes caused by the melanized fungi are classified as eumycetoma, Chromoblastomycosis, and Phaeohyphomycosis based on histological appearance. Eumycetoma and Chromoblastomycosis have a small number of fungi that result in deep tissue infection in lower extremities and tropical environments, respectively (Pang et al. 2004; Mcginnis 1983).

The phaeohyphomycoses are the fungal infections not covered by the earlier two (Revankar 2006). Melanized CNS fungal infections are covered under the umbrella term of Phaeohyphomycosis. Fungal agents cause Rhinocerebral Phaeomycosis under the Ascomycetes order Pleosporales, the pathogens on the grasses and spread by the air.

16.3 Melanized CNS Fungal Infections

Revankar et al. reported a review of 101 cases from 1966 to 2002 of melanized fungal CNS infections in 2004. They found that the brain abscess was the most common presentation; more than half had normal immunity, and the outcome was poor with mortality >50% irrespective of the individual's immune status (Revankar et al. 2004). In the review, the authors found that the most frequently isolated species for primary CNS melanized fungal infection was *Cladophialophora bantiana*, followed by *Ramichloridium mackenziei* (found exclusively in patients from the middle east) (Revankar et al. 2004). Other predominant species that cause primary CNS melanized fungal infection were *Ochroconis gallopava*, *Bipolaris spicifera*,

Exophiala dermatitidis, and *Chaetomium strumarium* (Revankar et al. 2004; Hong and Simon 2007; Chakrabarti 2007).

Secondary infections are due to grass-inhabiting species and secondary to chronic fungal sinusitis. They include *Bipolaris, Dissitimurus, Exserohilum, Curvularia lunata, Cladosporium cladosporioides, Nodulisporium species, and Pleosporaceae* (Hong and Simon 2007). The authors further found that though most of the cases occurred in the rural settings, there was no specific exposure in individual patients. Several other melanized fungal CNS infections were reported in the last two decades, and a new species, *Fonsecaea monophora,* was identified with a preference for the central nervous system.

16.4 Pathophysiology

There are animal models that illustrate CNS infection with melanized fungi in immunocompromised rats (Al-Abdely et al. 2005b; Dixon et al. 1987). However, it is not clear why some melanized fungi prefer the central nervous system in individuals with normal immunity. The route of CNS spread could be nasal and paranasal sinuses in immunocompromised individuals. In immunocompetent individuals, the primary route of CNS infection is thought to be due to the hematogenous spread from subclinical pulmonary infective foci.

The most common neurotropic melanized fungal species belong to the *Herpotrichiellaceae* and are acquired by inhalation and then rapidly disseminate to the CNS via the hematogenous spread. However, the exact route of CNS infection in immunocompetent individuals is difficult to describe because of the long asymptomatic period. The author's source is iatrogenic infection reported in contaminated steroid epidural injection with mortality>60%. Centers for Disease Control and Prevention (CDC) (2002) However, certain infections seem to have come about as a result of hematogenous spread, direct inoculation from penetrating head trauma, and through contaminated wounds.

Most CNS infections are believed to be secondary to extension from paranasal sinuses. (Skovrlj et al. 2014) Pleosporales has large conidia that cannot enter the lungs and get trapped in the sinuses and result in allergic sinusitis with gradual increase in the fungal ball and extended polyposis. As the conidia geminate in the sinus and due to pressure erosion of the bony wall of the sphenoid and ethmoid sinus due to chronic sinusitis, these are secondary cerebral infections. However, the above-mentioned route for the cerebral infection is rare; it has high fatality rate. Another difference between Rhinocerebral phaeomycosis and cerebral Phaeohyphomycosis is that the agents in the first one are grass pathogens and cause secondary cerebral infection. At the same time, in the latter, they are predominantly neurotropic fungi and are thought to reach the CNS via the hematogenous route. It is also believed that causative agents are fungi adapted to spend some part of their life cycle in the vertebrates in cases of cerebral Phaeohyphomycosis.

16.5 Clinical Manifestations

In the identified species causing the melanized fungal CNS infection, brain abscess was the most commonly identified presentation with standard treatment protocols. Excision tends to yield better results than the simple aspiration of the abscess (Revankar et al. 2004). The usual clinical presentation consisted of headache, seizures, and neurological deficits depending on the cortical location of the fungal lesion. In these cases, the mortality was >70% (Revankar et al. 2004). However, in patients with features of encephalitis and diffuse brain involvement, the mortality reported is as high as 100% (Abbott et al. 1995; Biggs et al. 1986). Meningitis is also reported in several reports of melanized fungal infection and is difficult to treat and associated with high mortality rates (Al-Aidaroos et al. 2007; Banerjee et al. 2002). Apart from this, there are reports of patients presenting with low backache (Shrivastava et al. 2017). Shrivastava et al. reported a middle-aged farmer who complained of low back ache and decreased urinary sensation. On neuroimaging, they found epidural collection with intrathecal extension from the D10-L2 vertebra causing spinal stenosis and compressing the cauda equina nerve roots. The authors identified the species as *Curvularia lunata* and successfully treated the patient with laminectomy and surgical excision and a combination of voriconazole and liposomal amphotericin B (Shrivastava et al. 2017).

16.6 Diagnosis

The diagnosis of Phaeohyphomycosis or neurological infections by melanized fungi depends on the clinical correlation with radiological evidences, sample as tissue or CSF analysis for direct microscopy, culture isolation, processing of culture for fungal genus, and species identification (Figs. 16.1, 16.2, 16.3, 16.4, 16.5 and 16.6). In case of difficulty in speciation by conventional methods, molecular confirmation is required for choice of treatment modality.

Molecular confirmation is the only and definitive diagnostic method for nonsporulating melanized fungus.

16.7 Medical Management

Combination therapy is more effective than monotherapy (Garg et al. 2007). There have been several reports on different treatment modules for melanized CNS infections; however, it is unclear if they are effective. There is no consensus on optimal antifungals to be used in these infections. The reported antifungals include amphotericin B, flucytosine, itraconazole, and Posaconazole (De Lastours et al. 2003; Nesky et al. 2000; Revankar et al. 2004). There are even reports of failure of medical management with the above-mentioned antifungals and it is suggested to use a combination of these antifungals (Levin et al. 2004; Lyons et al. 2005). In an animal study, it was found that the addition of flucytosine to the Posaconazole and

Fig. 16.1 Epidural abscess on MRI

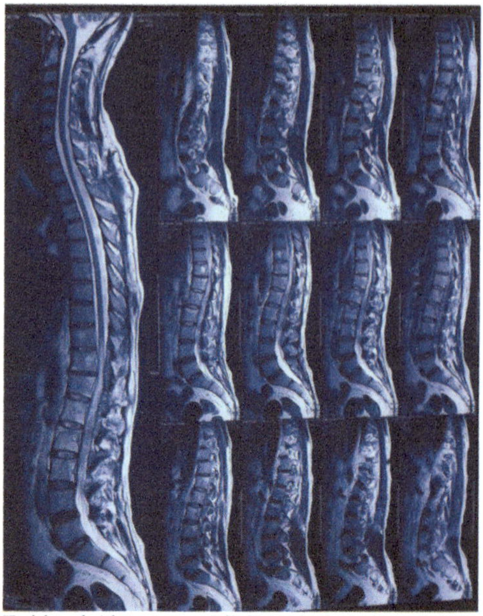

Fig. 16.2 Epidural abscess showing blackish soft material

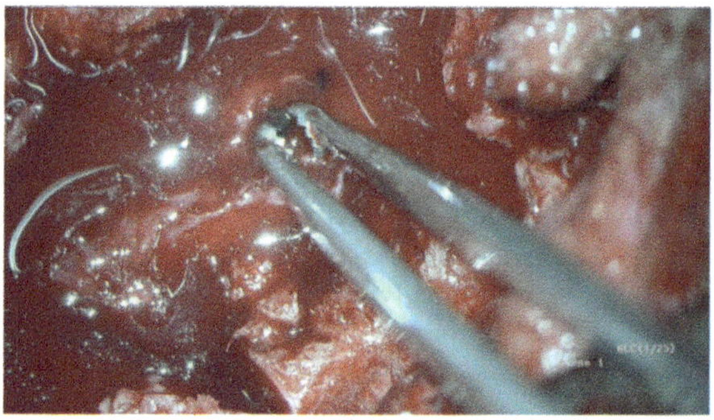

micafungin had better outcome than other antifungals in infection with *Cladosporium bantana* (Mariné et al. 2009). Studies with long-term follow-up of cerebral phaeohyphomycosis caused by *Ramichloridium mackenziei* species have shown 100% mortality. Al-abdely et al. reported successful treatment in a 62-year-old patient post-renal transplant and cerebral abscess by *Ramichloridium mackenziei* with posaconazole oral suspension, 800 mg/day, in divided doses after the failed treatment and recurrent lesion after therapy with liposomal amphotericin B, itraconazole, and 5-flucytosine (Al-Abdely et al. 2005a).

Fig. 16.3 Sabouraud dextrose agar from the black material with tissue from epidural space showing olive green growth with black pigment on reverse

Sabouraud dextrose agar from the black material with tissue from epidural space showing Olive green growth with black pigment on reverse.

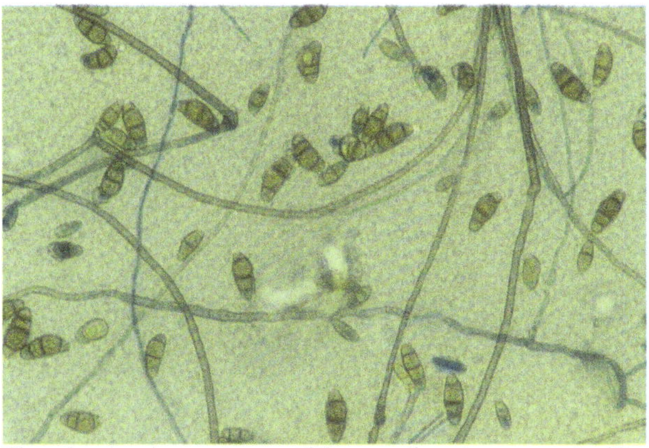

The microscopic appearance of the growth on tease mount with Lactophenol cotton blue shows melanized and hyaline fungal hyphae with brownish curved distoseptate conidia with three transverse septation and third cell enlarged suggestive of a melanized fungus *Curvularia lunata.*

Fig. 16.4 The microscopic appearance of the growth on tease mount with Lactophenol cotton blue shows melanized and hyaline fungal hyphae with brownish curved distoseptate conidia with three transverse septation and third cell enlarged suggestive of a melanized fungus *Curvularia lunata*

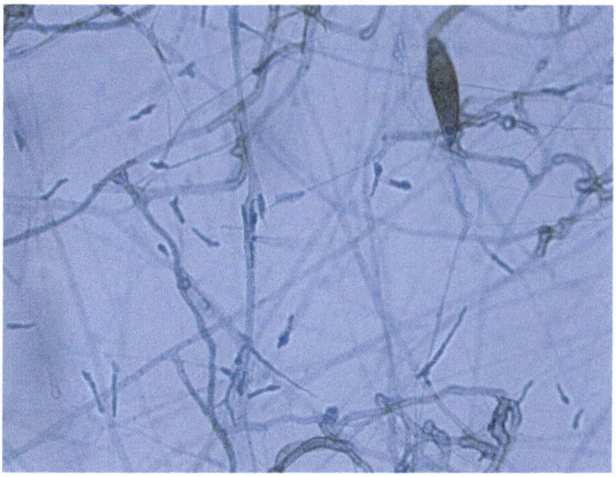

Fig. 16.5 Melanized fungus *Bipolaris cactivora* isolated from temporal lobe involvement

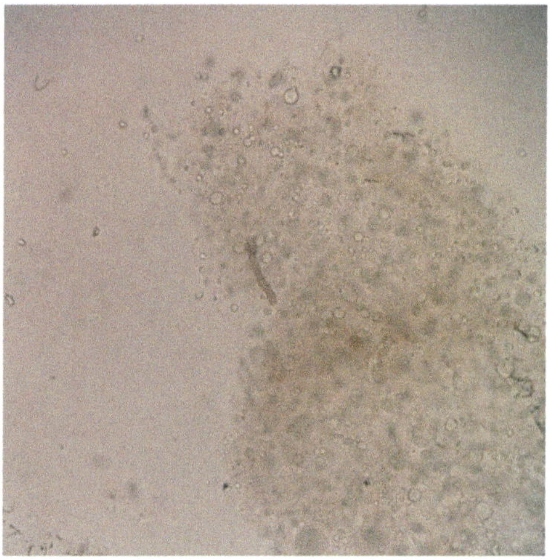

Fig. 16.6 Melanized hyphal fragment in tissue sample observed on direct microscopy by 20% potassium hydroxide

16.8 Surgical Management

According to the reported literature, complete excision of the abscess was more fruitful than partial excision and aspiration. A retrospective series on 10 cases of CNS cladosporiosis from south India spanning three decades in immunocompetent individuals found that nine had space-occupying mass lesions, while one patient presented with features of chronic meningitis (Garg et al. 2007). The authors further reported that the mortality was 50%, and patients who underwent burr hole and aspiration of abscess required excision of the abscess. The complete excision of the abscess had better outcomes than the aspiration or the partial excision of the abscess (Garg et al. 2007). In contrast, Delfino et al. reported a case of CNS cladosporiosis that was cured after repeat surgical excision of the abscess, despite the delayed and inappropriate primary therapy (Delfino et al. 2006).

References

Abbott SP, Sigler L, Mcaleer R, Mcgough DA, Rinaldi MG, Mizell G (1995) Fatal cerebral mycoses caused by the Ascomycete Chaetomium Strumarium. J Clin Microbiol 33:2692–2698

Al-Abdely HM, Alkhunaizi AM, Al-Tawfiq JA, Hassounah M, Rinaldi MG, Sutton DA (2005a) Successful therapy of cerebral phaeohyphomycosis due to Ramichloridium mackenziei with the new Triazole posaconazole. Med Mycol 43:91–95

Al-Abdely HM, Najvar LK, Bocanegra R, Graybill JR (2005b) Antifungal therapy of experimental cerebral phaeohyphomycosis due to Cladophialophora Bantiana. Antimicrob Agents Chemother 49:1701–1707

Al-Aidaroos A, Bin-Hussain I, El Solh H, Kofide A, Thawadi S, Belgaumi A, Al Ahmari A (2007) Invasive Chaetomium infection in two immunocompromised pediatric patients. Pediatr Infect Dis J 26:456–458

Banerjee TK, Patwari AK, Dutta R, Anand VK, Chabra A (2002) Cladosporium Bantianum meningitis in a neonate. Indian J Pediatr 69:721–723

Biggs PJ, Allen RL, Powers JM, Holley HP Jr (1986) Phaeohyphomycosis complicating compound skull fracture. Surg Neurol 25:393–396

Bloomfield BJ, Alexander M (1967) Melanins and resistance of fungi to lysis. J Bacteriol 93:1276–1280

Casadevall A, Rosas AL, Nosanchuk JD (2000) Melanin and virulence in Cryptococcus Neoformans. Curr Opin Microbiol 3:354–358

Centers for Disease Control and Prevention (CDC) (2002) Exophiala infection from contaminated injectable steroids prepared by a compounding pharmacy—United States, July–November 2002. MMWR Morb Mortal Wkly Rep 51:1109–1112

Chakrabarti AJNI (2007) Epidemiology of central nervous system. Mycoses 55:191

Cordero RJ, Robert V, Cardinali G, Arinze ES, Thon SM, Casadevall AJCB (2018) Impact of yeast pigmentation on heat capture and latitudinal distribution. Curr Biol 28(16):2657–2664.e3

De Lastours V, Lefort A, Zappa M, Dufour V, Belmatoug N, Fantin B (2003) Two cases of cerebral aspergillosis successfully treated with Voriconazole. Eur J Clin Microbiol Infect Dis 22:297–299

Delfino D, De Hoog S, Polonelli L, Benecchi M, Fanti F, Galatioto S, Manti G, Cusumano V (2006) Survival of a neglected case of brain abscess caused by Cladophialophora Bantiana. Med Mycol 44:651–654

Dixon DM, Merz WG, Elliott HL, Macleay S (1987) Experimental central nervous system phaeohyphomycosis following intranasal inoculation of Xylohypha Bantiana in cortisone-treated mice. Mycopathologia 100:145–153

Garg N, Devi IB, Vajramani GV, Nagarathna S, Sampath S, Chandramouli BA, Chandramuki A, Shankar SK (2007) Central nervous system cladosporiosis: an account of ten culture-proven cases. Neurol India 55:282–288

Gómez BL, Nosanchuk JD (2003) Melanin and fungi. Curr Opin Infect Dis 16:91–96

Hong L, Simon JD (2007) Current understanding of the binding sites, capacity, affinity, and biological significance of metals in melanin. J Phys Chem 111:7938

Jacobson ES (2000) Pathogenic roles for fungal melanins. Clin Microbiol Rev 13:708–717

Khajo A, Bryan RA, Friedman M, Burger RM, Levitsky Y, Casadevall A, Magliozzo RS, Dadachova E (2011) Protection of Melanized Cryptococcus Neoformans from lethal dose gamma irradiation involves changes in Melanin's chemical structure and paramagnetism. PLoS One 6:E25092

Levin TP, Baty DE, Fekete T, Truant AL, Suh B (2004) Cladophialophora Bantiana brain abscess in a solid-organ transplant recipient: case report and review of the literature. J Clin Microbiol 42: 4374–4378

Liu GY, Nizet V (2009) Color me bad: microbial pigments as virulence factors. Trends Microbiol 17:406–413

Lyons MK, Blair JE, Leslie KO (2005) Successful treatment with voriconazole of fungal cerebral abscess due to Cladophialophora Bantiana. Clin Neurol Neurosurg 107:532–534

Mariné M, Pastor FJ, Guarro J (2009) Combined antifungal therapy in a murine model of disseminated infection by Cladophialophora Bantiana. Med Mycol 47:45–49

Mcginnis MR (1983) Chromoblastomycosis and phaeohyphomycosis: new concepts, diagnosis, and mycology. J Am Acad Dermatol 8:1–16

Nesky MA, Mcdougal EC, Peacock JE Jr (2000) Pseudallescheria Boydii brain abscess successfully treated with voriconazole and surgical drainage: case report and literature review of central nervous system Pseudallescheriasis. Clin Infect Dis 31:673–677

Nosanchuk JD, Casadevall A (2003) The contribution of melanin to microbial pathogenesis. Cell Microbiol 5:203–223

Pacelli C, Bryan RA, Onofri S, Selbmann L, Shuryak I, Dadachova EJEM (2017) Melanin is effective in protecting fast and slow growing fungi from various types of ionizing radiation. Environ Microbiol 19:1612–1624

Pang KR, Wu JJ, Huang DB, Tyring SK (2004) Subcutaneous fungal infections. Dermatol Ther 17: 523–531

Pappagianis D, Ajello L (1994) Dematiaceous—a mycologic misnomer? J Med Vet Mycol 32:319–321

Perfect J, Durack DJ (1997) Fungal meningitis. 2:721–739

Revankar SG (2006) Phaeohyphomycosis. Infect Dis Clin N Am 20:609–620

Revankar SG, Patterson JE, Sutton DA, Pullen R, Rinaldi MG (2002) Disseminated phaeohyphomycosis: review of an emerging mycosis. Clin Infect Dis 34:467–476

Revankar SG, Sutton DA, Rinaldi MG (2004) Primary central nervous system phaeohyphomycosis: a review of 101 cases. Clin Infect Dis 38:206–216

Shrivastava A, Tadepalli K, Goel G, Gupta K, Kumar Gupta P (2017) Melanized fungus as an epidural abscess: a diagnostic and therapeutic challenge. Med Mycol Case Rep 16:20–24

Skovrlj B, Haghighi M, Smethurst ME, Caridi J, Bederson JB (2014) Curvularia abscess of the brainstem. World Neurosurg 82:241.E9–241.E13

Wang Y, Casadevall A (1994) Decreased susceptibility of Melanized Cryptococcus Neoformans to Uv light. Appl Environ Microbiol 60:3864–3866

Miscellaneous CNS Fungal Infections

17

Mohd Shadab, Ronni Mol Joji, Mohammad Shahid,
and Nadeem Rais

Abstract

The fungal involvement of central nervous system (CNS) poses formidable diagnostic and therapeutic challenges. Though, the *Cryptococcus neoformans*, *Histoplasma capsulatum*, *Coccidioides immitis* cause the CNS infections mainly, but other fungi like *Penicillium*, *Scedosporium*, *Fusarium* etc have also been recognized with some serious CNS problems. Fungi from *Penicillium* genera rarely cause infection, and if so, manifest as dermatitis, onychomycosis, or keratitis. However, there are reports of neuropenicilliosis linked to cerebral aneurysms and brain abscesses with *P. chrysogenum* isolated as the etiological agent. The *Trichosporon* genus of yeasts is a recognized cause of subcutaneous infection; however, there are reported cases of CNS mycosis due to *T. asahii* manifesting as meningitis in neutropenic patients. There is also an increase in cases reporting *Fusarium* isolates as etiological agents of invasive mold disease in profoundly immune-suppressed hosts. Cytotoxic fumonisin B1 products predominantly from *F. solani* have been linked to neuron demyelination, meningitis, and cerebral abscess. Another rarely encountered CNS mycoses is that associated with *Scedosporium apiospermum*. These species have potential to cause disseminated illness and neurological involvement manifesting as single or multiple brain abscesses, with mortality rates as high as 75%. Diabetics, transplant

M. Shadab (✉) · R. M. Joji · M. Shahid
Department of Microbiology, Immunology, and Infectious Diseases, College of Medicine and Medical Sciences, Arabian Gulf University, Manama, Kingdom of Bahrain
e-mail: shadab@agu.edu.bh; ronnimj@agu.edu.bh; mohammeds@agu.edu.bh

N. Rais
Department of Pharmacology, Al-Nakheel Medical Center (NMC Healthcare), Rabigh, Kingdom of Saudi Arabia

H. Sami et al. (eds.), *Viral and Fungal Infections of the Central Nervous System: A Microbiological Perspective*, https://doi.org/10.1007/978-981-99-6445-1_17

recipients, and leukemic patients are at highest risk. Here, we described the involvement of these four genera with central nervous system.

Keywords

Trichosporon · *Penicillium* · *Fusarium* · *Scedosporium* · CNS · Fungal infection

17.1 Introduction

Although infections of central nervous system (CNS) are rarely caused by fungi, once it crosses the blood-brain barrier, it is usually fatal and life threatening to patients. There are approximately 70,000 species well described to the world among 1.5 million existing fungal species. And nearly 300 of them are found virulent to humans (Goralska et al. 2018). 10–15% of virulent species could involve with central nervous system. The three most prevalent fungi that cause brain infections are *Candida, Aspergillus,* and *Cryptococcus*. But there are some other fungal species like *Trichosporon, Fusarium, Penicillium,* and *Scedosporium* which can cause a serious CNS infection, though the CNS infections of these are reported a few. CNS involvement of *Penicillium*, *Scedosporium*, *Trichosporon,* and *Fusarium* is found with a few underlying conditions like tissue transplant, immune status, malignancy, prolong steroid therapy, etc. Voriconazole and amphotericin B are found to be the most exploited drugs against these fungi. Their CNS infections are presented with several clinical presentations, mainly meningitis, meningospondylodiscitis, encephalitis, cerebritis, hydrocephalus, spinal cord abscess, cerebral abscess, meningoventriculitis, intraventricular fungal ball, and intracerebral lesion, etc. Common routes of transmission are inhalation, contamination of trauma or injury, and surgery. The diagnosis is always made according to the clinical presentation and clinical history as well. CSF, histology of incised tissue, and sometimes blood specimen are taken for culture and microscopy to examine the presence of fungi. MRI (magnetic resonance imaging) and CT (computed tomography) scan are crucial in the diagnosis of fungal CNS infections (Goralska et al. 2018).

17.2 Penicillium

Penicillium spp., which are Ascomycota, are frequently found in the environment, grow in decomposing organic materials, and they can also harm plants. About 200 species make up the genus. They may be linked to the formation of allergy symptoms because of their small size and high production of spores, and on extremely rare occasions, they can infect people (Noritomi et al. 2005). *Penicillium* species are rarely isolated from humans, and the recorded invasion cases mostly involved immunocompetent individuals. There have been occasional reports of *P. chrysogenum, P. commune, P. citrinum, P. decumbens, and P. brevicompactum* as causative agents of penicilliosis. The most frequent clinical manifestations are conjunctivitis, superficial mycosis, and otomycosis, while less frequently occurring are urinary tract infection, lung infection, endophthalmitis, endocarditis, esophagitis, and CNS infections. The progression of neuropenicilliosis included numerous brain edema, brain abscess, or mycotic cerebral aneurysm. *P. commune* (isolated from brain autopsy) and *P. chrysogenum* (isolated from CSF and brain biopsy) were shown to be associated with CNS illness, and in most cases, the identification was restricted to the genus (Kantarcıoğlu et al. 2004; Noritomi et al. 2005).

17.2.1 *P. commune*

A frequent mold utilized in the creation of soft cheese is *Penicillium commune*, which is regarded as an ancestral wild type of the fungus species *P. camemberti*. Both species have the ability to make cyclopiazonic acid, a metabolite that *Penicillium* species typically do not synthesize. *P. commune*, on the other hand, is a saprotroph that creates colonies that resemble soft, fluffy cotton (Pitt et al. 1986). It is common in warm, humid environments (soil, air). It may result in allergic responses, including rhinitis, coughing, rashes, or asthma. Cyclopaldinic acid and rugulovasin A and B are the main mycotoxins produced by *P. commune*. It can begin to create roquefortine, a neurotoxin, and penitrem A, a toxin that causes tremors under specific culture circumstances (Wagener et al. 1980). In the literature reports on humans, *P. commune* infections are very rare.

17.2.2 *P. chrysogenum*

P. chrysogenum has a widespread global distribution and can be found in a variety of habitats, including those with arid climates like the soil in Antarctica (De Sousa et al. 2017). These organisms can be found in sewage plants, soil, decomposing debris, and building sites (De Oliveira et al. 2023). *P. chrysogenum* is considered an opportunistic fungus and is rarely found to be the source of invasive infections, especially in people who are immunocompetent, because it has trouble growing at the temperature of the host. They can cause serious illness in immunodeficient individuals, whose risk factors include having HIV and tuberculosis concurrently, receiving stem cell transplantation therapy for a condition that requires immunomodulation, or having certain genetic flaws that produce primary immunodeficiency (Burgess et al. 2022). Despite their limited capacity to grow at elevated temperatures, which limits their ability to infect and thrive in mammals, fungi can be conditioned to develop thermal tolerance and progressive adaptation to rising temperatures brought on by climate change (Nnadi and Carter 2021). Virulence factors are by-products or a characteristic of the fungus that enhances the host-attacking potential. Proteases, which are degradative enzymes, are just one of the many virulence factors that *P. chrysogenum* is capable of producing. This fungus can also grow at 37 °C, has hemolytic activity, and synthesizes urease as additional characteristics (Jameel Far 2021).

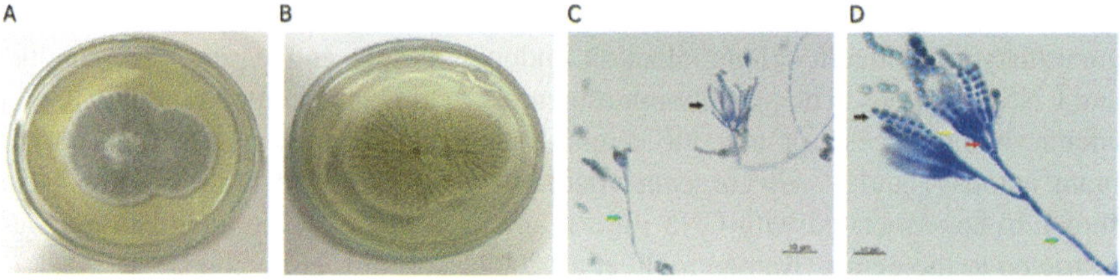

Fig. 17.1 *P. chrysogenum* in potato dextrose agar medium at 30 °C (after 7 days of culture). (**a**) Colony is grayish green and cotton texture on obverse side; (**b**) Reverse is yellow; (**c** and **d**) Cotton blue staining, showing globose conidia in chains shown by black arrows from the ends of the phialides represented by yellow arrow that emerge from metulae in a brush shown by red arrow. Metulae branched from conidiophores are marked by green arrows, (**c**) 400× (**d**) 1000× (De Oliveira et al. 2023). (Pictures are taken from de Oliveira et al. with permission (De Oliveira et al. 2023))

17.2.2.1 Growth Characteristics and Identification

P. chrysogenum typically reproduces by developing brush-shaped conidiophores into dried chains of spores (or conidia). To new colonization areas, the conidia are often carried by air currents. These organisms produce blue to blue-green conidia, and the mold can occasionally release a yellow color. Unfortunately, this species cannot be recognized just by its color. To validate its identity, observations of its morphological features and microscopic characteristics are required. DNA sequencing is also important to distinguish it from species that are closely related to it, like *P. rubens* (Böhm et al. 2013). *P. chrysogenum* was historically classified in the nebulous taxonomic group of "imperfect" fungi, or Deuteromycota, because it was believed to lack a reproductive cycle. The opportunistic pathogens like *Penicillium marneffei* (Talaromyces) and *Aspergillus fumigatus*, two close cousins of *P. chrysogenum*, had their genomes sequenced, and it was discovered that they had sexual reproduction-associated genes such as genes for pheromone generation and detection (Woo et al. 2006). Hoff et al. stated that *P. notatum* (NRRL 824) and *P. chrysogenum* (NRRL 1951) are two heterothallic species with different mating types (MAT1–1 and MAT1–2, respectively).

De Oliveira et al. reported a case of *P. chrysogenum* causing meningoencephalitis associated with brain injury in an immunocompetent patient. This case exemplifies the significance of a precise etiologic agent identification because, despite the fact that isolation and etiologic agent identification in culture (Fig. 17.1) are the gold standard for diagnosing fungal infections, molecular diagnosis is increasingly used, primarily to shorten treatment times, which are critical in the most severe disease conditions (De Oliveira et al. 2023).

17.2.2.2 Treatment

Penicillium species have been shown to be susceptible to the antifungal drugs posaconazole, voriconazole, and ravuconazole in in vitro tests. The in vitro activity of caspofungin against *Penicillium* spp. was just moderate. Although the recommended course of action for such species is not well understood, numerous authors have effectively used amphotericin B in conjunction with azoles, independent of the patient's immune status, in many cases (Swoboda-Kopec et al. 2003). There have been case reports of *Penicillium marneffei* and *P. chrysogenum*-related disseminated infections that responded effectively to amphotericin B intravenous treatment and were subsequently treated as outpatients with oral itraconazole for several weeks. Amphotericin B therapy failures have also been documented, though. Regrettably, in case reports of cerebral infection, the treatment options in every instance did not succeed in preventing mortality (De Oliveira et al. 2023; Chuah et al. 2020). Some of the case reports published in the PubMed database (1981–2022) are shown in Table 17.1.

There is no set timeframe for treatment.

17.3 Scedosporium

The genus *Scedosporium* contains two clinically significant species: *Scedosporium apiospermum* (*Pseudallescheria boydii*, its sexual or teleomorph state) and *Scedosporium prolificans* (formerly *S. inflatum*) (Cortez et al. 2008). *Scedosporium apiospermum* and *Scedosporium prolificans* are widely present in the ecosystem, especially in soil, aquatic, and decomposing organic materials. It is the most frequent cause of chronic subcutaneous white mycetoma, which is also called Madura foot. It can also result in endocarditis, endophthalmitis, lung abscess, pneumonia, and CNS infections (Buzina et al. 2006). Meningitis, mycotic aneurysms, and single or multiple brain abscesses (76.9%) are the CNS infections caused by *Scedosporium* species. The mortality rate among infected people is around 76%. The most vulnerable groups are those with leukemia and other malignancies, lupus erythematosus, insulin-dependent diabetes mellitus, transplant recipients, and near-drowning victims. Men are twice as likely as women to contract the disease, and adolescents and children make up the majority of those who suffer from it. Contact with polluted water is the most prevalent way for CNS infections to occur. In the scientific literature, a few cases of *S. apiospermum* CNS infection following near-drowning have been reported (Goralska et al. 2018).

Recent developments in molecular nomenclature have improved our understanding of the genus *Scedosporium* and revealed more species than previously thought. Studies of *Scedosporium* spp. pathology and immune response highlight the significance of innate host responses in defending against these pathogens. Microbiological identification of *Scedosporium* spp. now relies on morphological characterization and culture. The most helpful diagnostic method for *S. apiospermum* infections is culture because their clinical and histological manifestations often resemble those of other fungal infections. Scedosporiosis can be quickly, precisely, and quantitatively detected from clinical specimens using real-time polymerase chain reaction techniques (Sudke et al. 2020). Some of the case reports on human *Scedosporium* spp. CNS infections, published in the literature, are shown in Table 17.2.

Table 17.1 Some of the reported human *P. chrysogenum* CNS infections

Author	Place of study	Clinical case	Patient sex/age	Underlying condition	Treatment	Outcome
De Oliveira et al. (2023)	Brazil	Meningoencephalitis	F/ 14 year	None	Voriconazole, amphotericin B, emergency neurosurgical treatment	Died
Singh et al. (2022)	India	Cerebral abscess	F/ 63 year	Hypertension and type 2 diabetes mellitus	Liposomal amphotericin B	Improved
Kantarcıoğlu et al. (2004)	Turkiye	CNS infection	M/ 73 year	? Trauma	Fluconazole	Cured
Lyratzopoulos et al. (2002)	USA	Cerebral infection	M/ 51 year	None	Amphotericin B, amphotericin B + Flucytosin	Died

Table 17.2 Some of the reported human *Scedosporium* spp. CNS infections

Author	Place of study	Clinical case	Patient sex/age	Species	Underlying condition	Treatment	Outcome
Sudke et al. (2020)	India	Atypical fungal abscess	M/ 77 Year	*S. apiospermum*	None	Voriconazole	Deceased
Paajanen et al. (2019)	Finland	Brain and spinal cord abscess	F/ 18 Year	*S. apiospermum*	Cystic fibrosis underwent lung transplantation	Voriconazole, Miltefosin	Recovered
Wilson and Kennedy (2013)	Australia	Brain abscess	M/ 69 Year	*S. apiospermum*	COPD & Silicosis	Voriconazole, caspofungin	Deceased
Caggiano et al. (2011)	Italy	Cerebral abscess	F/ 58 Year	*S. apiospermum*	Chronic Liver disease and idiopathic pulmonary fibrosis	Liposomal Amphotericin B and posaconazole	Deceased
Sahi et al. (2007)	New Jersey	Hematogenous abscesses and cerebritis	F/ 57 Year	*S. apiospermum*	Lung Transplantation for chronic obstructive pulmonary disease	Voriconazole Posaconazole Terbinafine	Deceased

17.3.1 Growth Characteristics and Identification

Macroscopically, the colonies on Sabouraud dextrose agar grow quickly at a temperature of 25 °C. The fungus, however, may grow at 42 °C, low oxygen tension, and even at stringent anaerobic condition. When compared to sodium chloride, these isolates tolerate magnesium chloride (5%) better. Urea, potassium nitrate, asparagine, and ammonium nitrate can all be assimilated by these species (Cortez et al. 2008). They are also capable of assimilating mannitol, lactose, and maltose and grow in media containing cycloheximide (up to 8 mg/mL) (Cazin Jr. and Decker 1965). *P. boydii* (teleomorph state) forms floccose colonies with distinct appearances on both the obverse and reverse. The color is initially white on the obverse before changing to smoky brown or dark gray and it is pale on the reverse with zones of brownish black color (Hoog et al. 2000). The hyphae are hyaline, despite the fact that the growths are darkly colored because of pigments or the development of brown conidia. Whether *Scedosporium* spp. are hyaline or dematiaceous (pigmented) molds has been debated in the medical mycology literature. The absence of discernible pigment in *Scedosporium* spp. hyphae when stained histologically and the presence of nonpigmented grains in *Scedosporium* mycetoma cases support the idea that the mold is hyaline. The interpretation of a dematiaceous mold is favored by the presence of a diffusible pigment that resembles melanin on colonial morphology. The colored conidia are most likely the source of the diffusible melanin-like pigment, according to a closer investigation of this pigment (Cortez et al. 2008).

On Sabouraud agar, colonies of *S. prolificans* grow and mature in 5 days at 25 °C. Within a week, the colonies can grow up to 3 cm. The colony is flat and spreading, with a surface texture ranging from suede-like to downy and wet, and a white tint that eventually turns brownish olive-gray to black. The reverse takes on a light brown color. Additionally, *S. prolificans* develops more slowly on nutritional agar media and is unable to grow on cycloheximide containing media. In contrast to *S. apiospermum, S. prolificans* develops conidiophores with distinctly expanded bases. At 45 °C, the conidial mass forms apical clumps of conidia and demonstrates positive growth (Cortez et al. 2008).

Microscopically, Pseudallescheria is a homothallic fungus. On nutrient-poor medium such as cornmeal, pea agar, potato-carrot agar, potato dextrose agar, or plain water agar, many isolates generate brown cleistothecia with a diameter of 100 to 300 μm. Rarely, sexual reproductive structures are obtained from clinical samples, and cleistothecia (Fig. 17.2) are usually formed after 2 to 3 weeks of incubation. The cleistothecial forms (ascocarp) can be more numerous at the edge of an agar slant or on the edge of the culture plate. Coiled ascogonia are the first to create cleistothecia, which mature into fruiting bodies in 10 days. One layer of flat, thin, polygonal jigsaw-shaped brown cells makes up the ascocarp wall. The asci, which contain ascospores, are released when the cleistothecium breaks at maturity. Asci have eight ascospores inside and range in size from subglobose to globose. The ascospores easily break out from the ascus walls. Ascospores are smooth, ovoid to ellipsoidal, unicellular, and light golden brown to copper in color. They are roughly 4 to 5 by 7 to 9 m in size, and many of them contain an oil droplet. Sexual ascospores

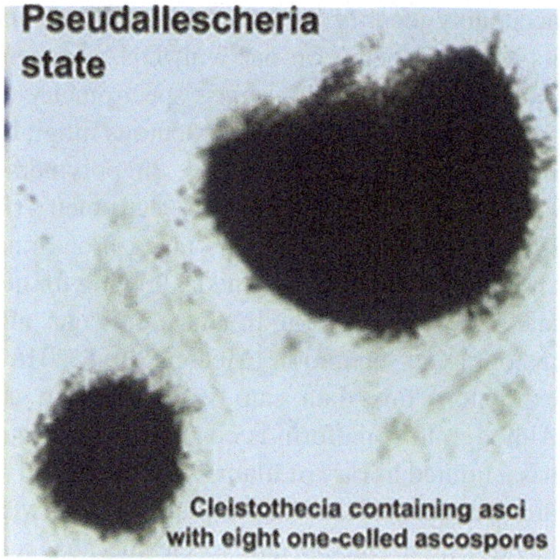

Fig. 17.2 Pseudallescheria state showing cleistothecia. (Picture is taken from Bouchara et al. with permission (Bouchara and Papon 2019))

and asexually produced conidia can be distinguished from one another by an interior oil droplet and the lack of a truncated base. *P. boydii* cleistothecium lacks ostioles and appendages (Hoog et al. 2000; Guarro et al. 2006).

There are numerous varieties of asexual reproduction. Almost always, a *Scedosporium* anamorph exists. This kind is distinguished by septate, cylindrical, and hyaline hyphae that range in diameter from 2-4um and give rise to conidiogenous cells. Anellidic conidiogenesis results in oval, brown, sticky conidia that range in size from 4-9x6-10um. Solitary annelloconidia are a characteristic of the *Scedosporium* type. The conidiophores of *S. apiospermum* are single. They produce unicellular, oval conidia (4-7x5-12um). At their base, they are often truncated (Caggiano et al. 2011; Cortez et al. 2008). In 1984, Malloch and Salkin (Malloch and Salkin 1984) initially identified *S. prolificans* in an osteomyelitis patient who was a child. It was then termed *S. inflatum*. Based on the morphology of the conidiogenous cells in culture, it was distinguished from *S. apiospermum* (Salkin et al. 1988). In addition to flask-shaped conidiophores that are basally swollen (inflated), *S. prolificans* also has septate hyaline hyphae and septate hyaline hyphae. The conidia have a narrower, truncated base, are hyaline to pale brown in color, ovoid to pyriform, and measure 2-5x3-13um (on average, 3.4-5.3um). Additionally, some isolates may result in the production of rounded, thick-walled conidia that grow directly from hyphae (Cortez et al. 2008). *Scedosporium* spp., however, may be difficult to recognize from *Fusarium or Aspergillus* species because they all have hyaline hyphae, dichotomous branching, and regular hyphal septation (Salehi et al. 2016). Due to such difficulties, newer strategies have been sought, including molecular techniques that make use of mass spectroscopy and nucleic acid sequencing.

The current gold standard for fungi identification is nucleotide sequence-based analysis; while partial-tubulin gene (BT2) sequencing can distinguish between closely related species, rDNA internal transcribed spacer (ITS) sequencing can accurately identify the principal *Scedosporium* species (Matray et al. 2015). Due to its accuracy being on par with DNA sequencing, matrix-assisted laser desorption ionization/time-of-flight mass spectrometry (MALDI-TOF MS) has been available for the initial detection of filamentous fungi; however, it is not frequently provided at many hospitals (Bader 2017). In polymerase chain reaction (PCR)/electrospray ionization time-of-flight mass spectrometry (ESI-TOF-MS), T2 magnetic resonance (T2MR) and 16 PCR assays using broad-range primers targeting nuclear or mitochondrial genes are combined, allowing for quick determination of molecular weight and base composition in the amplicons after electrospray ionization and chromatographic separation (Metzgar et al. 2016). Without the need for purification or extraction, this data can be matched to a database to identify *Scedosporium*. Although this platform is costly and not regularly utilized at most institutes, it also has a limited library of filamentous fungi and a potentially time-consuming specimen preparation method. These platforms showcase cutting-edge methods for quickly identifying *Scedosporium* species because a delay in diagnosis is frequently linked to inadequate care and unfavorable results (Yonetani et al. 2016).

Wide range of molecular methods, such as restriction fragment length polymorphism (RFLP) analysis, random amplification of polymorphic DNA (RAPD), amplified fragment-length polymorphism (AFLP), inter-simple-sequence-repeat PCR (ISSR-PCR), PCR fingerprinting, intergenic spacer region PCR (IGS-PCR), and multilocus enzyme electrophoresis (MLEE), have been used for discerning between strains and to identify possible sources of these infections. Along with these techniques, more current studies have concentrated on the development of a robust multilocus sequence typing (MLST) system to offer a trustworthy and repeatable way of strain differentiation (Harun et al. 2009).

Former investigations looked into using MLEE to type different strains of *Scedosporium apiospermum*. One study used 14 distinct enzymes to uncover a high level of variability in this species, with 27 polymorphic sites. (Zouhair et al. 2001). Several publications have used RAPD to distinguish between various *Scedosporium* isolates (Zouhair et al. 2001; Defontaine et al. 2002). Using 12 primers panel, San Millan et al. analyzed 17 isolates of *Scedosporium prolificans*. (San Millán et al. 1997). The genetic diversity of epidemiologically distinct strains of *S. apiospermum* and its previous teleomorph, *Pseudallescheria boydii*, has also been examined using RAPD analysis (Defontaine et al. 2002).

17.3.2 Treatment

At this time, there isn't a universally recognized standardized protocol for treating invasive Scedosporiosis (Sudke et al. 2020). Antifungal triazoles may be effective against *S. apiospermum* infections in humans and animals. In contrast, *S. prolificans* infections rarely heal with just medical treatment. For infections brought on by *S. prolificans*, surgery and immunosuppression reversal may be the only effective treatment choices (Cortez et al. 2008). *Scedosporium* spp. infections in immunocompromised hosts happen in the same clinical environment as aspergillus infections. Voriconazole, long regarded as the treatment of choice in this disease due to its greater blood-brain barrier penetration and a more robust response in vitro against *S. apiospermum*, is the mainstay of medical care. Posaconazole is a second choice behind voriconazole due to the dearth of data supporting its usage against *S. apiospermum*.

17.4 *Trichosporon*

Trichosporon species are yeast-like organisms that are commonly found in nature. They have been identified as opportunistic pathogens that can cause severe invasive diseases, particularly in individuals with weakened immune systems. However, cases of infection have also been reported in individuals with healthy immune systems. Biegel, in 1865, first discovered this organism from hair infection. Since then, there have been 50 species in the genus *Trichosporon*, 16 of which are of clinical significance (Colombo et al. 2011). Major pathogenic species of the genus *Trichosporon* include *T. asahii, T. asteroides, T. cutaneum, T. mucoides, T. inkin, T. ovoides, T. loubieri, and T. pullulans,* which have all been identified as significant sources of invasive infection (Mehta et al. 2021).

Although *Trichosporon* is found as a normal flora of the human skin, gastrointestinal tract, and vagina (Mehta et al. 2021), but it can cause various superficial infections (white piedra) of hair scalp, mustache, beard, axilla, eyebrows, and even genital hairs (Groll and Walsh 2001). Invasive infections caused by it are presented with brain abscess, endophthalmitis, meningitis, soft tissue lesions, endocarditis, arthritis, pneumonia, esophagitis, lymphadenopathy, liver, and splenic abscess, or sometimes uterine infections (Colombo et al. 2011). *Trichosporon* is also reported by Ando et al. as a cause of summer-type hypersensitivity (SHP) (Ando et al. 1991).

17.4.1 Pathogenesis

The virulence of this genus is attributed to the ability of *Trichosporon* strains to produce biofilms and enzymes like proteases and lipases, as well as the presence of glucuronoxylomannan in their cell walls. By avoiding both the host immune system's reaction and the impact of antifungal medications, biofilm aids in the invasion of the surface (Ramage et al. 2012). However, the proteases and lipases disrupt the proteins of the host's cell membrane allowing them to penetrate into it (Ghannoum 2000). These are common characteristics of *Trichosporon* species. *T. asahii* is found to be the most pathogenic one among all species. A review study done by de Almeida Júnior et al. found that *Trichosporon asahii* was the main etiologic agent (46.7%) responsible for various fungal infections (De Almeida Junior and Hennequin 2016). Another research work by Singh et al. shows that 79.16% of all the *Trichosopron* isolates were *T. asahii* from blood culture, nail, and pleural fluid. It is the most prevalent cause of disseminated trichosporonosis and the mortality rate is high even in those patients taking treatment (Singh et al. 2019).

17.4.2 Growth Characteristics and Identification

Like other basidiomycetous yeasts, *T. asahii* and other members of the genus produce true hyphae, pseudohyphae, arthroconidia, and blastoconidia (Heslop et al. 2011). Laboratory identification is carried out on the basis of phenotypic characteristics which are always not promising (Chagas-Neto et al. 2009) and require molecular identification. Sequencing of ITS, IGS, and D1/D2 regions of rRNA is required to confirm the species level identification (De Almeida Junior and Hennequin 2016; Mehta et al. 2021).

17.4.3 CNS Infections of *Trichosporon* spp.

The very first case of CNS infection caused by a *Trichosporon* was reported by Watson et al. in 1970 (Watson and Kallichurum 1970). Since then, there have been many cases reported of CNS infections caused mainly by *T. asahii* and other species like *T. inkin, T. cutaneum,* etc. The first reported case showed the CNS involvement of *T. cutaneum,* but later *T. asahii* emerged as the most involved species in CNS related morbidities starting with the first case in 2011 (Heslop et al. 2011).

Hundreds of disseminated cases of *T. asahii* infections in humans have been documented in the literature so far, despite the fact that these fungi are rarely encountered in human infections (Heslop et al. 2011). Invasive infections are often seen with neutropenia, immunodeficiency, an underlying malignancy, post-surgery, long time steroid therapy, etc. (Al Momani et al. 2021; Shibao et al. 2022). CNS infections are presented with different conditions as shown in Table 17.3. Here we have listed the cases reported around the world with different *Trichosporon* species involved in central nervous system infections. In this table, details of underlying conditions and diagnoses can be seen along with other related information. Disseminated trichosporonosis of the CNS is very rare and only seen in a few reported cases, as listed below.

17.4.4 Treatment

Trichosporon shows intrinsic resistance against echinocandins and polyenes (De Almeida Junior and Hennequin 2016). In most of the cases, azole drugs have been found promising to treat the patients (Thien et al. 2016; Ruan et al. 2009). Amphotericin B was also found to be less effective in invasive trichosporonosis (Ruan et al. 2009). Physicians have used different antifungal drugs like fluconazole, voriconazole, flucytosine, etc. to treat the patients (Ruan et al. 2009; Surmont et al. 1990; Thien et al. 2016). In particular, voriconazole showed a lower MIC in fungal CNS infections (Thien et al. 2016, Ruan et al. 2009).

Table 17.3 Reported cases of CNS infections caused by *Trichosporon*

Year	*Trichosporon* spp.	Underlying conditions/diagnosis	Specimen type	Country	Reference
2022	*Trichosporon* species	Intracerebral hemorrhage and aortic mechanical valve replacement surgery/intracranial fungal aneurysm	Blood from a microcatheter tip	Japan	Shibao et al. (2022)
2019	*T. asahii*	Blood cancer/invasive fungal infections of the central nervous system (IFI-CNS)	CSF/biopsy	Italy/Russia	Candoni et al. (2019)
2018	*T. inkin*	Neuroma and cerebrospinal fistula/meningoencephalitis	CSF	Brazil	Milan et al. (2018)
2016	*T. asahii*	Allo-HSCT and VP shunt infection/hydrocephalus	Blood, biopsy, CSF	Singapore	Thien et al. (2016)
2015	*T. asahii*	Immunocompetent/chronic meningo-ventriculitis and intra Ventricular fungal ball	CSF and intraventricular biopsy	India	Kumar et al. (2015)
2012	*T. asahii*	Autoimmune hepatitis and hypothyroidism/brain abscess	Brain abscess	Iran	Basiri et al. (2012)
2011	*T. asahii*	Diabetes, 50% burns/meningitis, and cerebral abscess	Sputum and facial wound, autopsy	Jamaica	Heslop et al. (2011)
2009	*T. montevideense*	Meningitis	CSF	Taiwan	Ruan et al. (2009)
2007	*T. asahii*	Pneumonia/meningoencephalitis	CSF & sputum	India	Rastogi and Nirwan (2007)
1995	*T. beigelii/T. asahii*	Chronic back pain/chronic meningitis	CSF	India	Mathews and Prabhakar (1995)
1990	*T. beigelii/T. asahii*	Acute lymphocytic leukaemia/chronic meningitis	CSF	Belgium	Surmont et al. (1990)
1970	*T. cutaneum*	Adenocarcinoma/brain abscess	Brain lesions	South Africa	Watson and Kallichurum (1970)

17.5 *Fusarium*

The genus *Fusarium* belongs to the Nectriaceae family of Hypocreales. This fungus is widely distributed in environment associated to mainly soil and plants. Most of the species do not cause any harm; nonetheless, they are important phytopathogens. Some of the species of *Fusarium* are also known as mycoparasites of smuts, rusts, etc. This is found in the environment as endophytes, soil saprobes, pathogens of plants, animals, and humans as well. *Fusarium* produces mycotoxins, which are found to be contaminating with foods and seeds (Torbati et al. 2021). A broad spectrum of human infections is reported by a few numbers of *Fusarium* species. Almost 70 different *Fusarium* species have been identified as opportunistic human infections due to rising infection rates in recent years. *Fusarium* disease has a wide range of clinical symptoms that are mostly influenced by the host's immune system, malignancies, neutropenia, etc. Onychomycosis and keratitis are the most common superficial infections caused by *Fusarium* species in immunocompetent patients, but invasive or disseminated infections typically affect critically ill or immunosuppressed patients and have a high fatality rate (Guo et al. 2022). *Fusaria* are classified into different species complexes (SC) based on their phylogeny derived from DNA sequences (Torbati et al. 2021). The six species complexes (SC), which include *F. solani* SC (FSSC), *F. oxysporum* SC (FOSC), *F. fujikuroi* SC (FFSC), *F. dimerum* SC (FDSC), *F. incarnatum-equiseti* SC (FIESC), and *F. chlamydosporum* SC (FCSC), comprise the majority of the clinically important *Fusarium* species (Guo et al. 2022). Details of the species under these complexes are given below.

***F. solani* SC (FSSC):** *F. keratoplasticum, F. cyanescens, F. falciforme, F. solani* sensu stricto, *F. ambrosium, F. petroliphilum, F. metavorans, F. lichenicola* (Guo et al. 2022).

***F. oxysporum* SC (FOSC):** *F. oxysporum, F. acutatum* (Guo et al. 2022).

***F. fujikuroi* SC (FFSC):** *F. proliferatum, F. bactridioides, F. sacchari, F. concentricum, F. udum, F. verticillioides, F. napiforme* (Guo et al. 2022; Torbati et al. 2021).

***F. dimerum* SC (FDSC):** *Fusarium biseptatum, F. dimerum, F. delphinoides, F. lunatum, F. domesticum* (Guo et al. 2022; Park et al. 2019).

***F. incarnatum-equiseti* SC (FIESC):** *F. incarnatum, F. compactum, F. pallidoroseum, F. clavum, F. semitectum, F. persianum* (Torbati et al. 2021).

***F. chlamydosporum* SC (FCSC):** *F. chlamydosporum, F. atrovinosum, F. sporodochiale, F. spinosum, F. peruvianum, F. microconidium, F. nelsonii, F. humicola* (Crous et al. 2022).

The most encountered *Fusaria* in human infections come from three main species complexes: FSSC, FOSC, and FFSC. FSSC is detected worldwide, causing superficial as well as invasive infections in humans, especially in Asia and Latin America (Guo et al. 2022). FIESC, FOSC, FFSC, and FSSC complexes have also been noted as the highest entomopathogenic *Fusarium* strains with biocontrol potential on various insect hosts (Torbati et al. 2021).

17.5.1 Growth Characteristics and Identification

Fusaria can grow on a variety of culture media, such as PDA, SDA, and SNA at 25 °C. Cultures show aerial cottony type colonies with pinkish mycelia on top and purple mycelia in the center when seen from the bottom of the media plate. Macroconidia are made up of conical or blunt apical cells and papillate basal cells with 3–5 septate. Similarly, microconidia are presented as oval, elongated oval, reinform with 0-septate. Hypha are seen 1.5–5 μm wide with long monophialides and 1–2 chlamydospores. FSSC strains appear in a shape of sickle macroconidia (Da Rosa et al. 2021).

As usual, the identification of *Fusarium* species starts from its culture characteristics. The morphology of colony and a wet mount help to identify the fungi in general (Da Rosa et al. 2021). Colony color, shape and length of macroconidia, and the arrangement of microconidia help in differentiating the species complexes of *Fusarium*. However, morphological and phenotypic methods alone are not promising for reaching species level identification. A few studies that use MALDI-TOF MS demonstrate that it is a reliable substitute for species complex level identification. (Guo et al. 2022; Da Rosa et al. 2021). Guo et al. successfully identified 91.6% isolates of *Fusaria* up to species complex (SC) level (Guo et al. 2022). The use of MLST for molecular identification might be considered the gold standard for precise species and strain level identification. For this, three genes have been widely recommended by mycologists, elongation factor 1a (EF-1α), the first and second biggest subunits of RNA polymerase II (RPB1 & RPB2), which demonstrate high discriminatory potential up to species and strain level identification. Most of the time, the EF-1α is the target of choice. But occasionally, further sequencing of the β-tubulin (TUB) and calmodulin (CaM) genes is also carried out in addition (Guo et al. 2022; Torbati et al. 2021; Da Rosa et al. 2021). Sequencing of 18S rRNA and internal transcribed spacer (ITS) is another option (Risum et al. 2022).

17.5.2 Pathogenesis

Fusarium spp. pathogenicity can take many different forms. This fungus may manifest as cutaneous infections, keratitis, onychomycosis, endophthalmitis, fungemia, septic otitis media, peritonitis, pneumonia, sinusitis, vertebral abscess, osteomyelitis, arthritis, brain abscess, and various infections of the central nervous system (CNS) (Peterson et al. 2014). Due to its affinity for blood vessels, *Fusarium* has a higher propensity to infiltrate people with T-cell immunodeficiency, hematological malignancies, solid organ and stem cell transplantation, steroid use, and persistent neutropenia (Goralska et al. 2018; Karthigeyan et al. 2022). The most common way to enter the human body is through the airway or a skin breakdown caused by burns or tissue trauma. However, foreign objects like contact lenses and intravenous catheters can also be colonized (Karthigeyan et al. 2022).

Numerous virulence characteristics of *Fusarium* species include the ability to produce mycotoxins such as fumonisins and trichothecenes, which suppress cellular and humoral defenses. The most common mycotoxin produced by *Fusarium* species, fumonisin B1, has been connected to neural axon demyelination and cerebral invasion. Fumonisin B1 harms microglia causing impaired function of mitochondria by altering cellular respiration and accumulating phospholipids in the cell membrane. In immunocompromised patients, Fusaria are the second most frequent cause of disseminated mold infections after Aspergillus spp. Death rates from fusariosis range from 50 to 80% (Goralska et al. 2018; Miceli 2019; Peterson et al. 2014).

17.5.3 CNS Infections of *Fusarium* spp.

The very first case of *Fusarium* causing CNS infection was reported by Carlos R. Abramowsky et al. in 1974 in a burnt child. The speciation was not performed in that study (Abramowsky et al. 1974). After a long period, Gary K. Steinberg et al. came up with second report of fusariosis in CNS in 1983. He reported the *F. oxysporum* to be the cause of meningitis in a patient suffering from immunodeficiency and chronic infectious mononucleosis syndrome (Steinberg et al. 1983). Since then, many cases have been reported of CNS infections by *F. oxysporum*. Here, we tried to enlist all the reports of CNS fusarioses and we found that *F. solani*, despite being the main cause of disseminated infections, *F. oxysporum* has been speciated as the main species behind most of the CNS infections. Table 17.4 shows the data of different *Fusaria* involved in CNS infections with their underlying conditions.

17.5.4 Treatment

Infections with *Fusarium* are usually treated empirically with different antimycotics. Many research studies showed different types of antifungal therapies against various species groups involved in disseminated infections. Guo et al. reported high MICs of >32 μg/mL for itraconazole (93.7%) and > 8 μg/mL for terbinafine (76.8%) in most species (Guo et al. 2022). However, Sun's research revealed that 50% of *Fusarium* strains were sensitive to the same antifungal compounds. He displayed a low MIC for terbinafine in FFSC (GM = 2.3 μg/mL) (Sun et al. 2015). Previously, Song et al. also presented good activities of terbinafine against FSSC and FOSC (GM = 2.4 μg/mL and 2.5 μg/mL, respectively) (Song et al. 2021). For amphotericin B, many studies revealed that their majority of isolates had low MICs against *Fusarium* (Al-Hatmi et al. 2015; Oliveira Dos Santos et al. 2019). Clinically, voriconazole is believed to be effective in invasive fusariosis (Guo et al. 2022; Sreedharan Namboothiri et al. 2014). A combination of voriconazole and amphotericin B lipid complex has been the drug of choice in many studies (Karthigeyan et al. 2022; Garcia et al. 2015). Although there has been surgical removal of infected tissues like brain abscess, cerebral lesion is also reported by some clinicians (Garcia et al. 2015).

Table 17.4 Case reports of CNS infections caused by *Fusarium*

Year	*Trichosporon* spp.	Underlying conditions/diagnosis	Specimen type	Country	Reference
2022	*F. proliferatum*	Rheumatoid arthritis, craniotomy/meningoencephalitis	Biopsy	Iran	Alavi Darazam et al. (2022)
2022	*F. dimerum*	AML/cerebral lesions	Sinus swab, sputum, biopsy, CSF	Denmark	Risum et al. (2022)
2021	*F. falciforme*	Tuberculous meningitis/brain abscess	Pus culture	India	Karthigeyan et al. (2022)
2019	*Fusarium* sp.	Blood cancer/invasive fungal infections of the central nervous system (IFI-CNS)	CSF/biopsy	Italy/Russia	Candoni et al. (2019)
2017	*Fusarium* sp.	Diabetes mellitus and liver Cirrhosis/brain abscess	Biopsy	Taiwan	Chen et al. (2017)
2016	*Fusarium* sp.	Tubercular Meningitis (TBM)/meningitis	CSF	India	Ravinder Kaur (2016)
2014	*F. solani*	Diabetes mellitus And acute lymphocytic leukemia/brain abscess	Biopsy	USA	Garcia et al. (2015)
2014	*Fusarium* sp.	Hypertension/brain abscess	Biopsy	USA	Peterson et al. (2014)
2014	*F. oxysporum*	Diabetes mellitus and systemic hypertension/ meningospondylodiscitis	Pus	India	Sreedharan Namboothiri et al. (2014)
2013	*Fusarium* sp.	Focal segmental glomerulosclerosis/meningitis	CSF	USA	Muhammed et al. (2013)
2009	*Fusarium* sp.	Pulmonary sarcoidosis, orthotopic lung transplant/ cerebritis	Autopsy	USA	Kleinschmidt-Demasters (2009)
2007	*F. oxysporum*	Promyelocytic leukemia/brain abscess	Aspirate	The Netherlands.	Anten et al. (2008)
2007	*F. oxysporum*	Acute promyelocytic leukemia/brain abscess	NA	NA	Njambi et al. (2007), Garcia et al. (2015)
2003	*F. solani*	Acute myelogenous leukemia/intracerebral lesion	Biopsy	USA	Vincent et al. (2003)
1998	*Fusarium* sp.	Acute lymphoblastic leukemia/brain abscess	NA	USA	Antunes et al. (1998)
1996	*Fusarium* sp.	Allogenic bone marrow transplant/encephalitis and cerebellitis	NA*	Brazil	Bleggi-Torres et al. (1996), Garcia et al. (2015)

(continued)

Table 17.4 (continued)

Year	*Trichosporon* spp.	Underlying conditions/diagnosis	Specimen type	Country	Reference
1991	*F. oxysporum*	Acute lymphoblastic leukemia (ALL)/ meningoencephalitis	CSF	USA	Agamanolis et al. (1991)
1986	*F. oxysporum*	Multiple myeloma/brain abscess and meningitis	Biopsy	USA	Anaissie et al. (1986)
1983	*F. oxysporum*	Chronic infectious mononucleosis syndrome, immunodeficiency/meningitis, and brain abscess	Aspirate	USA	Steinberg et al. (1983)
1974	*Fusarium* sp.	Second- and third-degree burns (60%)/brain abscess	Autopsy	USA	Abramowsky et al. (1974)

*NA=not available

References

Abramowsky CR, Quinn D, Bradford WD, Conant NF (1974) Systemic infection by Fusarium in a burned child. The emergence of a saprophytic strain. J Pediatr 84:561–564

Agamanolis DP, Kalwinsky DK, Krill CE Jr, Dasu S, Halasa B, Galloway PG (1991) Fusarium meningoencephalitis in a child with acute leukemia. Neuropediatrics 22:110–112

Al Momani M, Yusef DH, Hamasha D, Hamad MRA, Farran S (2021) Complicated Trichosporon Asahii mastoiditis in immunocompetent child. BMC Infect Dis 21:1229

Alavi Darazam I, Sharifi G, Jamali E, Khodavaisy S, Javandoust Gharehbagh F, Hakamifard A (2022) Meningoencephalitis caused by Fusarium Proliferatum: an unusual case. Infection 50: 1023–1027

Al-Hatmi AM, Van Diepeningen AD, Curfs-Breuker I, De Hoog GS, Meis JF (2015) Specific antifungal susceptibility profiles of opportunists in the Fusarium Fujikuroi complex. J Antimicrob Chemother 70:1068–1071

Anaissie E, Kantarjian H, Jones P, Barlogie B, Luna M, Lopez G-B, Bodey GP (1986) Fusarium. A newly recognized fungal pathogen in immunosuppressed patients. Cancer 57:2141–2145

Ando M, Arima K, Yoneda R, Tamura M (1991) Japanese summer-type hypersensitivity pneumonitis. Geographic distribution, home environment, and clinical characteristics of 621 cases. Am Rev Respir Dis 144:765–769

Anten S, Heddema ER, Visser O, Zweegman AS (2008) Images in haematology cerebral fungal abscess in a patient with acute Promyelocytic Leukaemia. Br J Haematol 140:253

Antunes NL, Hariharan S, Deangelis LM (1998) Brain abscesses in children with cancer. Med Pediatr Oncol 31:19–21

Bader O (2017) Fungal species identification by Maldi-Tof mass spectrometry. In: Human fungal pathogen identification: methods and protocols, pp 323–337

Basiri K, Meidani M, Rezaie F, Soheilnader S, Fatehi F (2012) A rare case of Trichosporon brain abscess, successfully treated with surgical excision and antifungal agents. Neurol Neurochir Pol 46:92–95

Bleggi-Torres LF, De Medeiros BC, Neto JZ, Loddo G, Telles FQ, De Medeiros CR, Pasquini R (1996) Disseminated Fusarium Sp. infection affecting the brain of a child after bone marrow transplantation. Bone Marrow Transplant 18:1013–1015

Böhm J, Hoff B, O'gorman CM, Wolfers S, Klix V, Binger D, Zadra I, Kürnsteiner H, Pöggeler S, Dyer PS, Kück U (2013) Sexual reproduction and mating-type-mediated strain development in the penicillin-producing fungus Penicillium Chrysogenum. Proc Natl Acad Sci U S A 110: 1476–1481

Bouchara J-P, Papon N (2019) Scedosporium apiospermum. Trends Microbiol 27:1045–1046

Burgess TB, Condliffe AM, Elks PM (2022) A fun-guide to innate immune responses to fungal infections. J Fungi 8:805

Buzina W, Feierl G, Haas D, Reinthaler F, Holl A, Kleinert R, Reichenpfader B, Roll P, Marth E (2006) Lethal brain abscess due to the fungus Scedosporium Apiospermum (teleomorph Pseudallescheria Boydii) after a near-drowning incident: case report and review of the literature. Med Mycol 44:473–477

Caggiano G, Cantisani P, Rolli M, Gianfreda CD, Pizzolante M, Montagna MT (2011) The importance of a proper Aetiological diagnosis in the management of patients with invasive mycoses: a case report of a brain abscess by Scedosporium Apiospermum. Mycopathologia 172: 317–322

Candoni A, Klimko N, Busca A, Di Blasi R, Shadrivova O, Cesaro S, Zannier ME, Verga L, Forghieri F, Calore E, Nadali G, Simonetti E, Muggeo P, Quinto AM, Castagnola C, Cellini M, Del Principe MI, Fracchiolla N, Melillo L, Piedimonte M, Zama D, Farina F, Giusti D, Mosna F, Capelli D, Delia M, Picardi M, Decembrino N, Perruccio K, Vallero S, Aversa F, Fanin R, Pagano L, Group, S (2019) Fungal infections of the central nervous system and paranasal sinuses in onco-haematologic patients. Epidemiological study reporting the diagnostic-therapeutic approach and outcome in 89 cases. Mycoses 62:252–260

Cazin J Jr, Decker DW (1965) Growth of Allescheria boydii in anitbiotic-containing media. J Bacteriol 90:1308–1313

Chagas-Neto TC, Chaves GM, Melo AS, Colombo AL (2009) Bloodstream infections due to Trichosporon spp.: species distribution, Trichosporon Asahii genotypes determined on the basis of ribosomal DNA intergenic spacer 1 sequencing, and antifungal susceptibility testing. J Clin Microbiol 47:1074–1081

Chen YJ, Chou CL, Lai KJ, Lin YL (2017) Fusarium brain abscess in a patient with diabetes mellitus and liver cirrhosis. Acta Neurol Taiwanica 26(3):128–132

Chuah CH, Ong YC, Kong BH, Woo YY, Wong PS, Leong KN, Chow TS (2020) Talaromyces (Penicillium) species infection in the central nervous system. J R Coll Physicians Edinb 50:138–140

Colombo AL, Padovan AC, Chaves GM (2011) Current knowledge of Trichosporon spp. and trichosporonosis. Clin Microbiol Rev 24:682–700

Cortez KJ, Roilides E, Quiroz-Telles F, Meletiadis J, Antachopoulos C, Knudsen T, Buchanan W, Milanovich J, Sutton DA, Fothergill A, Rinaldi MG, Shea YR, Zaoutis T, Kottilil S, Walsh TJ (2008) Infections caused by Scedosporium spp. Clin Microbiol Rev 21:157–197

Crous PW, Sandoval-Denis M, Costa MM, Groenewald JZ, Van Iperen AL, Starink-Willemse M, Hernandez-Restrepo M, Kandemir H, Ulaszewski B, De Boer W, Abdel-Azeem AM, Abdollahzadeh J, Akulov A, Bakhshi M, Bezerra JDP, Bhunjun CS, Camara MPS, Chaverri P, Vieira WAS, Decock CA, Gaya E, Gene J, Guarro J, Gramaje D, Grube M, Gupta VK, Guarnaccia V, Hill R, Hirooka Y, Hyde KD, Jayawardena RS, Jeewon R, Jurjevic Z, Korsten L, Lamprecht SC, Lombard L, Maharachchikumbura SSN, Polizzi G, Rajeshkumar KC, Salgado-Salazar C, Shang QJ, Shivas RG, Summerbell RC, Sun GY, Swart WJ, Tan YP, Vizzini A, Xia JW, Zare R, Gonzalez CD, Iturriaga T, Savary O, Coton M, Coton E, Jany JL, Liu C, Zeng ZQ, Zhuang WY, Yu ZH, Thines M (2022) Fusarium and allied fusarioid taxa (Fusa). 1. Fungal Syst Evol 9:161–200

Da Rosa PD, Aquino V, Fuentefria AM, Goldani LZ (2021) Diversity of Fusarium species causing invasive and disseminated infections. J Mycol Med 31:101137

De Almeida Junior JN, Hennequin C (2016) Invasive Trichosporon infection: a systematic review on a re-emerging fungal pathogen. Front Microbiol 7:1629

De Oliveira RVM, Corrêa-Moreira D, Mendes TV, Da Costa GL, Vieira RDM, Buchele CMN, Lins RS, Ferreira ABTBC, Veira DB, Pedroso RSDA, De Faria VTDP, Oliveira MME (2023) First report of fungal meningoencephalitis by Penicillium chrysogenum in Brazil. Int J Infect Dis 126:94–97

De Sousa JRP, Gonçalves VN, De Holanda RA, Santos DA, Bueloni CFLG, Costa AO, Petry MV, Rosa CA, Rosa LH (2017) Pathogenic potential of environmental resident fungi from Ornithogenic soils of Antarctica. Fungal Biol 121:991–1000

Defontaine A, Zouhair R, Cimon B, Carrère J, Bailly E, Symoens F, Diouri M, Hallet J-N, Bouchara J-P (2002) Genotyping study of Scedosporium Apiospermum isolates from patients with cystic fibrosis. J Clin Microbiol 40:2108–2114

Garcia RR, Min Z, Narasimhan S, Bhanot N (2015) Fusarium brain abscess: case report and literature review. Mycoses 58:22–26

Ghannoum MA (2000) Potential role of phospholipases in virulence and fungal pathogenesis. Clin Microbiol Rev 13:122–143, Table Of Contents, 13, 122

Goralska K, Blaszkowska J, Dzikowiec M (2018) Neuroinfections caused by fungi. Infection 46: 443–459

Groll AH, Walsh TJ (2001) Uncommon opportunistic fungi: new nosocomial threats. Clin Microbiol Infect 7(Suppl 2):8–24

Guarro J, Kantarcioglu AS, Horré R, Luis Rodriguez-Tudela J, Cuenca Estrella M, Berenguer J, Sybren De Hoog G (2006) Scedosporium Apiospermum: changing clinical spectrum of a therapy-refractory opportunist. Sabouraudia 44:295–327

Guo P, Chen J, Tan Y, Xia L, Zhang W, Li X, Jiang Y, Li R, Chen C, Liao K, Peng Y (2022) Comparison of molecular and Maldi-Tof Ms identification and antifungal susceptibility of clinical Fusarium isolates in Southern China. Front Microbiol 13:992582

Harun A, Perdomo H, Gilgado F, Chen SCA, Cano J, Guarro J, Meyer W (2009) Genotyping of Scedosporium species: a review of molecular approaches. Med Mycol 47:406–414

Heslop OD, Nyi Nyi MP, Abbott SP, Rainford LE, Castle DM, Coard KC (2011) Disseminated Trichosporonosis in a burn patient: meningitis and cerebral abscess due to Trichosporon Asahii. J Clin Microbiol 49:4405–4408

Hoog GD, Queiroz-Telles F, Haase G, Fernandez-Zeppenfeldt G, Angelis DA, Hg Gerrits Van Den Ende A, Matos T, Peltroche-Llacsahuanga H, Pizzirani-Kleiner A, Rainer J (2000) Black fungi: clinical and pathogenic approaches. Med Mycol 38:243–250

Jameel Far YS (2021) Virulence potential of Penicillium chrysogenum isolated from subclinical bovine mastitis. Iraqi J Sci 62:2131–2142

Kantarcıoğlu A, Apaydın H, Yücel A, De Hoog G, Samson R, Vural M, Özekmekçi S (2004) Central nervous system infection due to Penicillium chrysogenum. Mycoses 47:242–248

Karthigeyan M, Singh K, Kaur H, Salunke P, Pandey J, Nallasamy K (2022) Multiple Fusarium brain abscesses in a young child. Childs Nerv Syst 38:1017–1021

Kleinschmidt-Demasters BK (2009) Disseminated Fusarium infection with brain abscesses in a lung transplant recipient. Clin Neuropathol 28:417–421

Kumar A, Udayakumaran S, Babu R, Rajamma BM, Prakash A, Panikar D, Karim S, Chowdhary A (2015) Trichosporon Asahii infection presenting as chronic Meningo-Ventriculitis and intra ventricular fungal ball: a case report and literature review. Mycoses 58:99–103

Lyratzopoulos G, Ellis M, Nerringer R, Denning DW (2002) Invasive infection due to Penicillium species other than P. marneffei. J Infect 45:184–195

Malloch D, Salkin I (1984) A new species of scedosporium associated with osteomyelitis in humans. Mycotaxon 21:247–255

Mathews MS, Prabhakar S (1995) Chronic meningitis caused by Trichosporon beigelii in India. Mycoses 38:125–126

Matray O, Mouhajir A, Giraud S, Godon C, Gargala G, Labbe F, Rougeron A, Ballet J-J, Zouhair R, Bouchara J-P (2015) Semi-automated repetitive sequence-based Pcr amplification for species of the Scedosporium Apiospermum complex. Sabouraudia 54:409–419

Mehta V, Nayyar C, Gulati N, Singla N, Rai S, Chandar J (2021) A comprehensive review of Trichosporon spp.: an invasive and emerging fungus. Cureus 13:E17345

Metzgar D, Frinder MW, Rothman RE, Peterson S, Carroll KC, Zhang SX, Avornu GD, Rounds MA, Carolan HE, Toleno DM (2016) The Iridica bac Bsi assay: rapid, sensitive and culture-independent identification of bacteria and Candida in blood. PLoS One 11:E0158186

Miceli MH (2019) Central nervous system infections due to aspergillus and other hyaline molds. J Fungi (Basel) 5:79

Milan EP, Silva-Rocha WP, De Almeida JJS, Fernandes TUG, De Araujo Prudente AL, De Azevedo MF, Francisco EC, De Azevedo Melo AS, Colombo AL, Chaves GM (2018) Trichosporon Inkin meningitis in Northeast Brazil: first case report and review of the literature. BMC Infect Dis 18:470

Muhammed M, Anagnostou T, Desalermos A, Kourkoumpetis TK, Carneiro HA, Glavis-Bloom J, Coleman JJ, Mylonakis E (2013) Fusarium infection: report of 26 cases and review of 97 cases from the literature. Medicine (Baltimore) 92:305–316

Njambi S, Huttova M, Kovac M, Freybergh PF, Bauer F, Muli JM (2007) Fungal Neuroinfections: rare disease but unacceptably high mortality. Neuro Endocrinol Lett 28(Suppl 2):25–26

Nnadi NE, Carter DA (2021) Climate change and the emergence of fungal pathogens. PLoS Pathog 17:E1009503

Noritomi DT, Bub GL, Beer I, Da Silva ASF, De Cleva R, Gama-Rodrigues JJ (2005) Multiple brain abscesses due to Penicillium spp. infection. Crit Care 9:1–2

Oliveira Dos Santos C, Kolwijck E, Van Der Lee HA, Tehupeiory-Kooreman MC, Al-Hatmi AMS, Matayan E, Burton MJ, Eggink CA, Verweij PE (2019) In vitro activity of chlorhexidine compared with seven antifungal agents against 98 Fusarium isolates recovered from fungal keratitis patients. Antimicrob Agents Chemother 63:e02669

Paajanen J, Halme M, Palomäki M, Anttila V-J (2019) Disseminated Scedosporium Apiospermum central nervous system infection after lung transplantation: a case report with successful recovery. Med Mycol Case Rep 24:37–40

Park JH, Oh J, Song JS, Kim J, Sung GH (2019) Bisifusarium Delphinoides, an emerging opportunistic pathogen in a burn patient with diabetes mellitus. Mycobiol 47:340–345

Peterson A, Pham MH, Lee B, Commins D, Cadden J, Giannotta SL, Zada G (2014) Intracranial Fusarium fungal abscess in an immunocompetent patient: case report and review of the literature. J Neurol Surg Rep 75:E241–E245

Pitt JI, Cruickshank RH, Leistner L (1986) Penicillium commune, P. Camembertii, the origin of white cheese moulds, and the production of cyclopiazonic acid. Food Microbiol 3:363–371

Ramage G, Rajendran R, Sherry L, Williams C (2012) Fungal biofilm resistance. Int J Microbiol 2012:528521

Rastogi VL, Nirwan PS (2007) Invasive trichosporonosis due to Trichosporon Asahii in a non-immunocompromised host: a rare case report. Indian J Med Microbiol 25:59–61

Ravinder Kaur PSP (2016) Fusarium co-infection in a patient of tubercular meningitis. J Neuroanaesthesiol Crit Care 03:3

Risum M, Overgaard UM, Rubek N, Haastrup EK, Hare RK, Helweg-Larsen J (2022) Disseminated Fusariosis with cerebral involvement in a patient with acute myeloid leukemia: successful outcome with intrathecal-and systemic antifungal treatment. J Infect Chemother 28:1324–1328

Ruan SY, Chien JY, Hsueh PR (2009) Invasive Trichosporonosis caused by Trichosporon Asahii and other unusual Trichosporon species at a medical center in Taiwan. Clin Infect Dis 49:E11–E17

Sahi H, Avery RK, Minai OA, Hall G, Mehta AC, Raina P, Budev M (2007) Scedosporium Apiospermum (Pseudoallescheria Boydii) infection in lung transplant recipients. J Heart Lung Transplant 26:350–356

Salehi E, Hedayati MT, Zoll J, Rafati H, Ghasemi M, Doroudinia A, Abastabar M, Tolooe A, Snelders E, Lee HAVD, Rijs AJMM, Verweij PE, Seyedmousavi S, Melchers WJG (2016) Discrimination of aspergillosis, mucormycosis, fusariosis, and scedosporiosis in formalin-fixed paraffin-embedded tissue specimens by use of multiple real-time quantitative Pcr assays. J Clin Microbiol 54:2798–2803

Salkin I, Mcginnis M, Dykstra M, Rinaldi M (1988) Scedosporium inflatum, an emerging pathogen. J Clin Microbiol 26:498–503

San Millán R, Quindós G, Garaizar J, Salesa R, Guarro J, Ponton J (1997) Characterization of Scedosporium Prolificans clinical isolates by randomly amplified polymorphic DNA analysis. J Clin Microbiol 35:2270–2274

Shibao S, Kaburaki M, Saito K, Tomita H (2022) A case of subarachnoid and intracerebral hemorrhages complicated by trichosporonosis. Surg Neurol Int 13:472

Singh S, Capoor MR, Varshney S, Gupta DK, Verma PK, Ramesh V (2019) Epidemiology and antifungal susceptibility of infections caused by Trichosporon species: an emerging non-candida and non-cryptococcus yeast worldwide. Indian J Med Microbiol 37:536–541

Singh N, Anand K, Sondhi A (2022) Penicillium chrysogenum: a rare cause of cerebral abscess. D Y Patil J Health Sci 10:40–42

Song Y, Liu X, Yang Z, Meng X, Xue R, Yu J, Al-Hatmi AMS, De Hoog GS, Li R (2021) Molecular and Maldi-Tof Ms differentiation and antifungal susceptibility of prevalent clinical Fusarium species in China. Mycoses 64:1261–1271

Sreedharan Namboothiri P, Nair SN, Vijayan K, Visweswaran V (2014) Disseminated Fusarium Oxysporum neurospinal infection. Indian J Orthop 48:220–222

Steinberg GK, Britt RH, Enzmann DR, Finlay JL, Arvin AM (1983) Fusarium brain abscess. Case report. J Neurosurg 58:598–601

Sudke AY, Shaikh ST, Deopujari CE, Sakle AS (2020) Scedosporium apiospermum: rare cause of brain abscess in an immunocompetent patient. Neurol India 68:906–909

Sun S, Lyu Q, Han L, Ma Q, Hu H, He S, Tao S, Zhang J, Zhang H, Wang L (2015) Molecular identification and in vitro susceptibility of Fusarium from fungal keratitis in Central China. Zhonghua Yan Ke Za Zhi 51:660–667

Surmont I, Vergauwen B, Marcelis L, Verbist L, Verhoef G, Boogaerts M (1990) First report of chronic meningitis caused by Trichosporon Beigelii. Eur J Clin Microbiol Infect Dis 9:226–229

Swoboda-Kopec E, Wroblewska MM, Rokosz A, Luczak M (2003) Mixed bloodstream infection with staphylococcus aureus and Penicillium chrysogenum in an immunocompromised patient: case report and review of the literature. Clin Microbiol Infect 9:1116–1117

Thien SY, Chung SJ, Tan AL, Hwang WY, Tan BH, Tan TT (2016) Recurrent trichosporonosis with central nervous system involvement in an allogeneic hematopoietic stem cell transplant recipient. Transpl Infect Dis 18:768–772

Torbati M, Arzanlou M, Da Silva Santos AC (2021) Fungicolous Fusarium species: ecology, diversity, isolation, and identification. Curr Microbiol 78:2850–2859

Vincent AL, Cabrero JE, Greene JN, Sandin RL (2003) Successful voriconazole therapy of disseminated Fusarium solani in the brain of a neutropenic cancer patient. Cancer Control 10: 414–419

Wagener RE, Davis ND, Diener UL (1980) Penitrem A and roquefortine production by Penicillium commune. Appl Environ Microbiol 39:882–887

Watson KC, Kallichurum S (1970) Brain abscess due to Trichosporon cutaneum. J Med Microbiol 3:191–193

Wilson HL, Kennedy KJ (2013) Scedosporium apiospermum brain abscesses in an immunocompetent man with silicosis. Med Mycol Case Rep 2:75–78

Woo PC, Chong KT, Tse H, Cai JJ, Lau CC, Zhou AC, Lau SK, Yuen KY (2006) Genomic and experimental evidence for a potential sexual cycle in the pathogenic thermal dimorphic fungus Penicillium marneffei. FEBS Lett 580:3409–3416

Yonetani S, Ohnishi H, Ohkusu K, Matsumoto T, Watanabe T (2016) Direct identification of microorganisms from positive blood cultures by Maldi-Tof Ms using an in-house saponin method. Int J Infect Dis 52:37–42

Zouhair R, Defontaine A, Ollivier C, Cimon B, Symoens F, Hallet J-N, Deunff J, Bouchara J-P (2001) Typing of Scedosporium apiospermum by multilocus enzyme electrophoresis and random amplification of polymorphic DNA. J Med Microbiol 50:925–932

Part IV

Future Insights into Central Nervous System Infections

Advanced Microbiological Diagnostic Techniques in Viral Infections of the Central Nervous System

18

Mohd Shadab, Ronni Mol Joji, Hala Ali Ebrahim Almahmeed, and Mohammad Shahid

Abstract

Conventional diagnosis of viral infections includes culture, antigen detection, serology and nucleic acid detection but with time, new techniques are replacing the conventional methods of viral diagnosis. Polymerase chain reaction coupled with mass spectrometry and next-generation sequencing (NGS) are some of the most recent technological developments in viral detection. Various multiplex assays are now available which can simultaneously detect common viruses causing meningitis and encephalitis. For several viruses like herpes simplex virus, enterovirus, Epstein–Barr virus, varicella zoster virus and human herpesvirus-6, detection of viral nucleic acids from the cerebrospinal fluid by PCR or reverse-transcription PCR has supplanted culture and brain biopsy as the gold standard for diagnosing encephalitis. In various scientific and therapeutic settings, advances in genomic approaches-particularly in sequencing technologies-are being used. NGS technology, for example, has shown promising results in finding infections in clinical samples due to its ability to decode millions of DNA and RNA sequences at the same time. This chapter will provide significant information regarding current and potential advanced diagnostic techniques that could reform the diagnosis of CNS-viral infections in future. The chapter also

M. Shadab (✉) · R. M. Joji · M. Shahid
Department of Microbiology, Immunology, and Infectious Diseases, College of Medicine and Medical Sciences, Arabian Gulf University, Manama, Kingdom of Bahrain
e-mail: shadab@agu.edu.bh; ronnimj@agu.edu.bh; mohammeds@agu.edu.bh

H. A. E. Almahmeed
College of Medicine and Medical Sciences, Arabian Gulf University, Manama, Kingdom of Bahrain

Salmaniya Medical Complex, Manama, Kingdom of Bahrain

H. Sami et al. (eds.), *Viral and Fungal Infections of the Central Nervous System: A Microbiological Perspective*, https://doi.org/10.1007/978-981-99-6445-1_18

provides insight into some of the available diagnostic panels for diagnosing CNS infections.

Keywords

Virus identification · Diagnostic techniques · CNS · Cerebrospinal fluid · Meningitis · Encephalitis · Amplification · Sequencing

18.1 Introduction

Earth maintains the suitable environment for all living beings including the invisible organisms. Microbes are those invisible ubiquitous community cohabitated with us on Earth and viruses are a faction of it. Viruses affect nearly all living beings, i.e., humans, animals, plants, bacteria, etc. They are the tiniest particles thought to be between livings and non-livings and made up of only a group of proteins and nucleic acids requiring nothing to grow or reproduce but able to arrest the cellular mechanism for its own replication (Domingo 2020). Their small size helps them to avoid microscopic view, while their simple structure kept them unidentifiable for decades. Since the introduction of the field of virology by Dimitri Ivonovasky and Martinus Beijerinck in the late nineteenth century, initially, an infectious filtrate was used to infect a host to establish a virus infection (Lecoq 2001). Animal inoculation and egg inoculation techniques have been exploited for decades to cultivate the viruses. Later on, in the mid of the twentieth century, the development of cell culture techniques paved the way of culturing microbes in vitro (Payne 2017). With the advent of biological sciences and technologies, the complex structures of viruses are not only unveiled, almost all known viruses are now cultured in labs with the help of different cell lines. A plethora of detection methods are available nowadays which target the virus genome, over expressed antigens or the effects of viruses on host cells and tissues.

18.2 Diagnostic Approaches in Virology

Since the medical science is constantly evolving, new techniques are being discovered to help meet the easy and early diagnosis of diseases to fulfill the correct treatment needs. Central nervous system infections are one of those which require an urgent attention as the delay can reverberate into fatal impacts. Diagnostic virology, with the help of clinical picture of the patient, allows virologists to apply a cascade of laboratory tests to achieve a definitive diagnosis of CNS infections. Methodologies applied to study viruses and its diseases are mainly categorized into three types as follows (Abdullahi et al. 2020; Davis and Tyler 2005; Payne 2017):

1. Culture and isolation of virus from CSF, blood, brain tissue or sometimes stool and urine also.

2. Identifying the presence of virus particles or its components like antigen, nucleic acid, inclusion bodies in brain tissue, CSF, blood, etc.
3. A meaningful expression of intrathecal antibodies against the suspected virus.

18.3 Serving Purposes of Diagnostic Virology

Despite a handful of antiviral drugs available in pharmaceutical industry and their limited use or rather say, limited effects on viruses, we still want to identify a virus particle involved in an infection. Methods for detecting a virus in an infected patient should preferably be sensitive, specific, and quick to help the physician. But, once a patient has recovered or passed away, the diagnosis is less useful practically. Even so, a diagnosis could help scientists, members of the community, and family to act to stop its spread. And they can carry out preventative measures for this. The diagnostic virology actually serves the following purposes in spite of a mere diagnostic report (Payne 2017):

- Decide on treatment plans.
- Predict the course and expecting outcomes of the disease.
- Determine the potentiality of virus to spread ahead.
- Mark the individuals susceptible to it and their immunization.
- Pave the way of vaccine development.
- Trace the epidemiology of virus across a community, a nation or the world.
- Allow pharma industry to develop new antiviral drugs.

18.4 Advanced Diagnostic Techniques

Scientists have been using various conventional methods to detect and identify viruses causing meningitis, encephalitis and other neurotropic infections. These traditional methods include enzyme-linked immunosorbent assay (ELISA), antigen/antibody test cards, chemiluminescence immunoassay (CLIA), immunofluorescence, culture, polymerase chain reaction (PCR), etc. available in diagnostic as well as research industry. CSF-PCR and reverse transcription-PCR (RT-PCR) have already supplanted culture and brain biopsy as gold standard in meningitis and encephalitis diagnosis (Bookstaver et al. 2017). Whereas Multiplex PCR panels are available to simultaneously detect common viruses causing CNS infections. Nowadays, new molecular techniques are emerging promising better and early detection of several viruses like HSV, EV, EBV, VZV, HHV-6, etc. (Abdullahi et al. 2020; Mawuntu et al. 2018). New PCR mass spectrometry (PCR-MS) technology is emerging as a fast and more accurate method to detect a range of microbes in a given sample. Scientific and therapeutic settings have now started utilizing various genomic approaches particularly sanger sequencing, whole genome sequencing (WGS), in situ sequencing (ISS) and the last but not the least of course, shotgun metagenomic sequencing. Notably, these techniques are being used to diagnose a wide range of pathogens. Here, we tried to give the details of these advanced diagnostic techniques and how these are being used to detect the viruses causing CNS morbidities.

18.4.1 Advanced Nucleic Acid Amplification Methods

A crucial and widely used technique to amplify nucleic acid is the polymerase chain reaction (PCR) introduced in 1983 by Kary Mullis. PCR is actually based on utilizing multiple cycles of a set of temperatures and a polymerase enzyme (Mullis et al. 1986). Though, it only works with DNA but RNA can also be amplified by adding a reverse transcriptase enzyme (RT-PCR). Again, this does not complete the analyses, for which we have to run the amplicons with gel electrophoresis to visualize the nucleic acid for qualitative assessment. To solve this, a real-time quantitative PCR was introduced in 1993 by Higuch et al. (1993). This reaction utilizes fluorescence labelled probes or DNA intercalating fluorophores such as SYBR Green to give signals measuring amplification during the reaction and thus the quantity of reaction can be assessed in real time (Liu and Hong 2007). PCR methods are widely used by scientists and diagnostic labs, etc. This section will focus more on other advanced methods of nucleic acid amplification. PCR and RT-PCR techniques are being exploited in a vast area of scientific work. So is the diagnostic field, which has demonstrated the successful use of these techniques in the confirmation of various diseases. Use of PCR and RT-PCR in microbiology is widely established and after this casual remark, it requires to allude about some of the latest diagnostic panels available in market.

There are several multiplex panels available commercially which helps in identifying the related pathogens in a particular systemic infection. These panels are based on the PCR methods and utilize a number of primer sets able to amplify genes of causative agent, i.e., bacteria, viruses, fungi, etc.

18.4.1.1 Multiplex PCR Assays

One of the most used molecular biology techniques for the detection of pathogens is multiplex PCR, where multiple targets are amplified into a single PCR tube by adding various primer sets into it. This method has taken over other PCRs as it saves time and needs less effort. Multiplex PCR has two categories, PCR with a single template and PCR with multi-template. The first technique exploits a single template, which might be a piece of genomic DNA, along with a number of pairs of forward and reverse primers to amplify particular template areas. While in multi-template PCR, in a reaction tube, multiple DNA templates are added with multiple primer sets. Though the cross hybridization of multiple primers and mis-priming remains challenging with the multi-template technique.

The key to a successful multiplex reaction for a targeted amplification with high yield is the design of appropriate primer sets. The parameters that are used in multiplex PCR are comparable to conventional PCR primer design principles except for the primer length, melting temperature, specificity of designed primers to the target sequences, and to avoid primer dimer formation. Usually, short primers, with 18 to 22 bases, are employed. There are many advantages that made multiplex PCR superior to conventional PCR. It reveals false negative because each amplicon acts as an internal check on the other amplified segments. Secondly, it is more efficient as it reduces the expenses and the preparation time as well.

So far, a lot of multiplex PCR/RT-PCR panels have been developed to assist in diagnosing an infection in different systems such as respiratory panel, meningitis panel, gastrointestinal panel, blood panel, to name some of them. However, we tried to enlist a few of such panels which are used to detect various neurotropic viruses in cerebrospinal fluid collected by lumbar puncture:

18.4.1.1.1 Filmarray® ME Panel

This is a meningitis/encephalitis panel, product of BioFire Diagnostics, USA. It can identify 15 pathogens such as seven viruses, six bacteria, and two yeasts. Viruses included are enterovirus, human herpes virus 6, cytomegalovirus, varicella zoster virus, human par echovirus, herpes simplex virus-1 and herpes simplex virus-2. Only a tiny amount (0.2 mL) of CSF is required to run the test and it takes nearly 1 h to give the results.

The sensitivity for viruses is 90.1% which is slightly lower than that of bacteria (97.5%) (Graf et al. 2017). However, the company asserts that its total sensitivity is 94.2% and its specificity is 99.8% ("Biofiredx.com) (Biofire ME Panel n.d.). On contrary, the sensitivity of HSV genes is found very suboptimal, even false-negative results are also reported (Gomez et al. 2017; Graf et al. 2017; Piccirilli et al. 2018). The Filmarray® Meningitis/Encephalitis panel has established its accuracy for EV (Eichinger et al. 2019; Messacar et al. 2016). Although, Filmarray® ME is a widely used panel to detect the CNS infections but while analyzing the results of panel, the practitioners need to be always skeptical about the finding.

18.4.1.1.2 MeningoFinder® 2Smart

This is developed by PathoFinder, the Netherlands. It has the capacity to detect and differentiate 23 pathogens which includes 12 viruses, nine bacterial species and two fungi. Among the viruses, it can identify cytomegalovirus (CMV), herpes simplex virus 1, herpes simplex virus 2 (HSV1 & 2), enterovirus, human herpesvirus 6, human herpesvirus 7, human herpesvirus 8, Epstein-Barr virus (EBV), mumps virus, measles virus, par echovirus, and varicella zoster virus (VZV). It takes 2.5 h to produce results (MeningoFinder® 2Smart n.d.). In comparison to Filmarray® ME panel, its sensitivity is found lesser (Sall et al. 2019).

18.4.1.1.3 Fast Track Meningitis Kits

Fast Track Diagnostics has a high range of multiplex real-time PCR assays for syndromic screening of different systemic infections. For meningitis itself, they have developed four kits, two for viral and two for bacterial meningitis. These panels are "FTD Neuro 9 kit", "FTlyo Viral meningitis kit"(for viral meningitis), "FTlyo Bacterial meningitis kit", "FTD Neonatal meningitis kit" (Fast Track Kits n.d.).

FTD Neuro 9 kit can detect 11 viruses which includes enterovirus, EBV, herpes simplex virus 1, herpes simplex virus 2, adenovirus, cytomegalovirus, human herpesvirus 6, human herpesvirus 7, par echovirus, parvovirus b19, and varicella zoster virus.

FTlyo Viral meningitis kit identifies six viruses, i.e., enterovirus, HSV 1, HSV 2, par echovirus, mumps virus, and varicella zoster virus.

However, there is a paucity of validation studies on clinical samples employing these technologies (Ure et al. 2012).

18.4.1.1.4 Allplex™ Meningitis-V1 &V2 Assays

Seegene, Republic of Korea, has introduced two real-time PCR multiplex assays for the syndromic testing of viral meningitis. Allplex™ Meningitis-V1 assay includes varicella zoster virus (VZV) herpes simplex virus type 1 (HSV1), herpes simplex virus type 2 (HSV2), Epstein-Bart virus (EBV), human herpes virus 6 (HHV 6), human herpes virus 7 (HHV 7), cytomegalovirus (CMV),. While V2 assay includes only enterovirus (HEV), adenovirus (AdV), parvovirus B19 (B19V), mumps virus (MV), and human par echovirus (HPeV) (Allplex™ Assays n.d.; Cohen et al. 2023).

18.4.1.1.5 Viral CNS Flow Chip

This new one-step multiplex real-time PCR test is manufactured by Master Diagnostica. It can simultaneously detect eight DNA and RNA viruses causing meningitis as well as encephalitis. The viruses are HSV 1 & HSV 2, CMV, EV, EBV, VZV, HPeV, and Toscana virus (TOSV). The PCR is followed by flow-through hybridization technique. The sensitivity and specificity recorded was 95.9% and 99.9%, respectively. Inclusion of TOSV is an improvement of this kit specially for endemic areas (Perez-Ruiz et al. 2018).

18.4.1.2 PCR-Electrospray Ionization Mass Spectrometry (PCR/ESI-MS)

A new technique known as PCR/ESI-MS, which combines electrospray ionization-time of flight mass spectrometry (TOF MS) analysis with PCR amplification, offers the ability to directly identify a variety of human pathogens from clinical specimens as well as the potential to find genetic evidence of yet-to-be-discovered pathogens. A wider variety of bacteria, viruses and fungi can be identified using PCR/ESI-MS, which is designed for both broad-spectrum and target-specific identification of pathogens and their genetic features (Leveque et al. 2014).

Unlike MALDI-TOF MS, as aforementioned, it does not require culture and isolation of pathogens in order to identify them (Kaleta et al. 2011). In fact, the DNA or RNA can be extracted from the specimen and then amplified with various specific sets of primers and with an RT enzyme (RT-PCR/ESI-MS) if required (Deyde et al. 2011). Post amplification, desalting of the PCR product is done. Then, the amplicons are submitted to electrospray ionization-mass spectrometry analyzer for the assessment of genetic material. The analyzer provides the mass accuracy to discern the composition of base pairs. For this, electrospray ionization proceeds the charged PCR amplicons through high vacuum tube of mass spectrometer, and the TOF of charged molecules is measured according to the mass/charge ratio. For the purpose of quantitatively identifying bacterial species, base compositions are algorithmically predicted based on the obtained spectrum and then compared to a reference database (Wolk et al. 2012).

PCR/ESI-MS showed better detection for co-infections. In addition to giving a rapid detection and a broad-spectrum, PCR/ESI-MS can emerge as the first-line diagnostic tool in immunocompromised patients and patients in ICU suspected with multiple infections with a range of viral species from varicella zoster virus and herpes simplex virus to the rare ones such as Hendra virus, Nipah virus or par echoviruses, etc. Sensitivity and specificity of certain viruses involved in CNS infections have been assessed earlier by Leveque et al. Sensitivity and specificity values of HSV, VZV, EV, and EBV are showed as 92% and 97%, 94% and 99%, 84% and 99%, and 31% and 97%, respectively. PCR/ESI-MS showed lesser sensitivity to enteroviruses and EBV in comparison with routine PCR (Leveque et al. 2014).

In this regard, Abbott Molecular, USA, unveiled the commercial platform IRIDICA. In 2014, it debuted on the European market getting a CE mark (Tkadlec et al. 2020). This novel molecular system was earlier known as PLEX-ID system. This system utilizes PCR-based amplification of DNA/RNA and the electrospray ionization-TOF MS technology for the assessment. Different research studies demonstrated that PCR/ESI-MS can be utilized as a same-day diagnostic nucleic acid amplification technique (NAAT) to identify more pathogens than culture (Ozenci et al. 2018). But in 2017, despite of numerous testing and clinical trials, Abbott decided to withdraw the application from the US FDA and announced the discontinuation of an advanced technology-based instrument (Ozenci et al. 2018). No other system has been replaced so far to empower the field of microbiology based on the PCR/ESI-MS technique. This technology has performed at par with MALDI TOF MS reducing the labor and economic burden to a low level and with about a 6–8 h turnaround time (Kaleta et al. 2011).

18.4.1.3 Helicase-Dependent Amplification

Unlike a conventional method of denaturing the DNA strand with heat, helicase-dependent amplification (HDA) functions with a DNA-helicase enzyme which unwinds the two strands of template DNA. This development allows reaction to be thermally stable at 60–65 °C (Teo et al. 2015). To facilitate the RNA amplification, a reverse transcriptase enzyme can be incorporated (Goldmeyer et al. 2007). The primers used in HDA are longer than those generally employed in conventional PCR reaching about 30 bp (Vincent et al. 2004).

Since the main difference between a conventional PCR and HDA is it employs a helicase mediated rather than heat-based separation of DNA, the fluorescent probes of quantitative PCR (qPCR) can be applied to HDA in conjunction with quantitative real-time PCR to facilitate the quantitative assessment of amplicon (Kolm et al. 2019).

HDA offers a simplistic design of amplification assay in comparison to other techniques. Though it provides short and simple detection of pathogens, but this comes at the cost of less specificity. The issue of annealing of non-specific primers is the reason behind the reduced specificity of HDA (Barreda-Garcia et al. 2018). Teo et al. (2015) have performed the amplification of herpes simplex virus using the same technique, while Jevsnik et al. (2020) used it successfully to detect varicella zoster virus (Jevsnik et al. 2020; Teo et al. 2015).

18.4.1.4 Loop-Mediated Isothermal Amplification

Loop-mediated isothermal amplification (LAMP) is a low-cost amplification technique and a good alternative of simple PCR. LAMP employees 4–6 primer sets which operates at a temperature of 60–65 °C. Multiple primer sets confer it a high level of specificity (Parida et al. 2008). A DNA polymerase with strand displacement activity starts the reaction. Two specific primers form a loop to accelerate the amplification through extension on the loops allowing the primer annealing. Usually, four primer sets target six different areas of the target gene that increase reaction specificity (Notomi et al. 2015). For RNA amplification, reverse transcriptase enzyme facilitates the initial cDNA production. The rest is the same as aforementioned. This is simply termed as RT-LAMP (Wong et al. 2018).

LAMP is a single tube method which does not require sophisticated thermocycler machine (Tomita et al. 2008). That is why it can be used in a point of care testing setting or in a field laboratory (Moehling et al. 2021). The reaction time, typically takes less than an hour which can be reduced by incorporating additional primers yielding into a copious amount of amplified product (Nagamine et al. 2002). In comparison to classical PCR, LAMP provides highly sensitive, specific and rapid alternative to diagnose a range of viruses causing CNS infections.

SYBR green or other fluorescent dyes and probes can be used to quantify DNA in real time (Zhang et al. 2014). However, LAMP, as it implies a different polymerase enzyme with strand displacement activity, does not allow the same probes developed for a qPCR. Instead other alternative probes used in the LAMP reaction are:

- **QUASR**: Quenching of unincorporated amplification signal reporters (Tanner et al. 2012).
- **DARQ**: Detection of amplification by release of quenching (Ball et al. 2016).
- **MDP**: Mediator displacement probes (Becherer et al. 2018).
- **AuNPs**: Gold nanoparticles labelled with DNA (Arunrut et al. 2016).

Different LAMP, RT-LAMP and qRT-LAMP reactions have been utilized so far against several type of viruses, namely, lassa virus (Pemba et al. 2019), zika virus (Silva et al. 2019), dengue virus (Lopez-Jimena et al. 2018) and corona virus as well (Huang et al. 2020). Involvement of these viruses in different CNS-related diseases is well proven.

18.4.2 Advanced Nucleic Acid Sequencing Methods

Traditional microbiological techniques including culture, morphological identification, serological testing, and even PCR, according to researchers, which make it difficult to pinpoint the etiological agents (Xing et al. 2020). The choice of detection techniques is followed by the diagnosis, which is mostly based on the treating physician's clinical expertise and past knowledge in predicting the infections. This demonstrated the limitations of conventional approaches for the target-dependent detection of rare and new viruses (Carbo et al. 2021). The most recent method for identifying viruses that cause CNS disorders is known as next generation sequencing (NGS), also known as deep sequencing. This method analyzes samples and creates a single sequence from each DNA and cDNA present in the sample. Researchers can distinguish the origin of sequence fragments from a particular virus by analyzing the sequences (Brown et al. 2018). NGS applications can be categorized as follows:

1. Target amplicon sequencing, which involves the sequencing of targets that have been precisely enriched utilizing a variety of enrichment techniques, such as amplification or probe hybridization.
2. Whole genome sequencing, which brings together all DNA fragments from the pathogen's whole genome.
3. Metagenomic sequencing, which allows for massively parallel sequencing of all nucleic acids (DNA/RNA) producing millions to billions of reads and generating a plethora of genetic data for both pathogens and host within a specimen.

Here, we have given the details of how these methods can be utilized in detecting and identifying the CNS infections.

18.4.2.1 Target Amplicon Sequencing

Target amplicon sequencing allows you to identify a particular genetic segment of a virus or examine a genetic variation in certain genomic regions. Sanger sequencing is considered to be the first-generation DNA sequencing technique to sequence the short-read DNA segment. Sanger sequencing is frequently employed when specific genes or gene fragments need to be sequenced, such as for viral genotyping or for resistance testing where SNPs are linked to certain genome areas (Hagemann 2015). Next to it, targeted next-generation sequencing (NGS) is considered as the second-generation sequencing technique. It utilizes PCR to produce DNA amplicons, or sequences of DNA. Multiple samples can be sequenced during a single sequencing run by multiplexing, which involves barcoding samples so they can be combined into pools. But individual samples utilized for amplicon sequencing must first be converted into libraries by the addition of adapters and target region enrichment by PCR amplification before multiplexing (Bybee et al. 2011; Vincent et al. 2017). Sanger sequencing-derived information is exceedingly scarce, according to a study that was published in the Lancet in 2016 (Patel et al. 2016). While full-length viral gene sequences may influence the monitoring of medication resistance to provide the most effective therapeutic advice, sources of viral transmission within healthcare settings are identified and potential epidemics when coupled with clinical data is monitored (Patel et al. 2016). Sikazwe et al. (2022) performed an amplicon-based sequencing of Japanese encephalitis virus (JEV) (Sikazwe et al. 2022).

18.4.2.2 Whole Genome Sequencing (WGS)

Clinical medicine, encompassing both human and pathogens, is increasingly using whole-genome sequencing (WGS) techniques (Worthey et al. 2011). Particularly in clinical research and epidemiology, viral genome sequencing is becoming more and more significant. WGS of pathogens has the benefit of being able to identify all known drug-resistant variations in a single test, whereas deep sequencing (high coverage sequencing) can reveal low levels of drug-resistant variants to enable intervention before resistance that becomes clinically apparent (Witney et al. 2015). For the objectives of infection control and public health, whole genomes also offer useful information for identifying related infections. Viral WGS has, however, made only modest progress in clinical practice (Gire et al. 2014).

Viral WGS is becoming more clinically significant for illness management, diagnosis, molecular epidemiology, and infection prevention. Presently, the method of choice depends on both the virus and the clinical query. WGS of HSV 1 and HSV 2 strains has already been done by scientists (Chang et al. 2022; Saranathan et al. 2022). Lopez-Munoz et al. (2021) also sequenced an HSV 2 strain MS isolated from the midbrain of multiple sclerosis patient (Lopez-Munoz et al. 2021). Complete genome sequences of JEV were done by Taraphdar and Chatterjee (2015) from India (Taraphdar and Chatterjee 2015). Target enrichment is effective for all pathogen sizes but is particularly useful for large viruses and viruses with diverse but well-defined genomes. Metagenomic sequencing is best ideal for diagnostic sequencing of unknown or poorly characterized viruses (Houldcroft et al. 2017). As a laboratory considers a novel NGS program, practical factors including sequencing costs, data processing and maintenance, and data analysis complexity are crucial factors to consider. Because of the volume of data involved, these problems are exacerbated in WGS, and they have long been entrance obstacles for clinical laboratories wishing to implement WGS (Tsai et al. 2016).

WGS is very effective due to the depth of the sequencing data and the quality of the produced sequences. Nonetheless, despite the relatively tiny size of virus genomes, sequencing them is frequently still a challenge. Viral sequencing yield is reduced by the virus' limited genetic content compared to the host nucleic acid. The fact that multiple viral variants coexist in a single sample and show more or less varied sequences depending on the intrinsic mutation rate of the virus is another challenge that must be overcome (Maurier et al. 2019). NGS has been successfully used to sequence partial or WGS of the viruses in the clinical samples.

18.4.2.3 Metagenomic Sequencing

Metagenomics is another approach to find out the causative agent by sequencing from sample directly. It does not require culture and isolation of pathogens indeed. In addition to thoroughly analyzing all of the microbial and host genomic material present in clinical samples, metagenomic sequencing also significantly decreases the price and sequencing time (Gu et al. 2019). Moreover, it enables the genomic characterization and identification of various microbes including viruses, from clinical specimens without the need for prior knowledge of particular pathogens (Simner et al. 2017). The cause of unexplained or post-treatment infections can also

Table 18.1 Some case reports showing clinical implications of metagenomics in CNS viral infections

Authors	Metagenomic sequencing results	Diagnosis
Naccache et al. (2015)	Brain tissue: 12 of 134,068,968 reads aligned to astrovirus RNA	Neuroinvasive astrovirus infection
Wilson et al. (2017)	Urine sample positive for RNA JEV	JEV
Chiu et al. (2017)	CSF sample positive for St Louis encephalitis virus	St Louis encephalitis virus
Edridge et al. (2019)	CSF sample positive for novel Orthobunyavirus (Ntwetwe virus)	Encephalitis
Fang et al. (2020)	CSF sample: 97,248 sequence reads aligned to the VZV genome	Fulminant central nervous system varicella zoster virus (VZV) infection
Yan et al. (2021)	CSF sample showed 652 reads aligned to the Suid alpha herpesvirus	Pseudorabies virus (PRV) encephalitis and endophthalmitis
Liu et al. (2022)	CSF sample: Seoul orthohantavirus (106,253 reads)	Seoul Orthohantavirus CNS infection in a child after hematopoietic stem cell transplantation
Han et al. (2023)	>1 viral copy sequence found in CSF is defined as positive	VZV-associated Rhombencephalitis

be determined using metagenomic sequencing (Thoendel et al. 2016). It can be utilized to find latent pathogens even in the case of a negative cerebrospinal fluid culture to help in clinical diagnosis and quick treatment (Zhang et al. 2020). Some of the clinical implications of metagenomics in CNS infections are shown in Table 18.1.

Several investigations have demonstrated that there is frequently a discrepancy between metagenomic findings and clinical diagnosis (Miao et al. 2018; Qian et al. 2020). Moreover, many microorganisms are frequently found in a metagenomic sequencing result. There are currently no recommendations or studies that offer logical explanations for false-negative and false-positive results of metagenomic sequencing (Wang et al. 2022).

18.4.2.3.1 Shotgun Sequencing Method

Shotgun metagenomics refers to the untargeted, or "shotgun," sequencing of all the microbial genomes ("genomics") contained in a sample. Using shotgun sequencing, one may profile the taxonomic makeup and functional potential of microbiological communities and to retrieve whole genomic sequences. When profiling certain organisms or single marker genes, techniques like high-throughput 16S rRNA gene sequencing are frequently falsely referred to as metagenomic methods, although they are not metagenomic approaches because they do not aim to examine the complete genetic content of a sample (Quince et al. 2017).

This includes RNA and/or DNA approach separately. For the identification of RNA viruses, as well as the transcriptome-based study of other microbes and the host immune response, an RNA-based technique is necessary. Except for RNA viruses, all other microbes can be identified using a DNA-based technique. An RNA technique can further identify which organisms are transcriptionally active, whereas a DNA approach will only reveal what organisms are present (Simner et al. 2017).

Recently, metagenomics was employed to aid in the diagnosis of infectious diseases, including meningo–/encephalitis (Cordey et al. 2015). Shotgun metagenomics, which sequences every nucleic acid (NA) in a sample, has the potential to uncover unknown or mixed illnesses as well as known pathogens that have undergone mutations that prevent qPCR identification (Tan et al. 2013). Yet, given the high number of host sequences found in clinical samples, shotgun metagenomics must cope with this issue. Current host depletion methods to enhance pathogen sequences take a lot of effort, money, and/or specialization (Kohl et al. 2015). They either concentrate on bacterial or viral DNA or are solely focused on RNA sequencing that exclusively targets RNA viruses or 16S rRNA of bacteria (Lewandowska et al. 2015).

For a sample to undergo completely unbiased sequencing, effective extraction techniques are a crucial step. According to the organism type, reported metagenomic limit of detection from synthetic CSF matrix ranges from 9.4 copies/mL for cytomegalovirus to 100 copies/mL for HIV type 1 (Schlaberg et al. 2017). Moreover, it also differs depending on the type of the specimen used, such as the detection limit for hepatitis C and HIV virus from plasma was estimated to be 1×10^4 copies/mL (Kandathil et al. 2017). Although a negative metagenomic sequencing result does not completely rule out an infectious process, it can demonstrate the level of sensitivity reached by the test for that particular sample if the internal control is recovered sufficiently. Findings will still require clinical context for interpretation (Schlaberg et al. 2017).

The challenges faced in virus shotgun metagenomics is that this technique can identify viral species, but in order to access a larger variety of viruses, virome enrichment approaches are typically required. As there is a paucity of available viral genomes and inter-family phylogenetic signals, virome analysis remains computationally difficult (Quince et al. 2017). Shotgun metagenomics is a promising technique for aiding in CNS infection diagnosis, although it is not currently practical as a stand-alone approach (Oechslin et al. 2018). Workflow of metagenomic sequencing is shown in Fig. 18.1.

18.4.2.3.2 Limitations of Metagenomics

Despite its ground-breaking diagnostic potential, metagenomics has a number of significant limitations. They include the availability, cost, processing time, and restrictions of CSF testing for CNS infections. Access to metagenomic sequencing remains a hurdle despite growing awareness of this technology. Because most clinical microbiology laboratories lack the clinical expertise to undertake internal clinical metagenomic testing, samples must be sent to specific reference laboratories.

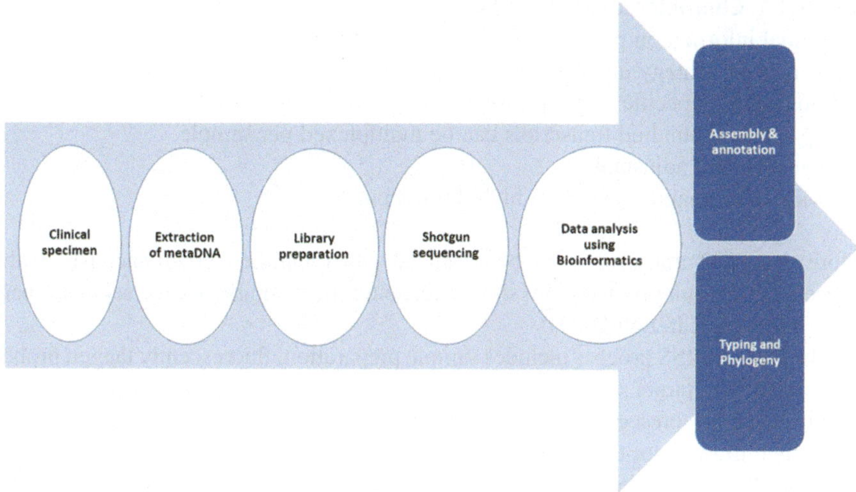

Fig. 18.1 Workflow of metagenomic sequencing in diagnostic microbiology lab, (modified from Simner et al. 2017)

Rapid diagnosis is severely hampered by the greater turnaround time. With additional transit time to reference laboratories, total turnaround times can reach beyond 10 days (Ramachandran and Wilson 2020). The utility of any test that looks for the presence of an organism at the time of sample collection is limited because despite being the most thorough test for CSF, many infectious organisms are still present in CSF but do not yet pose a threat to patients. Enterovirus D68, West Nile virus, California encephalitis virus, and other neuroinvasive arboviruses are some well-known viruses (Wilson et al. 2019). In order to determine host reaction and increased yield, metagenomic and serologic tests should be combined. When combined, these two complimentary methods may offer a more thorough diagnostic assessment for CNS infections (Graff et al. 2021).

18.4.2.4 In Situ Sequencing (ISS)

From single base pairs to whole chromosomes, genomes are physically organized. It is assumed that this structure, which differs between cells within animals, controls cellular activity and gene expression. The genome is covered with base-pair resolution using the current DNA sequencing techniques, but the spatial context is missing. Alternately, existing imaging-based techniques that capture spatial context are focused and lack base-pair resolution. Hence, a strategy for mapping genomic structure that bridges sequencing and imaging modalities is lacking (A. C. Payne et al. 2021). It is possible to simultaneously sequence and image the genomes of intact biological materials using in situ genome sequencing (IGS).

18.4.2.4.1 Characteristics of the ISS

- Spatial information preservation in a tissue setting.
- A focused strategy utilizing padlock probing.
- Molecularly specific amplification technique with high specificity.
- Up to a few hundred transcripts can be multiplexed per sample.
- Sensitive and adjustable.
- Wide-field imaging results in high throughput.

More than 100 transcripts can be localized and quantified simultaneously with subcellular resolution using in situ sequencing in a single tissue segment per experiment (SciLifeLab 2023).

The standard ISS process includes sample preparation, fluorescently tagged probe hybridization to target sequences, ligation to create longer templates, sequencing by synthesis, and fluorescent microscopy imaging to see how the target sequences are distributed throughout the tissue. For several target sequences, the entire procedure is repeated, enabling the spatial mapping of numerous genes or transcripts inside a single tissue sample. This can be applicable to both clinical and research settings (Omicsveu 2023). Further studies are required to explore the applications of ISS in the detection of CNS viral infections.

References

Abdullahi AM, Sarmast ST, Jahan N (2020) Viral infections of the central nervous system in children: a systematic review. Cureus 12(10):e11174. https://doi.org/10.7759/cureus.11174

Allplex™ Assays (n.d.). https://www.seegene.com/assays/allplex_meningitis_panel_assays

Arunrut N, Kampeera J, Sirithammajak S, Sanguanrut P, Proespraiwong P, Suebsing R, Kiatpathomchai W (2016) Sensitive visual detection of AHPND bacteria using loop-mediated isothermal amplification combined with DNA-functionalized gold nanoparticles as probes. PLoS One 11(3):e0151769. https://doi.org/10.1371/journal.pone.0151769

Ball CS, Light YK, Koh CY, Wheeler SS, Coffey LL, Meagher RJ (2016) Quenching of unincorporated amplification signal reporters in reverse-transcription loop-mediated isothermal amplification enabling bright, single-step, closed-tube, and multiplexed detection of RNA viruses. Anal Chem 88(7):3562–3568. https://doi.org/10.1021/acs.analchem.5b04054

Barreda-Garcia S, Miranda-Castro R, de Los Santos Alvarez N, Miranda-Ordieres AJ, Lobo-Castanon MJ (2018) Helicase-dependent isothermal amplification: a novel tool in the development of molecular-based analytical systems for rapid pathogen detection. Anal Bioanal Chem 410(3):679–693. https://doi.org/10.1007/s00216-017-0620-3

Becherer L, Bakheit M, Frischmann S, Stinco S, Borst N, Zengerle R, von Stetten F (2018) Simplified real-time multiplex detection of loop-mediated isothermal amplification using novel mediator displacement probes with universal reporters. Anal Chem 90(7):4741–4748. https://doi.org/10.1021/acs.analchem.7b05371

Biofire ME Panel (n.d.). https://www.biofiredx.com/products/the-filmarray-panels/filmarrayme/

Bookstaver PB, Mohorn PL, Shah A, Tesh LD, Quidley AM, Kothari R, Weissman S (2017) Management of viral central nervous system infections: a primer for clinicians. J Cent Nerv Syst Dis 9:1179573517703342. https://doi.org/10.1177/1179573517703342

Brown JR, Bharucha T, Breuer J (2018) Encephalitis diagnosis using metagenomics: application of next generation sequencing for undiagnosed cases. J Infect 76(3):225–240. https://doi.org/10.1016/j.jinf.2017.12.014

Bybee SM, Bracken-Grissom H, Haynes BD, Hermansen RA, Byers RL, Clement MJ, Crandall KA (2011) Targeted amplicon sequencing (TAS): a scalable next-gen approach to multilocus, multitaxa phylogenetics. Genome Biol Evol 3:1312–1323. https://doi.org/10.1093/gbe/evr106

Carbo EC, Blankenspoor I, Goeman JJ, Kroes ACM, Claas ECJ, De Vries JJC (2021) Viral metagenomic sequencing in the diagnosis of meningoencephalitis: a review of technical advances and diagnostic yield. Expert Rev Mol Diagn 21(11):1139–1146. https://doi.org/10.1080/14737159.2021.1985467

Chang W, Jiao X, Sui H, Goswami S, Sherman BT, Fromont C, Imamichi T (2022) Complete genome sequence of herpes simplex virus 2 strain G. Viruses 14(3):536. https://doi.org/10.3390/v14030536

Chiu CY, Coffey LL, Murkey J, Symmes K, Sample HA, Wilson MR, Klausner JD (2017) Diagnosis of fatal human case of St. Louis encephalitis virus infection by metagenomic sequencing, California, 2016. Emerg Infect Dis 23(10):1964–1968. https://doi.org/10.3201/eid2310.161986

Cohen R, Mahajnah M, Shlonsky Y, Golan-Shany O, Romem A, Halevy A, Genizi J (2023) Prospective, cross-sectional study finds no common viruses in cerebrospinal fluid of children with Pseudotumor cerebri. Brain Sci 13(2):361. https://doi.org/10.3390/brainsci13020361

Cordey S, Bel M, Petty T, Docquier M, Sacco L, Turin L, Zdobnov E (2015) Toscana virus meningitis case in Switzerland: an example of the ezVIR bioinformatics pipeline utility for the identification of emerging viruses. Clin Microbiol Infect 21(4):387.e381–387.e384

Davis LE, Tyler KL (2005) Molecular diagnosis of CNS viral infections. J Neurol Neurosurg Psychiatry 76(1):10. https://doi.org/10.1136/jnnp.2004.051698

Deyde VM, Sampath R, Gubareva LV (2011) RT-PCR/electrospray ionization mass spectrometry approach in detection and characterization of influenza viruses. Expert Rev Mol Diagn 11(1):41–52. https://doi.org/10.1586/erm.10.107

Domingo E (2020) Virus as populations. In: Domingo E (ed) Virus as populations, 2nd edn. Academic Press, Amsterdam

Edridge AW, Deijs M, Namazzi R, Cristella C, Jebbink MF, Maurer I, de Jong MD (2019) Novel orthobunyavirus identified in the cerebrospinal fluid of a Ugandan child with severe encephalopathy. Clin Infect Dis 68(1):139–142

Eichinger A, Hagen A, Meyer-Buhn M, Huebner J (2019) Clinical benefits of introducing real-time multiplex PCR for cerebrospinal fluid as routine diagnostic at a tertiary care pediatric center. Infection 47(1):51–58. https://doi.org/10.1007/s15010-018-1212-7

Fang M, Weng X, Chen L, Chen Y, Chi Y, Chen W, Hu Z (2020) Fulminant central nervous system varicella-zoster virus infection unexpectedly diagnosed by metagenomic next-generation sequencing in an HIV-infected patient: a case report. BMC Infect Dis 20(1):159. https://doi.org/10.1186/s12879-020-4872-8

Fast Track Kits (n.d.). https://marketing.webassets.siemens-healthineers.com/295496ad90a99ccb/95bdf9ac8890/30-21-DX-1054-76_FTD-RUO-Assays-EOL_Brochure_FINAL.pdf

Gire SK, Goba A, Andersen KG, Sealfon RS, Park DJ, Kanneh L, Dudas G (2014) Genomic surveillance elucidates Ebola virus origin and transmission during the 2014 outbreak. Science 345(6202):1369–1372

Goldmeyer J, Kong H, Tang W (2007) Development of a novel one-tube isothermal reverse transcription thermophilic helicase-dependent amplification platform for rapid RNA detection. J Mol Diagn 9(5):639–644. https://doi.org/10.2353/jmoldx.2007.070012

Gomez CA, Pinsky BA, Liu A, Banaei N (2017) Delayed diagnosis of tuberculous meningitis misdiagnosed as herpes simplex virus-1 encephalitis with the FilmArray syndromic polymerase chain reaction panel, vol 4. Paper presented at the open forum infectious diseases

Graf EH, Farquharson MV, Cardenas AM (2017) Comparative evaluation of the FilmArray meningitis/encephalitis molecular panel in a pediatric population. Diagn Microbiol Infect Dis 87(1):92–94. https://doi.org/10.1016/j.diagmicrobio.2016.09.022

Graff K, Dominguez SR, Messacar K (2021) Metagenomic next-generation sequencing for diagnosis of pediatric meningitis and encephalitis: a review. J Pediatr Infect Dis Soc 10(Supplement_4):S78–S87. https://doi.org/10.1093/jpids/piab067

Gu W, Miller S, Chiu CY (2019) Clinical metagenomic next-generation sequencing for pathogen detection. Ann Rev Pathol 14(1):319–338. https://doi.org/10.1146/annurev-pathmechdis-012418-012751

Hagemann IS (2015) Overview of technical aspects and chemistries of next-generation sequencing. In: Clinical genomics. Elsevier Inc., Amsterdam, pp 3–19. https://doi.org/10.1016/B978-0-12-404748-8.00001-0

Han J, Si Z, Wei N, Cao D, Ji Y, Kang Z, Zhu J (2023) Next-generation sequencing of cerebrospinal fluid for the diagnosis of VZV-associated rhombencephalitis. J Integr Neurosci 22(2):36. https://doi.org/10.31083/j.jin2202036

Higuchi R, Fockler C, Dollinger G, Watson R (1993) Kinetic PCR analysis: real-time monitoring of DNA amplification reactions. Biotechnology (N Y) 11(9):1026–1030. https://doi.org/10.1038/nbt0993-1026

Houldcroft CJ, Beale MA, Breuer J (2017) Clinical and biological insights from viral genome sequencing. Nat Rev Microbiol 15(3):183–192. https://doi.org/10.1038/nrmicro.2016.182

Huang WE, Lim B, Hsu CC, Xiong D, Wu W, Yu Y, Cui Z (2020) RT-LAMP for rapid diagnosis of coronavirus SARS-CoV-2. Microb Biotechnol 13(4):950–961. https://doi.org/10.1111/1751-7915.13586

Jevsnik M, Lusa L, Ursic T, Glinsek Biskup U, Petrovec M (2020) Detection of herpes simplex and varicella-zoster virus from skin lesions: comparison of RT-PCR and isothermal amplification for rapid identification. Diagn Microbiol Infect Dis 97(2):115015. https://doi.org/10.1016/j.diagmicrobio.2020.115015

Kaleta EJ, Clark AE, Cherkaoui A, Wysocki VH, Ingram EL, Schrenzel J, Wolk DM (2011) Comparative analysis of PCR-electrospray ionization/mass spectrometry (MS) and MALDI-TOF/MS for the identification of bacteria and yeast from positive blood culture bottles. Clin Chem 57(7):1057–1067. https://doi.org/10.1373/clinchem.2011.161968

Kandathil AJ, Breitwieser FP, Sachithanandham J, Robinson M, Mehta SH, Timp W, Balagopal A (2017) Presence of human Hepegivirus-1 in a cohort of people who inject drugs. Ann Intern Med 167(1):1–7. https://doi.org/10.7326/m17-0085

Kohl C, Brinkmann A, Dabrowski PW, Radonić A, Nitsche A, Kurth A (2015) Protocol for metagenomic virus detection in clinical specimens. Emerg Infect Dis 21(1):48

Kolm C, Martzy R, Fuhrer M, Mach RL, Krska R, Baumgartner S, Reischer GH (2019) Detection of a microbial source tracking marker by isothermal helicase-dependent amplification and a nucleic acid lateral-flow strip test. Sci Rep 9(1):393. https://doi.org/10.1038/s41598-018-36749-7

Lecoq H (2001) Discovery of the first virus, the tobacco mosaic virus: 1892 or 1898? C R Acad Sci III 324(10):929–933. https://doi.org/10.1016/s0764-4469(01)01368-3

Leveque N, Legoff J, Mengelle C, Mercier-Delarue S, N'Guyen Y, Renois F, Andreoletti L (2014) Virological diagnosis of central nervous system infections by use of PCR coupled with mass spectrometry analysis of cerebrospinal fluid samples. J Clin Microbiol 52(1):212–217. https://doi.org/10.1128/JCM.02270-13

Lewandowska DW, Zagordi O, Zbinden A, Schuurmans MM, Schreiber P, Geissberger F-D, Mueller NJ (2015) Unbiased metagenomic sequencing complements specific routine diagnostic methods and increases chances to detect rare viral strains. Diagn Microbiol Infect Dis 83(2):133–138

Liu XK, Hong Y (2007) Q-priming PCR: a quantitative real-time PCR system using a self-quenched BODIPY FL-labeled primer. Anal Biochem 360(1):154–156. https://doi.org/10.1016/j.ab.2006.10.011

Liu E, Lv S, Yi P, Feng L, Deng X, Xia H, Xu Y (2022) Central nervous system infection with Seoul orthohantavirus in a child after hematopoietic stem cell transplantation: a case report. Virol J 19(1):75. https://doi.org/10.1186/s12985-022-01766-6

Lopez-Jimena B, Bekaert M, Bakheit M, Frischmann S, Patel P, Simon-Loriere E, Weidmann M (2018) Development and validation of four one-step real-time RT-LAMP assays for specific detection of each dengue virus serotype. PLoS Negl Trop Dis 12(5):e0006381. https://doi.org/10.1371/journal.pntd.0006381

Lopez-Munoz AD, Rastrojo A, Kropp KA, Viejo-Borbolla A, Alcami A (2021) Combination of long- and short-read sequencing fully resolves complex repeats of herpes simplex virus 2 strain MS complete genome. Microb Genom 7(6):000586. https://doi.org/10.1099/mgen.0.000586

Maurier F, Beury D, Fléchon L, Varré J-S, Touzet H, Goffard A, Caboche S (2019) A complete protocol for whole-genome sequencing of virus from clinical samples: application to coronavirus OC43. Virology 531:141–148. https://doi.org/10.1016/j.virol.2019.03.006

Mawuntu AHP, Bernadus JBB, Dhenni R, Wiyatno A, Anggreani R, Feliana, Myint KSA (2018) Detection of central nervous system viral infections in adults in Manado, North Sulawesi, Indonesia. PLoS One 13(11):e0207440. https://doi.org/10.1371/journal.pone.0207440

MeningoFinder® 2Smart (n.d.). https://www.pathofinder.com/2smart-products/pf1200-2s

Messacar K, Breazeale G, Robinson CC, Dominguez SR (2016) Potential clinical impact of the film array meningitis encephalitis panel in children with suspected central nervous system infections. Diagn Microbiol Infect Dis 86(1):118–120. https://doi.org/10.1016/j.diagmicrobio.2016.05.020

Miao Q, Ma Y, Wang Q, Pan J, Zhang Y, Jin W, Hu B (2018) Microbiological diagnostic performance of metagenomic next-generation sequencing when applied to clinical practice. Clin Infect Dis 67(suppl_2):S231–s240. https://doi.org/10.1093/cid/ciy693

Moehling TJ, Choi G, Dugan LC, Salit M, Meagher RJ (2021) LAMP diagnostics at the point-of-care: emerging trends and perspectives for the developer community. Expert Rev Mol Diagn 21(1):43–61. https://doi.org/10.1080/14737159.2021.1873769

Mullis K, Faloona F, Scharf S, Saiki R, Horn G, Erlich H (1986) Specific enzymatic amplification of DNA in vitro: the polymerase chain reaction. Cold Spring Harb Symp Quant Biol 51(Pt 1): 263–273. https://doi.org/10.1101/sqb.1986.051.01.032

Naccache SN, Peggs KS, Mattes FM, Phadke R, Garson JA, Grant P, Chiu CY (2015) Diagnosis of neuroinvasive astrovirus infection in an immunocompromised adult with encephalitis by unbiased next-generation sequencing. Clin Infect Dis 60(6):919–923. https://doi.org/10.1093/cid/ciu912

Nagamine K, Hase T, Notomi T (2002) Accelerated reaction by loop-mediated isothermal amplification using loop primers. Mol Cell Probes 16(3):223–229. https://doi.org/10.1006/mcpr.2002.0415

Notomi T, Mori Y, Tomita N, Kanda H (2015) Loop-mediated isothermal amplification (LAMP): principle, features, and future prospects. J Microbiol 53(1):1–5. https://doi.org/10.1007/s12275-015-4656-9

Oechslin CP, Lenz N, Liechti N, Ryter S, Agyeman P, Bruggmann R, Beuret CM (2018) Limited correlation of shotgun metagenomics following host depletion and routine diagnostics for viruses and bacteria in low concentrated surrogate and clinical samples. Front Cell Infect Microbiol 8:375. https://doi.org/10.3389/fcimb.2018.00375

Omicsveu (2023) In situ sequencing workflow; https://omicsveu.com/

Ozenci V, Patel R, Ullberg M, Stralin K (2018) Demise of polymerase chain reaction/electrospray ionization-mass spectrometry as an infectious diseases diagnostic tool. Clin Infect Dis 66(3): 452–455. https://doi.org/10.1093/cid/cix743

Parida M, Sannarangaiah S, Dash PK, Rao PV, Morita K (2008) Loop mediated isothermal amplification (LAMP): a new generation of innovative gene amplification technique; perspectives in clinical diagnosis of infectious diseases. Rev Med Virol 18(6):407–421. https://doi.org/10.1002/rmv.593

Patel N, Ferns BR, Nastouli E, Kozlakidis Z, Kellam P, Morris S (2016) Cost analysis of standard sanger sequencing versus next generation sequencing in the ICONIC study. Lancet 388:S86. https://doi.org/10.1016/S0140-6736(16)32322-4

Payne S (2017) Methods to study viruses. In: Payne S (ed) Viruses. Academic Press, pp 37–52

Payne AC, Chiang ZD, Reginato PL, Mangiameli SM, Murray EM, Yao C-C, Chen F (2021) In situ genome sequencing resolves DNA sequence and structure in intact biological samples. Science 371(6532):eaay3446. https://doi.org/10.1126/science.aay3446

Pemba CM, Kurosaki Y, Yoshikawa R, Oloniniyi OK, Urata S, Sueyoshi M, Yasuda J (2019) Development of an RT-LAMP assay for the detection of Lassa viruses in Southeast and South-Central Nigeria. J Virol Methods 269:30–37. https://doi.org/10.1016/j.jviromet.2019.04.010

Perez-Ruiz M, Pedrosa-Corral I, Sanbonmatsu-Gamez S, Gomez-Camarasa C, Navarro-Mari JM (2018) Analytical validation of viral CNS flow chip kit for detection of acute meningitis and encephalitis. J Virol Methods 259:54–59. https://doi.org/10.1016/j.jviromet.2018.06.006

Piccirilli G, Chiereghin A, Gabrielli L, Giannella M, Squarzoni D, Turello G, Lazzarotto T (2018) Infectious meningitis/encephalitis: evaluation of a rapid and fully automated multiplex PCR in the microbiological diagnostic workup. New Microbiol 41(2):118–125

Qian L, Shi Y, Li F, Wang Y, Ma M, Zhang Y, Zhang G (2020) Metagenomic next-generation sequencing of cerebrospinal fluid for the diagnosis of external ventricular and lumbar drainage-associated ventriculitis and meningitis. Front Microbiol 11:596175

Quince C, Walker AW, Simpson JT, Loman NJ, Segata N (2017) Shotgun metagenomics, from sampling to analysis. Nat Biotechnol 35(9):833–844. https://doi.org/10.1038/nbt.3935

Ramachandran PS, Wilson MR (2020) Metagenomics for neurological infections—expanding our imagination. Nat Rev Neurol 16(10):547–556. https://doi.org/10.1038/s41582-020-0374-y

Sall O, Thulin Hedberg S, Neander M, Tiwari S, Dornon L, Bom R, Molling P (2019) Etiology of central nervous system infections in a rural area of Nepal using molecular approaches. Am J Trop Med Hyg 101(1):253–259. https://doi.org/10.4269/ajtmh.18-0434

Saranathan R, Asare E, Leung L, de Oliveira AP, Kaugars KE, Mulholland CV, Jacobs WR Jr (2022) Capturing structural variants of herpes simplex virus genome in full length by Oxford nanopore sequencing. Microbiol Spectr 10(5):e0228522. https://doi.org/10.1128/spectrum.02285-22

Schlaberg R, Chiu CY, Miller S, Procop GW, Weinstock G (2017) Validation of metagenomic next-generation sequencing tests for universal pathogen detection. Arch Pathol Lab Med 141(6):776–786. https://doi.org/10.5858/arpa.2016-0539-RA

SciLifeLab (2023) In situ sequencing; https://www.scilifelab.se/units/in-situ-sequencing/

Sikazwe C, Neave MJ, Michie A, Mileto P, Wang J, Cooper N, Williams DT (2022) Molecular detection and characterisation of the first Japanese encephalitis virus belonging to genotype IV acquired in Australia. PLoS Negl Trop Dis 16(11):e0010754. https://doi.org/10.1371/journal.pntd.0010754

Silva S, Pardee K, Pena L (2019) Loop-mediated isothermal amplification (LAMP) for the diagnosis of Zika virus: a review. Viruses 12(1):19. https://doi.org/10.3390/v12010019

Simner PJ, Miller S, Carroll KC (2017) Understanding the promises and hurdles of metagenomic next-generation sequencing as a diagnostic tool for infectious diseases. Clin Infect Dis 66(5):778–788. https://doi.org/10.1093/cid/cix881

Tan LV, van Doorn HR, Nghia HDT, Chau TTH, Tu LTP, de Vries M, Baker S (2013) Identification of a new cyclovirus in cerebrospinal fluid of patients with acute central nervous system infections. MBio 4(3):e00231–e00213

Tanner NA, Zhang Y, Evans TC Jr (2012) Simultaneous multiple target detection in real-time loop-mediated isothermal amplification. BioTechniques 53(2):81–89. https://doi.org/10.2144/000113902

Taraphdar D, Chatterjee S (2015) Complete genome sequence of two genotype III Japanese encephalitis virus isolates from West Bengal, India. Indian J Med Res 142(1):72–78. https://doi.org/10.4103/0971-5916.162122

Teo JW, Chiang D, Jureen R, Lin RT (2015) Clinical evaluation of a helicase-dependant amplification (HDA)-based commercial assay for the simultaneous detection of HSV-1 and HSV-2. Diagn Microbiol Infect Dis 83(3):261–262. https://doi.org/10.1016/j.diagmicrobio.2015.07.018

Thoendel M, Jeraldo PR, Greenwood-Quaintance KE, Yao JZ, Chia N, Hanssen AD, Patel R (2016) Comparison of microbial DNA enrichment tools for metagenomic whole genome sequencing. J Microbiol Methods 127:141–145. https://doi.org/10.1016/j.mimet.2016.05.022

Tkadlec J, Bebrova E, Berousek J, Vymazal T, Adamkova J, Martinkova V, Drevinek P (2020) Limited diagnostic possibilities for bloodstream infections with broad-range methods: a promising PCR/electrospray ionization-mass spectrometry platform is no longer available. Microbiology 9(5):e1007. https://doi.org/10.1002/mbo3.1007

Tomita N, Mori Y, Kanda H, Notomi T (2008) Loop-mediated isothermal amplification (LAMP) of gene sequences and simple visual detection of products. Nat Protoc 3(5):877–882. https://doi.org/10.1038/nprot.2008.57

Tsai EA, Shakbatyan R, Evans J, Rossetti P, Graham C, Sharma H, Lebo MS (2016) Bioinformatics workflow for clinical whole genome sequencing at partners HealthCare personalized medicine. J Pers Med 6(1):12. https://doi.org/10.3390/jpm6010012

Ure RLD, Edwards G, Dendzik T (2012) Evaluation of the FTD bacterial meningitis kit in comparison to in-house assays for the direct detection of N. *meningitidis, S. pneumoniae and H. influenzae in clinical specimens*. Paper Presented at the ECCMID, Basel

Vincent M, Xu Y, Kong H (2004) Helicase-dependent isothermal DNA amplification. EMBO Rep 5(8):795–800. https://doi.org/10.1038/sj.embor.7400200

Vincent AT, Derome N, Boyle B, Culley AI, Charette SJ (2017) Next-generation sequencing (NGS) in the microbiological world: how to make the most of your money. J Microbiol Methods 138: 60–71. https://doi.org/10.1016/j.mimet.2016.02.016

Wang J, Ye J, Yang L, Chen X, Fang H, Liu Z, Zhang Z (2022) Inconsistency analysis between metagenomic next-generation sequencing results of cerebrospinal fluid and clinical diagnosis with suspected central nervous system infection. BMC Infect Dis 22(1):764. https://doi.org/10.1186/s12879-022-07729-0

Wilson MR, Suan D, Duggins A, Schubert RD, Khan LM, Sample HA, DeRisi JL (2017) A novel cause of chronic viral meningoencephalitis: Cache Valley virus. Ann Neurol 82(1):105–114. https://doi.org/10.1002/ana.24982

Wilson MR, Sample HA, Zorn KC, Arevalo S, Yu G, Neuhaus J, Chiu CY (2019) Clinical metagenomic sequencing for diagnosis of meningitis and encephalitis. N Engl J Med 380(24): 2327–2340. https://doi.org/10.1056/NEJMoa1803396

Witney AA, Gould KA, Arnold A, Coleman D, Delgado R, Dhillon J, Stoker NG (2015) Clinical application of whole-genome sequencing to inform treatment for multidrug-resistant tuberculosis cases. J Clin Microbiol 53(5):1473–1483

Wolk DM, Kaleta EJ, Wysocki VH (2012) PCR-electrospray ionization mass spectrometry: the potential to change infectious disease diagnostics in clinical and public health laboratories. J Mol Diagn 14(4):295–304. https://doi.org/10.1016/j.jmoldx.2012.02.005

Wong YP, Othman S, Lau YL, Radu S, Chee HY (2018) Loop-mediated isothermal amplification (LAMP): a versatile technique for detection of micro-organisms. J Appl Microbiol 124(3): 626–643. https://doi.org/10.1111/jam.13647

Worthey EA, Mayer AN, Syverson GD, Helbling D, Bonacci BB, Decker B, Veith RL (2011) Making a definitive diagnosis: successful clinical application of whole exome sequencing in a child with intractable inflammatory bowel disease. Genet Med 13(3):255–262

Xing X-W, Zhang J-T, Ma Y-B, He M-W, Yao G-E, Wang W, Yu S-Y (2020) Metagenomic next-generation sequencing for diagnosis of infectious encephalitis and meningitis: a large, prospective case series of 213 patients. Front Cell Infect Microbiol 10:88. https://doi.org/10.3389/fcimb.2020.00088

Yan W, Hu Z, Zhang Y, Wu X, Zhang H (2021) Case report: metagenomic next-generation sequencing for diagnosis of human encephalitis and endophthalmitis caused by Pseudorabies virus. Front Med (Lausanne) 8:753988. https://doi.org/10.3389/fmed.2021.753988

Zhang X, Lowe SB, Gooding JJ (2014) Brief review of monitoring methods for loop-mediated isothermal amplification (LAMP). Biosens Bioelectron 61:491–499. https://doi.org/10.1016/j.bios.2014.05.039

Zhang Y, Cui P, Zhang H-C, Wu H-L, Ye M-Z, Zhu Y-M, Zhang W-H (2020) Clinical application and evaluation of metagenomic next-generation sequencing in suspected adult central nervous system infection. J Transl Med 18(1):199. https://doi.org/10.1186/s12967-020-02360-6

Advanced Microbiological Diagnostic Techniques in Fungal Infections of the Central Nervous System

19

Shalini Malhotra, Vikash Ranjan, C. Suman, S. Patil, A. Malhotra, and Nirmaljit Kaur Bhatia

Abstract

Conventional diagnostic approaches, for example, culture and histology, which are believed as gold standards, have a low detection rate, emphasising the need for novel ways to detect fungal infections. Novel serological and molecular approaches have been developed and are under constant evaluation. Serological tests for Cryptococcus spp., dimorphic fungi, and Pneumocystis spp., the galactomannan tests for aspergillus spp., and β-glucan assays for invasive Candida spp. and moulds are established diagnostic approaches and are used in routine diagnostic practise. PCR and other molecular methods, such as matrix-assisted laser desorption ionisation (MALDI) and fluorescence in situ hybridization (FISH), on the other hand, have shown promise in clinical trials but require standardisation before becoming widely used. These novel diagnostic technologies allow for the early detection of invasive fungal disease and the use of a preventative treatment strategy, which could lead to better treatment outcomes and lower toxicity.

Keywords

Molecular · Serological tests · Fungi · Novel diagnostic technologies

S. Malhotra (✉) · V. Ranjan · N. K. Bhatia
Department of Microbiology, Atal Bihari Vajpayee Institute of Medical Sciences & RML Hospital, New Delhi, India

C. Suman · S. Patil
ARMCH and Research Center Kumbhari, Solapur, Maharashtra, India

A. Malhotra
KIMSDU, Malkapur, Maharashtra, India

H. Sami et al. (eds.), *Viral and Fungal Infections of the Central Nervous System: A Microbiological Perspective*, https://doi.org/10.1007/978-981-99-6445-1_19

19.1 Introduction

With each passing year, the prevalence of fungal infections has increased due to improvements in the ability to identify these fungi as well as rising risk factors like immunocompromised states as more infections have been reported in patients who belong to high-risk groups, including HIV-positive and AIDS patients, transplant recipients, and immunosuppressed patients receiving chemotherapy or corticosteroids (Low and Rotstein 2011).

There are certain conditions that may make a patient more likely to develop a particular etiological agent. Disease/treatment associated factors like steroid therapy, long-term antibiotic therapy, transplantation, neutropenia, contaminated devices, neurosurgery, genetic factors like, CARD9 deficiency, chronic granulomatous disease, etc., may predispose to Candida infections (Góralska et al. 2018). Renal failure, necrotic burns, diabetic ketoacidosis and intravenous drug use may make patients prone to Mucoromycetes infection. Cryptococcus and Histoplasma may be transmitted via bird contact while iron load and deferoxamine therapy may likely cause Mucoromycetes infection. However, several fungi, including Cryptococcus, Coccidioides, and Histoplasma, can also infect patients who have a proper functioning immune system (Afroze et al. 2017; Góralska et al. 2018; Rodríguez-Cerdeira et al. 2014).

Cryptococcal meningoencephalitis is the most typical fungus infection of the central nervous system globally (Maziarz and Perfect 2016; Kauffman 2019). Invasive cryptococcosis patients are thought to develop CNS mycosis in about 67–84% of cases, invasive candidiasis in 3–64%, blastomycosis in 40%, disseminated coccidioidomycosis in 25%, disseminated histoplasmosis in 5–20%, Mucor mycosis in 12%, and invasive aspergillosis in 4–6% (Schmiedel and Zimmerli 2016). The results are typically less than satisfactory despite advancements in diagnostic techniques and the launch of several new and improved treatment in recent years (Fang et al. 2023). Several reviews on the subject of fungi that can infect the brain have been done in the recent times (Góralska et al. 2018).

Multiple different mechanisms allow fungi to enter the central nervous system (Chikley et al. 2019). Hematogenous spread is the most frequent cause of infection, with the lungs serving as the primary site of infection. Local extension might come from the orbits, ear, or paranasal sinuses (Archibald and Quisling 2013). A surgical operation, the placement of foreign objects such as ventriculoperitoneal shunts, the doing lumbar punctures, administering medications, or suffering head trauma are among the causes of traumatic introduction that can all result in infection (Fowler et al. 2023).

Fungal infections can cause a variety of various clinical syndromes; the particular syndrome may occasionally offer hints as to the relevant etiologic agents (Jain et al. 2010). The majority of endemic yeasts, including Candida, Coccidioides, Histoplasma, Blastomyces, Sporothrix, and Cryptococcus as well as those responsible for nosocomial infections, cause a meningitis syndrome, which is typically a subacute or chronic meningitis (Sharma and Anand 2002). Although it happens infrequently, filamentous fungi can cause meningitis. Some fungus, such as

Aspergillus, zygomycetes, phaeohyphomycotic etiologic agents, and occasionally Candida spp., are known to frequently cause abscesses (Góralska et al. 2018). Particularly in diabetic patients, the zygomycetes are infamous for causing rhino cerebral disease (Walsh and Dixon 1996). Vasculitis and stroke can be caused by fungi that affect the basilar meninges (Coccidioides, Cryptococcus, Histoplasma, and Candida) (Afroze et al. 2017). Sometimes, Candida and Histoplasma can infect the heart valves and result in embolic lesions that can lead to a stroke state (Sharma and Anand 2002). Additionally, there are some organisms (Aspergillus and zygomycetes) that are traditionally linked to angiotropism and frequently cause infarction and haemorrhagic necrosis. Although uncommon, spinal syndromes can happen in certain circumstances (Jabr and Hammoud 2020).

Given the high mortality rate, occasionally quick disease progression, and the fact that the brain is not a very forgiving organ, it is crucial that patients suspected of having a CNS fungal infection that gets evaluated as soon as possible (Gavito-Higuera et al. 2016). A comprehensive history may commonly reveal host factors, such as underlying conditions, immunosuppressive drugs (such as corticosteroids and biologic response modifiers), or interests (such as spelunking), that predispose to specific types of fungal infections (de Pauw 2011). Since many of the fungi that cause the disease have systemic effects, a physical examination may reveal signs in the mucous membranes, skin, prostate, or lungs (Kobayashi 1996). A hypothesis on probable etiologic agents should be used to guide the ordering of particular laboratory tests. It cannot be overstated how crucial it is to make a particular microbiologic diagnosis in order to recommend the best course of treatment due to the unique antimicrobial susceptibilities for the various fungus (Alastruey-Izquierdo et al. 2015). In this chapter, we will first talk about how to appropriately consider a diagnostic approach to distinct clinical presenting syndromes. Then, certain specific diagnostic techniques will be taken into account.

19.1.1 Clinical Picture of Fungal CNS Infections

Both Aspergillus spp. and Blastomyces dermatitidis have a similar clinical presentation that manifests as chronic meningitis and abscesses (McBride et al. 2017). Numerous microabscesses, macroscopic abscesses, and subacute meningitis are among the clinical symptoms of Candida spp. and can be epidemiologically linked to nosocomial settings, complicated ventricular shunts, or recent treatment for bacterial meningitis (Bays and Thompson 2021). Clinical signs of Coccidioides immitis and posadasii include subacute to chronic meningitis, focal intracerebral abscess(es), and cerebral vasculitis (Thakur and Wilson 2018). Epidemiologically, these conditions are associated with dark-skinned races and they can also be clinically linked to HIV infection and steroid therapy (Pappas 2013). As biologic response-modifying agents, Cryptococcus neoformans causes subacute to chronic meningoencephalitis, whereas Cryptococcus gattii is known to cause focal intracerebral abscess(es) (Guarner and Brandt 2011). Histoplasma capsulatum, Paracoccidioides brasiliensis, and Scedosporium/Pseudallescheria spp. can also result in focal intracerebral abscess(es) (Ines Andrade et al. 2014). Chronic meningitis is brought on by Sporothrix schenckii, which is typically linked to HIV infection, alcoholism, and occupational or environmental exposure (Scott et al. 1987). Zygomycetes' clinical symptoms include a rhino cerebral form, abscesses in the cerebrum and central nuclei, and they are frequently associated with acidosis, diabetes mellitus, and neutropenia (Ribes et al. 2000).

19.1.2 Etiological Agents Associated with Fungal Infections of the CNS

More than 100 different species of mould belong to the Aspergillus genus, which is ubiquitous and has a wide geographic range. Aspergillus fumigatus is the most common species to infect people (Sugui et al. 2015). Non-aspergillus moulds are a diverse group of common moulds that can be divided into three groups: hyalohyphomycetes (moulds without pigmentation), phaeohyphomycetes (melanin producing dark or dematiaceous moulds), and mucoromycetes (moulds of the order Mucorales with uniseptate or pauci-septate hyphae) (Juravel et al. 2023). Two important pathogenic yeast species make up the Cryptococcus genus: Cryptococcus neoformans, which is found around the world, and Cryptococcus gattii, which is native to Papua New Guinea, the Pacific Northwest of North America, and Australia (Kwon-Chung et al. 2014). A type of yeasts known as Candida are common commensal organisms but have the potential to infect people and spread invasive diseases (Talapko et al. 2021). The heterogeneous group of yeasts that seldom produce invasive infections, typically in individuals having accompanied (significant) immunosuppression, is composed of non-candida and non-cryptococcus yeasts. Dimorphic fungi are a class of fungi that depend on the in vivo or environmental conditions and can grow as either yeast or mould. The majority of dimorphic fungus are only found in endemic regions (Arya and Rafiq 2023).

19.1.3 Immunological States Predisposing to Various Fungal Infections of the CNS

There are several disease-induced immunosuppressed states that can make people more susceptible to fungus infections (Low and Rotstein 2011). Aspergillus spp. and non-aspergillus moulds in haematological malignancies, Aspergillus spp. in neutropenic conditions like aplastic anaemia, Candida spp. in neonate prematurity, and Mucorales in diabetes or iron overload are just a few examples of such fungal infections (Colombo et al. 2017).

19.1.3.1 Treatment-Induced Immunosuppression

Immunosuppression and other CNS-affected fungal infections are potential side effects of several treatments. In medical immunosuppression, haemopoietic stem cell transplantation, and solid organ transplantation, fungal infections such

Aspergillus spp. and non-aspergillus moulds are often encountered. Candida spp. is also linked to solid organ transplantation. While Cryptococcus and Aspergillus spp. are connected to the immunosuppression brought on by immunomodulatory drugs like ibrutinib, dimorphic fungus and mould infections are connected to the administration of biologicals like TNF (Rahi et al. 2021; Shoham and Marr 2012).

19.1.3.2 Immunodeficiency Syndromes

Patients with any immunodeficiency syndrome are usually at risk for CNS fungal infections that may include those with Aspergillus spp. in chronic granulomatous disease, Candida spp. in CARD9 (caspase recruitment domain-containing protein deficiency), and Cryptococcus gattii, which can cause autoantibodies against GM-CSF (granulocyte-macrophage colony-stimulating factor) (Kwon-Chung et al. 2014; Lanternier et al. 2013).

19.1.4 Clinical Syndromes

Meningitis/encephalitis—This section covers cases ranging from meningoencephalitis syndromes through subacute and chronic meningitis syndromes. Subacute hydrocephalus brought on by meningitis may manifest as normal-pressure hydrocephalus or obstructive hydrocephalus (caused by scarring of the arachnoid villus). These are typically caused by long-term infections with dimorphic fungi or yeasts (Mactier et al. 1998; Sapra and Singhal 2019).

Meningovascular syndromes—A variety of true moulds, especially Aspergillus species, can cause subacute meningitis, which can thereafter result in an angioinvasive disease affecting the arterial trunks close to the CSF cisterns at the base of the brain (Johnson 1996). Infarction of the brain tissue, particularly in a vascular region, may arise from the ensuing arteritis. Similar to how Mucorales in the sinuses or orbits are seen, invasive fungal infections at the base of the skull can affect nearby distal portions of the carotid arteries and their branches. These two infectious disorders typically manifest as complex strokes with meningitis-related symptoms and signs (Chen et al. 2023).

Focal infection/mass lesions—Patients with these infections frequently also experience headaches and focal neurological signs or seizures (Li et al. 2020). Clinical examinations of patients frequently reveal focal impairments, which calls for brain imaging with a CT or MRI (Tillema and Pirko 2013). This type of imaging typically identifies a distinct mass lesion, sometimes accompanied by contrast enhancement or not, and sporadically with displacement of nearby structures suggestive of a mass effect (Khan and Sepahdari 2012).

Spinal cord syndromes—Based on the particular level(s) at which the cord is affected and the degree of the cord's involvement, many clinical disorders appear differently (Beh et al. 2013). From focal spinal meningitis to frank fungal myelitis, infectious diseases that cause spinal cord disorders include epidural abscess (Archibald and Quisling 2013; Bhattacharya and Joshi 2011). The diagnostic strategy for focal lesions, like the previous two major categories of clinical syndromes, frequently involves a guided sample of tissue/abscess material, CSF sampling (preferably focal), and on occasion, tissue sampling (such as meningeal biopsy or cord biopsy) (Hrishi and Sethuraman 2019).

Foreign body-associated fungal infections—Although in theory, any foreign material that is in contact with the meninges, the brain tissue, or the CSF might cause a fungal infection, but these disorders are most frequently seen in the context of a ventricular drain, lumbar drain, ventricular shunt, or a pleural shunt (Gavito-Higuera et al. 2016). The wide range of ways in which infections associated with CSF shunts or drains manifest as well as the infection's localised nature can make them challenging to diagnose (Hanak et al. 2017).

Skull-based infections—These syndromes include fungi-related infections such as fungal sinusitis, osteomyelitis, and rhino cerebral mucormycosis (Chikley et al. 2019; Khan et al. 2018). These infections frequently spread to the central nervous system (CNS), sometimes with fatal results. Since these infections manifest without CNS involvement, the diagnostic strategy for them is primarily the same as that used for each of the distinct entities (Archibald and Quisling 2013; Li et al. 2020).

Peripheral nervous system syndromes—With concurrent or even primary peripheral nerve symptoms such as radiculopathy, plexopathy, and neuropathy, fungi infections of the CNS might present themselves. While many of these manifestations have been documented in the literature, multiple cranial palsies and lumbosacral nerve root involvement with cauda equina syndrome are more typical presentations (Brizzi and Lyons 2014).

19.2 Diagnostic Modalities

- Elucidation of clinical history
- Radiology (CT scan/MRI/other relevant radiological imaging)
- Laboratory investigations—microbiological/pathological/molecular

The bulk of infections that affect the CNS are brought on by viral or bacterial etiologies, but fungi are also becoming more and more recognised as serious pathogens, especially in individuals who have impaired immune systems. Certain radiological investigations may help in diagnosing different fungal CNS infections (Table 19.1) (Archibald and Quisling 2013; Gavito-Higuera et al. 2016). New fungal pathogens are being found, and for both established and emerging fungal pathogens, new symptoms are being reported. The development of diagnostic techniques to determine aetiology is a necessity for the clarification of the spectrum of disease linked to fungal infections of the CNS. CNS fungi infections can be grouped in a number of different ways (Arvanitis et al. 2014; Gavito-Higuera et al. 2016; Góralska et al. 2018; Guarner and Brandt 2011).

Table 19.1 Radiology (CT scan/MRI)

Fungus	CT scan	MRI
Aspergillus spp.	Superimposed haemorrhage	
	Hyperdense	T1WI—hyperintense
	Infarction	
	Hypodense	T2WI—hyperintense
Cryptococcus	Miliary lesions and cryptococcomas—variable density masses	T1WI—hypointense
		T2WI—hyperintense
Coccidiodes	Non-specific	Results are superior to CT scan
		Leptomeningeal enhancement of the basilar, sylvian, and interhemispheric cisterns
Histoplasma	Non-specific	Non-specific
Candidiasis	Microabscesses—hypodense	Non-specific
	Granulomas—nodular or ring appearance	

19.2.1 Microbiological Techniques for Isolating Fungi (Hage et al. 2019; Lockhart et al. 2021; Vainionpää and Leinikki 2008)

1. Methods for direct microscopic inspection
   India ink, KOH, and Calcofluor white for wet mounts
   Histopathological examinations
   Fluorescent antibody staining
2. Mycology culture
3. Non-culture techniques
   Serological techniques
   Tests for antibody detection
   Tests for antigen detection
   Immunohistochemical techniques
4. Examinations to look for cell-mediated immunity
5. Molecular techniques
   Polymerase chain reaction (PCR)
   Microarrays
6. Newer advanced diagnostic modalities

19.3 Microscopic Examination

Although this has been used for a long time, it still needs to be emphasised. When a patient has a fungal infection, this crucial process frequently offers the first microbiological evidence of the disease's aetiology and directs the choice of the best media to support growth. KOH, India ink, and calcofluor white are among the preparations for direct clinical specimen examination (Guarner and Brandt 2011). A few staining methods, including Giemsa and periodic acid Schiff (PAS), are also useful (Zaaroura and Bergman 2019).

19.3.1 KOH

An excellent tool for recognising the various fungal presenting patterns is a potassium hydroxide (KOH) mount (Ponka and Baddar 2014). Candida albicans is likely to be the culprit if yeasts and pseudohyphae are discovered in the same microscopic field (Mukaremera et al. 2017). The Aspergillus fungus usually appears as acutely angled, thin septate hyphae. In contrast, the hyphal structures of Mucorales fungi are non-septate, wide, and ribbon-like, with 90° branching. While Blastomyces is a yeast with a broad budding pattern in a figure-of-eight, Cryptococcus displays a capsule. The spherule with endospores is indicative of Coccidioides immitis while Histoplasma capsulatum is a tiny intracellular fungus (Guarner and Brandt 2011; Lee et al. 2015).

19.3.2 Calcofluor with KOH

Since Calcofluor white dye with KOH binds to the 1-3, 1-4 polysaccharides found in fungal cell walls, it is excellent for demonstrating the presence of fungal cells in clinical specimens (Baniya 2022). When exposed to UV radiation with shorter wavelengths, the dye fluoresces. For the purpose of finding fungal cells produced using Calcofluor White, a fluorescence microscope is required. When yeast cells, pseudohyphae, and hyphae are present, a chalk-white or vivid apple-green fluorescence is visible. This usually depends on the filter that is being used to isolate them from the background material. The drawback of utilising this method may be the requirement for a fluorescent microscope (Kumar et al. 2009; Punjabi et al. 2020).

19.3.3 Gram's Stain

In certain cases, fungi like Cryptococcus sp. simply exhibit stippling and barely stain. Some Nocardia spp. isolates either do not stain at all or stain very poorly (McMurray 1996; Sobottka et al. 1993; Soltani et al. 2013).

19.3.4 India Ink

When determining if extracellular polysaccharide capsules of fungi, particularly Cryptococcus neoformans, are present or absent in CSF, India ink (Fig. 19.1) is helpful. The yeast cells that have been encapsulated can easily be seen against the dark backdrop, thanks to this dye (Coradello and Tirelli 2021; Zaragoza et al. 2009).

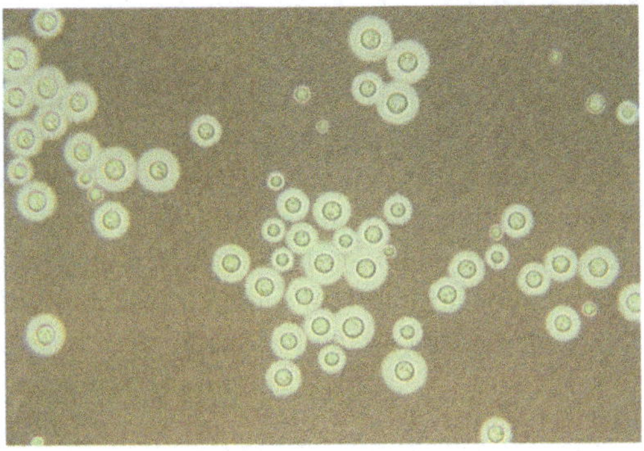

Fig. 19.1 India ink staining showing refractile capsules surrounding round budding yeast cells. (Image Source: Centers for Disease Control and Prevention—Public Health Image Library—ID#3771 (Haley 1969)—Copyright Restrictions: NONE; this image is in the public domain and thus free of any copyright restriction)

Lactophenol cotton or aniline blue wet mount—It is the approach that is employed the most. The fungal cell wall structures are preserved by lactic acid, allowing the slides to be rendered permanent (Leck 1999; Tankeshwar 2014).

19.3.5 Periodic Acid Schiff

PAS is one of the most often employed stains for fungal histopathology. Since glycogen is detected in tissues by PAS and is sufficiently present in the cell walls of fungi, it can be utilised to screen for the presence of fungi (Guarner and Brandt 2011).

19.3.6 Gomori's Methenamine Silver Stain: Grocott's Modification

By projecting mycological cell walls' polysaccharides, the reduced silver technique is used in this procedure to liberate and identify aldehyde groups. Aldehydes reduce the amount of methenamine silver nitrate complex, which causes the fungal cell wall to become stained brown-black. In comparison to PAS, which only stains live fungi, the Gomori's methenamine silver stain is superior since it can detect both live and dead fungi (Guarner and Brandt 2011; Wang et al. 2022).

19.3.7 Fluorescent Antibody Staining

This method can be applied to clinical samples such as pus, blood, CSF, tissue impression smears, paraffin sections of formalin-fixed tissues, and tissue impression smears to identify fungal antigen. The primary benefit of this method is the ability to identify fungi, even when only a small number of organisms are present, as seen in pus from sporotrichosis (Guarner and Brandt 2011; Kradin and Iafrate 2010).

19.3.8 Histopathology

To define the diagnostic relevance of positive culture isolates, including fungal invasion of tissue and vasculature as well as the host reaction to the fungus, histologic analysis of tissues to look for fungal cell walls remains a crucial technique. While awaiting the findings of a fungal culture, histopathology can also quickly presumptively diagnose the fungus or it may be the only source of material when no culture growth occurs or no cultures were ordered. H&E staining of the tissue should always come first in the histopathologic assessment of resection specimens from surgery, autopsy samples, and biopsy specimens. If a fungal etiology is suspected after evaluating tissue sections because of the presence of an inflammatory tissue response or when there is strong clinical suspicion, GMS and PAS staining should be carried out even if the H&E stain is equivocal. Additionally, mucin (mucicarmine) or melanin (Fontana-Masson) stains can be very helpful for identifying dematiaceous fungus, which may not produce an abundance of pigment (melanin), as well as for Cryptococcus identification (Guarner and Brandt 2011).

Advantages: Determining if an organism recovered in culture represents contamination, colonisation, or actual infection can be crucial in particular situations. Necrosis and tissue and vascular invasion are crucial histopathologic traits that can aid in the differentiation. Histopathologic presumptive diagnosis may be the only indication of a fungal infection because fungal cultures do not always grow. If the original specimen is ground too aggressively, the hyphal elements may be destroyed and the culture may not grow; for instance, Mucorales genera are moulds that typically grow within 24–48 h. Yeasts may be seen in tissue sections of immunocompetent patients with a persistent solitary nodule produced by an endemic mycosis, but the culture may not grow because the yeasts are not viable (Guarner and Brandt 2011; Zaragoza et al. 2009). The inability of Aspergillus spp. and other septate hyaline moulds to recover after therapy, the use of prophylactic antifungal drugs, or potential variations in the physiologic states of the mould in vivo and in vitro have all been reported (Jacobs and Walsh 2023).

When a small sample of biopsy material is sent to the microbiology lab with a request for several cultures, for instance, for aerobic microbes, anaerobic ones, acid-fast bacilli, and fungi, it may prove difficult to recover a microbe in culture despite the fact that a viable microbe initially existed in the specimen. This tiny amount of information must occasionally be split even more evenly between microbiology and histology (Bowler et al. 2001; Guarner and Brandt 2011).

Last but not least, some fungi, like Pneumocystis, cannot be grown using current microbiological methods and must instead be discovered using histopathologic or cytologic methods (Guarner and Brandt 2011).

A preliminary histopathologic diagnosis of a fungal infection may be ready before the culture in some circumstances, providing clinicians with enough information to begin therapy. In these situations, fungal cultures may take weeks to complete. Scedosporium, Sporothrix, Blastomyces, and Coccidioides, for instance, may take up to 3–4 days to grow, whereas Histoplasma and Paracoccidioides may take longer than 2 weeks. As a result, the histopathologic diagnosis may be available earlier than the results of a culture (Guarner and Brandt 2011).

Disadvantages: Pathologists must distinguish between stained typical human tissue structures and potentially fungal formations. Neurosecretory granules and melanin, in particular when using GMS stains, can be mistaken for yeasts, whereas hyphae need to be distinguished from collagen fibres, basement membranes, and other silver-staining filamentous structures. Pathologists should also evaluate whether or not fungi's internal features, such as their nuclei and cytoplasm, which stain with H&E but not GMS, are present (Guarner and Brandt 2011).

When the sample displays some transversally sliced hyphae, which subsequently seem as yeasts and may even appear to be budding, this might also be a disadvantage during the interpretation of particular stains. In these situations, it is crucial to look farther into the block to see whether the sample contains any more fungal elements sliced longitudinally. Typically, histopathology is unable to identify the fungal genus and species, which are crucial for treatment. Aspergillus spp., Fusarium, Scedosporium, and others are included in the differential for a case with hyaline septate hyphae for instance. Voriconazole, which is effective against all of these fungi, should thus be used as the primary treatment; amphotericin B should only be used as a secondary treatment for A. fumigatus and F. solani. For A. fumigatus, itraconazole or echinocandins might be employed, but they would not work for F. solani or Scedosporium. Fluconazole might be used to treat the majority of Candida spp. but would not work against C. krusei, C. glabrata, or the other septated hyaline moulds as the detection of yeasts with pseudohyphae suggests Candida spp. with a differential diagnosis that includes these moulds (Guarner and Brandt 2011; Herbrecht et al. 2002a; Wiederhold 2017).

19.3.9 Immunohistochemistry

In order to see the targets' (i.e., relevant antigens') morphologies in addition to the tissues around, an immunohistochemistry-based test may be undertaken. It involves the use of antibodies to target fungal antigens in tissue slices, particularly in extremely thin tissue specimens which are utilised in histopathology, and placed on a glass slide (Guarner and Brandt 2011). Both fresh frozen tissue and paraffin-embedded tissue are acceptable for this test. The embedded tissue is typically formalin fixed, which might distort antigens, especially if the fixing process has been prolonged. Formalin-fixed and paraffin-embedded (FFPE) tissues are typically preferred by pathologists because they can be easily preserved at room temperature and are a part of routine histopathologic procedure. Paraffin must be removed from the tissue in order to conduct the procedure, which is often accomplished using chemicals that dehydrate the tissue. Antifungal antibodies must first be given to the tissue after it has been rehydrated by treating it with enzymes or exposing to other antigen retrieval procedures to make fungal antigens accessible to the antibodies (Gao et al. 2020; Guarner and Brandt 2011; Rickerts 2016).

Advantages: It has numerous benefits, including the ability to precisely detect the organism using materials that have already undergone regular pathology lab processing to render them non-infectious and the combination of morphology (the fungal element itself, its position in the tissue, and the inflammatory reaction). For immunohistochemistry assays, a variety of automation platforms are commercially available, which lowers the cost and turnaround time (Guarner and Brandt 2011; Sugui et al. 2015). Last but not least, the reactions produced by enzyme-labeled antibodies are permanently recorded. Immunohistochemistry may offer a less expensive and quicker alternative to more expensive assays that do not integrate morphology with the detection of the specific fungus when new non-cross-reactive antibodies are created and evaluated using this method. Additionally, many fungi can be seen in tissue sections at the same time using double-staining immunohistochemical assays (Guarner and Brandt 2011; Hussain et al. 2020).

Disadvantages: Many of the currently available antibodies cross-react with other fungus, making them useless for the detection of specific organisms. Before results can be used for patient care, laboratory tests must be done on the antibodies and immunohistochemical assays to evaluate, verify, and validate them (Guarner and Brandt 2011; Hussain et al. 2020; Yeo and Wong 2002).

19.3.10 Laser Microdissection

It combines laser technology with microscopy to make it possible to examine particular cell types. After being separated, the cells of interest can be used for a number of research and tests, and the results will not be influenced or diluted by adjacent cells or tissue components in the tissue sample. There are two types of microdissection technologies: laser cutting and laser capture microdissection. Tissue sections are placed in specific slides that make it simple to separate them from the rest of the specimen. These tissue sections are typically thicker than the ones used for histopathology. The tissue can be fresh frozen or FFPE and stained with H&E or another chemical to enable bright-field microscope visualisation of the desired cells. The cells of interest are focused on (for laser capture) or cut around (for laser cutting) using a narrow-beam laser. A UV laser is used for cutting microdissection, whereas an infrared laser is used for capture microdissection. The selected or removed cells are collected in a plastic cap or tube. The preparation of the tissue before it is placed on the slides for microdissection must be specific to the secondary test. Fixation (or lack thereof) and staining are two aspects of preparation. The least amount of chemical exposure to the tissues ensures the least degree of sample modification for later testing. Nucleic acid detection via PCR, electron microscopic visualization, identification by means of mass spectrometry, or two-dimensional polyacrylamide gel electrophoresis (for proteins) are examples of potential secondary testing. The micro-dissected cells may even be cultured if properly prepared (Guarner and Brandt 2011; Liu 2010).

Advantages: The ability to selectively choose the components to be evaluated is laser microdissection's biggest benefit. Thus, it is possible to study in-depth dual infections and the local environment in which they take place. The material obtained won't be contaminated with non-fungal tissues, which is an additional benefit (Guarner and Brandt 2011; Liu 2010).

Disadvantages: Because they are expensive, many laboratories might not have access to or be able to afford laser capture microdissection equipment. From a technological standpoint, it is possible that some components won't be preserved because the laser's heat could damage the components chosen for secondary testing (Gavito-Higuera et al. 2016; Góralska et al. 2018; Guarner and Brandt 2011). If the cells must be chosen using specific tools in a microdissection system as opposed to being placed in the tube where the second test will be conducted, contamination may result. It has been noted that obtaining complete fungal nuclei is necessary for DNA extraction, which is another drawback. It is especially possible to successfully collect hyphal material from mucoromycete organisms that are pauciseptate; however, it might not contain nuclei (Guarner and Brandt 2011; Skiada et al. 2018).

19.4 Fungal Culture

19.4.1 Culture Methods

Fungal infections have always been diagnosed primarily through tissue culture. Culture also enables susceptibility testing. Potato dextrose agar, or the slightly modified potato flakes agar, Sabouraud dextrose agar (SDA), SDA with antimicrobial agents, and BHI agar loaded with blood and antimicrobial agents are some examples of fungi cultures. The antimicrobials that are typically included with the fungal medium are either gentamycin or chloramphenicol and cycloheximide (Liu et al. 2002). The first two prevent the growth of bacteria, whereas cycloheximide also prevents the growth of numerous environmental fungi that are normally regarded as contaminants (Basu et al. 2015). Petri dishes or large test tubes (Fig. 19.2) can be used for culture. Although the former has a bigger surface area, they are more susceptible to dehydration. The latter is less drying-prone and safer to handle. Because many fungi that cause disease can thrive at temperatures below 37 °C, cultures are incubated at 25, 30, and 37 °C. After recording colony traits including colour and growth texture, slide mounts should be made with lactophenol cotton blue dye to analyse the morphological details. One of the dimorphic species should be taken into account for moulds that grow in 7–14 days or those have an aerial mycelium that resembles a web. A direct mount of the fungal isolate is the most typical approach for microscopic examination of fungal cultures. These mount or cellophane tape mount are prepared to achieve this (Kozel and Wickes 2014). These two techniques can identify a wide variety of fungus; however, when atypical mycological forms may be involved, slide culture techniques should additionally be carried out (Guarner and Brandt 2011).

Aspergillus—At 25 and 37 °C, Aspergillus is inoculated on Sabouraud dextrose agar with antibiotics but without cycloheximide. The culture is inspected twice a

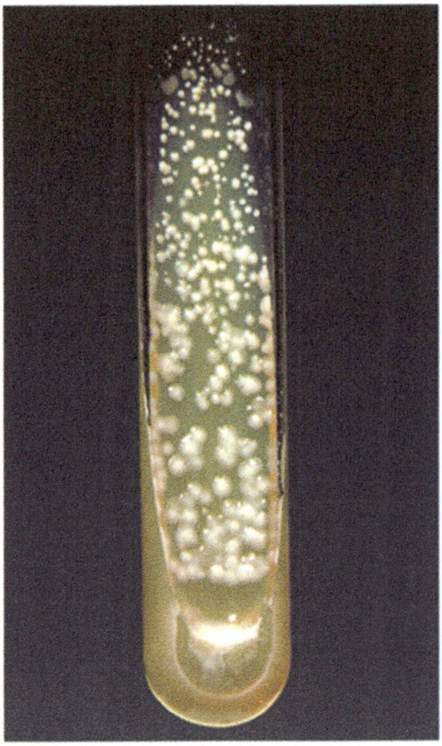

Fig. 19.2 Moist colonies of Cryptococcus neoformans on Sabouraud dextrose agar test tube. (Image Source: Centers for Disease Control and Prevention—Public Health Image Library—ID#3199 (Kaplan 1969)—Copyright Restrictions: NONE; this image is in the public domain and thus free of any copyright restriction)

week for the following 4 weeks after the first week's daily inspection (Das et al. 2010).

Mucorales—The majority of the species are inhibited by cycloheximide; thus, they can only grow on common fungal culture media without it (Dal Pizzol et al. 2021).

Histoplasma—It is inoculated onto SDA, treated with actidione and antimicrobial agents, and then incubated at 25 and 37 °C. The same antimicrobials are also used to inoculate it on BHI agar, which is then incubated at 25 °C (Kandi et al. 2016).

Blastomyces—At 25 and 37 °C, it may be inoculated on a variety of media, including blood agar, SDA, BHI agar, and blood glucose cysteine agar. The growing process typically takes 2–4 weeks (Kane 1984).

19.5 Non-culture Methods

19.5.1 Histology and Special Stains

It is the gold standard method for identifying IFI. Differentiation of organisms into broad groups of potential pathogens is made possible. Genus- and species-level identification, however, has a limited degree of specificity. It takes invasive methods to obtain a specimen, and the sensitivity depends on how well the sample was collected (Arvanitis et al. 2014; Kozel and Wickes 2014).

19.5.2 Nucleic Acid Detection/Biomarkers

Using a variety of specimen types has the ability to produce results with high sensitivity and specificity. It delivers outcomes faster than conventional approaches. There is a lack of uniformity among assays developed in laboratories. It cannot distinguish between invasive infection and contamination or colonisation, and sequencing-based assays can be pricey (Maziarz and Perfect 2016; Terrero-Salcedo and Powers-Fletcher 2020).

Cryptococcal meningitis—On histological analysis, pleomorphic narrow-based budding yeast cells with thick capsules are visible. A lateral flow assay (LFA) test can find the Cryptococcal antigen. It is possible to make 1,3-β-d-glucan. As a molecular technique, pan fungal DNA sequencing is carried out (Guarner and Brandt 2011; Kozel and Wickes 2014).

Cerebral histoplasmosis—On histological examination, small budding yeast cells (2–5 μm) are observed. Antigen from the histoplasma is detected. However, 1,3-β-d-glucan and antibody detection are done. Real-time PCR is the one molecular technique performed for its detection (Guarner and Brandt 2011; Stevens et al. 2016).

Cerebral coccidioidomycosis—On histological examination, spherules with sizes between 20 and 50 μm are visible. However, 1,3-β-d-glucan, antibody, and the Coccidioides antigen are all detectable. The molecular technique can be real-time PCR (Guarner and Brandt 2011).

Cerebral blastomycosis—On histological analysis, broad-based budding yeast cells (5–10 μm) are seen. Along with the detection of 1,3-β-d-glucan and antibodies, Blastomyces antigen is also found. The molecular technique can be real-time PCR (Arvanitis et al. 2014; Guarner and Brandt 2011).

Cerebral aspergillosis—On histological examination, septate hyphae with acute branching are visible. There is evidence of the Galactomannan antigen. However, 1,3-β-d-glucan and antibody detection is done. Real-time PCR is the one molecular technique performed (Guarner and Brandt 2011; Lehrnbecher et al. 2019).

Cerebral phaeohyphomycosis (e.g., Exserohilum)—On histological examination, brown-pigmented septate hyphae well-stained by Fontana Masson are observed. There is no antigen or antibody detection. However, 1,3-β-d-glucan is done. Real-time PCR and pan fungal DNA sequencing are used as the molecular techniques (Arvanitis et al. 2014; Guarner and Brandt 2011).

Cerebral mucormycosis (zygomycosis)—On histological examination, ribbon-like, deformed aseptate hyphae with nearly 90-degree branching are visible. Along with 1,3-β-d-glucan, antigen and antibody detection are not done. As a molecular technique, pan fungal DNA sequencing is carried out (Arvanitis et al. 2014; Guarner and Brandt 2011; Lockhart et al. 2021).

19.6 Serology

The method of serological diagnostics, which is advantageous in many ways, is based on the identification of antibodies or antigens from serum samples of suspected individuals. It can nevertheless give results with incredibly small sample quantities and is, to start with, more accurate than culture-based approaches (Vainionpää and Leinikki 2008). Furthermore, it is barely invasive. Once serological assays are positive, the need for culturing of a potentially dangerous fungus, like Coccidioides species, for example, may be reduced (Arvanitis et al. 2014; Kozel and Wickes 2014). The drawbacks of serology are its low sensitivity and specificity levels (Guarner and Brandt 2011; Richardson and Page 2018).

A fungal infection should not be ruled out by a negative serologic test. The limitation of serology is its inability to distinguish between current and previous infections (Hussain et al. 2020). In addition, the time of testing in relation to the disease process—for instance, early vs. late—affects sensitivity. The many serological methods include immunodiffusion, countercurrent immunoelectrophoresis, immunofluorescent antibody, ELISA, and western blotting/immunoblotting (Richardson and Page 2018). Measuring organism-specific antigens like Galactomannans, which are thought to be specific for the diagnosis of aspergillosis, is one of the tests that is particularly helpful. With the exception of Mucorales and Cr. Neoformans, serum glucan detection is utilised as a diagnostic technique for the detection of a wide range of fungal diseases (Arvanitis et al. 2014; Hussain et al. 2020).

19.6.1 Detection of Fungal Antibodies

It identifies patient populations at risk and offers proof of prior exposure. It aids in the diagnosis of specific illnesses, such as chronic and allergic aspergillosis, histoplasmosis, and coccidioidomycosis, and it is helpful for monitoring the effectiveness of treatment. It does not consistently distinguish between past exposure and present illness (Arvanitis et al. 2014; Guarner and Brandt 2011).

19.6.2 Complement Fixation

The foundation of CF is free complement's capacity to lyse sensitised RBCs (Arthur et al. 2019; Fida et al. 2022; Han et al. 2001).

Positive result: The patient sample's Ab interacts favourably with the fungal Ag in the reaction well. The free complement binds to the Ab-Ag complex. The cells sink to the bottom of the well and there is no RBC lysis.

Negative result Lack of Ab-Ag complexes leaves the complement free. When RBCs are lysed, a red pigment is spread across the well. The patient sample is titrated to ascertain the antibody titre.

It is typically done for Coccidioidomycosis, Blastomycosis, and Histoplasmosis. When diagnosing Histoplasma, yeast and mycelia are utilised as two distinct Ag, and a single titre of 1:32 or higher is suggestive of active infection but not definitive. The diagnosis is predicated on a fourfold increase in Ab titre. About 16–25% of instances of pulmonary Blastomycosis have CF findings that are positive, according to the Blastomycosis database. A titre of 1:32 or higher for Coccidioides may indicate a disseminated disease, whereas a titre of less than 1:32 may indicate a self-limited or previous infection. Positive results should be further characterised by immunodiffusion (Guimarães et al. 2006; Hage et al. 2019; Miceli and Krishnamurthy 2023; Saubolle et al. 2007).

19.6.3 Immunodiffusion

It is based on the precipitate that looks like a band that is formed when patient Abs and test Ag are mixed together (Coleman and Kaufman 1972). Separate wells are used to allow soluble fungal Ag and Abs (control and patient sample) to diffuse into the test medium (Gavito-Higuera et al. 2016; Zaragoza et al. 2009). Visible precipitate line formation that correlates with the positive control suggests the presence of target Abs (Brown et al. 2018).

Usually, it is done to treat cases of histoplasmosis, blastomycosis, and Coccidioidomycosis. Abs to M and H glycoprotein are found in cases of histoplasmosis, where the M band is present in both acute and chronic cases and can last for months to years after infection (Guarner and Brandt 2011). In contrast, the H band is less frequent and is only seldom seen without the M band (McBride et al. 2017). Compared to CF, it is less sensitive but more specific. Partially purified A antigen is used for Blastomyces (Navalkar et al. 2014). In between 28 and 40% of instances with pulmonary blastomycosis, ID results are positive (Gavito-Higuera et al. 2016; Guarner and Brandt 2011). Using an ID-CF antigen or an ID-TP antigen, ID can identify Abs that correlate with CF or a tube precipitin in Coccidioidomycosis can help as ID-TP looks for IgM antibodies, whereas ID-CF looks for IgG ones (Góralska et al. 2018; Jackson et al. 2019).

Within 1–3 weeks following the start of the primary infection, serum precipitins can be found, although they are rarely found 6 months after the infection. When a disease has spread widely, precipitins may persist or recur (Góralska et al. 2018).

19.6.4 Detection of Fungal Antigens

Serial monitoring of patients who are at risk for fungal infections may be useful for early detection and preventative therapy approaches. It might be helpful for predicting outcomes and monitoring the response to treatment. The diagnostic value of a single test result is limited, and the confirmation of a positive result necessitates multiple positive results. Specificity is limited by cross reactivity between related organisms, and false-positive tests are frequent (Arvanitis et al. 2014; Guarner and Brandt 2011; Riwes and Wingard 2012).

19.6.5 1,3-β-d-Glucan Assay

Fungal wall component (1,3-β-d-glucan) authorised by the U.S. Food & Drug Administration (FDA), a brand-new fungal surrogate marker called the D-glucan assay, provides a non-invasive way to potentially monitor and identify invasive fungal infections. A minimally intrusive sample is needed for 1,3-β-d-glucan testing, which can be used to both diagnose an invasive fungal infection and track the effectiveness of treatment.

Testing for 1,3-β-d-glucan has the drawback that a positive result on its own does not have enough sensitivity and specificity to make a conclusive diagnosis. Although there are no official guidelines for the use of 1,3-β-d-glucan testing, this chromogenic assay offers a new way to evaluate populations that are at risk. Polysaccharides are the primary structural elements of the fungal cell wall. Glucan, chitin, and mannan make up the structural core polysaccharides in the majority of fungi. The most significant and prevalent polysaccharide in the cell wall of most fungus is glucan. Glucans are a significant component of the cell walls of various fungi; exceptions being Cryptococcus, Blastomyces, Mucor, Rhizopus species, etc. (Garcia-Rubio et al. 2020; Theel and Doern 2013). It has been proposed that yeast cell walls contain less glucan than filamentous moulds. The 1,3-β-d-glucan backbone is made mostly of glucose polymers linked in a linear pattern by carbons 1 and 3 by a glycosidic bond with a beta conformation. Transporting glucose subunits to the plasma membrane, where they are then carried across the plasma membrane and organised by a linear (1-3) glycosidic bond, is a step in the production of this polysaccharide backbone. The enzyme 1,3-β-d-glucan synthase is in charge of creating the polysaccharide backbone (Douglas 2001; Theel and Doern 2013). There are 1500 glucose subunits that make up the polysaccharide backbone branch inside each chain using either a (1-4) or (1-6) glycosidic link. The branching assignments are extremely diverse and unique to the fungi species. D-glucan generally appears as an insoluble molecule when absorbed into the fungal cell wall (1-3). Blood or other bodily fluids cause 1,3-β-d-glucan to change into soluble single helix, triple helix (the most common), or random coil forms. The immune system may be able to be modulated by this soluble 1,3-β-d-glucan by preventing leukocyte phagocytosis (Theel and Doern 2013).

The following are five beta-glucan assays (Theel and Doern 2013) that are currently in use:

1. Beta-Glucan Test (Waco Pure Chemical Industries, Osaka, Japan)
2. BGSTAR-Glucan Test (Maruha, Tokyo, Japan)
3. Endosafe-PTS (Charles River Laboratories, Charleston, SC)
4. Fungitec-G (Seikagaku Biobusiness, Tokyo, Japan)
5. Fungitell (Associates of Cape Cod, East Falmouth, MA)

The Endosafe—PTS and beta-Glucan Test Kits should not be used to diagnose invasive fungal infection in clinical samples; they are solely meant for research use (Parveen and Bhattacharya 2023).

The FDA classified Glucatell as a class II medical device in May 2004 to enable its introduction into national commercial sale as a serological compound for the determination of 1,3-β-d-glucan. This instrument, which is now known as Fungitell, is the only one that the FDA has approved for use in detecting serum 1,3-β-d-glucan as a supplementary indicator of invasive fungal infections (Arvanitis et al. 2014; Theel and Doern 2013).

Individual fungal pathogens have a restricted range of specificity, and their effectiveness varies depending on the organism. For Pneumocystis pneumonia, the highest sensitivity (96%) and specificity (84%) were observed. For invasive candidiasis, increased sensitivity is observed as compared to blood culture, with appropriate specificity (63–73%) (Arvanitis et al. 2014; Theel and Doern 2013; Thompson et al. 2022).

Similar sensitivity to serum galactomannan is observed for Aspergillus species (Theel and Doern 2013). Although no official recommendations for the use of 1,3-β-d-glucan testing have been made, the proposed testing recommendations are listed below based on the available data. According to the available data, a negative test result does not completely rule out the presence of an invasive fungal infection (Arvanitis et al. 2014; Riwes and Wingard 2012). For the diagnosis of invasive fungal infections, testing should be done in conjunction with other techniques and undertaken before antifungal treatment. In at-risk individuals, testing should typically be done twice weekly to monitor infections; once weekly testing should be done to monitor treatment effectiveness (de Pauw 2011; Kozel and Wickes 2014).

Before being accepted as true positive tests that are positive ought to be repeated from the original specimen or confirmed using an additional new, second specimen. The value of testing on samples other than serum needs to be investigated further. For paediatric patients, the ideal cut-off to determine a positive test has not been established (Parveen and Bhattacharya 2023; Theel and Doern 2013).

Assay Procedure: The examples are mislabeled or unlabeled samples, bacterial or fungal-infected equipment, and uncalibrated spectrophotometers.

False-positive reactions: Among them are concurrent bacteremia (most frequently caused by Streptococcus species), cellulose membranes and filters used in haemodialysis, immunoglobulin products (such as IVIG), human serum (due to its naturally yellow colour), surgical sponges and gauze products, lung transplantation, and medications like Crestin, Lentinan, Schizophyllan, and Scleroglucan. Beta-lactam antibiotics might not be as harmful as originally believed.

False-negative reactions: These comprise lipemic blood samples, haemolyzed blood samples, and specific fungi like Zygomycetes, Cryptococcus neoformans, and Blastomyces dermatitidis that are deficient in 1,3-β-d-glucan (Theel and Doern 2013).

19.6.6 Fully Automatic Chemiluminescence Immunoassay System

It is an open system experiment that can be carried out for a quick, precise diagnosis using either chemiluminescence or photometry. The FACIS-1 model from Bio-state Inc. guarantees a 42-min detection time for samples like serum, BAL fluid, and CSF. According to the literature, the test has a sensitivity of 86.36% and a specificity of 100%. Additionally, the accuracy is found to be 89.3%, with 100% PPV and 66.7% NPV (Gavito-Higuera et al. 2016; Infantino et al. 2020).

19.6.7 Galactomannan

This diagnostic toolbox includes the use of ELISA or immune-chromatographic approaches (e.g., lateral flow tests) to detect galactomannan as well as galactomannan-like biomarkers. Galactomannan (GM) is a part of the cell wall of Aspergillus species. The range of serum testing's sensitivity is between 22 and 90%, and its specificity is between 80 and 90% (Herbrecht et al. 2002b; Serin and Dogu 2021). Serum sensitivity is significantly higher in individuals who are highly neutropenic (such as those with haematological malignancies) and not on mould-active prophylaxis than in people who are not neutropenic. Since GM triggers the production of antibodies, IA cases that develop more slowly are more likely to be negative due to antibody binding. In practically all patient categories, bronchoalveolar lavage GM performs better than serum GM; however, sensitivity is decreased in patients receiving empiric antifungal medication and anti-mould prophylaxis (de Heer et al. 2019; Fontaine and Latgé 2020; Mikulska et al. 2021; Susianti et al. 2021). The presence of GM in BAL does not distinguish between invasive illness and colonisation.

A combination of baseline GM and 1-week decay has been demonstrated to be predictive of all-cause death in IA patients with detectable GM in serum. "False-positive" results may be brought about by pharmacological and dietary contamination, as well as infection from Histoplasma and Fusarium, two other fungal diseases (D'Haese et al. 2012; Zhou et al. 2017). Other "false-positives" have been observed to be associated with the co-administration of specific antibiotics (although piperacillin/tazobactam may no longer be of concern), specific foods (such as milk formulae), and Bifidobacterium (i.e., translocated from newborns' stomachs) (Patterson et al. 2016).

Comparable to Aspergillus PCR in terms of diagnostic performance in IA, the GM test also has a higher level of specificity (Hsu et al. 2021). For the diagnosis of IA and for screening for IA in high-risk patients, the GM assay is advised.

For the detection of GM, there are two commercial ELISA tests available (Bio-Rad and Dynamiker) (Arvanitis et al. 2015). There are two freshly released point-of-care LFD Aspergillus antigen tests on the market that are already in extensive usage (White et al. 2013).

19.6.8 Lateral Flow Assay

Typically, this is done to find the Cryptococcus antigen. Before the onset of clinical signs, Cryptococcus spp. shed their polysaccharide capsular antigen (CrAg) extremely early into the bloodstream during dissemination. The diagnosis of cryptococcal meningitis and asymptomatic cryptococcal antigenemia has been transformed by CrAg testing (Boulware et al. 2014; Liu et al. 2022; Maziarz and Perfect 2016).

In comparison to urine samples (sensitivity 85%), the CrAg-LFA performed on CSF, or on plasma, serum or blood samples produces very high sensitivity (>95%) and specificity (99%) (Boulware et al. 2014).

The CrAg test performs better than both India ink preparation (sensitivity 70–80% in CSF) and Cryptococcus culture (sensitivity 90% in CSF sample). Bronchoalveolar lavage (BAL) samples show a higher yield in isolated pulmonary cryptococcosis compared to serum samples (sensitivity of 83% vs. 74%, respectively) (Vidal and Boulware 2015; Zeng et al. 2021).

A CrAg titre >1:8 in BAL samples from HIV-positive patients yields a sensitivity of 100% and a specificity of 98% (Vidal and Boulware 2015).

19.6.9 Latex Agglutination

With quick findings (10–30 min, depending on the specimen type), high sensitivity and specificity for meningeal and disseminated cases (>90%), and a decreased sensitivity for CrAg of serotype C, polyclonal antibodies are employed to identify polysaccharide antigen (Fang et al. 2023; Grundy et al. 1995). Cross-reactivity has been linked to Trichosporon beigelii, Capnocytophaga canimorsus, and Stomatococcus mucilaginosus as well as cancer, bacterial septicaemia, hydroxyethyl starch, soaps, disinfectants, sample transport vials, and inadequate protein treatment (Murex Diagnostics, Norcross, Ga.) (Chanock et al. 1993; Tee et al. 1987). However, using the Mouse monoclonal IgM-based LA eliminates these false-positive (Arvanitis et al. 2014; Guimarães et al. 2006). Here, a direct antigen detection assay is carried out on blood, cerebrospinal fluid (CSF), or bodily fluid samples such as pleural fluid and bronchoalveolar lavage in which the latex particles are coated with polyclonal cryptococcal capsular antibodies. The sensitivity of CrAg-latex serum is between 83 and 97%. Pronase was not used on serum specimens in the tests with reduced sensitivity (Dominic et al. 2009; Vidal and Boulware 2015). The range of the CrAg-latex specificity on serum is 93–100%. The CrAg-latex test had high CSF sensitivity, ranging from 93 to 100%, and specificity, 93 to 98% (Hevey et al. 2020).

There are several Latex agglutination kits currently on the market, including Crypto-La (International Biologicals NJ) polyclonal Ab, Myco-immune 9 American microscan NJ) polyclonal Ab, IMMY Latex-Crypto® test (Immuno-Mycologics, Norman, OK, USA) polyclonal Ab, CALAS (Meridian diagnostics, Ohio) mAbs, Eiken test (Eiken Co Remel mAbs for the Remel Cryptococcus Antigen Test Kit, Lenexa, Kansas, USA (Babady et al. 2009; Coovadia and Solwa 1987; Tanner et al. 1994).

19.6.10 Enzyme Immunoassay

It is utilised to find antigens for Candida (Mannan & Enolase), Histoplasma, Aspergillus (galactomannan or galactomannan-like biomarkers), and Cryptoccus (CrAg). For Candida, it does not need samples to be pre-treated, but it has lower sensitivity for CrAg of serotypes C and D and is not appropriate for use in environments with little to no infrastructure. It costs more money overall and takes longer to complete (35–45 min) (Forster et al. 2022; Sendid et al. 1999). The EIA on serum is found to be 96% specific, and has a sensitivity 94%. On CSF EIA testing, the sensitivity and specificity were 100% and 98%, respectively (Vijayan et al. 2023). It is a Biotin amplified sandwich ELISA based on the idea of polyclonal capture and monoclonal detection, with a limit of detection for A–D serotypes of 0.63, 0.63, 7.8, and 62 ng/ml (McHugh et al. 2019; Nath et al. 2021; Toscanini et al. 2021). Monoclonal antibodies can recognise the 34–38 and 115 kDa non-capsular culture filtrate antigens. It has been demonstrated to be helpful in identifying noncapsular forms of Cryptococcosis in AIDS patients.

There are two commercial ELISA tests (Bio-Radand Dynamiker) for the detection of GM in Aspergillus. Two point-of-care Aspergillus antigen assays that were recently commercialised in an LFD format may drastically change how and where these tests are conducted, particularly in LMICs (Binnicker et al. 2012; Jarvis et al. 2011).

19.6.11 Mannan Detection

It is a significant part of the Candida cell wall and causes a potent antibody response. As a result, various authors claim that the simultaneous detection of mannan (Platelia Candida Ab, Bio-Rad Laboratories) significantly enhances the diagnosis of candidemia. While each of these tests has a low sensitivity (50%), when detected together, the sensitivity rises to 60–80% (Fang et al. 2023; Hussain et al. 2020).

19.6.12 Enolase

Except for C. glabrata, all Candida spp. produce it. However, it is not commercially available and only has a 54% sensitivity (Pasqualotto and Denning 2005; Malhotra et al. 2014).

19.6.13 Secreted Aspartyl Proteinase

Sap (extracellularly secreted enzyme) production by Candida is induced because it has been demonstrated that Sap is produced during active tissue invasion. Therefore, rather than sample colonisation, its extracellular concentration is correlated with invasive illness. It has a sensitivity of 93.9% and a specificity of 96% when using an

ELISA inhibition approach based on monoclonal antibodies to detect C. albicans (Naglik et al. 2003).

19.7 Tests for Detection of Metabolites

Finding a characteristic microbial metabolic product in the bodily fluids of an infected host is an entirely distinct way of identifying an infectious disease. The discovery of various fungal metabolites in the bodily fluids or tissues of infected or colonised hosts serves as an example of the sustainability of this strategy. Examples include ethanol in human CSF positive for C. neoformans, aflatoxin B1 in urine taken from frogs with intraperitoneal implantation of Aspergillus flavus mycelial mats, and pulmonary oxalic acid crystals in Aspergillus fungal ball patients (Cary et al. 2018; Kanaujia et al. 2023; Rajarajan et al. 2021).

19.7.1 D-Arabinitol

It is applied to species of Candida. The five-carbon acyclic polyol D-arabinitol was the first microbial metabolite that was effectively utilised as a diagnostic marker.

The following are discovered by some researchers (Christensson et al. 1999; Yeo et al. 2006; Yeo and Wong 2002):

- Compared to uninfected/colonised controls, animals/humans with invasive variety of Candidal infections showed greater serum arabinitol concentrations.
- Regardless of the presence of an invasive variety of Candidal infection, C. albicans produced large levels of D-arabinitol in culture.
- Serum-arabinitol levels were higher in people with impaired kidney functions than in individuals having optimal kidney function.

Despite the fact that these outcomes were positive, they did not prove the efficacy of arabinitol as an indicator of diagnosis for candidiasis (Yeo et al. 2006) because of the following reasons:

1. The frequency of considerable arabinitol production by clinical isolates of the medically relevant Candida species was relatively unknown.
2. The exact impact of renal failure on serum arabinitol concentrations remained unknown.
3. It was unknown whether the excess arabinitol seen in serum was produced by microbial or host metabolism.
4. Whether the presence of D-arabinitol-producing Candida species could likewise result in higher serum levels of arabinitol was still a question.

Therefore, Bernard et al. demonstrated that while no strain of Candida glabrata, Candida krusei, or Candida neoformans tested produced D-arabinitol, whereas all strains of Candida albicans, C. tropicalis, C. parapsilosis, and Candida pseudotropicalis did. Gas chromatography can measure D-arabinitol production in serum and urine, or it can be quantified enzymatically and detected spectrophotometrically (Silva et al. 2012; Turner and Butler 2014).

19.7.2 Mannitol in Aspergillus and Cryptococcus Infections

It is used for the infections caused by Aspergillus and Cryptococcus. Mannitol is a six-carbon acyclic polyol that is abundantly produced by a variety of fungi. Mannitol is also excreted through the urine at a rate that is equivalent to glomerular filtration rate in healthy adults. Additionally, mannitol has been researched as a potential diagnostic indicator for pathogenic fungi that produce mannitol (Góralska et al. 2018; Guimarães et al. 2006). These characteristics all help to make D-arabinitol valuable as a diagnostic marker. Multiple clinical isolates of Aspergillus fumigatus, Aspergillus flavus, and Aspergillus niger (B. Wong, unpublished observations), as well as C. neoformans, all produced significant levels of mannitol in culture, according to early research. Additionally, two studies have demonstrated that experimental A. fumigatus infections in animals resulted in higher levels of mannitol in body fluid and/or tissue than did uninfected controls, while a third study demonstrated that experimental cryptococcal meningitis in animals resulted in increased CSF-mannitol content compared to unaffected controls. The existing data fail to back up the utility of mannitol as an absolute indicator for aspergillosis or cryptococcosis, despite animal research showing that the two clinically relevant fungi create considerable levels of mannitol in affected mammalian tissues. This is due to the fact that mannitol has not been examined in any published trials as a diagnostic indicator for aspergillosis in people (Góralska et al. 2018; Guimarães et al. 2006; Latgé 1999).

19.8 Tests for Detection of CMI

19.8.1 Skin Testing

Skin testing can be used to establish an etiological diagnosis, particularly in nonendemic areas, for epidemiological surveys, and as a prognostic indicator because a positive skin test may turn negative with severe or widespread disease (Arvanitis et al. 2014; Kozel and Wickes 2014). Aspergillus fumigatus, Histoplasma capsulatum, Sprothrix schenkii, and other organisms are subjected to skin testing. However, due to cross reactivity brought on by the sharing of antigens and endemic areas, it can result in false-positive reactions. Early infection in a CMI with a flaw can result with a false-negative (Terrero-Salcedo and Powers-Fletcher 2020; Yeo and Wong 2002).

19.9 Matrix-Assisted Laser Desorption/Ionization-Time of Flight Mass Spectrometry (MALDI-TOF MS)

One of the proteomic techniques that has completely altered clinical microbiology labs around the world is MALDI-TOF MS. The ability to quickly and accurately identify yeasts and moulds using MALDI-TOF fingerprinting is increasingly gaining popularity. The resulting mass spectral fingerprints serve as each microorganism's relative unique signature, making them perfect for precise species identification (Ghosh et al. 2015; Oviaño and Bou 2018).

19.9.1 Commercially Available Are as Follows

- Andromas (Andromas, Paris, France)
- AXIMA SARAMIS database (AnagnosTec, Potsdam, Germany and Shimadzu, Duisburg, Germany)
- MALDI Biotyper (Bruker Daltonics, Bremen, Germany)
- VITEK MS systems (bio Merieux, Marcy l'Etoile, France)

A wide range of clinical isolates simultaneously with approved standards are yet to be evaluated in this technique. For 95.9% of Candida albicans and 86.5% of non-albicans Candida species, the identification outcomes by MALDI-TOFMS were consistent with those of the traditional culture-based approach (Alizadeh et al. 2017). This method provides a reliable, time-saving tool for routine identification of Candida species that cause bloodstream infection, according to the 30-min turnaround time of the results. When utilising this method to distinguish between mould species like Aspergillus, Fusarium, and dermatophytes, there were initially few data available. However, many species have since been identified utilising as a diagnostic method (Góralska et al. 2018; Hussain et al. 2020).

19.10 Molecular Diagnostics

Frequently used molecular targets for species identification—Numerous molecular targets are frequently employed for species identification (Brizuela et al. 2020; De Carolis et al. 2012; Landlinger et al. 2009; Lockhart et al. 2021; Miceli 2019), including:

- ACT—species discrimination [Aspergillus, Cladosporium, Coniochaeta, Verticillium, Verruconis]
- Cal—Alternaria
- Chi18-5—species discrimination [Trichoderma]
- D1/D2—Mucorales
- EF-1—Fusarium
- GPDH—species discrimination [Bipolaris, Curvularia and Verticillium]

- ITS—species-level identification [wide range of fungi]
- RPB1—complex identification [Fusarium, Penicillium, Talaromyces]
- β -tubulin—Aspergillus spp.

19.10.1 Multiplex PCR

In comparison to traditional PCR, this assay may identify a wide range of fungus concurrently in the same samples, saving time and money (Carvalho-Pereira et al. 2020). There are two rounds of amplification on the platform (Elnifro et al. 2000). To enable target DNA enrichment without causing competition amongst amplicons, a multiplex PCR is carried out at 10–15 cycles in the first stage. This result is thinned out and utilised as a template for the second amplification, which entails numerous separate quantitative PCR reactions using primers that are nested within those used in the multiplex PCR (Carvalho-Pereira et al. 2020). Multiplexed PCR can run up to 72 distinct PCR reactions simultaneously (Souza et al. 2013). At the conclusion of each extension cycle, SYBR green technology measures fluorescence, and melt-curve analysis identifies species or genes. Each target is amplified and detected correctly, thanks to the use of two sets of species-specific primers, saving the cost of DNA probes. SYBR green detection significantly expands multiplexing and quantitative capabilities of real-time PCR systems, which are often constrained by the accessibility of fluorescent channels and the requirement to optimise each multiplex PCR individually (Carvalho-Pereira et al. 2020).

19.10.2 Panfungal PCR Assays

These assays use all-inclusive fungal primers to identify "all" fungal DNA contained in a clinical specimen (Lau et al. 2007). The internal transcribed spacers 1 and 2 (ITS1 and ITS2) and the D1/D2 sections of the 28S rRNA gene are the preferred target(s) for the rRNA gene cluster. DNA sequencing is most frequently done after amplification; however, a high-resolution melt curve analysis in real-time PCR techniques is becoming more popular (Valero et al. 2016). Together, these assays have been effective in detecting and identifying fungi from a variety of samples. Blood, bronchoalveolar lavage fluid (BALF), formalin-fixed paraffin-embedded (FFPE) tissue, cerebrospinal fluid (CSF), vitreous fluid, and other specimen types can all be detected with reasonable specificity and accuracy; however, sensitivity varies between specimen types. The absence of standardisation of assays for comparability of data in the clinical trial or study context was a possible downside of panfungal PCR techniques (Srinivas et al. 2021). The International Society for Human and Animal Mycology (ISHAM) Fungal PCR Initiative (FPCRI) working group is presently developing and refining protocols to assess PCR techniques to identify fungi from tissue specimens (Van Burik et al. 1998).

19.10.3 Syndromic Testing

Syndromic testing PCR panels are being used more frequently to diagnose different infections (Dien Bard and McElvania 2020). There are a number of FDA- or equivalent regulatory agency-approved panels for IFDs. The BioFire Film Array Meningitis/Encephalitis panel (bioMérieux, Marcy l'Etoile, France) identifies one fungus target, Cryptococcus neoformans/Cryptococcus gattii, along with bacterial and viral targets in CSF in roughly 1 h as part of the target menu for CNS infections (Trujillo-Gómez et al. 2022; Van et al. 2020). Although quick, this test cannot distinguish between C. neoformans and C. gattii, and subsequent investigations have found that it performs far worse at detecting cryptococcal infection than tests that use the cryptococcal antigen (Maziarz and Perfect 2016; Posnakoglou et al. 2020).

19.10.4 Multiplex Tandem PCR (MT-PCR)

The highly multiplexed gene expression profiling method known as MT-PCR is used to quickly identify infections. It is advised to employ MT-PCR to quickly identify fungal components straight from specimens (Lau et al. 2013).

19.10.5 PCR-ELISA

Amplification, immobilisation, and detection are the three immunological phases of PCR-ELISA operation. The PCR-ELISA approach has a faster time to result and is around 100 times more specific than traditional PCR. It excludes the use of mutagen-staining materials and permits the testing of many samples. It is also capable of performing quantitative and qualitative tests with a lower risk of contamination. Given that it just requires the use of basic lab tools, it is a simple procedure (Arvanitis et al. 2014; Latgé 1999; Löffler et al. 1998; Sue et al. 2014).

19.10.6 Fluorescent In Situ Hybridization (FISH)

AdvanDx (Woburn, Massachusetts) has created a brand-new, quick molecular diagnostic platform based on its own PNA probe technology. It is an effective probe-based technique for detecting living organisms in environmental samples in situ. It has the ability to pinpoint the precise position of specific DNA or RNA sequences in the organelles, cytoplasm, or nucleus of biological materials. In less than 3 h, PNA probes enable the identification of candida species from positive blood cultures. Biological samples or environmental samples are prepared, and a nucleic acid sequence is tagged (incorporated with a fluorescent label or marker) to generate a probe constitute the main steps in FISH (Rigby et al. 2002). The probe is then hybridized to the DNA or RNA in biological materials to create a double-stranded molecule under carefully regulated experimental circumstances. The sites of hybridization are then found and made visible. It makes use of several currently authorised probes, including those for Candida albicans, glabrata, krusei, parapsilosis, tropicalis, as well as different Staphylococcus species, Enterococcus species, Klebsiella pneumoniae, Escherichia coli, and Pseudomonads (AdvanDx). Recently, new QuickFISH, created by AdvanDx, which enables very quick (20 min) detection of bacteria or yeasts from positive blood samples, has been introduced, further improving this technology (Da Silva et al. 2015; Rigby et al. 2002).

19.10.7 Microarray

A collection of organised probes on a stable support is what is known as a microarray. These arrayed probes have been used since Gergen et al. produced the first known ordered arrays of E. coli carrying individual plasmids in the middle to late 1970s. These arrays served as the forerunner of the ordered cDNA libraries contained in E. coli in the 1980s. At that time, radioactively labelled nucleic acids that hybridised to the target clone were used to identify the DNAs of interest (Spiess et al. 2007). The advancement of DNA synthesis, the availability of sequenced genomes, and the development of fluorescence-based detection techniques allowed for the creation of a unique genome-wide technology based on microarrays.

A ground-breaking technique for transcriptional analysis using many cDNAs spotted in high density onto glass slides was described in 1995 by Schena and colleagues (Leinberger et al. 2005). Modern microarray technologies may be regarded as having their origins in this groundbreaking breakthrough. It made it possible to identify or genotype various species as well as assess differential expression using simultaneous, two-color fluorescence hybridization (Govindarajan et al. 2012). This progress is inextricably linked to the genome-wide sequencing initiatives that began in the late 1980s and led to the publishing of the first eukaryotic genomes—S. cerevisiae—in the middle of the 1990s. Only 5 years later, as a result of this, was the human genome decoded (Gresham et al. 2008). The ability to represent each anticipated gene or transcriptional unit of a genome as a probe on a microarray was made possible by the understanding of the genomes. Genome-wide transcriptional profiling was not feasible before microarray technology (Rupp 2017). In the past 20 years, array technologies have rapidly evolved from being initially developed for multiparametric nucleic acid detection to becoming a tool for the detection of all types of biological compounds (DNA, RNA, proteins, cells, nucleic acids, carbohydrates, etc.) or their modifications (methylation, phosphorylation, etc.) (Leinberger et al. 2005). The fundamental idea, as mentioned before, is straightforward. An array is made up of a specified number of ordered probes that are immobilised on a solid support and are directed at specific molecules, such as lectins, proteins, mRNA, or genomic DNA. The interaction of these molecules with their corresponding, typically labelled ligands at the designated spot on the array can now be used to detect their biological interaction, such as the interaction of complementary bases of nucleic acids, antibody-antigen interactions, or the interaction of carbohydrates with lectins (Goyal et al. 2018).

Suspension bead arrays, which are based, for example, on the immobilisation of the probes on tiny polystyrene beads as the solid support and flow cytometry for bead and target detection, have been created in addition to these two dimensional or planar arrays (Dunbar 2006). By doing so, they adhere to a notion that is very dissimilar from planar arrays. Here, rather than on a planar surface, reactions take happen in solution; nevertheless, the multiplexing capacity is constrained (Rupp 2017).

Here, multiplexing is achieved, for instance, by utilising various microsphere sets depending on a combination of two colours. This makes it possible to produce a set of around 100 distinct microbeads that can be easily distinguished from one another and then examined separately after being coupled to the appropriate probe (Vafajoo et al. 2018). Today, a number of nucleic acid-based tests for pathogen identification are FDA-cleared and were originally developed for the detection of antibodies and antigens. Microarrays, in contrast to the majority of conventional biological tests, enable the concurrent investigation of up to many thousand analytes, spanning from proteins, nucleic acids, carbohydrates and others. The potential uses for nucleic acid-based arrays already extend from the focused analysis of a small number of genes for diagnostic purposes, such as the detection of pathogens, to the analysis of SNPs (single nucleotide polymorphisms) to identify resistance phenotypes, and even to the transcriptomic profiling or resequencing of entire genomes. mRNA, miRNA, or genomic DNA complementary oligonucleotides are immobilised or synthesised directly on solid supports like glass, silicon, nylon, or other polymers in an orderly manner to set up a microarray. These oligonucleotides may match a variety of established diagnostic targets, including all of an organism's open reading frames to enable transcriptional profiling or the pathogen identification target (Govindarajan et al. 2012). The so-called tiling arrays for tracking transcriptional activities across entire chromosomes or identifying interactions between transcription factors and DNA (Chromatin IP, Chip-on-chip studies) may even comprise overlapping sections of a full genome (Spiess et al. 2007). The most popular technique for detecting the hybridization event involves labelling the RNA using nucleic acid polymerizing enzymes and fluorescently labelled nucleotides (Miller and Tang 2009). Following hybridization with the microarray, a certain molecular interaction occurs where the complimentary strand is immobilised, which results in fluorescence tagging of the corresponding site. A semi-quantitative readout employing a fluorescence scanner plus photomultiplier tube, or CCD camera-based imaging can provide the necessary information to determine if the target molecule is present in the analysed sample or not. Techniques for indirect labelling have been employed in addition to direct labelling. For instance, during PCR, biotin-labeled nucleotides are added and then identified using streptavidin or antibody conjugates. Another example is the DualChip® microarrays produced by Eppendorf using the colorimetric Silverquant® technology (Bumgarner 2013; Rupp 2017; Spiess et al. 2007). The PCR reaction uses nucleotides that have been biotin-tagged. Following hybridization, gold-coupled antibodies against biotin are added, and the colorimetric reaction is triggered by the addition of silver nitrate and a reducing agent. Silver then precipitates at the gold particles as a result. Additionally, streptavidin-horseradish peroxidase conjugate has been utilised for colorimetric detection (Pavlickova and Hug 2004; Spiess et al. 2007).

19.10.8 DNA Array Hybridization

Reverse Dot Blot Hybridization (RDBH), also known as DNA array hybridization or microarray, is a hybridization technique that is regarded as a useful way for locating and identifying fungus and other microbes in culture (Arvanitis et al. 2014; Spiess et al. 2007; Zeng et al. 2007).

19.10.9 Electrophoretic Karyotyping (EK)

In this method, whole DNA molecules move over an agarose gel matrix while being affected by pulsed fields, making it simple to separate DNA molecules with several megabases. EK analysis, which moves intact chromosomes through an agarose gel matrix using electric fields of alternate orientation, assesses chromosome length polymorphism. The examination of chromosomal binding patterns, also known as electrophoretic karyotypes, and the identification of karyotypic variants within the species are both facilitated by electrophoretic karyotyping. EK has been widely used to identify C. albicans and other Candida species by fingerprinting. It has a fair amount of discriminating power yet decent repeatability (Beadle et al. 2003; Magee and Magee 1987).

19.10.10 Multilocus Enzyme Electrophoresis (MLEE)

The polymorphism of the isolates' isoenzymes or alloenzymes is assessed using MLEE. Under natural circumstances, proteins from cell extracts are separated by electrophoresis, and the enzymes are identified using particular enzyme-staining techniques. When a large enough number of enzymes are assessed, this method's key benefits include its strong discriminatory power and the extremely low possibility of homoplasy in clonal organisms (Caugant et al. 1986; El-Gendi et al. 2021; Santos et al. 2012).

19.10.11 Light Cycler Septi Fast (LC-SF)

This new commercial multiplex PCR method, which is intended for pathogen in patients with sepsis, may identify several pathogens responsible for bloodstream infections in few hours. By utilising PCR probes which may target internal transcribed spacers (ITS) in-between 18S and 5.6S fungal ribosomal RNA (rRNA), it is feasible to distinguish between Candida albicans, glabrate, krusei, parapsilosis, tropicalis, and Aspergillus fumigatus (Chang et al. 2013; Fang et al. 2023; White et al. 2003).

19.10.12 Multianalyte Profiling (xMAP) System

This assay makes use of a cutting-edge flow cytometer and tiny beads that can be identified from one another in the laser reader thanks to their distinct colours and color-coding (Dunbar 2006).

19.10.13 Loop Mediated Isothermal Amplification

It is the isothermal amplification of a small number of copies of the target DNA. Pathogenic fungi like Penicillium marneffei and Cryptococcus species can be quickly identified with this method. It is an efficient and straightforward amplification technique that may be applied in developing nations without the need for expensive machinery or highly trained personnel. In a PCR ELISA test, the steps of PCR amplification, hybridization with a complementary-coloured probe, and EIA reaction product detection are all carried out. It can distinguish between several dermatophytes' species. When compared to dermatophyte cultures, it is more sensitive (Sun et al. 2010).

19.10.14 Nanotechnology Approaches

Metallic nanoparticles, such as those made of gold or silver, exhibit a wide range of electrical, optical, and chemical characteristics. For instance, colour shifts can be used to identify gold nanoparticles coupled with particular oligonucleotides. A number of complicated matrices, including culture media, cell lysate, urine, serum, plasma, sputum, fine-needle aspirates, and blood, have been employed for the detection of a wide range of target types, including proteins, drugs, pathogens, enzymes, and cells using T2 magnetic resonance (T2MR)-based biosensing. As a diagnostic tool for candidemia, an automated T2 Candida assay is created employing T2 magnetic resonance technology, a hybridization method based on nanoparticles and PCR amplification (Hussain et al. 2020; Mikhailova 2021). By creating a combination of two superparamagnetic particle populations, each one equipped with a unique oligonucleotide capture probe [one connected to a capture probe that hybridizes the single-stranded DNA target's 5′ end, and the other to the DNA's 3′ end] T2MR may be utilized in nucleic acid detection. Candida albicans, glabrata, krusei, parapsilosis and tropicalis, may all be distinguished using this method from whole blood. In comparison to FISH or MALDITOF, T2MR also offers quicker turnaround times for results. In clinical studies, it produces results that are 98% positive and 100% negative in concordance with blood culture data (Fang et al. 2023).

19.11 Genome-Based Fungal Identification and Characterization

19.11.1 Direct Clinical Specimen-Based Precision Diagnosis of IFDs Using DNA Metabarcoding

DNA barcoding's guiding premise is that intra-species variance that must be smaller than inter-species variation in order to create a "barcoding gap." The major fungal DNA barcode has been recommended as the ribosomal TS1/2 region in an effort to standardise sequence-based identification of fungus. The absence of reference sequence databases that have been quality-controlled; however, it limits its widespread application. The ISHAM working group for barcoding medical fungus was created in 2010 to address this issue, and in 2015 the ISHAM ITS bar code database for human and animal harmful fungi was created. The hunt for a secondary fungal DNA barcode was sparked by research into the intra- and interspecies polymorphism of the ITS1/2 region in human pathogenic fungi, which supported prior results that only around 75% among all species of fungi could be successfully identified by this genomic locus at the species level (Xu 2016).

The translocation elongation factor 1 alpha (TEF1a) gene was proposed as the secondary fungal DNA barcode in a global study examining the genetic variability and ability to construct universal primers for a number of genetic loci. Again, the widespread use of this technique in routine fungal diagnosis has been hampered by the absence of a quality-controlled reference database. As a result, the seventh iBOL Conference in South Africa in 2017 produced a supplemental reference database for TEF1a. The ISHAM Barcoding database contains quality-controlled reference sequences for the targets ITS1/2 ($n = 4200$) and TEF1a ($n = 2432$), representing 645 and 346 species of human and animal pathogenic fungus, respectively. The majority of human and animal harmful fungi may be identified thanks to the Dual DNA barcoding technique for fungi. Using either Illumina or Oxford Nanopore Technology, the introduction of NGS has made it possible to simultaneously sequence mixed microbial communities directly from a range of clinical samples (such as blood, sputum, BALF, tissue, and faeces). Then, using sequence alignment tools like BLAST or WIMP against the appropriate, publicly available, quality-controlled reference sequence databases, such as the ISHAM barcoding database UNITE, RefSeq, and BOLD, it is possible to identify sequences from clinical samples down to the genus and species level. However, there are currently a number of significant limitations in this technology that could result in inaccurate or insufficient identification of the fungal pathogen (Meyer et al. 2019; Xu 2016).

The limitations are as follows:

1. pre-PCR biases—for example, handling of specimens, contamination at time of collection, aliquoting, nucleic acid extraction, library preparation, or pooling

DNA extraction practices, plus selecting storage buffer and extraction kits, host DNA's quantity (that may be reduced due to a variety of different reagents).
2. PCR biases—such as, primer mismatches and varying amplicon length.
3. The present NGS technologies' significant sequencing error rates, particularly for long read sequencing.
4. The absence of comprehensive reference sequencing databases with accurate taxonomic tagging.
5. A lack of relevant bioinformatic tools, such as cross-talk between fungal sequences and alignment methods

The clinical context of the disease symptoms should be used to interpret and review any DNA metabarcoding-based pathogen identification. It is envisaged that these problems would be solved with technical development and growing shifts to metagenomics techniques for clinical diagnosis.

Using high-quality reference sequence databases and NGS-based DNA metabarcoding, it is possible to diagnose IFDs in as little as 24–48 h. However, the expense of clinical metagenomics is now its biggest downside, coming in at a minimum of $1 million after taking into account the sequencing facility, computing equipment, and staff needed to execute it (Khot and Fredricks 2009; Kidd et al. 2020; Salem-Bango et al. 2023; Xu 2016).

19.12 Future Prospects in Diagnosis of Fungal Infection

19.12.1 Artificial Intelligence Methods

Filamentous fungi have high metabolic adaptability that can be used in bioprocesses to achieve circular economy. The interest in automating fungi-based systems has grown as a result of the present digital transformation occurring within the biomanufacturing industry. Automation tries to replace manual labour with automated equipment. On the other hand, a branch of computer science called artificial intelligence tries to create tools and machines that can carry out tasks that ordinarily need human intelligence. One advantage of incorporating automation and artificial intelligence in fungi-based bioprocesses is greater productivity. Other advantages include improved data dependability, flexible processes, and better production. The employment of such data-based technologies in the creation of food from fungi is one of the current gaps that needs more research (Wainaina and Taherzadeh 2023).

Fungi-based bioprocesses can benefit from AI and automation to boost production yields, while soft sensors make it possible to comprehend the intricate dynamics of bioprocesses. MATLAB for evaluating fungal morphology, robotic systems made up of clever sensors, actuators, and appropriate software for material handling are just a few instances of how automation and AI are being used in bioprocesses, including those based on fungi (Picek et al. 2022; Wainaina and Taherzadeh 2023).

19.13 Conclusion

The frequency of CNS fungal infections has sharply increased recently. The decision to treat a patient with CNS fungal infections is primarily based on clinical and mycological data, and early and accurate diagnosis is essential (Jia and Thakur 2019). However, conventional techniques that are now employed in everyday practise for the diagnosis of CNS fungal infections may be somewhat generic and insensitive. New methods to diagnose CNS fungal infections are being developed based on a deeper comprehension of the biology of infections caused by fungal species and the factors that determine the virulence of such fungal diseases (Góralska et al. 2018; Panackal and Williamson 2015).

It is obvious that fungal infection diagnosis has advanced in recent years, and some of these recently discovered technologies are expected to achieve earlier infection identification. However, it has been reported that the results were poor in terms of sensitivity and specificity (Gavito-Higuera et al. 2016). Additionally, the majority of the data from the various tests examined here are insufficient on their own to help determine when to start antifungal medication, which antifungal agent or drugs to use, or when therapy should be terminated. Additionally, some recently created techniques can be too laborious for regular use (Riwes and Wingard 2012). Therefore, these procedures need to be further refined to make them easier to use and to increase their sensitivity and/or specificity so that they might be useful for directing clinical treatment decisions (Raman Sharma 2010). Traditional approaches may be insensitive and/or nonspecific, yet they are not completely ineffective. They are now the best techniques used in the typical clinical laboratory. Therefore, it is likely that the most information will be obtained by combining both conventional morphological and cultural approaches with one or more of the more recent diagnostic modalities discussed here. Sadly, very few research studies have systematically and prospectively explored this multimodality method. It would be helpful to define the function of these tests in the context of clinical investigations by prospective studies of mixed modalities, culture and morphology, and nonculture techniques (Kozel and Wickes 2014). Due to the significant morbidity and mortality that CNS fungal infections are linked to, it is crucial to achieve positive outcomes by making an early diagnosis using both traditional and cutting-edge methods. Artificial intelligence would therefore be the method of the future for diagnosis, and greater emphasis should be placed on exploring newer and more sophisticated modalities for diagnosis.

References+

Afroze SN, Korlepara R, Rao GV, Madala J (2017) Mucormycosis in a diabetic patient: a case report with an insight into its pathophysiology. Contemp Clin Dent 8:662–666. https://doi.org/10.4103/ccd.ccd_558_17

Alastruey-Izquierdo A, Melhem MSC, Bonfietti LX, Rodriguez-Tudela JL (2015) Susceptibility test for fungi: clinical and laboratorial correlations in medical mycology. Rev Inst Med Trop Sao Paulo 57:57–64. https://doi.org/10.1590/S0036-46652015000700011

Alizadeh M, Kolecka A, Boekhout T, Zarrinfar H, Nahzag G, Mohamad A, Badiee P, Rezaei-Matehkolaei A, Fata A, Dolatabadi S, Najafzadeh MJ (2017) Identification of Candida species isolated from vulvovaginitis using matrix assisted laser desorption ionization-time of flight mass spectrometry. Curr Med Mycol 3:21–25. https://doi.org/10.29252/cmm.3.4.21

Archibald LK, Quisling RG (2013) Central nervous system infections. In: Textbook of neurointensive care, pp 427–517. https://doi.org/10.1007/978-1-4471-5226-2_22

Arthur CM, Chonat S, Fasano R, Yee MEM, Josephson CD, Roback JD, Stowell SR (2019) Examining the role of complement in predicting, preventing, and treating hemolytic transfusion reactions. Transfus Med Rev 33:217–224. https://doi.org/10.1016/j.tmrv.2019.09.006

Arvanitis M, Anagnostou T, Fuchs BB, Caliendo AM, Mylonakis E (2014) Molecular and nonmolecular diagnostic methods for invasive fungal infections. Clin Microbiol Rev 27:490–526. https://doi.org/10.1128/CMR.00091-13

Arvanitis M, Anagnostou T, Mylonakis E (2015) Galactomannan and polymerase chain reaction-based screening for invasive aspergillosis among high-risk hematology patients: a diagnostic meta-analysis. Clin Infect Dis 61:1263–1272. https://doi.org/10.1093/cid/civ555

Arya NR, Rafiq NB (2023) Candidiasis. In: StatPearls. StatPearls Publishing, Treasure Island

Babady NE, Bestrom JE, Jespersen DJ, Jones MF, Beito EM, Binnicker MJ, Wengenack NL (2009) Evaluation of three commercial latex agglutination kits and a commercial enzyme immunoassay for the detection of cryptococcal antigen. Med Mycol 47:336–338. https://doi.org/10.1080/13693780802607400

Baniya S (2022) Calcofluor white staining: principle, procedure, and application. Microbe [WWW Document]. https://microbeonline.com/calcofluor-white-staining-principle-procedure-and-application/. Accessed 16 July 23

Basu S, Bose C, Ojha N, Das N, Das J, Pal M, Khurana S (2015) Evolution of bacterial and fungal growth media. Bioinformation 11:182–184. https://doi.org/10.6026/97320630011182

Bays DJ, Thompson GR (2021) Coccidioidomycosis. Infect Dis Clin North Am 35:453–469. https://doi.org/10.1016/j.idc.2021.03.010

Beadle J, Wright M, McNeely L, Bennett JW (2003) Electrophoretic karyotype analysis in fungi. Adv Appl Microbiol 53:243–270. https://doi.org/10.1016/s0065-2164(03)53007-6

Beh SC, Greenberg BM, Frohman T, Frohman EM (2013) Transverse myelitis. Neurol Clin 31:79–138. https://doi.org/10.1016/j.ncl.2012.09.008

Bhattacharya M, Joshi N (2011) Spinal epidural abscess with myelitis and meningitis caused by Streptococcus pneumoniae in a young child. J Spinal Cord Med 34:340–343. https://doi.org/10.1179/107902610x12883422813507

Binnicker MJ, Jespersen DJ, Bestrom JE, Rollins LO (2012) Comparison of four assays for the detection of Cryptococcal antigen. Clin Vaccine Immunol 19:1988–1990. https://doi.org/10.1128/CVI.00446-12

Boulware DR, Rolfes MA, Rajasingham R, von Hohenberg M, Qin Z, Taseera K, Schutz C, Kwizera R, Butler EK, Meintjes G, Muzoora C, Bischof JC, Meya DB (2014) Multisite validation of Cryptococcal antigen lateral flow assay and quantification by laser thermal contrast. Emerg Infect Dis 20:45–53. https://doi.org/10.3201/eid2001.130906

Bowler PG, Duerden BI, Armstrong DG (2001) Wound microbiology and associated approaches to wound management. Clin Microbiol Rev 14:244–269. https://doi.org/10.1128/CMR.14.2.244-269.2001

Brizuela AM, De la Lastra E, Marín-Guirao JI, Gálvez L, de Cara-García M, Capote N, Palmero D (2020) Fusarium Consortium populations associated with Asparagus crop in spain and their role on field decline syndrome. J Fungi 6:336. https://doi.org/10.3390/jof6040336

Brizzi KT, Lyons JL (2014) Peripheral nervous system manifestations of infectious diseases. Neurohospitalist 4:230–240. https://doi.org/10.1177/1941874414535215

Brown EM, McTaggart LR, Dunn D, Pszczolko E, Tsui KG, Morris SK, Stephens D, Kus JV, Richardson SE (2018) Epidemiology and geographic distribution of blastomycosis, histoplasmosis, and coccidioidomycosis, Ontario, Canada, 1990–2015. Emerg Infect Dis 24:1257–1266. https://doi.org/10.3201/eid2407.172063

Bumgarner R (2013) DNA microarrays: types, applications and their future. Curr Protoc Mol Biol 22:Unit-22.1. https://doi.org/10.1002/0471142727.mb2201s101

Carvalho-Pereira J, Fernandes F, Araújo R, Springer J, Loeffler J, Buitrago MJ, Pais C, Sampaio P (2020) Multiplex PCR based strategy for detection of fungal pathogen DNA in patients with suspected invasive fungal infections. J Fungi (Basel) 6:308. https://doi.org/10.3390/jof6040308

Cary JW, Gilbert MK, Lebar MD, Majumdar R, Calvo AM (2018) Aspergillus flavus secondary metabolites: more than just aflatoxins. Food Saf (Tokyo) 6:7–32. https://doi.org/10.14252/foodsafetyfscj.2017024

Caugant DA, Bøvre K, Gaustad P, Bryn K, Holten E, Høiby EA, Frøholm LO (1986) Multilocus genotypes determined by enzyme electrophoresis of Neisseria meningitidis isolated from patients with systemic disease and from healthy carriers. J Gen Microbiol 132:641–652. https://doi.org/10.1099/00221287-132-3-641

Chang S-S, Hsieh W-H, Liu T-S, Lee S-H, Wang C-H, Chou H-C, Yeo YH, Tseng C-P, Lee C-C (2013) Multiplex PCR system for rapid detection of pathogens in patients with presumed sepsis – a systemic review and meta-analysis. PloS One 8:e62323. https://doi.org/10.1371/journal.pone.0062323

Chanock SJ, Toltzis P, Wilson C (1993) Cross-reactivity between Stomatococcus mucilaginosus and latex agglutination for cryptococcal antigen. Lancet 342:1119–1120. https://doi.org/10.1016/0140-6736(93)92106-4

Chen C-H, Chen J-N, Du H-G, Guo D-L (2023) Isolated cerebral mucormycosis that looks like stroke and brain abscess: a case report and review of the literature. World J Clin Cases 11:1560–1568. https://doi.org/10.12998/wjcc.v11.i7.1560

Chikley A, Ben-Ami R, Kontoyiannis DP (2019) Mucormycosis of the central nervous system. J Fungi (Basel) 5:59. https://doi.org/10.3390/jof5030059

Christensson B, Sigmundsdottir G, Larsson L (1999) D-arabinitol--a marker for invasive candidiasis. Med Mycol 37:391–396. https://doi.org/10.1046/j.1365-280x.1999.00249.x

Coleman RM, Kaufman L (1972) Use of the immunodiffusion test in the serodiagnosis of aspergillosis. Appl Microbiol 23:301–308. https://doi.org/10.1128/am.23.2.301-308.1972

Colombo AL, de Almeida Júnior JN, Slavin MA, Chen SC-A, Sorrell TC (2017) Candida and invasive mould diseases in non-neutropenic critically ill patients and patients with haematological cancer. Lancet Infect Dis 17:e344–e356. https://doi.org/10.1016/S1473-3099(17)30304-3

Coovadia YM, Solwa Z (1987) Sensitivity and specificity of a latex agglutination test for detection of cryptococcal antigen in meningitis. S Afr Med J 71:510–512

Coradello G, Tirelli N (2021) Yeast cells in microencapsulation. general features and controlling factors of the encapsulation process. Molecules 26:3123. https://doi.org/10.3390/molecules26113123

D'Haese J, Theunissen K, Vermeulen E, Schoemans H, De Vlieger G, Lammertijn L, Meersseman P, Meersseman W, Lagrou K, Maertens J (2012) Detection of galactomannan in bronchoalveolar lavage fluid samples of patients at risk for invasive pulmonary Aspergillosis: analytical and clinical validity. J Clin Microbiol 50:1258–1263. https://doi.org/10.1128/JCM.06423-11

Da Silva RM, Da Silva Neto JR, Santos CS, Frickmann H, Poppert S, Cruz KS, Koshikene D, De Souza JVB (2015) Evaluation of fluorescence in situ hybridisation (FISH) for the detection of fungi directly from blood cultures and cerebrospinal fluid from patients with suspected invasive mycoses. Ann Clin Microbiol Antimicrob 14:6. https://doi.org/10.1186/s12941-015-0065-5

Dal Pizzol M, Freitas EC, Locatelli C, Guareze F, Reginatto P, Machado G, Fuentefria A, Marinho D (2021) Antifungal efficacy and safety of cycloheximide as a supplement in optisol-GS. Drug Des Devel Ther 15:2091–2098. https://doi.org/10.2147/DDDT.S298059

Das S, Sharma S, Kar S, Sahu SK, Samal B, Mallick A (2010) Is inclusion of Sabouraud dextrose agar essential for the laboratory diagnosis of fungal keratitis? Indian J Ophthalmol 58:281–285. https://doi.org/10.4103/0301-4738.64122

De Carolis E, Posteraro B, Lass-Flörl C, Vella A, Florio AR, Torelli R, Girmenia C, Colozza C, Tortorano AM, Sanguinetti M, Fadda G (2012) Species identification of Aspergillus, Fusarium and Mucorales with direct surface analysis by matrix-assisted laser desorption ionization time-of-flight mass spectrometry. Clin Microbiol Infect 18:475–484. https://doi.org/10.1111/j.1469-0691.2011.03599.x

Dien Bard J, McElvania E (2020) Panels and syndromic testing in clinical microbiology. Clin Lab Med 40:393–420. https://doi.org/10.1016/j.cll.2020.08.001

Dominic RS, Prashanth H, Shenoy S, Baliga S (2009) Diagnostic value of latex agglutination in Cryptococcal meningitis. J Lab Physicians 1:67–68. https://doi.org/10.4103/0974-2727.59702

Douglas CM (2001) Fungal beta(1,3)-D-glucan synthesis. Med Mycol 39(suppl 1):55–66. https://doi.org/10.1080/mmy.39.1.55.66

Dunbar SA (2006) Applications of Luminex® xMAP™ technology for rapid, high-throughput multiplexed nucleic acid detection. Clin Chim Acta 363:71–82. https://doi.org/10.1016/j.cccn.2005.06.023

El-Gendi H, Saleh AK, Badierah R, Redwan EM, El-Maradny YA, El-Fakharany EM (2021) A comprehensive insight into fungal enzymes: structure, classification, and their role in mankind's challenges. J Fungi (Basel) 8:23. https://doi.org/10.3390/jof8010023

Elnifro EM, Ashshi AM, Cooper RJ, Klapper PE (2000) Multiplex PCR: optimization and application in diagnostic virology. Clin Microbiol Rev 13:559–570

Fang W, Wu J, Cheng M, Zhu X, Du M, Chen C, Liao W, Zhi K, Pan W (2023) Diagnosis of invasive fungal infections: challenges and recent developments. J Biomed Sci 30:42. https://doi.org/10.1186/s12929-023-00926-2

Fida M, Misra A, Harring JA, Kubbara A, Theel ES (2022) Histoplasma capsulatum complement fixation and immunodiffusion assay sensitivity in culture-confirmed cases of Histoplasmosis: a 10-year retrospective review (2011 to 2020). J Clin Microbiol 60:e01057–e01022. https://doi.org/10.1128/jcm.01057-22

Fontaine T, Latgé J-P (2020) Galactomannan produced by Aspergillus fumigatus: an update on the structure, biosynthesis and biological functions of an emblematic fungal biomarker. J Fungi (Basel) 6:283. https://doi.org/10.3390/jof6040283

Forster J, Hoenigl M, Suerbaum S, Wagener J, Dichtl K (2022) Serologic biomarkers in Candida and Aspergillus infections of the central nervous system: a comparison of galactomannan, mannan and β-1,3-D-gucan testing from serum and cerebrospinal fluid. Mycoses 65:709–714. https://doi.org/10.1111/myc.13451

Fowler JB, De Jesus O, Mesfin FB (2023) Ventriculoperitoneal shunt. In: StatPearls. StatPearls Publishing, Treasure Island

Gao XH, Li J, Gong HF, Yu GY, Liu P, Hao LQ, Liu LJ, Bai CG, Zhang W (2020) Comparison of fresh frozen tissue with formalin-fixed paraffin-embedded tissue for mutation analysis using a multi-gene panel in patients with colorectal cancer. Front Oncol 10:310. https://doi.org/10.3389/fonc.2020.00310

Garcia-Rubio R, de Oliveira HC, Rivera J, Trevijano-Contador N (2020) The fungal cell wall: Candida, Cryptococcus, and Aspergillus species. Front Microbiol 10:2993

Gavito-Higuera J, Mullins CB, Ramos-Duran L, Olivas Chacon CI, Hakim N, Palacios E (2016) Fungal infections of the central nervous system: a pictorial review. J Clin Imaging Sci 6:24. https://doi.org/10.4103/2156-7514.184244

Ghosh AK, Paul S, Sood P, Rudramurthy SM, Rajbanshi A, Jillwin TJ, Chakrabarti A (2015) Matrix-assisted laser desorption ionization time-of-flight mass spectrometry for the rapid identification of yeasts causing bloodstream infections. Clin Microbiol Infect 21:372–378. https://doi.org/10.1016/j.cmi.2014.11.009

Góralska K, Blaszkowska J, Dzikowiec M (2018) Neuroinfections caused by fungi. Infection 46: 443–459. https://doi.org/10.1007/s15010-018-1152-2

Govindarajan R, Duraiyan J, Kaliyappan K, Palanisamy M (2012) Microarray and its applications. J Pharm Bioallied Sci 4:S310–S312. https://doi.org/10.4103/0975-7406.100283

Goyal S, Castrillón-Betancur JC, Klaile E, Slevogt H (2018) The interaction of human pathogenic fungi with C-type lectin receptors. Front Immunol 9:1261. https://doi.org/10.3389/fimmu.2018.01261

Gresham D, Dunham MJ, Botstein D (2008) Comparing whole genomes using DNA microarrays. Nat Rev Genet 9:291–302. https://doi.org/10.1038/nrg2335

Grundy MA, Barnes RA, Coakley WT (1995) Highly sensitive detection of fungal antigens by ultrasound-enhanced latex agglutination. J Med Vet Mycol 33:201–203. https://doi.org/10.1080/02681219580000411

Guarner J, Brandt ME (2011) Histopathologic diagnosis of fungal infections in the 21st century. Clin Microbiol Rev 24:247–280. https://doi.org/10.1128/CMR.00053-10

Guimarães AJ, Nosanchuk JD, Zancopé-Oliveira RM (2006) Diagnosis of histoplasmosis. Braz J Microbiol 37:1–13. https://doi.org/10.1590/S1517-83822006000100001

Hage CA, Carmona EM, Epelbaum O, Evans SE, Gabe LM, Haydour Q, Knox KS, Kolls JK, Murad MH, Wengenack NL (2019) Microbiological laboratory testing in the diagnosis of fungal infections in pulmonary and critical care practice. an official American Thoracic Society Clinical Practice Guideline. Am J Respir Crit Care Med 200:535–550

Haley L (1969) ID#3771 - Public Health Image Library (PHIL). CDC

Han Y, Kozel TR, Zhang MX, MacGill RS, Carroll MC, Cutler JE (2001) Complement is essential for protection by an IgM and an IgG3 monoclonal antibody against experimental, hematogenously disseminated candidiasis. J Immunol 167:1550–1557. https://doi.org/10.4049/jimmunol.167.3.1550

Hanak BW, Bonow RH, Harris CA, Browd SR (2017) Cerebrospinal fluid shunting complications in children. Pediatr Neurosurg 52:381–400. https://doi.org/10.1159/000452840

de Heer K, Gerritsen MG, Visser CE, Leeflang MM (2019) Galactomannan detection in broncho-alveolar lavage fluid for invasive aspergillosis in immunocompromised patients. Cochrane Database Syst Rev 2019:CD012399. https://doi.org/10.1002/14651858.CD012399.pub2

Herbrecht R, Denning DW, Patterson TF, Bennett JE, Greene RE, Oestmann J-W, Kern WV, Marr KA, Ribaud P, Lortholary O, Sylvester R, Rubin RH, Wingard JR, Stark P, Durand C, Caillot D, Thiel E, Chandrasekar PH, Hodges MR, Schlamm HT, Troke PF, de Pauw B, Invasive Fungal Infections Group of the European Organisation for Research and Treatment of Cancer and the Global Aspergillus Study Group (2002a) Voriconazole versus amphotericin B for primary therapy of invasive aspergillosis. N Engl J Med 347:408–415. https://doi.org/10.1056/NEJMoa020191

Herbrecht R, Letscher-Bru V, Oprea C, Lioure B, Waller J, Campos F, Villard O, Liu K-L, Natarajan-Amé S, Lutz P, Dufour P, Bergerat J-P, Candolfi E (2002b) Aspergillus galactomannan detection in the diagnosis of invasive aspergillosis in cancer patients. J Clin Oncol 20:1898–1906. https://doi.org/10.1200/JCO.2002.07.004

Hevey MA, George IA, Rauseo AM, Larson L, Powderly W, Spec A (2020) Performance of the lateral flow assay and the latex agglutination serum cryptococcal antigen test in cryptococcal disease in patients with and without HIV. J Clin Microbiol 58:e01563–e01520. https://doi.org/10.1128/JCM.01563-20

Hrishi AP, Sethuraman M (2019) Cerebrospinal fluid (CSF) analysis and interpretation in neurocritical care for acute neurological conditions. Indian J Crit Care Med 23:S115–S119. https://doi.org/10.5005/jp-journals-10071-23187

Hsu AJ, Tamma PD, Zhang SX (2021) Challenges with utilizing the 1,3-beta-d-glucan and galactomannan assays to diagnose invasive mold infections in immunocompromised children. J Clin Microbiol 59:e03276–e03220. https://doi.org/10.1128/JCM.03276-20

Hussain KK, Malavia D, Johnson EM, Littlechild J, Winlove CP, Vollmer F, Gow NAR (2020) Biosensors and diagnostics for fungal detection. J Fungi (Basel) 6:349. https://doi.org/10.3390/jof6040349

Ines Andrade A, Donato M, Previgliano C, Hardjasudarma M (2014) Histoplasmosis brain abscesses in an immunocompetent adult. Neuroradiol J 27:334–338. https://doi.org/10.15274/NRJ-2014-10038

Infantino M, Grossi V, Lari B, Bambi R, Perri A, Manneschi M, Terenzi G, Liotti I, Ciotta G, Taddei C, Benucci M, Casprini P, Veneziani F, Fabbri S, Pompetti A, Manfredi M (2020) Diagnostic accuracy of an automated chemiluminescent immunoassay for anti-SARS-CoV-2 IgM and IgG antibodies: an Italian experience. J Med Virol 92:1671–1675. https://doi.org/10.1002/jmv.25932

Jabr R, Hammoud K (2020) CNS infection mimicking stroke. Kans J Med 13:127–128

Jackson NR, Blair JE, Ampel NM (2019) Central nervous system infections due to coccidioidomycosis. J Fungi (Basel) 5:54. https://doi.org/10.3390/jof5030054

Jacobs SE, Walsh TJ (2023) Non-Aspergillus hyaline molds: a host-based perspective of emerging pathogenic fungi causing sinopulmonary diseases. J Fungi 9:212. https://doi.org/10.3390/jof9020212

Jain A, Jain S, Rawat S (2010) Emerging fungal infections among children: A review on its clinical manifestations, diagnosis, and prevention. J Pharm Bioallied Sci 2:314–320. https://doi.org/10.4103/0975-7406.72131

Jarvis JN, Percival A, Bauman S, Pelfrey J, Meintjes G, Williams GN, Longley N, Harrison TS, Kozel TR (2011) Evaluation of a novel point-of-care cryptococcal antigen test on serum, plasma, and urine from patients with HIV-associated cryptococcal meningitis. Clin Infect Dis 53:1019–1023. https://doi.org/10.1093/cid/cir613

Jia DT, Thakur K (2019) Fungal infections of the central nervous system. Semin Neurol 39:343–357. https://doi.org/10.1055/s-0039-1688916

Johnson RT (1996) Microbiology of the nervous system. In: Baron S (ed) Medical microbiology. University of Texas Medical Branch at Galveston, Galveston

Juravel E, Polacheck I, Isaacson B, Dagan A, Korem M (2023) The distinction between dematiaceous molds and non-dematiaceous fungi in clinical and spiked samples treated with hydrogen peroxide using direct fluorescence microscopy. J Fungi (Basel) 9:227. https://doi.org/10.3390/jof9020227

Kanaujia R, Singh S, Rudramurthy SM (2023) Aspergillosis: an update on clinical spectrum, diagnostic schemes, and management. Curr Fungal Infect Rep 1–12. https://doi.org/10.1007/s12281-023-00461-5

Kandi V, Vaish R, Palange P, Bhoomagiri MR (2016) Chronic pulmonary histoplasmosis and its clinical significance: an under-reported systemic fungal disease. Cureus 8:e751

Kane J (1984) Conversion of Blastomyces dermatitidis to the yeast form at 37 degrees C and 26 degrees C. J Clin Microbiol 20:594–596

Kaplan W (1969) ID#3199 - Public Health Image Library (PHIL). CDC

Kauffman CA (2019) Central nervous system infection with other endemic mycoses: rare manifestation of blastomycosis, paracoccidioidomycosis, talaromycosis, and sporotrichosis. J Fungi (Basel) 5:64. https://doi.org/10.3390/jof5030064

Khan SN, Sepahdari AR (2012) Orbital masses: CT and MRI of common vascular lesions, benign tumors, and malignancies. Saudi J Ophthalmol 26:373–383. https://doi.org/10.1016/j.sjopt.2012.08.001

Khan MA, Quadri SAQ, Kazmi AS, Kwatra V, Ramachandran A, Gustin A, Farooqui M, Suriya SS, Zafar A (2018) A comprehensive review of skull base osteomyelitis: diagnostic and therapeutic challenges among various presentations. Asian J Neurosurg 13:959–970. https://doi.org/10.4103/ajns.AJNS_90_17

Khot PD, Fredricks DN (2009) PCR-based diagnosis of human fungal infections. Expert Rev Anti Infect Ther 7:1201–1221. https://doi.org/10.1586/eri.09.104

Kidd SE, Chen SC-A, Meyer W, Halliday CL (2020) A new age in molecular diagnostics for invasive fungal disease: are we ready? Front Microbiol 10:2903

Kobayashi GS (1996) Disease mechanisms of fungi. In: Baron S (ed) Medical microbiology. University of Texas Medical Branch at Galveston, Galveston

Kozel TR, Wickes B (2014) Fungal diagnostics. Cold Spring Harb Perspect Med 4:a019299. https://doi.org/10.1101/cshperspect.a019299

Kradin RL, Iafrate AJ (2010) General principles in the diagnosis of infection. In: Diagnostic pathology of infectious disease, pp 3–15. https://doi.org/10.1016/B978-1-4160-3429-2.00002-X

Kumar RS, Ganvir S, Hazarey V (2009) Candida and calcofluor white: study in precancer and cancer. J Oral Maxillofac Pathol 13:2–8. https://doi.org/10.4103/0973-029X.44575

Kwon-Chung KJ, Fraser JA, Doering TL, Wang Z, Janbon G, Idnurm A, Bahn Y-S (2014) Cryptococcus neoformans and Cryptococcus gattii, the etiologic agents of Cryptococcosis. Cold Spring Harb Perspect Med 4:a019760. https://doi.org/10.1101/cshperspect.a019760

Landlinger C, Preuner S, Willinger B, Haberpursch B, Racil Z, Mayer J, Lion T (2009) Species-specific identification of a wide range of clinically relevant fungal pathogens by use of Luminex xMAP technology. J Clin Microbiol 47:1063–1073. https://doi.org/10.1128/JCM.01558-08

Lanternier F, Cypowyj S, Picard C, Bustamante J, Lortholary O, Casanova J-L, Puel A (2013) Primary immunodeficiencies underlying fungal infections. Curr Opin Pediatr 25:736–747. https://doi.org/10.1097/MOP.0000000000000031

Latgé J-P (1999) Aspergillus fumigatus and Aspergillosis. Clin Microbiol Rev 12:310–350

Lau A, Chen S, Sorrell T, Carter D, Malik R, Martin P, Halliday C (2007) Development and clinical application of a panfungal PCR assay to detect and identify fungal DNA in tissue specimens. J Clin Microbiol 45:380–385. https://doi.org/10.1128/JCM.01862-06

Lau A, Stanley K, Sorrell T (2013) Multiplex-tandem PCR for fungal diagnostics. Methods Mol Biol 968:195–201. https://doi.org/10.1007/978-1-62703-257-5_15

Leck A (1999) Preparation of lactophenol cotton blue slide mounts. Community Eye Health 12:24

Lee C-Y, Thompson GR III, Hastey CJ, Hodge GC, Lunetta JM, Pappagianis D, Heinrich V (2015) Coccidioides endospores and spherules draw strong chemotactic, adhesive, and phagocytic responses by individual human neutrophils. PloS One 10:e0129522. https://doi.org/10.1371/journal.pone.0129522

Lehrnbecher T, Rath PM, Attarbaschi A, Cario G, Döring M, Moser O, Mücke U, Poyer F, Rieken S, Temme C, Voigt S, Groll AH, Lauten M (2019) Galactomannan and PCR in the central nervous system to detect invasive mold disease - a retrospective analysis in immunocompromised children. Sci Rep 9:12950. https://doi.org/10.1038/s41598-019-49426-0

Leinberger DM, Schumacher U, Autenrieth IB, Bachmann TT (2005) Development of a DNA microarray for detection and identification of fungal pathogens involved in invasive mycoses. J Clin Microbiol 43:4943–4953. https://doi.org/10.1128/JCM.43.10.4943-4953.2005

Li S, Nguyen IP, Urbanczyk K (2020) Common infectious diseases of the central nervous system—clinical features and imaging characteristics. Quant Imaging Med Surg 10:2227–2259. https://doi.org/10.21037/qims-20-886

Liu A (2010) Laser capture microdissection in the tissue biorepository. J Biomol Tech 21:120–125

Liu Y, Tortora G, Ryan ME, Lee H-M, Golub LM (2002) Potato dextrose agar antifungal susceptibility testing for yeasts and molds: evaluation of phosphate effect on antifungal activity of CMT-3. Antimicrob Agents Chemother 46:1455–1461. https://doi.org/10.1128/AAC.46.5.1455-1461.2002

Liu Y, Kang M, Wu S-Y, Wu L-J, He L, Xiao Y-L, Zhang W-L, Liao Q-F, Deng J, Chen Z-X, Ma Y (2022) Evaluation of a Cryptococcus capsular polysaccharide detection FungiXpert LFA (lateral flow assay) for the rapid diagnosis of Cryptococcosis. Med Mycol 60:myac020. https://doi.org/10.1093/mmy/myac020

Lockhart SR, Bialek R, Kibbler CC, Cuenca-Estrella M, Jensen HE, Kontoyiannis DP (2021) Molecular techniques for genus and species determination of fungi from fresh and paraffin-embedded formalin-fixed tissue in the revised EORTC/MSGERC Definitions of Invasive Fungal Infection. Clin Infect Dis 72:S109–S113. https://doi.org/10.1093/cid/ciaa1836

Löffler J, Hebart H, Sepe S, Schumcher U, Klingebiel T, Einsele H (1998) Detection of PCR-amplified fungal DNA by using a PCR-ELISA system. Med Mycol 36:275–279. https://doi.org/10.1080/02681219880000441

Low C-Y, Rotstein C (2011) Emerging fungal infections in immunocompromised patients. F1000 Med Rep 3:14. https://doi.org/10.3410/M3-14

Mactier H, Galea P, McWilliam R (1998) Acute obstructive hydrocephalus complicating bacterial meningitis in childhood. BMJ 316:1887–1889

Magee BB, Magee PT (1987) Electrophoretic karyotypes and chromosome numbers in Candida species. J Gen Microbiol 133:425–430. https://doi.org/10.1099/00221287-133-2-425

Malhotra S et al (2014) Recent diagnostic techniques in mycology. Med Microbiol Diagn 3:1–6. https://doi.org/10.4172/2161-0703.1000146

Maziarz EK, Perfect JR (2016) Cryptococcosis. Infect Dis Clin North Am 30:179–206. https://doi.org/10.1016/j.idc.2015.10.006

McBride JA, Gauthier GM, Klein BS (2017) Clinical manifestations and treatment of blastomycosis. Clin Chest Med 38:435–449. https://doi.org/10.1016/j.ccm.2017.04.006

McHugh KE, Gersey M, Rhoads DD, Procop GW, Zhang Y, Booth CN, Sturgis CD (2019) Sensitivity of cerebrospinal fluid cytology for the diagnosis of cryptococcal infections: a 21-year single-institution retrospective review. Am J Clin Pathol 151:198–204. https://doi.org/10.1093/ajcp/aqy133

McMurray DN (1996) Mycobacteria and Nocardia. In: Baron S (ed) Medical microbiology. University of Texas Medical Branch at Galveston, Galveston

Meyer W, Irinyi L, Hoang MTV, Robert V, Garcia-Hermoso D, Desnos-Ollivier M, Yurayart C, Tsang C-C, Lee C-Y, Woo PCY, Pchelin IM, Uhrlaß S, Nenoff P, Chindamporn A, Chen S, Hebert PDN, Sorrell TC, ISHAM Barcoding of Pathogenic Fungi Working Group (2019) Database establishment for the secondary fungal DNA barcode translational elongation factor 1α (TEF1α) 1. Genome 62:160–169. https://doi.org/10.1139/gen-2018-0083

Miceli MH (2019) Central nervous system infections due to Aspergillus and other hyaline molds. J Fungi (Basel) 5:79. https://doi.org/10.3390/jof5030079

Miceli A, Krishnamurthy K (2023) Blastomycosis. In: StatPearls. StatPearls Publishing, Treasure Island

Mikhailova EO (2021) Gold nanoparticles: biosynthesis and potential of biomedical application. J Funct Biomater 12:70. https://doi.org/10.3390/jfb12040070

Mikulska M, Balletto E, Castagnola E, Mularoni A (2021) Beta-D-glucan in patients with haematological malignancies. J Fungi (Basel) 7:1046. https://doi.org/10.3390/jof7121046

Miller MB, Tang Y-W (2009) Basic concepts of microarrays and potential applications in clinical microbiology. Clin Microbiol Rev 22:611–633. https://doi.org/10.1128/CMR.00019-09

Mukaremera L, Lee KK, Mora-Montes HM, Gow NAR (2017) Candida albicans yeast, pseudohyphal, and hyphal morphogenesis differentially affects immune recognition. Front Immunol 8:629. https://doi.org/10.3389/fimmu.2017.00629

Naglik JR, Challacombe SJ, Hube B (2003) Candida albicans secreted aspartyl proteinases in virulence and pathogenesis. Microbiol Mol Biol Rev 67:400–428. https://doi.org/10.1128/MMBR.67.3.400-428.2003

Nath P, Kabir MA, Doust SK, Ray A (2021) Diagnosis of herpes simplex virus: laboratory and point-of-care techniques. Infect Dis Rep 13:518–539. https://doi.org/10.3390/idr13020049

Navalkar KA, Johnston SA, Woodbury N, Galgiani JN, Magee DM, Chicacz Z, Stafford P (2014) Application of immunosignatures for diagnosis of valley fever. Clin Vaccine Immunol 21: 1169–1177. https://doi.org/10.1128/CVI.00228-14

Oviaño M, Bou G (2018) Matrix-assisted laser desorption ionization–time of flight mass spectrometry for the rapid detection of antimicrobial resistance mechanisms and beyond. Clin Microbiol Rev 32:e00037–e00018. https://doi.org/10.1128/CMR.00037-18

Panackal AA, Williamson PR (2015) Fungal infections of the central nervous system. Continuum (Minneap Minn) 21:1662–1678. https://doi.org/10.1212/CON.0000000000000241

Pappas PG (2013) Cryptococcal infections in non-HIV-infected patients. Trans Am Clin Climatol Assoc 124:61–79

Parveen S, Bhattacharya S (2023) Options appraisal of various β–D–glucan assay systems for a diagnostic microbiology laboratory. J Acad Clin Microbiol 25:1. https://doi.org/10.4103/jacm.jacm_28_22

Pasqualotto A, Denning D (2005) Diagnosis of invasive fungal infections – current limitations of classical and new diagnostic methods. Eur Oncol Haematol 1. https://doi.org/10.17925/EOH.2005.0.0.1p

Patterson TF, Thompson GR, Denning DW, Fishman JA, Hadley S, Herbrecht R, Kontoyiannis DP, Marr KA, Morrison VA, Nguyen MH, Segal BH, Steinbach WJ, Stevens DA, Walsh TJ, Wingard JR, Young J-AH, Bennett JE (2016) Practice guidelines for the diagnosis and management of Aspergillosis: 2016 update by the Infectious Diseases Society of America. Clin Infect Dis 63:e1–e60. https://doi.org/10.1093/cid/ciw326

de Pauw BE (2011) What are fungal infections? Mediterr J Hematol Infect Dis 3:e2011001. https://doi.org/10.4084/MJHID.2011.001

Pavlickova P, Hug H (2004) A streptavidin-biotin-based microarray platform for immunoassays. Methods Mol Biol 264:73–83. https://doi.org/10.1385/1-59259-759-9:073

Picek L, Šulc M, Matas J, Heilmann-Clausen J, Jeppesen TS, Lind E (2022) Automatic fungi recognition: deep learning meets mycology. Sensors (Basel) 22:633. https://doi.org/10.3390/s22020633

Ponka D, Baddar F (2014) Microscopic potassium hydroxide preparation. Can Fam Physician 60: 57

Posnakoglou L, Siahanidou T, Syriopoulou V, Michos A (2020) Impact of cerebrospinal fluid syndromic testing in the management of children with suspected central nervous system infection. Eur J Clin Microbiol Infect Dis 39:2379–2386. https://doi.org/10.1007/s10096-020-03986-6

Punjabi V, Patel S, Pathak J, Swain N (2020) Comparative evaluation of staining efficacy of calcofluor white and acridine orange for detection of Candida species using fluorescence microscopy – a prospective microbiological study. J Oral Maxillofac Pathol 24:81–86. https://doi.org/10.4103/jomfp.JOMFP_315_18

Rahi MS, Jindal V, Pednekar P, Parekh J, Gunasekaran K, Sharma S, Stender M, Jaiyesimi IA (2021) Fungal infections in hematopoietic stem-cell transplant patients: a review of epidemiology, diagnosis, and management. Ther Adv Infect Dis 8:20499361211039050. https://doi.org/10.1177/20499361211039050

Rajarajan P, Sylvia K, Periasamy MP, Subramanian M (2021) Detection of aflatoxin producing Aspergillus flavus from animal feed in Karnataka, India. Environ Anal Health Toxicol 36: e2021017. https://doi.org/10.5620/eaht.2021017

Raman Sharma R (2010) Fungal infections of the nervous system: current perspective and controversies in management. Int J Surg 8:591–601. https://doi.org/10.1016/j.ijsu.2010.07.293

Ribes JA, Vanover-Sams CL, Baker DJ (2000) Zygomycetes in human disease. Clin Microbiol Rev 13:236–301

Richardson M, Page I (2018) Role of serological tests in the diagnosis of mold infections. Curr Fungal Infect Rep 12:127–136. https://doi.org/10.1007/s12281-018-0321-1

Rickerts V (2016) Identification of fungal pathogens in formalin-fixed, paraffin-embedded tissue samples by molecular methods. Fungal Biol 120:279–287. https://doi.org/10.1016/j.funbio.2015.07.002

Rigby S, Procop GW, Haase G, Wilson D, Hall G, Kurtzman C, Oliveira K, Von Oy S, Hyldig-Nielsen JJ, Coull J, Stender H (2002) Fluorescence in situ hybridization with peptide nucleic acid probes for rapid identification of Candida albicans directly from blood culture bottles. J Clin Microbiol 40:2182–2186. https://doi.org/10.1128/JCM.40.6.2182-2186.2002

Riwes MM, Wingard JR (2012) Diagnostic methods for invasive fungal diseases in patients with hematologic malignancies. Expert Rev Hematol 5:661–669. https://doi.org/10.1586/ehm.12.53

Rodríguez-Cerdeira C, Arenas R, Moreno-Coutiño G, Vásquez E, Fernández R, Chang P (2014) Systemic fungal infections in patients with human inmunodeficiency virus. Actas Dermosifiliogr 105:5–17. https://doi.org/10.1016/j.adengl.2012.06.032

Rupp S (2017) Microarray technologies in fungal diagnostics. Methods Mol Biol 1508:385–409. https://doi.org/10.1007/978-1-4939-6515-1_22

Salem-Bango Z, Price TK, Chan JL, Chandrasekaran S, Garner OB, Yang S (2023) Fungal whole-genome sequencing for species identification: from test development to clinical utilization. J Fungi (Basel) 9:183. https://doi.org/10.3390/jof9020183

Santos PO, Melo JO, Ponzzes CMPBS, Alves JAB, de Melo DLFM, Botelho N d S, Yamada-Ogatta SF, Mann RS, de Cássia Trindade R (2012) Multilocus enzyme electrophoresis analysis and exoenzymatic activity of Candida albicans strains isolated from women with vaginal candidiasis. Mycoses 55:64–72. https://doi.org/10.1111/j.1439-0507.2011.02043.x

Sapra H, Singhal V (2019) Managing Meningoencephalitis in Indian ICU. Indian J Crit Care Med 23:S124–S128. https://doi.org/10.5005/jp-journals-10071-23189

Saubolle MA, McKellar PP, Sussland D (2007) Epidemiologic, clinical, and diagnostic aspects of coccidioidomycosis. J Clin Microbiol 45:26–30. https://doi.org/10.1128/JCM.02230-06

Schmiedel Y, Zimmerli S (2016) Common invasive fungal diseases: an overview of invasive candidiasis, aspergillosis, cryptococcosis, and Pneumocystis pneumonia. Swiss Med Wkly 146:w14281. https://doi.org/10.4414/smw.2016.14281

Scott EN, Kaufman L, Brown AC, Muchmore HG (1987) Serologic studies in the diagnosis and management of meningitis due to Sporothrix schenckii. N Engl J Med 317:935–940. https://doi.org/10.1056/NEJM198710083171505

Sendid B, Tabouret M, Poirot JL, Mathieu D, Fruit J, Poulain D (1999) New enzyme immunoassays for sensitive detection of circulating Candida albicans mannan and antimannan antibodies: useful combined test for diagnosis of systemic candidiasis. J Clin Microbiol 37:1510–1517

Serin I, Dogu MH (2021) Serum Aspergillus galactomannan lateral flow assay for the diagnosis of invasive aspergillosis: a single-centre study. Mycoses 64:678–683. https://doi.org/10.1111/myc.13265

Sharma M, Anand SK (2002) Characterization of constitutive microflora of biofilms in dairy processing lines. Food Microbiol 19:627–636. https://doi.org/10.1006/fmic.2002.0472

Shoham S, Marr KA (2012) Invasive fungal infections in solid organ transplant recipients. Future Microbiol 7:639–655. https://doi.org/10.2217/fmb.12.28

Silva S, Negri M, Henriques M, Oliveira R, Williams DW, Azeredo J (2012) Candida glabrata, Candida parapsilosis and Candida tropicalis: biology, epidemiology, pathogenicity and antifungal resistance. FEMS Microbiol Rev 36:288–305. https://doi.org/10.1111/j.1574-6976.2011.00278.x

Skiada A, Lass-Floerl C, Klimko N, Ibrahim A, Roilides E, Petrikkos G (2018) Challenges in the diagnosis and treatment of mucormycosis. Med Mycol 56:S93–S101. https://doi.org/10.1093/mmy/myx101

Sobottka I, Freiesleben H, Laufs R (1993) Simple, rapid stain technique for diagnosis of cryptococcosis. Eur J Clin Microbiol Infect Dis 12:479–481. https://doi.org/10.1007/BF01967452

Soltani M, Bayat M, Hashemi SJ, Zia M, Pestechian N (2013) Isolation of Cryptococcus neoformans and other opportunistic fungi from pigeon droppings. J Res Med Sci 18:56–60

Souza TB, Lozer DM, Kitagawa SMS, Spano LC, Silva NP, Scaletsky ICA (2013) Real-time multiplex PCR assay and melting curve analysis for identifying diarrheagenic Escherichia coli. J Clin Microbiol 51:1031–1033. https://doi.org/10.1128/JCM.02478-12

Spiess B, Seifarth W, Hummel M, Frank O, Fabarius A, Zheng C, Mörz H, Hehlmann R, Buchheidt D (2007) DNA microarray-based detection and identification of fungal pathogens in clinical samples from neutropenic patients. J Clin Microbiol 45:3743–3753. https://doi.org/10.1128/JCM.00942-07

Srinivas S, Kumari P, Gupta DK (2021) Utility of Panfungal PCR in the diagnosis of invasive fungal infections in febrile neutropenia. J Family Med Prim Care 10:2533–2540. https://doi.org/10.4103/jfmpc.jfmpc_2325_20

Stevens DA, Zhang Y, Finkelman MA, Pappagianis D, Clemons KV, Martinez M (2016) Cerebrospinal fluid (1,3)-beta-d-glucan testing is useful in diagnosis of coccidioidal meningitis. J Clin Microbiol 54:2707–2710. https://doi.org/10.1128/JCM.01224-16

Sue MJ, Yeap SK, Omar AR, Tan SW (2014) Application of PCR-ELISA in molecular diagnosis. Biomed Res Int 2014:653014. https://doi.org/10.1155/2014/653014

Sugui JA, Kwon-Chung KJ, Juvvadi PR, Latgé J-P, Steinbach WJ (2015) Aspergillus fumigatus and related species. Cold Spring Harb Perspect Med 5:a019786. https://doi.org/10.1101/cshperspect.a019786

Sun J, Li X, Zeng H, Xie Z, Lu C, Xi L, de Hoog GS (2010) Development and evaluation of loop-mediated isothermal amplification (LAMP) for the rapid diagnosis of Penicillium marneffei in archived tissue samples. FEMS Immunol Med Microbiol 58:381–388. https://doi.org/10.1111/j.1574-695X.2009.00647.x

Susianti H, Parmadi L, Firani NK, Setyawan UA, Sartono TR (2021) Diagnostic value of serum human Galactomannan aspergillus antigen and 1,3-beta-D-glucan in immunocompromised patient suspected fungal infection. J Clin Lab Anal 35:e23806. https://doi.org/10.1002/jcla.23806

Talapko J, Juzbašić M, Matijević T, Pustijanac E, Bekić S, Kotris I, Škrlec I (2021) Candida albicans—the virulence factors and clinical manifestations of infection. J Fungi (Basel) 7:79. https://doi.org/10.3390/jof7020079

Tankeshwar A (2014) Lactophenol Cotton Blue (LPCB) mounts. Microbe Online [WWW Document]. https://microbeonline.com/lactophenol-cotton-blue-lpcb-mounts-principle-staining-protocol/. Accessed 16 July 2023

Tanner DC, Weinstein MP, Fedorciw B, Joho KL, Thorpe JJ, Reller L (1994) Comparison of commercial kits for detection of cryptococcal antigen. J Clin Microbiol 32:1680–1684. https://doi.org/10.1128/jcm.32.7.1680-1684.1994

Tee RD, Gordon DJ, Taylor AJ (1987) Cross-reactivity between antigens of fungal extracts studied by RAST inhibition and immunoblot technique. J Allergy Clin Immunol 79:627–633. https://doi.org/10.1016/s0091-6749(87)80159-8

Terrero-Salcedo D, Powers-Fletcher MV (2020) Updates in laboratory diagnostics for invasive fungal infections. J Clin Microbiol 58:e01487–e01419. https://doi.org/10.1128/JCM.01487-19

Thakur KT, Wilson MR (2018) Chronic meningitis. Continuum (Minneap Minn) 24:1298–1326. https://doi.org/10.1212/CON.0000000000000664

Theel ES, Doern CD (2013) β-d-Glucan testing is important for diagnosis of invasive fungal infections. J Clin Microbiol 51:3478–3483. https://doi.org/10.1128/JCM.01737-13

Thompson GR, Boulware DR, Bahr NC, Clancy CJ, Harrison TS, Kauffman CA, Le T, Miceli MH, Mylonakis E, Nguyen MH, Ostrosky-Zeichner L, Patterson TF, Perfect JR, Spec A, Kontoyiannis DP, Pappas PG (2022) Noninvasive testing and surrogate markers in invasive fungal diseases. Open Forum Infect Dis 9:ofac112. https://doi.org/10.1093/ofid/ofac112

Tillema J-M, Pirko I (2013) Neuroradiological evaluation of demyelinating disease. Ther Adv Neurol Disord 6:249–268. https://doi.org/10.1177/1756285613478870

Toscanini M, Nusblat A, Cuestas M (2021) Diagnosis of histoplasmosis: current status and perspectives. Appl Microbiol Biotechnol 105:1–23. https://doi.org/10.1007/s00253-021-11170-9

Trujillo-Gómez J, Tsokani S, Arango-Ferreira C, Atehortúa-Muñoz S, Jimenez-Villegas MJ, Serrano-Tabares C, Veroniki A-A, Florez ID (2022) Biofire FilmArray Meningitis/Encephalitis panel for the aetiological diagnosis of central nervous system infections: a systematic review and diagnostic test accuracy meta-analysis. EClinicalMedicine 44:101275. https://doi.org/10.1016/j.eclinm.2022.101275

Turner SA, Butler G (2014) The Candida pathogenic species complex. Cold Spring Harb Perspect Med 4:a019778. https://doi.org/10.1101/cshperspect.a019778

Vafajoo A, Rostami A, Parsa SF, Salarian R, Rabiee N, Rabiee G, Rabiee M, Tahriri M, Vashaee D, Tayebi L, Hamblin MR (2018) Multiplexed microarrays based on optically encoded microbeads. Biomed Microdevices 20:66. https://doi.org/10.1007/s10544-018-0314-4

Vainionpää R, Leinikki P (2008) Diagnostic techniques: serological and molecular approaches. In: Encyclopedia of virology, pp 29–37. https://doi.org/10.1016/B978-012374410-4.00585-9

Valero C, de la Cruz-Villar L, Zaragoza Ó, Buitrago MJ (2016) New panfungal real-time PCR assay for diagnosis of invasive fungal infections. J Clin Microbiol 54:2910–2918. https://doi.org/10.1128/JCM.01580-16

Van Burik J-A, Myerson D, Schreckhise RW, Bowden RA (1998) Panfungal PCR assay for detection of fungal infection in human blood specimens. J Clin Microbiol 36:1169–1175
Van TT, Kim TH, Butler-Wu SM (2020) Evaluation of the Biofire FilmArray meningitis/encephalitis assay for the detection of Cryptococcus neoformans/gattii. Clin Microbiol Infect 26:1375–1379. https://doi.org/10.1016/j.cmi.2020.01.007
Vidal JE, Boulware DR (2015) Lateral flow assay for cryptococcal antigen: an important advance to improve the continuum of HIV care and reduce cryptococcal meningitis-related mortality. Rev Inst Med Trop Sao Paulo 57:38–45. https://doi.org/10.1590/S0036-46652015000700008
Vijayan T, Chiller T, Klausner JD (2023) Sensitivity and specificity of a new cryptococcal antigen lateral flow assay in serum and cerebrospinal fluid. MLO Med Lab Obs 45(3):16, 18, 20. PMID: 23822028; PMCID: PMC4119400
Wainaina S, Taherzadeh MJ (2023) Automation and artificial intelligence in filamentous fungi-based bioprocesses: a review. Bioresour Technol 369:128421. https://doi.org/10.1016/j.biortech.2022.128421
Walsh TJ, Dixon DM (1996) Spectrum of mycoses. In: Baron S (ed) Medical microbiology. University of Texas Medical Branch at Galveston, Galveston
Wang S, Lai J, Wu R, Zhang L, Huang M, Xiao Y, Zhang X, Chen J (2022) Grocott methenamine silver staining is the optimal approach to histological diagnosis of pulmonary cryptococcosis. Front Microbiol 13:885511. https://doi.org/10.3389/fmicb.2022.885511
White PL, Shetty A, Barnes RA (2003) Detection of seven Candida species using the Light-Cycler system. J Med Microbiol 52:229–238. https://doi.org/10.1099/jmm.0.05049-0
White PL, Parr C, Thornton C, Barnes RA (2013) Evaluation of real-time PCR, galactomannan enzyme-linked immunosorbent assay (ELISA), and a novel lateral-flow device for diagnosis of invasive aspergillosis. J Clin Microbiol 51:1510–1516. https://doi.org/10.1128/JCM.03189-12
Wiederhold NP (2017) Antifungal resistance: current trends and future strategies to combat. Infect Drug Resist 10:249–259. https://doi.org/10.2147/IDR.S124918
Xu J (2016) Fungal DNA barcoding. Genome 59:913–932. https://doi.org/10.1139/gen-2016-0046
Yeo SF, Wong B (2002) Current status of nonculture methods for diagnosis of invasive fungal infections. Clin Microbiol Rev 15:465–484. https://doi.org/10.1128/CMR.15.3.465-484.2002
Yeo SF, Huie S, Sofair AN, Campbell S, Durante A, Wong B (2006) Measurement of serum d-arabinitol/creatinine ratios for initial diagnosis and for predicting outcome in an unselected, population-based sample of patients with Candida fungemia. J Clin Microbiol 44:3894–3899. https://doi.org/10.1128/JCM.01045-06
Zaaroura H, Bergman R (2019) How useful is periodic acid-Schiff stain to detect fungi in biopsies from dermatoses of the palms and soles? J Cutan Pathol 46:418–420. https://doi.org/10.1111/cup.13451
Zaragoza O, Rodrigues ML, De Jesus M, Frases S, Dadachova E, Casadevall A (2009) The capsule of the fungal pathogen Cryptococcus neoformans. Adv Appl Microbiol 68:133–216. https://doi.org/10.1016/S0065-2164(09)01204-0
Zeng X, Kong F, Halliday C, Chen S, Lau A, Playford G, Sorrell TC (2007) Reverse line blot hybridization assay for identification of medically important fungi from culture and clinical specimens. J Clin Microbiol 45:2872–2880. https://doi.org/10.1128/JCM.00687-07
Zeng H-Q, Zhang X-B, Cai X-Y, Yang D-Y, Lin L, Chen M-J, Guo W-F, Luo X (2021) Diagnostic value of bronchoalveolar lavage fluid cryptococcal antigen-lateral flow immunochromatographic assay for pulmonary cryptococcosis in non-HIV patients. Diagn Microbiol Infect Dis 99:115276. https://doi.org/10.1016/j.diagmicrobio.2020.115276
Zhou W, Li H, Zhang Y, Huang M, He Q, Li P, Zhang F, Shi Y, Su X (2017) Diagnostic value of galactomannan antigen test in serum and bronchoalveolar lavage fluid samples from patients with nonneutropenic invasive pulmonary Aspergillosis. J Clin Microbiol 55:2153–2161. https://doi.org/10.1128/JCM.00345-17

Association of Viral and Fungal Infections of the CNS During Immunosuppression

20

Reba Kanungo and Shampa Anupurba

Abstract

A major concern of immunodeficiency due to any cause is secondary infection by microorganisms, including virus and fungi. These microorganisms are generally of low virulence or are commensals found ubiquitously in and around human habitation, posing a major threat to immunocompromised individuals. The central nervous system infections unlike other systems are secondary to a primary site of infection elsewhere in the body. It is acquired through blood stream, peripheral nerves or breach in the blood-brain and blood cerebrospinal fluid barriers. The risk of viral and fungal infections of CNS depends on two primary determinants, namely, exposures and the net state of immunosuppression. Specific virus belonging to the family "Herpesviridae", Adenoviruses "Adenoviridae" and JC virus "Papovaviridae" are commonly associated with infections in CNS common fungi involved in infection include Cryptococcus, Histoplasma, and Candida. Infection with Mucorales occurs by contiguous spread via the rhino-orbital pathway. Clinical manifestation may be acute as meningitis, or encephalitis caused mainly by virus or chronic as in cryptococcal meningitis. Symptoms are overlapping, making specific clinical diagnosis difficult. Imaging techniques, CSF analysis, and rapid molecular and antigen detection techniques have eased confirmatory diagnosis of viral and fungal infections of CNS in immunocompromised individuals. Treatment options are limited due to narrow range of antivirals and anti-fungals that act against the infecting agents. This chapter discusses

R. Kanungo (✉)
Department of Microbiology, Pondicherry Institute of Medical Sciences, Pondicherry, India

S. Anupurba
Department of Microbiology, Institute of Medical Sciences, Banaras Hindu University, Varanasi, India

H. Sami et al. (eds.), *Viral and Fungal Infections of the Central Nervous System: A Microbiological Perspective*, https://doi.org/10.1007/978-981-99-6445-1_20

the virus and fungi causing secondary infections of the CNS in the immunocompromised individuals.

Keywords

Immunocompromised · Central nervous system · Viral infections · Fungal infections

20.1 Introduction

Defective components of innate or adaptive immunity leading to immunodeficiency is either congenital (primary) or acquired. The deficiency may be due to genetic abnormality, or loss of immune function as a result of damage to immune cells by drugs, or physical and chemical agents, leading to immunodeficiency in the host. Aging also contributes to a weakened immune system. The host is then referred to as **immunocompromised** with decreased or absence of the ability to fight off infection. Congenital immunodeficiencies resulting from genetic abnormalities can be divided into three categories: humoral, cellular, and combination immunodeficiency. More often, acquired immunodeficiencies can be brought on by cytotoxic medication, other immunosuppressive therapy, steroid use, treatment with radiation, severe nutritional deficiencies, or fatal burns. A decrease in the efficiency of the immune system is also associated with hematological malignancies like leukemia, solid organ transplant (SOT) as well as hematopoietic stem cell transplantation (HSCT) recipients and autoimmune diseases like systemic lupus erythematosus (SLE). Chronically ill patients are also prone to secondary viral and fungal infections. Certain infections themselves can also lead to immunodeficiency, e.g., infection with human immunodeficiency virus (HIV), which leads to acquired immunodeficiency syndrome (AIDS).

A major concern of immunodeficiency is secondary infection by microorganisms. Bacteria, viruses, fungi, and parasites can cause severe infections in these individuals. These microorganisms are generally of low virulence or are commensals found ubiquitously in and around human habitation. In this chapter, two important groups, namely, the virus and fungi, causing secondary infection are discussed.

20.2 Viral Infections in Immunocompromised Individuals

The risk for central nervous system (CNS) infection rests on two fundamental determinants, exposures and the net state of immunosuppression (Castro et al. 2018). Following viruses are predominantly associated with infections in immunocompromised hosts, namely, Herpesviridae, Adenoviruses (Adenoviridae), and JC virus (Papovaviridae) (Table 20.1). The presentation and course of CNS infections in cancer and immunosuppressed patients differ from those without cancer or are immunocompetent.

Table 20.1 Common viruses associated with CNS infection in the immunocompromised individuals

Virus	Disease
Herpes simplex virus	Encephalitis
Epstein-Barr virus	CNS lymphoma
Cytomegalovirus/varicella-zoster	Disseminated infections
JC virus	Progressive multifocal leukoencephalopathy

20.2.1 Pathogenesis

There are essentially two major methods that viruses enter the CNS, despite the fact that they can penetrate defenses in other ways. The blood supply is one means to get in. The blood-brain and blood cerebrospinal fluid barriers form a complex barrier network that guards the sensitive CNS parenchyma from hazardous blood-borne pathogens. However, viruses have evolved one or more strategies to get beyond this barrier. Some viruses are capable of infecting vascular endothelial cells, allowing them to penetrate the blood-brain barrier (BBB) and enter the CNS. Additionally, there are areas of CNS that are not completely protected by the BBB. These are the choroid plexus and circumventricular organs, that serve as entry points for several viruses. "Trojan horse" phenomenon is adapted by the infected hematopoietic cells, which transport the virus into the CNS via the blood supply (Swanson II and McGavern 2015).

Migration through peripheral nerves is the second main CNS entry point. Once viruses reach the CNS, viral tropism and the ensuing immune response combine to shape the resulting disease. Meningitis is frequently caused by viruses that remain inside meningeal cells or the ventricular lining, whereas meningoencephalitis, encephalitis, or myelitis are caused by viruses that infect the CNS parenchyma.

20.2.2 Common Viruses Associated with CNS Infection

20.2.2.1 Herpesviridae

Following the primary infection by the virus, it can progress to latent infection. Reactivations of the infection are likely to occur during periods of immunosuppression. Both primary infection and reactivations often become more serious in immunocompromised patients.

Human herpes virus group include the following virus:

- Herpes simplex virus—type 1, i.e., HSV-1.
- Herpes simplex virus—type 2, i.e., HSV-2.
- Varicella zoster virus, i.e., VZV.
- Epstein-Barr virus, i.e., EBV.
- Cytomegalovirus, i.e., CMV.
- Human herpesvirus 6, i.e., HHV-6.

- Human herpesvirus 7, i.e., HHV-7.
- Human herpesvirus 8, i.e., HHV-8.

20.2.2.1.1 Herpes Simplex Viruses (HSV)

Craniospinal ganglia is the primary site for latency by HSV. It may then be reactivated from time to time by several triggers which include immunosuppression, stress, any other infection, and sunlight. Severe HSV infection in the immunocompromised individuals is associated with AIDS, those receiving cytotoxic drugs or are recipients of organ graft.

Herpes simplex encephalitis (HSE) occurs as a serious complication in neonates with the global involvement of the brain leading to liquefaction with 100% mortality as the worst case scenario. The other is a focal disease with the temporal lobe most commonly affected. This appears both in children and adults arising from the reactivation of latent virus and has a mortality of 70% without treatment.

Disseminated herpes simplex infection is the other manifestation that may occur in immunocompromised individuals. Widespread vesicular rash might appear which resembles the rash of chickenpox. Other than the skin, liver, spleen, lungs, and the central nervous system (CNS) may be involved.

20.2.2.1.2 Varicella-Zoster Virus (VZV)

VZV causes chicken pox (varicella) and reactivates to produce zoster or shingles. This neurotropic virus can cause aseptic meningitis, encephalomyelitis, multifocal polyradiculitis, cranial neuritis, and a vasculopathy affecting both small and big cerebral arteries that can result in unifocal and multifocal strokes. The virus goes into a latent phase along the entire neuroaxis which include the cranial nerve ganglia, dorsal root ganglia, and autonomic ganglia. The latent VZV reactivates to cause a variety of neurologic diseases in aged people who have had a loss in cell-mediated immunity over time or in those who have impaired immune systems. Patients with impaired immune systems are more likely to get an extensive infection (Bookstaver et al. 2017).

VZV infection is linked to aneurysm, subarachnoid and cerebral hemorrhage, carotid dissection, and, in rare cases, peripheral artery disease in addition to ischemic infarction of the brain and spinal cord. Typically, VZV vasculopathy involves both large and small arteries (Gilden et al. 2009), which frequently accompanies VZV-meningitis, myelitis, or radiculitis.

20.2.2.1.3 Cytomegalovirus (CMV)

Disease may manifest with symptoms in the immunocompromised individuals due to both primary and recurrent infections by the cytomegalovirus. Encephalitis, polymyeloradiculitis, ventriculitis, Guillain–Barre syndrome (GBS), and inflammatory polyneuropathy are some of the predominant manifestations in immunosuppressed individuals including those with HIV/AIDS. Transplant recipients and AIDS patients are prone to severe CMV disease such as pneumonitis, colitis, encephalopathy, and retinitis. Inpatients with HIV, retinitis is the most common presentation accounting for 85% of CMV disease. Patients undergoing hematopoietic allogeneic stem cell transplants (HSCT) are more likely to get primary CMV infection and experience latent illness reactivation (Bookstaver et al. 2017).

CMV can be transmitted vertically or horizontally. Transmission may occur vertically, in utero or horizontally during the perinatal period. Once infection is acquired, it persists lifelong. The virions may appear in urine and saliva, following activation of the virus, which may occur from time to time. Infection acquired during the perinatal period is said to be approximately 10 times more common than the congenital infection. Postnatal infection is frequently transmitted through saliva. Other modes of transmission are through sexual contact, through blood and blood products and transplanted organs.

20.2.2.1.4 Epstein-Barr Virus (EBV)

Acute and subacute infections due to EBV are seen in immunosuppressed patients. The virus causes meningitis, encephalitis, and AIDS-related lymphoma of the CNS in the immunocompromised as well as immunocompetent individuals. Other diseases are EBV-associated post-transplant lymphoproliferative disease (PTLD) of the CNS, Guillain–Barre syndrome (GBS), myeloradiculitis, and encephalomyeloradiculitis. The most common form is the lymphoproliferative disorder.

20.2.2.1.5 HHV-6

HHV-6A and HHV-6B are two different genetic variations of the human herpesvirus 6. Together with CMV, the virus is a member of the beta herpesvirus subfamily. They are responsible for a wide range of illnesses, including meningoencephalitis, recurrent febrile seizures in children, and encephalitis in transplant patients. Infants between the ages of 6 months and 2 years old can be exposed to HHV-6 at a young age. In this age, it is prevalent. Small newborns with the clinical condition roseola infantum are thought to have an exanthematous infection. Immunocompromised hosts experience reactivation of the virus after it goes latent.

20.2.2.2 Adenoviridae

Adenoviruses are now more well acknowledged as important viral infections among immunocompromised people, where they cause considerable morbidity and mortality. This may be associated with the increasing number of immunocompromised populations, increased knowledge of this virus as a pathogen, more aggressive therapies, the development of more sensitive diagnostic techniques, and patients with acquired immunodeficiency in particular. After an acute infection, T-cell-mediated immunity plays a critical role in healing. Patients with impaired immune systems who lack functional cellular immunity are more vulnerable to viral infection. After an acute infection, T-cell-mediated immunity plays a critical role in recovery. Patients with impaired immune systems who lack functional cellular immunity are more vulnerable to viral infection (Echavarrı'a 2008).

Human adenoviruses (HAdVs) belong to the family Adenoviridae and the genus Mastadenovirus. They have been divided into seven different species, numbered from A to G, with species B being further classified into B1 and B2 based on immunologic, biologic, and biochemical traits. However, 88 distinct HAdV types are currently recognised, and new adenovirus types are constantly emerging. Up to type 51, types were only defined as serotypes (by cross-neutralization), but since 2007, new genotype types have been discovered through genomic and bioinformatic investigations (Dhingra et al. 2019).

All adenoviruses except B use the coxsackie-adenovirus receptor (CAR). Group B adenoviruses have been shown to bind to CD46, a complement-related protein. Following primary infection, HAdVs can persist in a latent state in different tissues and may be reactivated in the presence of immunosuppression. A significant portion of their genomes are involved in gene products which modulate the host immune responses (Echavarrı'a 2008). The best-documented sites of HAdV persistence include tonsillar and adenoidal T lymphocytes, which appear to represent a sanctuary for adenoviral latency (Garnett et al. 2009). Latency is characterized by the expression of viral proteins by the host cell without the replication of a complete virus. In addition to intestinal lymphocytes which represent a reservoir for HAdV persistence and reactivation, the central nervous system is also a sanctuary for persistence of adenovirus (Kosulin et al. 2007), which is particularly seen with species C. It is yet unknown whether adenovirusspecies other than species C are capable of inducing a persistent or latent infection. Species C of the adenoviruses can establish latent infections in mucosal lymphocytes. Stimulation of these mucosal lymphocytes leads to reactivation of the virus resulting in RNA transcription, DNA replication, leading to release of infectious virus (Garnett et al. 2009). The virus must avert immune tracking, which is carried out by a variety of processes, in order to establish an ongoing infection at susceptible, cellular, and tissue levels.

Human adenoviruses cause subacute or chronic meningoencephalitis in immunosuppressed individuals. More than 50% of adenovirus infections in the immunocompromised hosts are associated with species C type 1 and 2. In patients with transplants, some of the most commonly reported adenovirus types include HAdV-C1, -C2, -C5, -A12, -A31, -B3, -B11, -B16, -B34, and -B35, with a strong predominance of species C (Al Qurashi et al. 2011).

20.2.2.3 Papovaviridae

20.2.2.3.1 John Cunningham Virus (JCV)

JCV can result in progressive multifocal leukoencephalopathy (PML) when the immune system is suppressed. However, most other people continue to have no symptoms. Oligodendrocyte infection that results in several brain foci of demyelinated neurons is a hallmark of PML. In patients with AIDS and those receiving various immunosuppressive medications for autoimmune illnesses, the JC virus is now more common. The current hypothesis is that neurotropic JCV is reactivated from a peripheral latent reservoir after immunosuppression, mostly of CD4 T cells, even if the exact mechanisms that lead to PML are not fully known. JCV then travels to the CNS via B cells and breaches the BBB. The virus then infects oligodendrocytes leading to destruction of these cells and the breakdown of the myelin sheath (Swanson II and McGavern 2015). Another possibility is that HHV-6 and JCV co-infect patients, leading to demyelination of the brain and contributing significantly to the pathophysiology. PML pathology comprises of many demyelination foci of varying sizes ranging from pinpoint lesions to areas of several centimeters. These are located usually in the cerebral hemispheres. PML develops gradually, causing deterioration in mental capacity as well as speech and vision disturbances. Movement may then be impacted as well. The illness advances quickly, causing severe impairment and, finally, dementia, paralysis, coma, and death. Improved long-term survival of transplant recipients and of HIV patients has shown a change in recent years, in the viral etiology of the illnesses. According to reports, EBV and JCV both contribute to immunoproliferative illness and increased PML incidence.

There is no specific treatment for PML. However, in individuals receiving combination antiretroviral therapy (ART), immunosuppression has been reversed and there is evidence of JCV removal from CS. This has led to clinical and MRI stabilization in 50–60% of patients receiving ART.

20.2.3 Laboratory Diagnosis of Viral CNS Infections

Neurological consultants face a tremendous challenge when it comes to the identification and treatment of CNS infections in cancer or immunosuppressed patients. Patients with impaired immune systems should constantly be on surveillance for an opportunistic CNS infection when they come with new neurological symptoms. A thorough diagnostic workup is required because there may be several opportunistic infections present. The diagnostic process should involve blood and CSF laboratory tests, blood and CSF microbiology, contrast brain imaging (MRI if practical), and should be designed to identify the causal agent as well as determine the kind and degree of immunosuppression.

20.2.3.1 Abnormalities Detected in the CSF Following Viral Infections in the Immunocompromised

CSF analysis confirms that there is CNS inflammation. White blood cells are increased, though typically not above 250/L. Although lymphocytes predominate in the differential diagnosis, early infection might also be accompanied by neutrophils. In two-thirds of individuals with VZV vasculopathy, a mild pleocytosis is seen, usually consisting of less than 100 mononuclear cells. Concentrations of CSF protein are mildly elevated (< 150 mg/dL) in viral encephalitis whereas those of glucose are normal.

20.2.3.2 Virus Isolation

Virus isolation in cell culture is possible; however, due to low yield and the cumbersome technique, it is rarely attempted as a diagnostic tool. The adenoviruses (except serotypes 40 and 41) replicate well and produce obvious cytopathology characterized by clumping, cell rounding, and refractile intranuclear inclusion bodies in primary human embryonic or neonatal kidney (HEK, HNK), continuous human cell lines of epithelial origin (HeLa, HEp-2, KB), and other cell lines like A549.The virus is multiplied in cell culture before being serotyped using hemagglutination inhibition and neutralization tests using hyperimmune type-specific animal antisera.

20.2.3.3 Antigen Detection

CMV-pp65 test allows for the rapid recognition of infection in patients with a compromised immune system, by detecting antigenemia. Clinical CMV disease is possible by the detection of CMV from blood or bronchoalveolar lavage is more prognostic than the detection of the virus from urine or saliva.

Adenoviruses can be detected directly in clinical specimens by rapid tests that measure their common, group specific hexon antigen. This is widely used for the diagnosis of respiratory and gastrointestinal infections.

20.2.3.3.1 In Situ Hybridization

VZV-specific RNA probes have been used to study the pathogenesis of VZV-associated neurological syndromes showing localization of the genome in a neuronal cell.

20.2.3.4 Serology: Antibodies in CSF or Serum

During the first week of symptoms, HSV-1 DNA can be detected by molecular test as antibodies to HSV-1 are negative in this phase. The antibodies to HSV-1 begin to appear in the second week when DNA levels vanish (Aurelius et al. 1991).

The detection of anti-VZV IgG antibody in the CSF has a higher diagnostic value than the detection of VZV DNA. The envelope glycoproteins of VZV are highly immunogenic and VZV glycoprotein G have been used in VZV-specific EIAs.

Acute CMV infection is indicated by four-fold increase in the IgG level (sero-conversion) or detection of IgM in the serum.

The most common ways to diagnose acute EBV-infections are by looking for heterophile antibodies or anti-EBV VCA IgM.

20.2.3.5 PCR

Conventional techniques have some drawbacks, including the potential for lengthy culture and the possibility for inhibition by neutralizing antibodies or other interfering compounds, as well as the potential for insensitivity of electron microscopy and antigen detection techniques. PCR analysis of CSF has revolutionized the diagnosis viral infections of the nervous system, particularly those caused by human herpesviruses (HHV). PCR has now replaced brain biopsy as the gold standard for the diagnosis of encephalitis due to HSV (Kleinschmidt-De Masters et al. 2001). Regardless of high specificity and sensitivity of the HSV CSF PCR, the results of the test should be correlated clinically. Clinical judgment must be used both in determining when a diagnostic CSF PCR assay needs to be ordered and in the interpretation of the results.

False-negative CSF PCR results may occur when specimens are collected either too early or too late. Davies et al., found that the prevalence of positive CSF PCR results was maximal when specimens were obtained at 3–14 days after onset of symptoms. It was significantly lower when specimens were obtained earlier or later (Davies et al. 2005).

Only negative results in both VZV PCR and anti-VZV IgG antibody testing in the CSF can reliably exclude the diagnosis of VZV vasculopathy, even though a positive PCR for VZV DNA in CSF is beneficial.

Despite PCR-based techniques being the technique of choice for suspected infection with HHV6, positive tissue PCR results must be interpreted cautiously, especially in cases that have insufficient correlating clinical and neuropathologic evidence of infection. Multiplex PCRs are efficient and cost-effective and are presently being evaluated in multiple centers. It is being evaluated to detect opportunistic infections caused by a large number of herpes viruses, including, infections due to, HSV-1, HSV-2, VZV, CMV, EBV, and HHV-6 (Markoulatos et al. 2001; Ramamurthy et al. 2012).

Another advancement in the technique for detection of viral infections is the Film Array Meningitis/Encephalitis panel (BioFire Diagnostics) (FA/ME). This technique seems better for ruling in, than ruling out, the disease. Biofire for viral infections has an overall high sensitivity and specificity for the diagnosis of meningoencephalitis with a fast turnaround time. However, the fact that all of the herpesviruses on the ME Panel [namely, HSV-1; HSV-2; CMV; VZV; HHV-6] can cause latent infections, it should also be asserted. As a result, a positive ME Panel outcome could indicate a latent infection, or it could indicate the initial infection. The ME Panel should not be used as the sole diagnostic tool and clinicians should be very cautious and try to interpret ME Panel results in combination with clinical, epidemiological, and laboratory data.

Diagnosis of HAdV infections has become easier with highly sensitive techniques, based primarily on PCR, and have become the gold standard for the detection and monitoring of the infection. Monitoring of patients in the immunocompromised setting requires tests permitting reliable detection of all potentially relevant HAdV types in different clinical specimens, as well as accurate quantitative assessment of viral loads (Lion 2014). The use of PCR assays on blood samples from patients has greatly aided in identifying those who have disseminated adenovirus illness. PCR primers for the hexon gene, fiber gene, or virally associated RNA I and II regions are usually preferred because they contain some sequences that are highly conserved across serotypes. The wide range of heterogeneity among the numerous serotypes is one of the main obstacles to the creation of a sensitive generic PCR. A number of commercial kits are available, including the Adenovirus R-gene kit (bioMérieux, Lyon, France), ELITeMGB kit (ELITech Group Molecular Diagnostics, Puteaux, France), FilmArray RP kit (BioFire Diagnostics, Inc., Salt Lake City, UT), sensor RVP kit (GenMark Diagnostics, Carlsbad, CA), xTAG RVP Fast and xTAG RVPv1 kits (Luminex Molecular Diagnostics, Toronto, Canada), Prodesse ProAdeno assay (Hologic Gen-Probe, San Diego, CA), and Anyplex II RV16 kit (Seegene, South Korea), etc.

Developing laboratory criteria that differentiate between active virus replication and latent virus will be useful in deciding preemptive therapy. Documentation of DNA by qualitative assays which reflect latent virus presence especially with

Herpesviridae members is less indicative of virus replication than the viral mRNA demonstration in whole blood which is a better indicator of virus replication (Fleischer and Aronson 2020).

20.2.3.6 MRI

The initial stage in the investigation of a possible CNS illness is brain imaging. Brain imaging is mostly used to rule out lesions that take up a lot of space, like abscesses, tumours, edema, or hydrocephalus, which might result in brain herniation during lumbar puncture (LP). Hence, neuro-imaging prior to lumbar puncture is recommended for all patients suffering from severe immunosuppression who come having indications of neuro-infection. Magnetic resonance imaging (MRI) is now widely accepted as the preferred test, to detect early changes in the CNS, in suspected viral infections (Sonneville et al. 2017).

HSV patients develop tempero-basal or fronto-basal focal encephalitis with hemorrhagic necrotic lesions and significant cerebral edema. The hemorrhagic necroses can be visualized in MRI images (Arendt and Maschke 2017).

In CNS infections due to VZV, abnormalities are cortical and deep. They occur in both the grey and white matter and at grey–white matter junctions. VZV vasculopathy should be included in the differential diagnosis of patients with lesions at grey–white matter junctions. Concomitant intracranial vasculopathy and brainstem infarction may occur in Ramsay Hunt syndrome (Sun et al. 2022).

20.2.3.7 Angiography

Characteristic angiographic changes include segmental constriction, often with post-stenotic dilatation. A negative angiography does not rule out the diagnosis of VZV vasculopathy, most likely because disease in tiny arteries is harder to detect than in large arteries, even though the presence of stenosis or occlusion aids in the diagnosis of VZV vasculopathy.

20.2.4 Treatment

Early empirical treatment should target the most likely infections. In some circumstances, stopping or reducing immunosuppressive medication may be an option (Sonneville et al. 2017).

20.2.4.1 HSV

Acyclovir may be used to treat HSV infection, but the duration of therapy is longer in immunocompromised (21 days) patients compared to immunocompetent (14 days) patients. Acyclovir resistance may develop during the course of prolonged treatment. It has previously been demonstrated that acyclovir resistance in HSV is linked to mutations in the thymidine kinase (TK) gene (Frobert et al. 2005). Drug-resistant virus variants are found in 3.8–15.7% of immunocompromised patients, as compared to immunocompetent patients (Frobert et al. 2014). However, for people receiving organ graft transplants, acyclovir is increasingly frequently used as prophylactic, even before herpetic encephalitis is confirmed, and HSV is no longer a major problem in these patients.

20.2.4.2 VZV

Acyclovir, valaciclovir, famciclovir, or combination of ganciclovir and foscarnet are treatment options available for Varicella and zoster cases. VZIG is indicated in immunocompromised patients.

20.2.4.3 CMV

Ganciclovir, forscarnet, and cidofovir are anti-CMV medications currently in use. The medication of choice is ganciclovir. Treatment options for CMV infection in immunocompromised individuals today include intravenous ganciclovir, foscarnet, or a combination of the two, as well as cidofovir. The prodrug valganciclovir (vGCV) has taken the place of oral ganciclovir (GCV) for CMV prevention in recipients of kidney transplants.

20.2.4.4 EBV

Acyclovir and other more recent medications have demonstrated effective therapy of the illness if it is identified early. It has been demonstrated that ganciclovir, valaciclovir, famciclovir, and cidofovir have anti-EBV action in vitro.

20.2.4.5 HHV 6

It has been demonstrated that cidofovir, ganciclovir, and valaciclovir or famciclovir have in vitro action against HHV-6.

20.2.4.6 Adenoviruses

The most frequently utilized medication currently among stem cell transplant (SCT) recipients is cidofovir. Preemptive medication and active viral surveillance are strongly considered, especially for pediatric SCT patients who are at high risk of contracting the illness, until enough T-cell activity is restored. The most crucial source for clinical HAdV infection surveillance in immunocompromised environments is peripheral blood. Adult organ transplant recipients do not seem to need routine HAdV surveillance by PCR.

20.3 Fungal Infections in the Immunocompromised Host

With the increasing use of immunosuppressive therapy in various diseases and other causes of immunosuppression, there has been an increased incidence of fungal infections globally. Bone marrow transplants, hematological malignancies, solid organ transplants, and acquired immunodeficiency syndromes (AIDS) have been associated with secondary fungal infections (Reyes and Shinohara 2022). Fungal cells which infiltrate into the central nervous system parenchyma in the absence of an intact immune system cannot be cleared adequately. On accessing the microcirculation of brain, fungal cells infect the astrocytes, local endothelial cells, and microglia, leading to lepto-meningitis, as well as encephalitis.

20.3.1 Mechanisms of CNS Fungal Invasion in the Immunocompromised Host

Neural parenchyma becomes penetrated by fungal elements when the immune response in an immunocompromised host is unable to remove the microbial threat at the initial infection site, which is typically the gastrointestinal or pulmonary system. To establish the infection, fungal elements need to cross the BBB which consists of endothelial cells, pericytes, astrocytes, and microglia. Pathogenic fungi on the other hand produce various molecules that induce host cell death, thereby damaging the barriers. Fungi such as Cryptococcus can enter into the host macrophages by phagocytosis. Given their capsulated nature, they remain alive inside the cells and instead use camouflage to pass over the BBB to enter the central nervous system. This process is referred to as the "Trojan Horse" phenomenon. The fungal cells can also enter by transcellular migration, by attaching ligands to the cells and crossing the barrier, or paracellular migration by squeezing through the cells without damaging them (Góralska et al. 2018).

Candida, Cryptococcus, and Aspergillus are the three major groups of fungi that are associated with infections in the CNS of an immunosuppressed host (Table 20.2). Other major categories of fungi include Mucorales, and dimorphic fungi, such as Blastomyces, Coccidioides, and Histoplasma, black fungi *Cladophialophora bantiana, Exophiala dermatitidis*. Rarely encountered fungi include Blastomyces dermatitidis, Coccidioides spp. The fungi listed above cause meningoencephalitis, hydrocephalus, brain or epidural abscesses, and spinal cord lesions.

Table 20.2 Common fungi associated with CNS infection in the immunocompromised individuals

Fungus	Clinical presentation	Predisposing condition
Candida	Chronic meningitis, Brain abscess	Prolonged steroid therapy, AIDS
Cryptococcus	Chronic meningitis	AIDS and immunosuppression
Histoplasma capsulatum	Chronic meningitis	AIDS and immunosuppression
Aspergillus	Brain abscess	Granulocytopenia and immunosuppression
Mucorales	Rhinocerebrorbital zygomycosis	Immunosuppresssion, prolonged steroid thrapy

20.3.2 Candida

Candida species are yeast like fungi, found as commensals on mucosal surfaces of the oropharynx, gastrointestinal, reproductive tracts, skin, and the respiratory tract. They become pathogenic when the mucosal or skin barriers are breached in the immunocompromised individual. Candida infection is also seen in individuals implanted with CNS shunts. Most other Candida-related CNS infections are a result of hematogenous spread from a primary site, elsewhere in the body. They can cause chronic meningitis, brain abscesses, vasculitis with cerebral infarctions, spinal infections, ventriculitis, and mycotic aneurysms. The most common species associated are Candida *albicans* followed by *C. tropicalis*. Other species of Candida have also been encountered infrequently.

20.3.3 Cryptococcus

C. neoformans which includes three serotypes (A, D, and AD) and *C. gattii* (two serotypes: B and C) are the two main human pathogens. Two varieties of C. neoformans have been distinguished: C. neoformans var. neoformans and C. neoformans var. grubii. Fungal virulence factors include melanin production and the presence of a polysaccharide capsule (glucuronoxylomannan). The capsule enables Cryptococcus to evade immune recognition. Cryptococcus is the most common opportunistic CNS fungal pathogen observed in HIV-positive patients. It causes chronic meningitis and meningoencephalitis. The infection often arises from a primary lung infection.

20.3.4 Aspergillus

Aspergillus is a mold found ubiquitously in nature. It is an abundantly sporulating fungus. The spores are disseminated by air, primarily causing pulmonary infections and invasive aspergillosis in immunocompromised individuals. Patients with tuberculosis, neutropenia, chronic granulomatous disease, cancer, and those taking prolonged immunosuppressant medications are most susceptible to this infection. Following inhalation, the fungal spores establish themselves in the respiratory tract. Hyphal forms disseminate to the CNS either through the bloodstream or by contiguous spread from the orbits, periorbital regions, middle ear, or paranasal sinuses, crossing the BBB. Pathogenesis depends on the gliotoxin of aspergillus which can damage and kill microglial cells, astrocytes, neurons by inducing apoptosis (Koutsouras et al. 2017). The commonest species causing CNS infections in the immunocompromised host is *A. fumigatus*.

20.3.5 Mucorales (Zygomycetes)

Zygomycetes which include Mucor, Rhizopus, and Rhizomucor are the most commonly encountered pathogens. They occur in the environment, with high humidity. Infection can develop as a result of skin damage or mucous membrane damage, or by inhaling airborne spores. Patients with immunosuppression, with suppressed inflammatory response due to impaired function of monocyte and granulocyte, are prone for infections by the mucorales. The infection disseminates, to the CNS through the rhino-orbital route. A common contiguous invasion is rhino-orbital-cerebral mucormycosis. Most frequently associated species in rhino-orbital-cerebral mucormycosis are *R. arrhizus* and *R. microsporus*. This form of infection was seen as a complication in a large number of patients, in the recent Covid-19 pandemic. Prolonged steroid therapy, in addition to T-Cell dysfunction, has been attributed to this complication. Rhino-orbital-cerebral mucormycosis begins with the fungal infection spreading contiguously from maxillary and ethmoid sinus, orbit, extending to the brain. The most common symptoms include headache, rhinitis, periorbital edema, diplopia or loss of vision, fever, and decreased mental function. Mortality is between 30 and 90%, depending on the time of diagnosis and the progression of the disease. Black discharge and crusts from the nose, involvement of the turbinates, and palatal perforation may be seen. Brain abscess, granulomatous inflammation, meningitis, myelitis, or encephalitis are seen less commonly in these patients; however, those affected have a high mortality (Gavito-Higuera et al. 2016).

20.3.6 Histoplasma

This is a dimorphic fungus, which is primarily found in Africa and Latin America and Asia. It primarily affects the respiratory tract. It occurs in 3–5% of immunosuppressed individuals including AIDS patients, transplant recipients, patients on corticosteroid and tumor necrosis factor antagonists, and patients with ventriculoperitoneal shunts. Histoplasmosis develops in 3–5% immunosuppressed individuals. Among whom, 10–20% individuals experience CNS involvement, with mortality. Histoplasmosis typically begins with lung infection brought on by spore inhalation. Macrophages phagocytize yeast-phase cells, which is followed by the spread of intracellular forms to the central nervous system. CNS infections have also been noted with delayed presentation following a protracted latent period or a recurrence. Histoplasma-related CNS lesions are non-specific and can resemble illnesses caused by other causative agents. Hydrocephalus, along with acute or chronic meningitis, is one of the most recognizable signs of CNS involvement.

20.3.7 Miscellaneous Fungi Causing Infection in the Immunocompromised Individual

There are several other fungi that are less commonly encountered as pathogens causing CNS infections in the immunocompromise host. Dimorphic endemic fungi, such as Blastomyces dermatitidis, and Coccidioides spp., dematiaceous fungi which include Cladosporium, Cladophialophora, and Foncescaea, and molds including Fusarium, Scedosporium species, and the yeast Sporothrix can also be associated with CNS infections in the immunocompromised patients.

20.3.8 Common CNS Manifestations of Fungal Infections Are as Follows

- **Brain abscess:** This leads to specific focal neurologic deficits and increased intracranial pressure. They are multiple and affect the basal ganglia.
- **Cryptococcoma**: This is a form of a granuloma, due to cryptococcal infection.
- **Fungal cerebritis:** This is brought on by fungi that enter the parenchyma of the brain.
- **Fungal meningitis**: It is commonly brought on by the yeasts, Cryptococcus, and Candida; they can have acute/sub-acute presentations.
- **Meningeal vasculitis:** These can occur with vessel thrombosis and localized brain infarctions. They are often due to infections with Aspergillus species and Zygomycetes.

20.3.9 Common Clinical Features

The patients complain of one or more of the following symptoms: chronic headache, neck or back pain, facial weakness, double vision, or visual loss. Other complaints may be weakness of arms and legs, numbness, and sphincter dysfunction, which may be due to myelopathy or radiculopathy, and frontal dysfunction. The above symptoms may be accompanied by altered mental status, drowsiness, disorientation, and memory loss. Signs of meningeal irritation may be elicited by Brudzinski's and Kernig's signs.

20.3.10 Laboratory Diagnosis of Fungal CNS Infections

Due to an insidious onset, generally secondary to a primary infection, fungal CNS infections present diagnostic challenges. Recognition of the organism ought to be taken into account alongside the patient's condition, including any underlying illnesses, manifestations, and findings. Necessary imaging along with microscopic examination, histopathological findings, and staining relevant for fungi are all crucial for the diagnosis of fungal infections. Final identification depends on the culture. Additionally, some of the recent molecular techniques and automated systems have improved fungal diagnostics.

20.3.10.1 Imaging

Recognition of typical imaging patterns on CT and MRI helps in the differential diagnosis and helps in initiating early treatment.

20.3.10.2 CSF Examination

Microscopic examination of centrifuged deposits of CSF provides a rapid and presumptive diagnosis. A large volume of CSF needs to be tapped for satisfactory laboratory tests. **Wet mount** for fungal filaments can help differentiate a mold by its characteristic shape and color. **Fluorochrome stains** like acridine orange may also be used to stain the fungal filaments for ease of detection. **Gram's/Leishman's** stain and other fungal-specific staining of the CSF reveal the presence of inflammatory cells. Generally, mononuclear cells predominate. Yeast cells can also be visualized. The shape of the cells and the presence of budding or pseudo-hyphae may help in provisional identification. **India ink** is used to demonstrate capsules of Cryptococcus. This is one of the rapid methods for confirmatory identification of Cryptococcal meningitis.

20.3.10.3 Antigen Detection

Several rapid antigen detection kits have now come into the market, to detect specific fungal antigens in cerebrospinal fluid, urine, and other body fluids. A **lateral flow immunoassay** (dipstick) assay detects glucuronoxylomannan (GXM), the major capsular polysaccharide of *C. neoformans* major serotypes: A, B, C, and D and a hybrid serotype AD (Kozel and Wickes 2014). **Galactomannan enzyme immunoassay** (EIA) is useful in the diagnosis of invasive aspergillosis in the immunocompromised neutropenic patients. *H. capsulatum* polysaccharide antigen is detected in urine, of patients with diagnosis of disseminated histoplasmosis. **Assay for Beta D Glucan** can be used to screen for presumptive diagnosis of invasive fungal infection.

20.3.10.4 PCR

Species-specific assays using multiplex polymerase chain reaction (PCR) in conjunction with metagenomics and next-generation sequencing have recently emerged.

20.3.10.5 Other Rapid Automated Diagnostic Tests

Lateral flow immunoassays (LFA) for cryptococcosis and invasive candidiasis matrix-assisted laser desorption ionization-time of flight mass spectrometry (MALDI-TOF MS) have been used to detect yeasts from clinical samples. However, these tests have not come widely into clinical practice.

20.3.10.6 Culture

The gold standard for determining whether a patient has a fungal infection is culture from a clinical sample. If positive, it offers the benefit of revealing the precise etiological agent. Testing for susceptibility is also possible. However, the drawbacks of typical culture systems include reporting delays brought on by the time it takes for fungi to grow, particularly dimorphic fungi. Automated methods, such as yeast identification systems from VITEK and the MALDI TOF mass spectrometry, may speed up identification.

20.3.11 Treatment

Treatment of fungal infections of CNS is particularly difficult due to a lack of knowledge about the pharmacokinetics and dynamics of the nervous system, and the penetration of the drug into the brain parenchyma (Dodds and Ashley 2019).

The cornerstone of treatment for many CNS fungal infections is amphotericin B and related lipid formulations. When treating CNS infections, lipidosomal amphotericin B (L-Amb) is the preferable formulation.

The other antifungal agent used widely is flucytosine which has very good action in the CNS. However, this drug needs to be given in combination with other antifungal agents, due to its propensity to develop resistance if given as monotherapy. Voriconazole has also emerged as a successful treatment option for aspergillus infection of the CNS.

It has been tried to directly administer antifungal medications into the nervous system due to the slim chance that systemic medications will enter the CNS. The administration of amphotericin B deoxycholate via the intrathecal route is one of the preferred routes to treat fungal CNS infections.

References

Al Qurashi YM, Guiver M, Cooper RJ (2011) Sequence typing of adenovirus from samples from hematological stem cell transplant recipients. J Med Virol 83:1951–1958

Arendt G, Maschke M (2017) Update opportunistic infections of the central nervous system in patients with iatrogenic immunosuppression. Neurol Int Open 1:E316–E325. https://doi.org/10.1055/s-0043-118924

Aurelius E, Johansson B, Skoldenberg B, Staland A, Forsgren M (1991) Rapid diagnosis of herpes simplex encephalitis by nested polymerase chain reaction assay of cerebrospinal fluid. Lancet 337(8735):189–192. https://doi.org/10.1016/0140-6736(91)92155-u

Bookstaver PB, Mohorn PL, Shah A, Tesh LD, Quidley AM, Kothari R, Bland CM, Weissman S (2017) Management of viral central nervous system infections: a primer for clinicians. J Cent Nerv Syst Dis 9:1–12. https://doi.org/10.1177/1179573517703342

Castro I, Ruiz J, Tasias M, Montero M, Salavert M (2018) Central nervous system infections in immunocompromised patients. Rev Esp Quimioter 31(Suppl. 1):56–61

Davies NWS, Brown LJ, Gonde J, Swan AV, Banatwala J, Howard RS, Shareif MK, Muir P (2005) Factors influencing PCR detection of viruses in cerebrospinal fluid of patients with suspected viral infections. J Neurol Neurosurg Psychiatry 76:82–87. https://doi.org/10.1136/jnnp.2004.045336

Dhingra A, Hage E, Ganzenmueller T, Böttcher S, Hofmann J, Hamprecht K, Obermeier P, Rath B, Hausmann F, Dobner T, Heim A (2019) Molecular evolution of human adenovirus (HAdV) species C. Sci Rep 9:1039. https://doi.org/10.1038/s41598-018-37249-4

Dodds E, Ashley J (2019) Antifungal drugs: special problems treating central nervous system infections. J Fungi (Basel) 5:97. https://doi.org/10.3390/jof5040097

Echavarrı'a M (2008) Adenoviruses in immunocompromised hosts. Clin Microbiol Rev 21(4): 704–715. https://doi.org/10.1128/CMR.00052-07

Fleischer E, Aronson PL (2020) Rapid diagnostic tests for meningitis and encephalitis—BioFire. Pediatr Emer Care 36:397–403

Frobert E, Ooka T, Cortray JC, Lina B, Thouvenot D, Morfin F (2005) Herpes simplex virus thymidine kinase mutations associated with resistance to acyclovir: a site-directed mutagenesis study. Antimicrob Agents Chemother 49(3):1055–1059. https://doi.org/10.1128/AAC.49.3.1055-1059.2005

Frobert E, Burrel S, Ducastelle-Lepretre S, Billaud G, Adar F, Casalegno JS, Nave V, Boutolleau D, Michallet M, Lina B, Morfin F (2014) Resistance of herpes simplex viruses to acyclovir: an update from a ten-year survey in France. Antivir Res 111:36–41. https://doi.org/10.1016/j.antiviral.2014.08.013; Epub 2014 Sep 8

Garnett CT, Talekar G, Mahr JA, Huang W, Zhang Y, Ornelles DA, Gooding LR (2009) Latent species C adenoviruses in human tonsil tissues. J Virol 83:2417–2428. https://doi.org/10.1128/JVI.02392-08

Gavito-Higuera J, Mullins CB, Ramos-Duran L, Olivas Chacon CI, Hakim N, Palacios E (2016) Fungal infections of the central nervous system: a pictorial review. J Clin Imaging Sci 6:24

Gilden D, Cohrs RJ, Mahalingam R, Nagel MA (2009) Varicella zoster virus vasculopathies: diverse clinical manifestations, laboratory features, pathogenesis, and treatment. Lancet Neurol 8(8):731–740. https://doi.org/10.1016/S1474-4422(09)70134-6

Góralska K, Blaszkowska J, Dzikowiec M (2018) Neuroinfections caused by fungi. Infection 46: 443–459. https://doi.org/10.1007/s15010-018-1152-2

Kleinschmidt-De Masters BK, RL DB, Tyler KL (2001) Polymerase chain reaction as a diagnostic adjunct in herpesvirus infections of the nervous system. Brain Pathol 11(4):452–464. https://doi.org/10.1111/j.1750-3639.2001.tb00414.x

Kosulin K, Haberler C, Hainfellner JA, Amann G, Lang S, Lion T (2007) Investigation of adenovirus occurrence in pediatric tumor entities. J Virol 81:7629–7635. https://doi.org/10.1128/JVI.00355-07

Koutsouras GW, Ramos RL, Martinez LR (2017) Role of microglia in fungal infections of the central nervous system. Virulence 8(6):705–718. https://doi.org/10.1080/21505594.2016.1261789

Kozel TR, Wickes B (2014) Fungal diagnostics. Cold Spring Harb Perspect Med 4:a019299

Lion T (2014) Adenovirus infections in immunocompetent and immunocompromised patients. Clin Microbiol Rev 27(3):441–462. https://doi.org/10.1128/CMR.00116-13

Markoulatos P, Georgopoulou A, Siafakas N, Elias PE, Tzanakaki G, Kourea-Kremastinou J (2001) Laboratory diagnosis of common herpesvirus infections of the central nervous system by a multiplex PCR assay. J Clin Microbiol 39(12):4426–4432. https://doi.org/10.1128/JCM.39.12.4426-4432.2001

Ramamurthy M, Kannangai R, Abraham AM, Sridharan G (2012) Viral infections in immunocompromised hosts. Proc Natl Acad Sci Biol Sci 82(1):95–109. https://doi.org/10.1007/s40011-011-008-7

Reyes EY, Shinohara ML (2022) Host immune responses in the central nervous system during fungal infections. Immunol Rev 311(1):50–74. https://doi.org/10.1111/imr.13101

Sonneville R, Magalhaes E, Meyfroidt G (2017) Central nervous system infections in immunocompromised patients. Curr Opin Crit Care 23(2):128–133. https://doi.org/10.1097/MCC.0000000000000397

Sun H, You H, Wu H (2022) Brainstem infarction and vertebral artery vasculopathy after ramsay Hunt syndrome. Neurology 98:890–891. https://doi.org/10.1212/WNL.0000000000200522

Swanson PA II, McGavern DB (2015) Viral diseases of the central nervous system. Curr Opin Virol 11:44–54. https://doi.org/10.1016/j.coviro.2014.12.009

SPRINGER NATURE

GPSR Compliance

The European Union's (EU) General Product Safety Regulation (GPSR) is a set of rules that requires consumer products to be safe and our obligations to ensure this.

If you have any concerns about our products, you can contact us on ProductSafety@springernature.com

In case Publisher is established outside the EU, the EU authorized representative is:

Springer Nature Customer Service Center GmbH
Europaplatz 3
69115 Heidelberg, Germany

The manufacturer's authorised representative in the EU is Springer Nature Customer Service Centre GmbH, Europaplatz 3, 69115 Heidelberg, Germany. If you have any concerns regarding our products, please contact ProductSafety@springernature.com

Printed and bound by CPI Group (UK) Ltd, Croydon, CR0 4YY
30/09/2026
02225322-0020